Reproductive physiology of marsupials

MONOGRAPHS ON MARSUPIAL BIOLOGY

Reproductive physiology of marsupials

HUGH TYNDALE-BISCOE

Chief Research Scientist, CSIRO Division of Wildlife and Rangelands Research, Canberra, Australia

MARILYN RENFREE

NHMRC Principal Research Fellow, Department of Anatomy Monash University, Melbourne, Australia

CAMBRIDGE UNIVERSITY PRESS

Cambridge

London New York New Rochelle

Melbourne Sydney

Published by the Press Syndicate of the University of Cambridge
The Pitt Building, Trumpington Street, Cambridge CB2 1RP
32 East 57th Street, New York, NY 10022, USA
10 Stamford Road, Oakleigh, Melbourne 3166, Australia

First published 1987

Printed in Great Britain by the University Press, Cambridge

British Library cataloguing in publication data

Tyndale-Biscoe, C. H.

Reproductive Physiology of Marsupials.

(Monographs on marsupial biology)

1. Marsupialia – Reproduction
2. Mammals – Reproduction
I. Title II. Renfree, Marilyn B.
III. Series
599.2′0416 QL737.M3

Library of Congress cataloguing in publication data

Tyndale-Biscoe, C. H.

Reproductive Physiology of Marsupials.

(Monographs on marsupial biology)

Bibliography.
Includes index.
1. Marsupialia – Reproduction. I. Renfree, M. B.
II. Title. III. Series.
QL737.M3T97 1986 599.2′016 86-2251

ISBN 0 521 25285 7 hard covers
ISBN 0 521 33792 5 paperback

UP

To three pioneers who recognised the special role of marsupials for the understanding of mammalian reproduction

J. P. Hill   C. G. Hartman   G. B. Sharman

Contents

	Preface	xiii
1	**Historical introduction**	1
2	**Breeding biology of marsupials by family**	14
	Didelphidae	15
	Oestrous cycle and pregnancy	20
	Parturition and lactation	25
	Breeding season and annual productivity	26
	Other didelphids	29
	Microbiotheriidae	30
	Caenolestidae	31
	Dasyuridae	31
	Oestrous cycle and pregnancy	33
	Parturition and lactation	37
	Breeding seasons	40
	Thylacinidae	47
	Myrmecobiidae	47
	Perameloidea	48
	Oestrous cycle and pregnancy	49
	Parturition and lactation	51
	Breeding season and annual productivity	52
	Thylacomyidae	53
	Herbivorous marsupials – Diprotodonta	53
	Phalangeridae and Petauridae	54
	Trichosurus vulpecula	55
	Other species of Phalangeridae	57
	Petauridae	58
	Burramyidae and Tarsipedidae	60

Tarsipedidae 62
Phascolarctidae 64
Vombatidae 65
Macropodidae 66
Oestrous cycle and pregnancy 69
Parturition in Macropodidae 76
Lactation 81
Breeding seasons 83
Facultative breeding macropodids 83
Seasonal breeding macropodids 85
Conclusions 88
Reproductive cycles 88
Classification of reproductive patterns 90
Embryonic diapause 91
Cost of lactation and the timing of the breeding season 92
3 Sexual differentiation and development 95
Sex chromosomes 95
Sex-linked inheritance and dosage compensation for X chromosomes in the Macropodidae 98
X inactivation in other marsupials 103
Sex differentiation 105
Primordial germ cells 105
Differentiation of testis and ovary 108
Differentiation of the urogenital system 114
Effects of gonadectomy and gonadotrophin treatment 116
Effects of testosterone and oestradiol on the urogenital system 117
Differentiation of pouch and scrotum 121
Secondary sex characters 123
4 Male anatomy and spermatogenesis 124
Anatomy and physiology of the male genitalia 124
The scrotum, testes and epididymides 124
Testicular secretions 135
Rete testis fluid 135
Testicular endocrinology 135
Penis morphology 145
Accessory gland structure 147
Prostatic secretions 149
Spermatozoa 151
Sperm morphology 151
Spermatogenesis 156
Post-testicular maturation of the spermatozoa 163
Sperm pairing 168
Conclusions 171

5 The female urogenital tract and oogenesis 172
Anatomy of the urogenital tract 172
The oviduct 172
The uterus 175
The vaginal complex and birth canal 175
Histology and functional aspects of the uterus and vaginal canals 181
The proliferative phase 181
The luteal phase 184
The post-luteal phase 186
Steroid receptors in the urogenital tract 188
Vascular anatomy of the urogenital tract 190
Oocyte and follicular growth and development 195
Folliculogenesis 196
Follicular atresia and the origin of interstitial tissue 201
6 Ovarian function and control 203
Oestrus and ovulation 204
Formation and development of the corpus luteum 208
Type 1: short gestation, short luteal phase 212
Type 2: short gestation, prolonged luteal phase 224
Type 3: long gestation, delayed luteal phase 230
The oestrous cycle and pregnancy of M. eugenii, *uninterrupted by lactation* 231
The delayed oestrous cycle and pregnancy 232
Growth and development of the corpus luteum after removal of pouch young 233
Other species of macropodid 240
Role of the corpus luteum in follicular growth, ovulation and oestrus 242
Influence of the corpus luteum on the male 248
Endocrine control of the corpus luteum 249
Demise of the corpus luteum 254
Conclusions 256
7 Pregnancy and parturition 258
Sperm transport and fertilisation 258
Sperm transport 260
Fertilisation 261
Egg membranes 265
The primary vitelline membrane 265
The secondary egg membrane – the zona pellucida 265
The inner tertiary egg membrane 267
The shell membrane 269
'Yolk' extrusion, cleavage and blastocyst formation 274

The unilaminar blastocyst and the phenomenon of diapause 278
Embryonic diapause 279
Reactivation after diapause 282
Role of the corpus luteum 282
First changes in the blastocyst 286
The role of uterine secretions 291
Conclusions concerning reactivation 293
Primary endoderm and formation of the bilaminar blastocyst 294
The medullary plate, primitive streak and embryogenesis 299
Organogenesis 304
The excretory system 305
Derivatives of the endoderm 307
Endocrine organs 308
The marsupial placenta 310
Placentation 311
Feto-maternal contact 314
Placental functions 323
Biochemical functions of the placenta 324
Maternal recognition of pregnancy 327
Morphogenetic effects of the placenta 328
Endocrine functions of the placenta 330
Immunosuppressive function of the placenta 333
Parturition 333
Myometrial activity during pregnancy in M. eugenii 334
Preparation of the birth canal 334
Endocrine changes at parturition 337
Conclusions 341
8 Lactation 343
Mammary gland development and growth 344
Differentiation of the mammary gland 344
Milk composition 352
Mammary regression and successive lactation 355
Hormonal control of lactation 356
Control of mammary development before lactation 356
Initiation of lactation 358
Galactopoiesis and the maintenance of lactation 361
Control of milk secretion 364
Role of striated muscle 364
Oxytocin and milk ejection 365
Conclusions 371
9 Neuroendocrine control of seasonal breeding 373
Seasonal change in pituitary and hypothalamus 374
The control of lactational quiescence 377

The control of seasonal quiescence 382
Influence of daylength on seasonal quiescence 384
Response to experimental change in photoperiod 387
Role of the pineal gland 389
Stage of development of photosensitivity in the female 393
Conclusions 394
10 Marsupials and the evolution of mammalian reproduction 396
The palaeontological record of mammals 398
Comparison of reproduction in living mammals 400
Sex chromosomes 400
Male anatomy 400
Female anatomy 401
Follicle and corpus luteum 402
Intrauterine development 404
Fetal membranes 407
Size and development at birth 409
Mammary glands 410
Pouch and epipubic bones 411
Lactation 411
Conclusions 414
The evolution of mammalian reproduction 415

References 424
Index 470

Preface

The three sub-classes of living mammals are distinguished most clearly by their modes of reproduction, which have developed independently during the past 100 million years of their separate evolution. Study of the differences in reproduction between monotremes, marsupials and eutherians helps to throw light on the origins of mammalian reproduction, especially the origin of mammalian viviparity. More importantly, comparisons between these three groups elucidates general principles by disclosing the different means that each has used to achieve the same ends. The central control of ovarian function, the nature and the role of the corpus luteum, the transport and fate of the germ cells at fertilisation, maternal recognition of pregnancy, the initiation and control of lactation, and the neuroendocrine control of seasonal reproduction are all topics that interest the student of reproduction and to which the results of studies on marsupials are contributing new insights.

This is the first attempt to bring the wealth of material, old and new, on marsupial reproduction within one book. In attempting to do this we have been greatly helped by our many colleagues who have generously read those parts that cover their speciality and given us the benefit of their criticism. While we accept responsibility for what is written we thank Des Cooper, Brian Green, Mervyn Griffiths, Lyn Hinds, Jim Kenagy, Russell Jones, Steve McConnell, Kevin Nicholas, Jeremy O'Shea, John Rodger, Lynne Selwood, Brian Setchell, Geoff Sharman, Geoff Shaw and Peter Temple-Smith for reading particular chapters, and we especially thank John Calaby and Roger Short for giving us the benefit of their wide experience on all aspects of the book.

We also thank the following for allowing us full access to their unpublished work: Geoff Alcorn, Jon Curlewis, Leigh Findlay, Terry

Fletcher, Lyn Hinds, Steve McConnell, Helen McCracken, Jim Merchant, Geoff Shaw and Marcus Walker, and to those many others who provided the original plates for those figures acknowledged in the appropriate captions. To Graeme Chapman, who prepared all the half-tone figures, and to Frank Knight, who prepared all the line drawings, our special thanks. Finally, we thank those people who helped in various ways with the production of the text: Dani Blanden, Lyn Hinds, Janice Rudd and Beverley White.

C.H.T-B, M.B.R.

1

Historical introduction

> Their manner of generation or procreation is exceeding strange and highly worth observing; below the belly the female carries a pouch, into which you may put your hand; inside this pouch are her nipples, and we have found that the young ones grow up in this pouch with the nipples in their mouths. We have seen some young ones lying there, which were only the size of a bean, though at the same time perfectly proportioned, so that it seems certain that they grow there out of the nipples of the mammae, from which they draw their food, until they are grown up and are able to walk.
>
> *Francisco Pelsaert from the translation of Heeres* (1899)

The three sub-classes of living mammals are distinguished most clearly by their modes of reproduction. Many other features set the oviparous Monotremata apart (see Griffiths, 1978) but no other function so distinguishes the Marsupialia from the Eutheria as the manner of their reproduction. This has been recognised since the first marsupials came to the attention of scientists almost 500 years ago and has remained the predominant interest in their study ever since.

Although the original inhabitants of South America and of Australia and the New Guinea islands knew of them, scientific interest began on 8 February 1500 when Vincente Yanez Pinzon collected a female opossum during his first voyage to Brazil. He was so impressed by its pouch and contained young that he took it back to Spain and presented it to King Ferdinand and Queen Isabella at Granada. By then the young were lost and the opossum dead but the description of its presentation was published by Peter Martyr and from thence was republished many times in Europe. In his description Pinzon implies that the young must leave the pouch to suck, which led to the idea that the mammary glands and teats were elsewhere than in the pouch.

A contemporary description, had it been published, would have cleared up this error and established the presence of similar animals in the East Indies, but accident and the rivalry of Spain and Portugal prevented it. Antonio Galvao was Station Captain of the Portuguese settlement at Ternate in the Moluccas from 1536 to 1540 and brought back to Lisbon extensive notes from which he wrote a treatise on the Moluccas. However, the final manuscript was lost before being published and the rough draft remained in the Jesuit library at Seville until translated and published by Father Hubert Jacobs in 1971. Galvao wrote as follows:

> Some animals resemble ferrets, only a little bigger. They are called kusus...On their belly they have a pocket like an intermediate balcony; as soon as they give birth to a young one they grow it inside there at a nipple until it does not need nursing anymore. As soon as she has borne and nourished it, the mother becomes pregnant again.

This is a clear reference to the only species of marsupial that lives on Ternate, *Phalanger orientalis*, which is still known as the cuscus.

Some knowledge of Galvao's observations may have persisted, or else other reports from the Dutch and Portuguese settlements in the East Indies reached scientists in Europe, because writers in the seventeenth century such as Nierembergius (1635) and Piso (1648), referred to animals, similar to the opossum of South America, being found in the Moluccas and Ambon, which were known as cous cous.

Another accurate account, which apparently did not reach the world of science was Pelsaert's description, quoted at the head of this chapter, of reproduction in the tammar wallaby, *Macropus eugenii*, on the Houtmans Albrolhos Islands off Western Australia. His report to the Commissioners of the Dutch East India Company was published in 1648 but was not referred to by later writers, most notably Tyson (1698), who remained sceptical of the reports of marsupials from the East Indies and believed the opossum's pouch to be unique among quadrupeds.

A persistent idea, from the earliest writings except Galvao's, was that the young grows out of the teat instead of being born *per vaginum*. This idea is implicit in Pelsaert's account and was first given explicit form by Piso (1648). Piso spent some time as surgeon to the Governor of the Dutch settlement in Brazil and there observed opossums which, from his description, were *Didelphis albiventris*. He stated firmly that 'the pouch is the uterus of the animal, it has no other as I have determined by dissection. Into this pouch the semen is received and the young formed therein'.

Tyson (1698), however, was scornful of Piso's claims of dissection and, on the basis of his own thorough dissection of a nulliparous Virginian opossum (*Didelphis virginiana*), firmly refuted the notion that the young is formed in the pouch. He described the internal anatomy and showed clearly that the genital tract was double from the ovaries to the opening of the urethra where the two lateral vaginae joined to form a common urogenital canal. He emphasised that there was no direct passage from the 'uterus' (= median vagina) to the common canal and concluded that the young traversed the lateral canals at parturition. He surmised that the penis of the male might be divided to enter the two vaginae, and in this he was confirmed by his colleague William Cowper a few years later. Cowper (1704) dissected a male of the same species (Fig. 1.1) and these two studies laid the foundations of marsupial reproductive biology and can be read today with profit for their accuracy and lucidity. Cowper used the evidence of the complicated but complementary anatomies of the male and female genital tracts to refute the prevailing view that in all mammals the *aura semenalis* of the male passes by way of the blood of the female to fecundate the ova. As he says:

> For to what end has Nature been at the trouble of making double emissaries for the semen of the male opossum, [at the time] she designed the impregnation of a double uterus of the female? Certainly one passage in the glans penis would have been sufficient to convey the semen masculinum to the mass of blood of the female in the manner they conceive. Nature would never have been at the trouble of all this clutter in this animal, in making a double glans, and contriving two distinct apertures in the glans, when its penis is erected, if the propagation of the species had not depended on't: doubtless 'twas for that end chiefly, that the penis of this animal differs so much from what we meet in other Creatures.

Notwithstanding these studies, the idea that the young grow out of the teats persisted during the eighteenth century. Francçis Valentyn (1726), in describing the filander (*Thylogale brunii*) from Ambon in the Moluccas, concluded that the young develop on the teats because bleeding follows their forcible removal.

As late as 1840 Surgeon Bynoe was still perpetuating the error in his observations on the Marsupiata (Stokes, 1846) relating to *Macropus eugenii* on the Albrolhos Islands and kangaroos elsewhere. He was

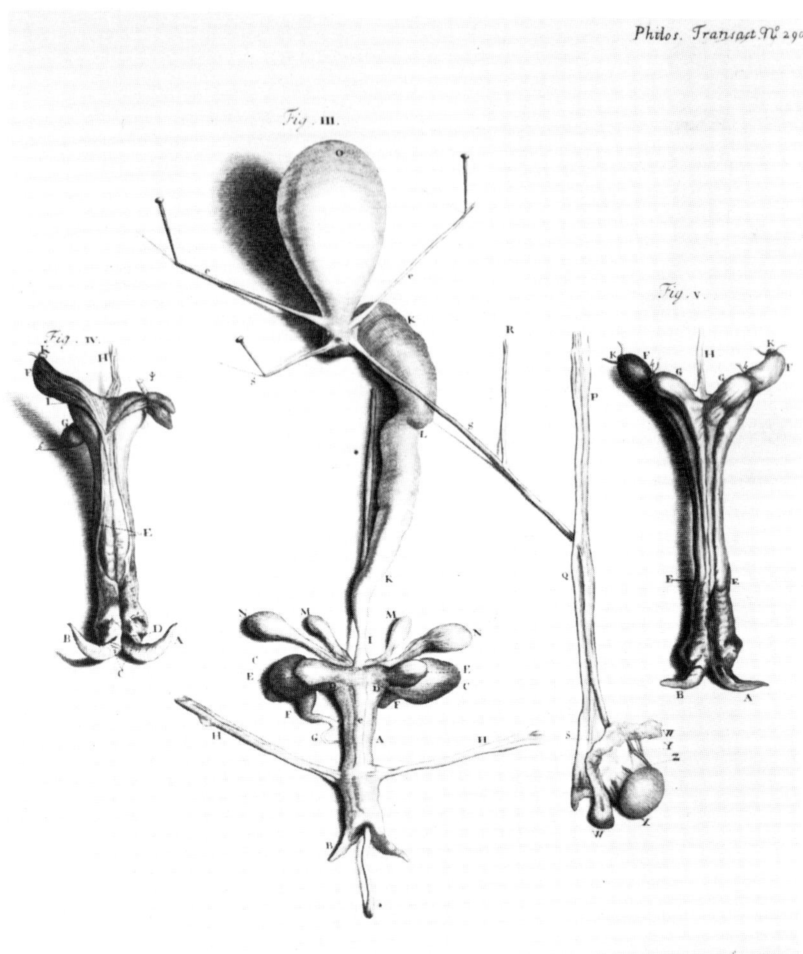

Fig. 1.1. The urogenital tract of a male *Didelphis* as depicted in 1704 by William Cowper, showing the characteristic carrot-shaped prostate and the Cowper's glands. Cowper's key reads as follows:

Fig. III. The backside of the genitals of the male opossum. A the body of the *Penis*; B its glans; CC the *Bulbi* of the *Corpora Cavernosa Penis* covered with their muscles; DD the *Corpora Cavernosa Penis*; EE the two distinct bulbs of the cavernous body of the *Urethra*, inclosed with their particular muscles; FFG parts of the muscles exprest on the fore part of the penis; HH the other pairs of muscles springing from the *Rectum* and inserted to the sides of the *Corpora Cavernosa Penis*; IKL the *Urethra* covered with the prostate KLK; MN the two mucous bags on each side [now called Cowper's glands]; O the bladder of urine; P the *Musculus Cremaster*; Q the *Tunica Vaginalis* opened; R *Vasa Praeparantia* cut from the great trunks [blood vessels supplying the vas]; SS the *Vas Deferens* on each side; WXYZ the left Testicle as it appeared [on

convinced that he had discovered the direct connection from the uterus to the pouch, when he thought he saw a young one in the process of passing to the teat.

By the end of the eighteenth century the discovery of the east coast of Australia, with its variety of marsupials and monotremes, stimulated further interest in marsupial reproduction. Home (1795) dissected a female kangaroo and recognised that it had the same anatomy as Tyson (1698) had described in the opossum, but he was confused in his interpretation of the parts because the specimen he examined was recently mated and had a characteristically enlarged median vagina, filled with semen. He thought this large organ was the gravid uterus and unsuccessfully attempted to find a fetus within it. However, he did observe a small but open canal between this chamber and the urogenital sinus, posterior to the bladder and concluded, correctly, that this was the route taken by the fetus at birth. Others followed him and while some were unable to confirm the presence of the birth canal others did.

Controversy, sometimes quite heated, continued through most of the nineteenth century on this matter. Seiler (1828) examined a lactating kangaroo and attempted to test Home's idea by filling the chamber from

Fig. 1.1. *cont.*

opening the *Tunica Vaginalis*]; W its *Epididymis*; X the body of the Testicle; Y the spermatick vein and artery, as they pass to and from the testicle; Z the excretory duct of the testicle [ductulus efferentis], which could be distinctly seen arising from the *Testes* and marching to the *Epididymis*; W, where it is folded up and constitutes that body, whence it is continued to the bladder of urine and called *Vas Deferens* SS; ee parts of the ureter; **a probe inserted into part of the *Urethra*.

Fig. IV. The fore part of the *Penis*, as it appears when its *Corpora Cavernosa* are filled with mercury and dried; figured big as the life; AB its forked *Glans*; CC the two distinct apertures that appear in the distention on erection of its *Corpora Cavernosa*; D the middle part of the orifice of the urethra which is occluded on the intumescence or erection of the *Penis*; E the two veins of the *Glans*, which are compressed by the two *Sphincter Muscles* of the *Male* and *Female* in coition; F the bulbs of one of the cavernous bodies of the *Penis* distended; G one of the bulbs of the cavernous body of the *Urethra* also distended. These *Bulbi* were opened on the other side to fill the cavernous bodies with quick-silver, but are all exprest as they ought to appear on both sides in the following figure. H the *Urethra*; I the muscles dried, exprest in Fig. III FFG; Kk the veins tied up to keep in the mercury, as they pass the muscles of the bulbi.

Fig. V. The back part of the *Penis* exprest in the preceding Fig.; AB its forked *Glans*; EE parts of the veins arising from the *Glans*; FF the bulbs of the cavernous bodies of the *Penis*; GG the two bulbs the cavernous body of the *Urethra*; H the *Urethra*; KKkk the veins tied up, as they pass out of the *Bulbi* to keep in the mercury.

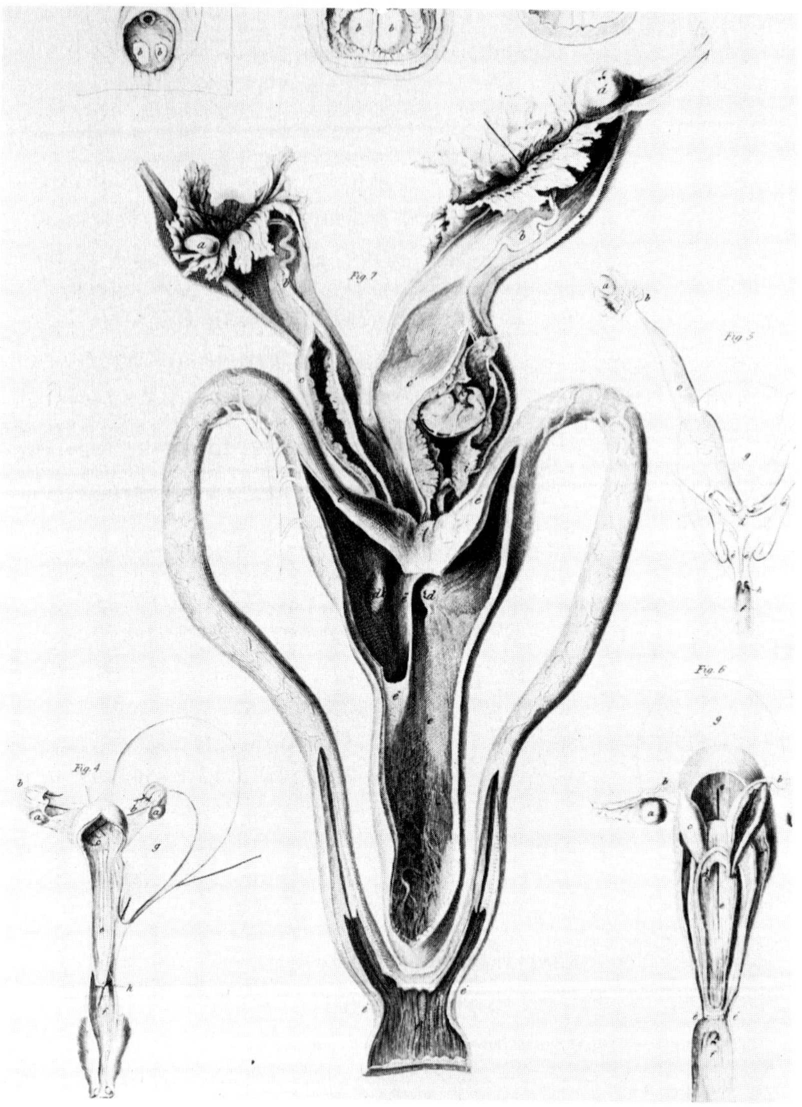

Fig. 1.2. A part of Plate VI from Richard Owen's (1834) paper on the generation of the marsupial animals showing the female organs of *Dasypus novemcinctus* (4), *Didelphis dorsigera* (= *Marmosa murina*) (5), *Hypsiprymnus whitei* (= *Bettongia gaimardi*) (6) and *Macropus major* (= *M. giganteus*) (7). Part of Owen's legend reads as follows: The letters indicate the same parts in each of the figures.

a. Ovaries; b. Fallopian tubes; c. Uteri; (*Cornua uteri*, Tyson); d. Os tincae; e. Mesial cul de sac of the vagina; é, é. Divided portion of the vagina. (*Uteri reduplicati*, and *Vaginae*, Tyson; *Vaginae*, Geoffroy; *Lateral uterine canals*,

one fallopian tube with mercury to such an extent that it was tightly distended, but could find no opening. Owen (1834) did find an open canal in parous females of the grey kangaroo, *Macropus giganteus* (Fig. 1.2); Owen (1852) and Poelman (1851) found the same in parous, but not in nulliparous, Bennett's wallabies, *M. rufogriseus*, but the significance of their findings eluded them. Not until 1881 was the matter satisfactorily resolved when Lister & Fletcher (1881) and Fletcher (1881, 1883) reviewed the existing literature and examined a large series of several species of Macropodidae in which the previous reproductive condition was known. They established that a pseudovaginal canal forms before parturition and in some species, such as *M. rufogriseus*, it remains open thereafter and in others, as *M. giganteus*, it closes after each birth. Stirling (1889) confirmed this in a female euro (*M. robustus*) taken in the act of parturition with the fetal membranes passing from the uterus through the median vagina to the fetus in the urogenital sinus, about to be delivered.

In North America on the other hand, the start of the nineteenth century saw serious attempts being made to observe the birth of opossums and to determine to what extent the female participates. Although Miegs (1847) in Philadelphia observed the birth of opossum young he believed that the mother participated in its transport to the pouch and Owen (1834) also considered that the young would need to be conveyed to the pouch by the mother. The matter was finally resolved by Carl Hartman (1920*a*) and his wife, who witnessed the whole event in *Didelphis virginiana* and confirmed that the young travel unaided to the pouch. Many subsequent observers of this and other marsupials have confirmed the Hartmans' conclusion (see Chapter 2).

Fig. 1.2. *cont.*

Home); f. Urethro-sexual canal. ('*Canalis communis*', or common passage from the *urethra and the two vaginae*, Tyson; *Canal urétro-sexuel*, Geoffroy; *Vagina*, Home); g. Urinary bladder; h. Urethra.
Fig. 7. The impregnated female organs of the Kangaroo (*Macropus major*, Shaw). The gravid uterus *c′* is laid open, and also the chorion *i*, or membrana corticalis of the foetus, showing the latter suspended from *k*, the umbilical chord. In addition to the letters above explained, *a′* is the left ovary, with a large corpus luteum showing the orifice from which the ovulum escaped not yet cicatrized. *The ovarian ligaments. Bristles are inserted into the Fallopian tubes. The vaginal apparatus *e*, *e′ e′*, not having been preserved along with the impregnated uterus, is here added from another specimen, in which the imperfect septum of the mesial cul de sac
(*e″*) did not extend to the lower end of that cavity, as is usual in the Kangaroo. The cellular membrane which connects the vaginal cul de sac with the urethro-sexual passage has been removed.

During the nineteenth century an equally vigorous debate was pursued about the nature of reproduction in monotremes and of their relationship to marsupials. The two main protagonists in this debate were Etienne Geoffroy-Saint Hilaire and Richard Owen. Neither of them could reconcile the idea that monotremes could be oviparous and lactate, because these were mutually exclusive characters of birds and reptiles on the one hand and of mammals on the other. Geoffroy (1833) accepted the widely held belief of people in Australia that the platypus, *Ornithorhynchus anatinus*, laid eggs and argued that, this being so, the abdominal glands were not mammary glands but modified odoriferous or mucus-secreting glands involved in maintaining the condition of the fur in water. Richard Owen (1832) drew the opposite conclusion: being convinced that the glands were anatomically the same as mammary glands, albeit without teats, were only to be found in females, became enlarged during the breeding season and secreted milk, he concluded that the monotremes lactated; therefore he concluded that they could not be oviparous but must be ovo-viviparous like marsupials. He held to this view against all contrary evidence for 50 years (Owen, 1880). Finally Caldwell (1884) established the oviparity of the monotremes when he collected an echidna with an egg in its pouch and a platypus that had laid one egg and had a second in the uterus. In the same month Haacke (1885) found an echidna with an egg in the pouch and Owen (1887) then somewhat grudgingly conceded that monotremes are oviparous and lactiferous.

Caldwell's specimen was obtained during an expedition in which over 1400 specimens of platypus and echidna were collected for him. Only one preliminary paper (Caldwell, 1887) resulted from this 'fantastic slaughter' (Griffiths, 1978) but in it he established the manner of formation of the egg in the follicle, the meroblastic nature of egg cleavage, and the stage of development reached at egg laying.

Caldwell's other contribution was as the co-inventor of the automatic microtome, which enabled serial sectioning of tissue and embryos and so set the stage for the much more thorough and extensive embryological studies of the late nineteenth and early twentieth centuries. Richard Semon visited Australia to collect monotreme and marsupial embryos, which he described in beautiful detail (Semon, 1894) (Fig. 1.3) and Selenka (1887, 1892) imported opossums from America to breed in Germany, so that he could obtain dated embryos for the first detailed study of a marsupial.

In 1894 J. P. Hill arrived in Sydney to work with J. T. Wilson and began one of the most fruitful periods in the study of marsupial and monotreme reproduction. He discovered that the Peramelidae, unlike all other marsupials, have an allantoic as well as a yolk sac placenta (Hill, 1895, 1899, 1900*a*); that the pseudovaginal canal is a transient structure in the Peramelidae (Hill, 1899, 1900*a*), *Trichosurus vulpecula* (Hill, 1900*c*), *Dasyurus viverrinus* (Hill, 1900*b*) and, by inference, in all marsupials. He and Wilson collected material of *Ornithorhynchus* and *Tachyglossus* with which, during the next 30 years, they published a detailed description of the ovary, genital tract and embryo of both these species of monotreme (summarised by Griffiths, 1978). Hill's most enduring achievement however was to be, with his students, among the first anywhere to appreciate the endocrine relationship between the corpus luteum and the uterus, embryo and mammary gland in mammals. In a paper, based on Hill's material, Sandes (1903) wrote '...it may be stated as probable, firstly, that the corpus luteum is a glandular structure with an internal secretion; and secondly, that it influences the genital organs and the organism generally and prevents ovulation during pregnancy, and temporarily if pregnancy does not occur'.

Later, in their study of *Dasyurus viverrinus*, Hill & O'Donoghue (1913) observed that after ovulation similar changes occurred in the uteri and mammary glands of unmated females as occurred in mated females and they coined the term pseudo-pregnancy for this condition. The term was subsequently adopted by Long & Evans (1922) and Hammond & Marshall (1925) to describe the condition that follows sterile mating in the rat and the rabbit, and has now been widely assumed for this phenomenon. Because this meaning has changed from the original one, Sharman (1959) recommended that it be no longer used in referring to the marsupial condition, which has no reference to the occurrence of sterile mating. The main deficiency of Hill's work was that little of his material could accurately be referred to the stage of the reproductive cycle. This criticism does not apply to Carl Hartman's great study on reproduction and development of *Didelphis virginiana* (Fig. 1.4) done contemporaneously in Texas, from 1913 to 1952. Hartman was quick to adopt Stockard & Papanicolau's (1917) technique of the vaginal smear for detecting the stage of the oestrous cycle (Hartman, 1923*a*) and was able to use it to examine experimentally the endocrine role of the ovary (Hartman, 1925*b*), the corpus luteum (Hartman, 1927) and the uterus (Hartman, 1925*a*) in opossum reproduction. He was the first to recognise the importance of marsupials for comparative studies on reproduction and to pioneer the experimental approach to the endocrinology of reproduction. He also recognised the potential that the monovular marsupials of Australia offered for this, but more than 30 years elapsed before this began to be realised.

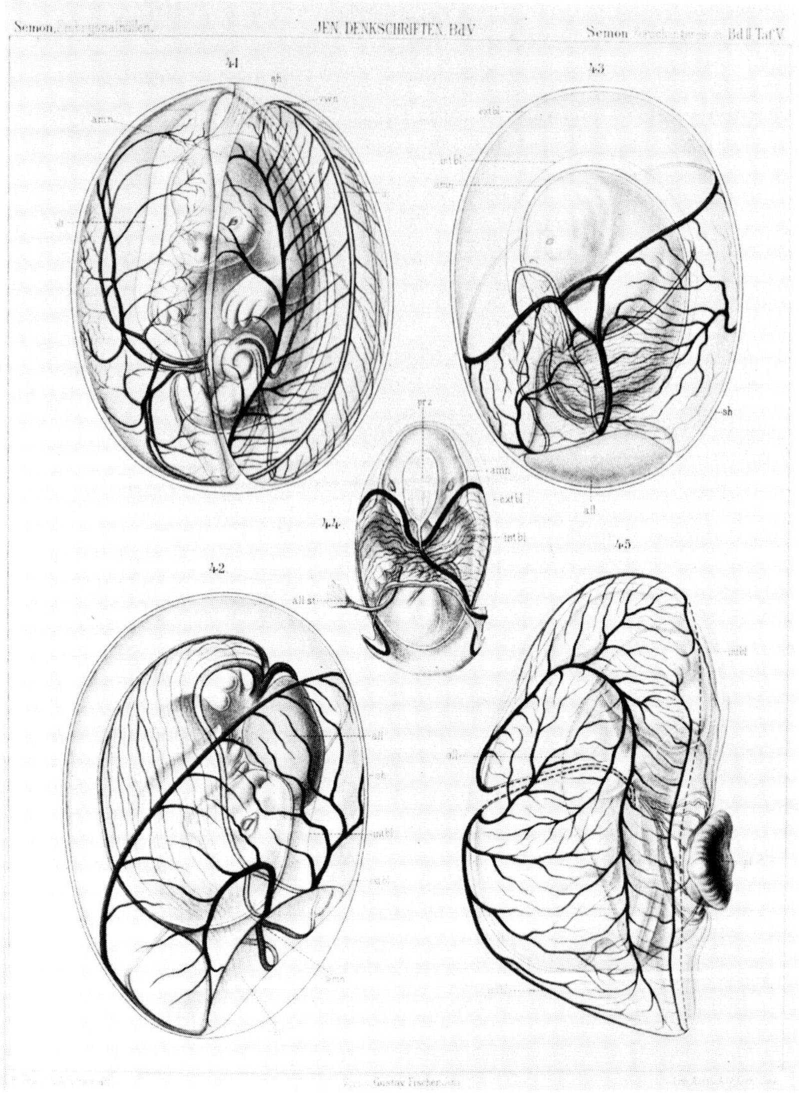

Fig. 1.3. The arrangement of the embryonic membranes of monotreme, marsupial amd eutherian mammals as depicted by Semon (1894). Semon's key reads as follows:

Fig. 41: Embryo of *Echidna aculeata* [= *Tachyglossus aculeatus*] in its fetal membranes.

Fig. 42: Embryo of *Aepyprymnus rufescens* in its membranes.

Fig. 43: Embryo of *Phascolarctos cinereus*. The same is in a shifted position inside the yolk sac.

Fig. 44: Another *Phascolarctos* which is in the normal position, but less

The modern period in marsupial reproductive biology can be dated from Sharman's (1954, 1955*b*) discovery of delayed implantation, or embryonic diapause, in *Setonix brachyurus* and his confirmation of the phenomenon in *Macropus eugenii* (Sharman, 1955*c*), the species Pelsaert had described 300 years before. His study of these monovular diprotodont marsupials reawakened interest in marsupial reproduction and, in the following decade, the reproductive cycles of many marsupials were described (Sharman, Calaby & Poole, 1966).

The discovery of embryonic diapause also stimulated the first experiments on the hormonal control of reproduction in monovular marsupials and, less directly, the first studies on lactation and pouch young (Waring, Moir & Tyndale-Biscoe, 1966). Two trends in research have developed subsequently, the one ecological and the other physiological. There is a burgeoning interest in the ecological significance of reproductive patterns and strategies of the different species of marsupials, made possible by the ever-increasing number of species that are being studied in Australasia and in South America (see Hunsaker & Shupe, 1977; Nelson, 1978; Archer, 1982; Charles-Dominique, 1983; Smith & Hume, 1984). These studies are now enabling much more critical analyses of reproductive strategies (Parker, 1977; Low, 1978; Morton *et al.*, 1982; Russell, 1982*a*, *b*; Cockburn, Lee & Martin, 1983). In Chapter 2 the diversity of marsupial reproduction is considered systematically, because we find that there are common patterns within families, but the larger evolutionary and ecological implications of this are not developed because they are the subject of the companion volume in this series by Lee & Cockburn (1985).

The remainder of this book treats each of the main areas of reproductive physiology that have been actively studied in the past two decades. Most of this work has been motivated by interest in marsupials for their own sake, but increasingly, as the approaches become more critical, the special features of marsupial reproduction are providing new understanding of reproductive process in other mammals as well.

Fig. 1.3. *cont.*

enlarged. The chorion and outer wall of the yolk sac are only indicated by a dotted line, and the allantois is shown cut off at the end of the stalk.
Fig. 45: Older rabbit embryo in its yolk sac membranes. The outer mesoderm and avascular yolk sac, which are closely connected to the mucous membrane of the uterus, is left out. all, allantois; allst, allantoic stalk; amn, amnion; ds, yolk sac; extbl, outer (non-inverted) yolk sac wall; intbl, inner (inverted) yolk sac wall; kr, cavity formed in the yolk sac, in which the embryo lies; pl, placenta; prz, fusion between the amnion and the yolk sac (pro-amnion remnant); sh, chorion; vwn, contact between the amnion and chorion.

Marsupials have a similar primary sex-determing mechanism as other mammals but different mechanisms for dosage compensation of sex chromosomes. The low number and large size of the chromosomes has enabled these to be identified more readily and the mechanism examined. The young are born at the indifferent stage of sexual differentiation, so that the process is readily accessible for experimental investigation. These aspects are reviewed in Chapter 3. The maturation of the spermatozoa and the role of the epididymis in this are different from eutherian patterns (Chapter 4). While oogenesis and folliculogenesis provide no unusual features, the female genital tract, as already mentioned, is very unusual and the function of the several parts and their endocrine control are reviewed in Chapter 5.

The many functions of the corpus luteum and the unusual manner of its control by the pituitary have been intensively studied for many years and form the substance of Chapter 6. In Chapter 7 several aspects of pregnancy and intrauterine development are considered from fertilisation, through blastocyst formation and the control of embryonic diapause, to placental function and parturition. In the beautiful and close relationship

Fig. 1.4. A female Virginian opossum, *Didelphis virginiana*, with a litter of six young about 90 days old. It occurs in Central and North America and was the subject of the classical studies of Tyson, Cowper, Selenka, Hartman and McCrady.

throughout lactation between composition and production of milk, on the one hand, and development of the pouch young on the other, marsupials provide an unique system of profound interest for understanding mammalian development and evolution. The physiology of lactation is discussed in Chapter 8 but it is expected that the growth and development of the pouch young will form the main subject of a companion volume later.

In Chapter 9 the central control of reproduction in *Macropus eugenii* is reviewed. No other marsupial has been investigated to anywhere near the same extent as this species and its remarkable sensitivity to photoperiod change makes it an exceptionally useful species in which to investigate the interdependent roles of the pineal, the pituitary and the ovary in regulating reproductive activity.

In all these respects we hope to show that marsupials have much to offer to the general understanding of reproductive physiology. This change in perception of the value of marsupial studies has occurred at the same time as palaeontology is providing a clearer idea about the the origin and evolution of mammals. We now know that marsupials and eutherian mammals have been separated for more than 100 million years, so that much of the diversity in reproductive processes in both groups has evolved since the dichotomy and comparisons can provide deeper understanding of how they came about. These matters are discussed in the final chapter.

2

Breeding biology of marsupials by family

The most recent checklist of living marsupials is that of Kirsch & Calaby (1977) and their nomenclature is used throughout this book except for more recent revisions in the Dasyuridae reported by Archer (1982). Kirsch & Calaby (1977) recognise 249 species in 16 families. Sixty species live in the Americas and the rest in Australia or the New Guinea islands. They range in size from 7 g to 90 kg and occupy rainforest, dry woodlands, open grasslands and semi-desert from sea level to altitudes of 3000 m in New Guinea and South America. Some are carnivorous, some insectivorous, others are arboreal folivores and others grazers or browsers (see Hume, 1982). Of these species there is now substantial information about the reproduction of 32 species representing 6 families and moderate or slight information about a further 45 species (Tables 2.1 and 2.2). In compiling the material for this chapter we have drawn extensively on the reviews of Collins (1973), Hunsaker (1977), Hunsaker & Shupe (1977) and Charles-Dominique (1983) for neotropical marsupials, Strahan (1983) for Australian marsupials, Archer (1982) for Dasyuridae, Smith & Hume (1984) for Phalangeroidea and Tyndale-Biscoe, Hearn & Renfree (1974) and Tyndale-Biscoe (1984) for the Macropodidae.

Russell (1982*a*) has provided a fine review of the maternal contribution to reproduction in marsupials, which emphasises the importance of adult body size in reproduction and the constraints it imposes on the strategies adopted by various species. For marsupials there are other constraints as well, which need to be borne in mind when reviewing reproductive patterns systematically. The young are born at a very early stage of development, are very small and are wholly dependent on the mother for a certain time after birth. During this period each young is permanently attached to one teat and has exclusive use of the associated mammary gland, so that the maximum litter size is determined absolutely by the number of functional mammary glands the female possesses. During lactation the young undergo a major part of their growth and become homeothermic so that at weaning the litter may weigh, in total, considerably more than the mother (see Fig. 2.5). For these reasons lactation is generally of long duration and the major investment in reproduction is made by a female then, rather than in gestation. Furthermore the young is most vulnerable, not at birth but at the time of emergence from the nest or pouch. Seasonal breeding reflects this: in places where the food resources or climate vary through the year, pouch emergence (not parturition) is synchronised to the most favourable period of the year.

Nevertheless, each major family grouping of marsupials shares a common and distinctive reproductive pattern, with each species displaying modifications of the basic pattern in relation to its own constraints of size, diet and habitat (see Lee & Cockburn, 1985). The treatment in this chapter reflects these constraints. For each of the larger families the common features of distribution and habits are mentioned, followed by a review of oestrous behaviour and pregnancy, parturition and lactation for the best-known representatives and concluding with a review of breeding in these and other species in the family. Data for the best known species are summarised in Tables 2.1 and 2.2, following the schemes of Russell (1982*a*), Tyndale-Biscoe (1984) and Lee & Cockburn (1985).

Didelphidae

The Didelphidae comprise the predominant family of American marsupials but, of the 60 species, the reproduction of only three has been investigated in some detail (Table 2.2). The best-known species is *Didelphis virginiana* from North and Central America, which has been the subject of numerous laboratory and field studies. Two smaller species from South America, *Marmosa robinsoni* (= *mitis*) and *Monodelphis domestica*, have been studied in captivity with a view to their use as laboratory animals. In addition there is scattered or slight information about several other South American species: *D. marsupialis*, *D. albiventris*, *Philander opossum*, *Caluromys derbianus* and *C. philander*. The basic reproductive pattern of all these species is very similar and *D. virginiana* will serve as the main example.

D. virginiana is a terrestrial omnivore that ranges in body weight from 1 to 4 kg, the heaviest animals being found in northern latitudes and the lightest nearer the equator. It is a seasonal breeder with the onset of reproduction occurring soon after the winter solstice. In northern parts of

Table 2.1. *Summary of data on reproduction and development for marsupial species, arranged by family and within families by increasing adult female body weight*

Species	Adult female weight (g)	Reproductive group	Mammary area type	Number of teats	Litter size	Neonate weight (mg)
Didelphidae						
Marmosa robinsoni	40–50	1	1	13	8	100
Monodelphis domestica	80–100	1	1	22	3–14	100
Philander opossum	250–400	1	5		4–6	—
Caluromys derbianus	330	1		6	3–4	—
Caluromys philander	285–340	1		6	4.1	—
Didelphis marsupialis	1200	1	5	9	2–9	—
Didelphis virginiana	1000–4000	1	5	11–17	3–13	130
Didelphis albiventris	2000–4000	1	5	—	4–5	—
Caenolestidae						
Caenolestes obscurus	20–30	?	1	4	3–4	—
Dasyuridae						
Ningaui ridei	8–10	1	1		7	5
Planigale ingrami	6–9	1	2		4–12	—
Planigale tenuirostris	6–9	1	2	—	4–8	—
Planigale maculatus	9–16	1	2	8–13	4–12	—
Planigale gilesi	6–10	1	2	12	7	—
Sminthopsis crassicaudata	12–18	1	3	8–10	7.5	10
Sminthopsis macroura	16–24	1	3	8	1–8	10
Sminthopsis murina	21	1	3	—	4–10	—
Antechinomys laniger	63		2	—	5	—
Antechinus stuartii	26–30	M	1	6–10	6–10	16
Antechinus flavipes	30	M	1	—	12	—
Antechinus swainsonii	50	M	1	—	8–10	—
Parantechinus bilarni	12–34	1	1	6	4–5	—
Parantechinus apicalis	68	M	1	—	—	—
Phascogale tapoatafa	140–180	M		8	1–8	—
Dasycercus cristicauda	60–130	M	1	6	6	—
Dasyuroides byrnei	90–150	1	1	6	4.8	—
Dasyurus hallucatus	300–500	—	1	—	6.4	—
Dasyurus viverrinus	1350	1	1	5–8	5.8	12.5
Dasyurus maculatus	2000–4000	—	2	—	5	—
Sarcophilus harrisii	6700–12000	1	2	4	2.9	18
Myrmecobius fasciatus	500	1	1	—	2.4	—
Notoryctes typhlops	40–70	—	6	2	—	—

…chment/ off (…s)	Permanent pouch exit/ left in nest (days)	Weaning (days)	Sexual maturity (months)		Birth season	References
			Female	Male		
	35–40	60–70	10	12	Mar–Sept	Barnes & Barthold (1969); Collins (1973); Godfrey (1975); Hunsaker & Shupe (1977)
	—	50	—	—	All year	Fadem *et al.* (1982); VandeBerg (1983*b*)
	—	90	—	—	Feb–Nov	Hunsaker (1977)
	—	—	—	—	Feb–Dec	Phillips & Jones (1968)
	75–80	100–125	—	—	Oct–Dec	Atramentovicz (1982)
	—	100		6–8	Feb–Aug	Fleming (1973); Tyndale-Biscoe & Mackenzie (1976); Hunsaker (1977)
	70	110	5	?	Jan–Aug	Hartman (1921); Reynolds (1952)
	—	—	—	—	May–Aug	Tyndale-Biscoe & Mackenzie (1976); Streilein (1982)
	—	—	—	—	Feb–May	Osgood (1921); Kirsch & Waller (1979); Tyndale-Biscoe (1980*a*)
2–44	—	90	—	—	Sept–Jan	Fanning (1982)
5–40	—	90	—	—	Dec–Mar	Fleay (1965); Heinsohn (1970); Woolley (1974); Archer (1976)
	—	—	—	—	Summer	Denny (1982)
8	45	70	10	—	July–Jan	Aslin (1975); Van Dyck (1979)
7	37	65–70	—	—	July–Jan	Denny (1982); Whitford *et al.* (1982)
3	59–63	65–68	4	5	June–Feb	Godfrey & Crowcroft (1971); Woolley (1973); Morton (1978)
0	—	70	4	7	July–Feb	Godfrey (1969*a*)
5	—	65	—	—	Aug–Jan	Fox & Whitford (1982)
8	—	80	—	—	July–Nov	Woolley (1973); Lee & Cockburn (1985)
40–45	75	90–110	9–10	—	July–Aug	Marlow (1961); Woolley (1966*a*) Wood (1970); Selwood (1982*a*)
42	—	90–120	—	—	—	Woolley (1973)
33–43	—	90–95	—	—	—	Dickman (1982)
–	—	*c.* 90	—	—	Aug–Sept	Begg (1981*a*)
–	—	–120	10–11	—	—	Woolley (1971*b*; 1973)
54	—	120	7.5	—	July–Aug	Cuttle (1982)
55–60	88	100–120	12	12	May–July	Michener (1969); Sorenson (1970); Woolley (1971*a*)
56	70–78	100–120	12	12	June–Nov	Mack (1961); Woolley (1971*a*); Aslin (1974; 1980); Fletcher (1983)
60–70	—	125	10–11	—	July–Aug	Nelson & Smith (1971); Woolley (1973); Begg (1981*b*)
49–56	91	135–140	12	12	May–Aug	Hill & O'Donoghue (1913); Hill & Hill (1955); Green (1967); Nelson & Smith (1971); Woolley (1971*a*)
49	96	150	12	12	July–Aug	Collins (1973); Settle (1978)
90–105	105	150–240	24	—	Mar–July	Green (1967); Guiler (1970)
–	160	180	—	—		Calaby (1960); Friend & Burrows (1983)
–	—	—	—	—	—	Johnson in Strahan (1983)

Table 2.1. cont.

Species	Adult female weight (g)	Reproductive group	Mammary area type	Number of teats	Litter size	Neonatal weight (mg)
Peramelidae						
Perameles gunnii	816	2	6	8	2.3	250
Perameles nasuta	860	2	6	8	2.4	237
Isoodon macrourus	1130	2	6	8	3.4	180
Isoodon obesulus	766	2	6	8	2.8	350
Thylacomyidae						
Macrotis lagotis	800–1100	2	6	8	2	—
Burramyidae & Tarsipedidae						
Cercartetus concinnus	50	4	5	6	4	12
Cercartetus nanus	24	4	5	—	4–5	—
Burramys parvus	40	1	5	4	4	—
Acrobates pygmaeus	11–17	4	5	4	2	—
Tarsipes rostratus	10–12	4	5	4	2.4	3–6
Petauridae						
Gymnobelideus leadbeateri	122–133	1	5	?	1.6	—
Petaurus breviceps	150–200	1	5	4	1.6	194
Petaurus australis	450–700	1	4	2	1	—
Pseudocheirus peregrinus	700–1000	1	5	4	2	300
Petauroides volans	1000–1250	1	5	2	1	273
Phalangeridae						
Trichosurus vulpecula	1500–3500	1	5	2	1	200
Trichosurus caninus	2500–4500	1	5	2	1	—
Phascolarctidae						
Phascolarctos cinereus	4500–7900	?	6	2	1	—
Vombatidae						
Vombatus ursinus	26000	1	6	2	1	500
Lasiorhinus latifrons	25000	1	6	?	1	500
Potoroidae						
Hypsiprymnodon moschatus	510	?	4	4	1–2	—
Potorous tridactylus	660–1000	3	5	4	1	330
Bettongia lesueur	1100	3	5	4	1	317
Bettongia penicillata	1000–1600	3	5	4	1	290
Bettongia gaimardi	1800	3	5	4	1	300
Aepyprymnus rufescens	1000–3500	3	5	4	1	—

achment/ off /s)	Permanent pouch exit/ left in nest (days)	Weaning (days)	Sexual maturity (months) Female	Male	Birth season	References
	48–53	59–61	3	4–6	Jun–Dec	Lyne (1964); Heinsohn (1966)
	62–63	62–68	4	5	All year	Lyne (1964); Stodart (1966*a*); Close (1977)
	50	60	3.4	—	All year	Lyne (1964); Gordon (1974); Gemmell (1982)
	53	58	4–5	6	July–Dec	Lyne (1964); Heinsohn (1966); Stoddart & Braithwaite (1979)
	75	90	—	—	All year	Hulbert (1972); Johnson in Strahan (1983)
25	30	50	12–15	—	Nov–Jan	Bowley (1939); Casanova (1958); Clark (1967)
	42	50–60	5	—	Sept–Apr	Turner in Strahan (1983)
	33–37	70–75	12	—	Nov	Dimpel & Calaby (1972); Kerle (1984)
	50	90–95	6–8	—	July–Jan	Fleming & Frey (1984)
	63–70	90	6	—	All year	Wooller *et al.* (1981); Renfree *et al.* (1984*a*)
	87	120	12	—	Apr–Jun Oct–Dec	Smith (1984*a*)
	70–74	120	8–15	12	Jun–Nov	Smith (1971; 1973; 1979*b*); Suckling (1984)
	100	180–240	24	18	Nov–May	Russell in Strahan (1983); Russell (1984*b*)
2–49	120	180–210	12	12	May–Aug	Thomson & Owen (1964); How *et al.* (1984)
75	*c.* 150	*c.* 240	24	24	Apr–Jun	Smith (1969); Tyndale-Biscoe & Smith (1969); Bancroft (1973); Henry (1984)
4	140–150	230	12–24	24	Mar–Nov	Lyne & Verhagen (1957); Dunnet (1964); Smith, Brown & Frith, (1969); Crawley (1973)
2	175–200	275	22–36	36	Apr–July	Smith & How (1973); How (1976)
	240–270	360–380	36	—	Oct–Mar	Smith (1979*a*); Martin & Lee (1984)
	> 150	400	24	—	All year	McIlroy (1973); G. E. Young & G. D. Brown (personal communication)
	300	400	—	—	Nov–Jan	Crowcroft & Sonderland (1977) Gaughwin & Wells (1978)
	147	—	> 12	—	Feb–July	Johnson in Strahan (1983)
0	130	147	12	—	June–Aug	Hughes (1962*a*); Shaw & Rose (1979); Heinsohn (1968)
1	115	165	7	14	All year	Tyndale-Biscoe (1968)
	100	130	10	—	All year	Parker (1977)
	109	160	8–11	—	All year	Rose (1978)
37	114	155	10	12	All year	Johnson (1978)

Table 2.1. cont.

Species	Adult female weight (g)	Reproductive group	Mammary area type	Number of teats	Litter size	Neonatal weight (mg)
Macropodidae						
Peradorcas concinna	1400	?	?	?	1	—
Setonix brachyurus	2750	3	5	4	1	350
Macropus parma	3500	3i	5	4	1	510
Petrogale penicillata	4000	3	5	4	1	—
Thylogale thetis	3900	3	5	4	1	—
Thylogale billardierii	3900	3	5	4	1	400
Macropus eugenii	5000	3	5	4	1	370
Wallabia bicolor	11 500	3	5	4	1	610
Macropus rufogriseus	14 000	3	5	4	1	450
Macropus parryi	14 000	3i	5	4	1	—
Macropus agilis	12 000	3	5	4	1	630
Macropus robustus	16 000	3	5	4	1	—
Macropus rufus	27 300	3	5	4	1	817
Macropus giganteus	27 600	3i	5	4	1	740
Macropus fuliginosus	27 600	3i	5	4	1	828

3i = intermediate form of Group 3.

its range the males as well as the females show a recrudescence after the solstice (Chase, 1939; Biggers, 1966) but, in the tropics, Biggers (1966) found no change in testis size or in spermatogenesis through the year.

Oestrous cycle and pregnancy

The females are polyoestrous and, in captivity, will undergo up to 7 cycles in a year (Jurgelski & Porter, 1974). Hartman (1923*a*) considered the mean to be 28 days (22–34) but Jurgelski & Porter (1974) found it to be 25.5 days (17–34). Some of this variation may be due to a seasonal shift in cycle length as Reynolds (1952) observed a shift in mean cycle length of 10 females from 24.9 days for the first cycle to 35.3 days for the last cycle. Some females also exhibit much shorter anovulatory cycles, which were ascribed to dietary deficiency (Hartman, 1923*b*; Martinez-Esteve, 1937). The cycles of captive *Monodelphis domestica* and *Marmosa robinsoni* are similar (Table 2.2); Godfrey (1975) and Fadem & Rayve (1985) observed some females of each species to undergo short anovulatory cycles of 16 and 14.4 days respectively, compared to ovulatory cycles of 25.5 and 32.3 days. These short cycles, probably represent the follicular phase of

achment/ : off ys)	Permanent pouch exit/ left in nest (days)	Weaning (days)	Sexual maturity (months) Female	Male	Birth season	References
	180	360	12–24		All year	G. Sanson in Lee & Cockburn (1985)
	190	240	9–12	13	Jan–Mar	Shield (1968)
	212	300	16	22	All year	Maynes (1973*a*)
	204	290	18	20	All year	Johnson (1979)
	181	210	26	20	?	Johnson (1977)
	200	300	14	—	Apr–June	Rose & McCartney (1982*b*)
	250	270	8	24	Jan–June	Andrewartha & Barker (1969); Murphy & Smith (1970)
	256	—	15	—	All year	Calaby & Poole (1971)
	270	360	11–21	13	Jan–Aug	Catt (1977); Merchant & Calaby (1981)
	275	420	24	18–24	—	Maynes (1973*b*)
	219	328	12	14	All year	Kirkpatrick & Johnson (1969); Merchant (1976); Bolton, Newsome & Merchant (1982)
	256	380	27	—	All year	Ealey (1967)
	235	360	14–20	24–36	All year	Sharman & Calaby (1964); Sharman & Pilton (1964)
	319	540	18	48	Oct–Mar	Kirkpatrick (1965); Poole (1973; 1975); Poole & Catling (1974)
	310	540	14	31	Oct–Mar	Poole (1973; 1975; 1976); Poole & Catling (1974)

the oestrous cycle (Table 2.2). In captive females the oestrous cycle can be followed by monitoring the changes in cells collected from the urogenital sinus or posterior vaginal sinus (see Chapter 5). The method and interpretation has been described by Jurgelski & Porter (1974).

Behavioural oestrus in all three species lasts for up to 36 h, and copulatory behaviour is similar in the three species. During the initial approach, the male *Didelphis* utters a characteristic clicking sound and alternately attempts to mount the female and, if repulsed, adopts a submissive posture. Oestrous females are passive and allow the male to approach and initiate copulation. The male approaches the female from behind and grasps each of her limbs with his feet and may also seize the nape of her neck in his mouth. Males of *Marmosa* use the prehensile tail to hold onto some firm object such as the cage wall (Fig. 2.1*b*) and, unless this is achieved, penile erection does not occur. *Didelphis* and *Monodelphis* couples generally fall over, usually to the right side when copulation takes place (Fig. 2.1*a*). Copulation lasts 5–30 min (Hunsaker & Shupe, 1977). Although copulation may take place at any time during the 36 h, conception is more likely to occur if it takes place early in the oestrous

Table 2.2. *Duration of the oestrous cycle, gestation and the follicular phase, and average ovulation rate for the best-studied species of marsupial*

Species	Oestrous cycle	Gestation	Ratio	Post-partum oestrus	Diapause	Follicular phase of oestrous cycle	Removal of pouch young to birth	Ovulation number	References
Group 1									
Didelphis virginiana	25.5 (22–34)	13	0.51	N	N	7–17	26	22	Hartman (1923*a*); Reynolds (1952)
Marmosa robinsoni	25.5	13.5	0.53	N	N	16.0	—	20	Godfrey (1975)
Monodelphis domestica	32.3	13.5	0.42	N	N	14.4	—	?	Fadem *et al.* (1982); Fadem & Rayve (1985)
Dasyurus viverrinus	37	19	0.51	N	N	20	?	7–35	Hill & O'Donoghue (1913); Fletcher (1985); J. C. Merchant (personal communication)
Dasyuroides byrnei	60	30–31	0.51	N	N	—	—	11	Woolley (1971*a*); Aslin (1980); Close (1983); Fletcher (1983)
Sminthopsis crassicaudata	31	13–16	0.47	N	N	14–16	30	14	Smith & Godfrey (1970); Godfrey & Crowcroft (1971)
Sminthopsis macroura	26	12.5	0.48	N	N	—	—	30–40	Godfrey (1969*a*)
Antechinus stuartii	Monoestrous	27	—	N	Y	N	N	11–19	Selwood (1982*a*, *b*; 1983)
Gymnobelideus leadbeateri	< 30	< 20	0.60	N	N	—	—	12	Smith (1984*a*)
Petaurus breviceps	29	16	0.55	N	N	12	28	> 2	Smith (1971)
Trichosurus vulpecula	25.7	17.5	0.68	N	N	8	26	1	Pilton & Sharman (1962)
Trichosurus caninus	26.4	16.2	0.61	N	N	10–11	26	1	Smith & How (1973)
Group 2									
Perameles nasuta	21 (10–34)	12.5	0.60	N	N	5–10	—	3.3	Stodart (1966*a*); Lyne (1976); Close (1977); Lyne & Hollis (1979)
Isoodon macrourus	20.5 (9–34)	12.5	0.61	N	N	10–20	—	5.1	Lyne (1976); Lyne & Hollis (1979) Gemmell (1981)
Macrotis lagotis	20.4 (12–37)	14 (13–16)	0.69	N	N	—	—	—	McCracken (1986)

Group 3 (Intermediate)									
Macropus parma	41.8±0.7	34.5±0.1	0.83	N	Y	6–15 (17%) > 45 (83%)	31.2	1	Maynes (1973*a*)
Macropus parryi	42.2	36.3	0.86	N	Y	6	?	1	Calaby & Poole (1971); Maynes (1973*b*)
Macropus giganteus	45.6±9.8	36.4±1.6	0.80	N	Y	10.9±4.8	28–32	1	Kirkpatrick (1965); Poole & Catling (1974); Poole (1975)
Macropus fuliginosus	34.9±4.4	30.6±2.6	0.88	N	N	8.3±5.8	Never	1	Poole & Catling (1974); Poole (1975)
Group 3									
Bettongia lesueur	23	21	0.91	Y	Y	22	18	1	Tyndale-Biscoe (1968)
Bettongia gaimardi	23.2	21.1	0.91	Y	Y	—	18	1	Rose (1978)
Bettongia penicillata	22–23	21	0.93	Y	Y	—	21	1	Parker (1977)
Potorous tridactylus	42	38	0.90	Y	Y	—	29	1	Shaw & Rose (1979)
Aepyprymnus rufescens	21–36	21–30	—	Y	Y	20.6	18.7	1	Moors (1975); Johnson (1978)
Setonix brachyurus	28	27	0.96	Y	Y	26–27	25	1	Sharman (1955*a*, *b*); Tyndale-Biscoe 1963*a*); Shield (1968)
Thylogale billardierii	30.3	30	0.99	Y	Y	—	28.7	1	Rose & McCartney (1982*a*)
Petrogale penicillata	32	31	0.97	Y	Y	31	29	1	Johnson (1979)
Macropus eugenii	30.6	29.3	0.96	Y	Y	30.4	26.2±0.7	1	Merchant (1979)
Macropus rufogriseus	31.9	29.4	0.92	Y	Y	29.6	27.8	1	Merchant & Calaby (1981)
Macropus agilis	32.4	29.4	0.91	Y	Y	29.8	26.5	1	Merchant (1976)
Macropus rufus	34.8+0.6	33.2+0.2	0.95	Y	Y	34.7+0.3	31.3+0.4	1	Sharman (1963); Calaby & Poole (1971)
Wallabia bicolor	32.6±3.6	35.5±2.3	1.09	N	Y	24.8±3.4	29.7±1.1	1	Calaby & Poole (1971)

Fig. 2.1. Copulation in Didelphidae (*a*) *Monodelphis domestica:* the male has immobilised the female's rear legs by grasping her ankles with his hind feet, while clasping her just anterior to the pelvis with his forepaws. Copulation usually occurs with the animals lying upon their right sides. Redrawn from Trupin & Fadem (1982). (*b*) *Marmosa robinsoni:* in this species the tail is used by the male to obtain support while all four feet are used to grasp and immobilise the female. The partial withdrawal of the penis was caused by disturbance during photography but note the relative position of the penis and pre-penial scrotum. Drawn from a photograph in Barnes & Barthold (1969).

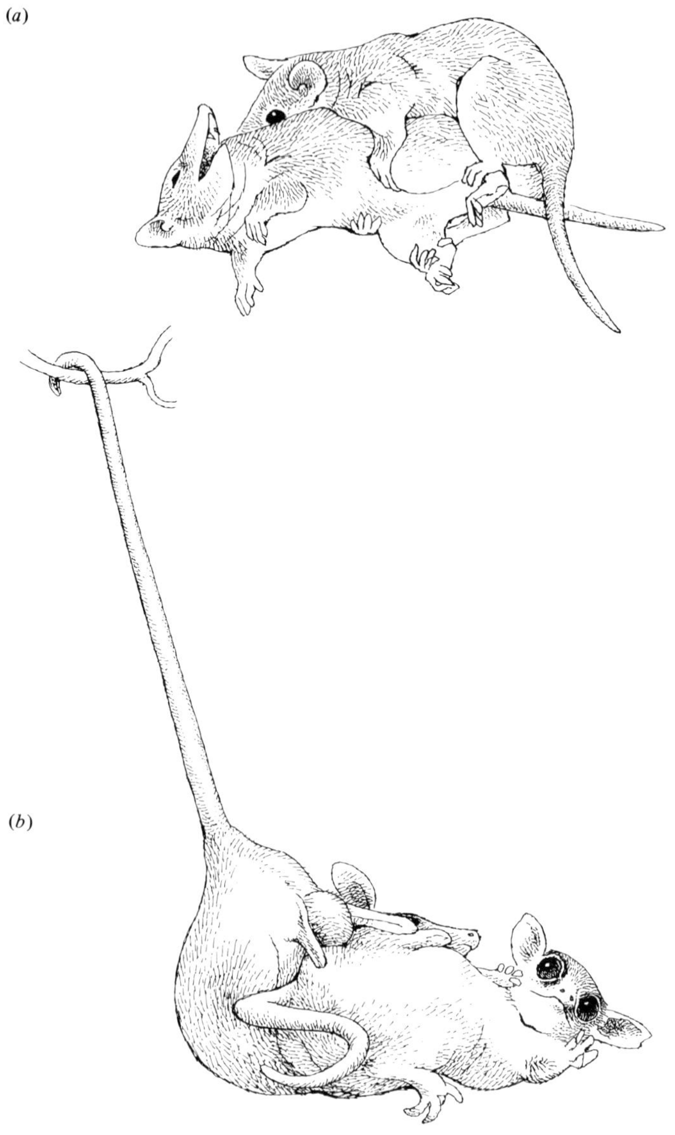

period. This is because ovulation, which is spontaneous, occurs within 24 h of the onset of oestrus and the ova pass through the oviduct in less than 24 h when they acquire the mucoid coat (Hartman, 1923*a*) (see p. 264 and Fig. 7.3). Accurately timed intervals from observed copulation to birth in *Didelphis virginiana* range from 12 days 19 h (McCrady, 1938), to 13 days ± 6 h (Reynolds, 1952). The gestation period for *Marmosa* is either 13.5 or 14.5 days depending on whether the females were paired with males 48 h or 24 h after the beginning of oestrus, as determined by vaginal smear (Godfrey, 1975). For *Monodelphis* gestation is given as 14–15 days by Fadem *et al.*, (1982) but, as females would mate on two consecutive nights, this figure is probably not different from that for *Marmosa*. For *D. albiventris* the gestation is also reported to be 13 days (Talice & Lagomarsino, 1961).

The gestation period in *Didelphis* and in *Marmosa* is of the same length as the luteal phase of the oestrous cycle and the subsequent follicular phase is inhibited by the presence of the young (Hartman, 1923*a*; Godfrey, 1975). If the young are removed during the first week, oestrus occurs 8–17 days later but if they are removed more than 21 days after birth the interval to next oestrus is 7 days (Feldman & Ross, 1975). *D. marsupialis* and *D. albiventris* display the same pattern, returning to oestrus 7–14 days after young were removed (Tyndale-Biscoe & Mackenzie, 1976). In the normal course of events, oestrus occurs at the end of lactation, about 97 days in *Didelphis*, and 65 days in *Marmosa*.

Parturition and lactation

Parturition in *D. virginiana* was first fully observed and described by Hartman (1920*a*) and subsequently by Reynolds (1952), who observed the birth of four entire litters, one of which included 25 young. The first change in behaviour that could be related to parturition occurred during the last few hours before birth and comprised more active grooming of the pouch and ventral surface. Fifteen minutes before birth the animals showed signs of pain and sat in a hunched position and underwent abdominal contractions about once every 3–4 s. Just before birth, a sitting position was adopted with the tail drawn forward, so that the distance from vulva to pouch was reduced to 4 to 5 cm. As fetal fluid appeared, the female licked it up; this action may have freed the fetuses from their membranes. As they emerged, the young travelled rapidly to the pouch. Reynolds (1952) recorded one that took 16.5 s to cover the distance and the time for the whole litter to be born and reach the pouch varied from 2.5 to 12 min. Not all the young reached the pouch; of those that did, only those that attached to a teat survived.

The young of *Didelphis virginiana* are accommodated in a well-formed pouch which normally contains 13 teats (Table 2.1); however, the anterior one or two pairs are not always functional (Reynolds, 1952; Burns & Burns, 1957), whereas the young on the most posterior teats have been observed to be the heaviest in the litter (Cutts, Krause & Leeson, 1978). These observations suggest that there is a gradient in functional capacity of the mammary glands, which may be the means of effecting the variation in mean litter size from high to low latitude (see below). The young are continuously attached for about 48 days after which they begin to release the teat occasionally. At 60 days they may emerge from the pouch but can remain attached to the now much-lengthened teat. From about 80 days the young are left in a den to which the mother returns and, at 86 days, they begin to eat solid food and ride on the mother's back (Fig. 1.4). Weaning occurs at 100–10 days and, if conception occurred at the first oestrus after lactation, the next litter may be born about this time. About 6 days elapse between cessation of lactation and birth of the next litter and, in that time, the mammary glands regress and the teats shorten and become firm (see p. 355). The teats do not retun to the size they were at the start of the season, however, but a small papilla develops on the end of the teat, which the newborn young is able to grasp in its mouth. Neither *Marmosa* nor *Monodelphis* have pouches, but the young remain continuously attached to a teat for the first 20 days or 14 days respectively. This is described as Type 1 mammary area by Russell (1982*a*) as shown in Fig. 2.8. After this time they are left in a nest to which the mother returns after foraging. At 34 days *Marmosa* young accompany the mother clinging to her back and are weaned at 70 days (Barnes & Wolf, 1971). *Monodelphis* young may be weaned at 50 days (Fadem *et al.*, 1982).

Breeding season and annual productivity

The three species of *Didelphis* together cover a huge range of latitudes and climates and their reproductive strategies reflect this, in particular the litter size, the number of litters and the onset and close of breeding (Hunsaker, 1977). *D. albiventris* is restricted to the cooler climate of South America above 1500 m and as far south as 39° S. Over part of its range it is sympatric with *D. marsupialis*, which occurs in warm humid habitats of the Brazilian subregion and extends as far as northeastern Mexico (Hershkovitz, 1972). In Central America it is sympatric with *D. virginiana* (Gardner, 1973) but the latter species, which can occupy cooler temperate habitats, extends as far as the Great Lakes (45° N).

The mean litter size of *D. virginiana* is 8.4 in New York State (44–42° N) (Hamilton, 1958) and about the same in other localities north of latitude

39° N but 6.2 or 6.8 in Texas (Hartman, 1928; Lay, 1942) and 6.3 in Florida (Burns & Burns, 1957). The mean litter size of *D. marsupialis* is the same (6.0) in Panama (Fleming, 1973) and in eastern Colombia but at higher altitude near Cali, at 3° N, it is 4.5 (Tyndale-Biscoe & Mackenzie, 1976) and the same at 5° N in French Guyana (Charles-Dominique, 1983). At 23° S the mean litter sizes of two series were 8.5 (Davis, 1945) and 7.1 (Hill, 1918; Tyndale-Biscoe & Mackenzie, 1976). In Colombia the litter size of *D. albiventris* was 4.2 (Tyndale-Biscoe & Mackenzie, 1976) and in Brazil (12° S) was 5 (Cerquiera, 1984).

In *D. virginiana* breeding commences in January in lower latitudes such as in Florida and Texas, but in February or even in March further north (Hamilton, 1958). In the zone of sympatry of the two species in Nicaragua, breeding also commences in January (Biggers, 1966) as it does for *D. marsupialis* in Panama and Colombia. However, south of the Equator, in Brazil, this species commences to breed in June or July (Hill, 1918, Davis, 1945), which would seem to implicate photoperiod as the proximate factor for the initiation of breeding in both species. Farris (1950) induced *D. virginiana* females to give birth in December by increasing photoperiod from November.

For *D. virginiana* two litters are the rule, although a third litter may be possible in Texas (Hartman, 1928), while in New York State even a second litter is uncommon (Hamilton, 1958). The close of breeding in North America is probably due to diminishing food resources and cooling temperatures in the autumn. Fleming, Harder & Wuckie (1981) found that the resting metabolic rate of lactating *D. virginiana* was 92% higher than that of summer-acclimated non-reproductive animals measured by Lustick & Lustick (1972), so that females bearing litters late in the year would be less able to nourish their young.

At lower latitudes, where low temperatures are not a factor, *D. marsupialis* nevertheless displays similar patterns with two litters a year in Panama (Fleming, 1973), eastern Colombia (Tyndale-Biscoe & Mackenzie, 1976), Guyana (Charles-Dominique *et al.*, 1981) and Brazil (Davis, 1945). At each of these sites the second litters are weaned or lost towards the end of a marked wet season and the females enter anoestrus (Fig. 2.2). In Panama, Fleming (1973) considered that the breeding season of *D. marsupialis* may be so timed that weaning coincides with the main fruiting times from May to September (Fig. 2.2). In Nicaragua and western Colombia, where there is no marked dry and rainy season, three peaks of births occurred in February, May and August (Biggers, 1966, Tyndale-Biscoe & Mackenzie, 1976).

Fig. 2.2. Breeding seasons of Didelphidae in Panama. (*a*) *Marmosa robinsoni*, (*b*) *Didelphis marsupialis*, (*c*) annual rainfall. □ no reproductive activity; ▨ lactating, pouch empty; ■ young present or pregnant. Redrawn from Fleming (1973).

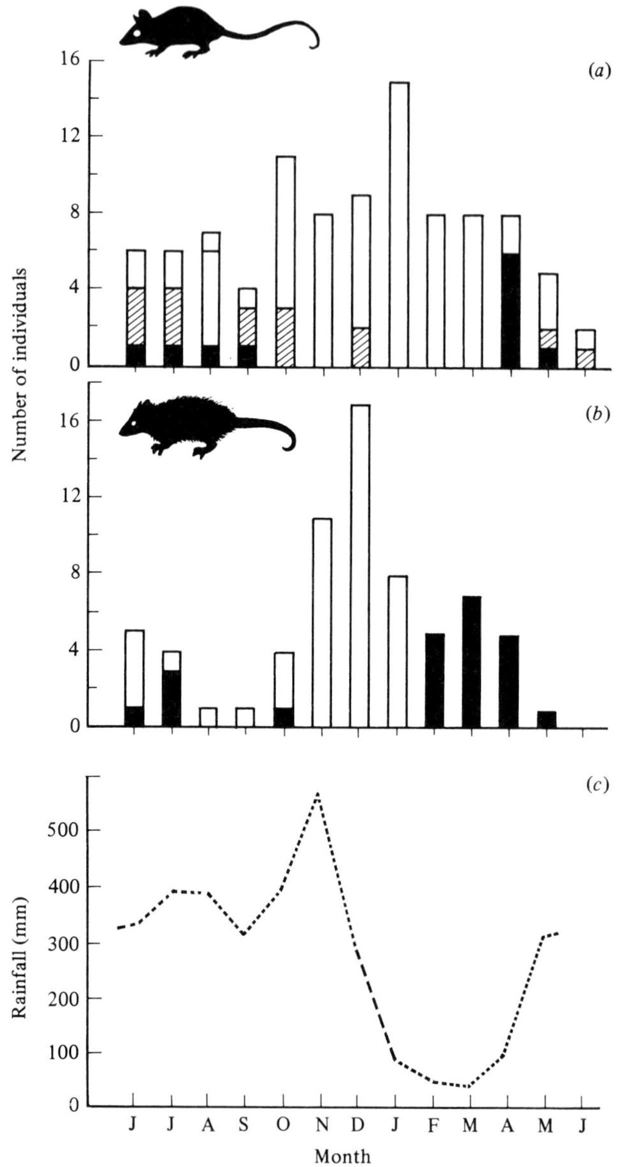

Other didelphids

In Central and South America the breeding seasons of three other species of Didelphidae, like *D. marsupialis*, are closely related to the annual wet season from April to November. *Caluromys derbianus* begins to breed in February (Enders, 1966; Biggers, 1967) and may continue until December (Phillips & Jones, 1968) (Table 2.1). The average litter size is 3–4, maximum of 6, but according to Enders (1966) up to 50% of litters

Fig. 2.3. (*a*) Breeding pattern of *Caluromys philander* related to the annual changes in abundance of food plants. (*b*) Number of births recorded in the study area each month. (*c*) Number of plants bearing fruit each month. Redrawn from Atramentowicz (1982).

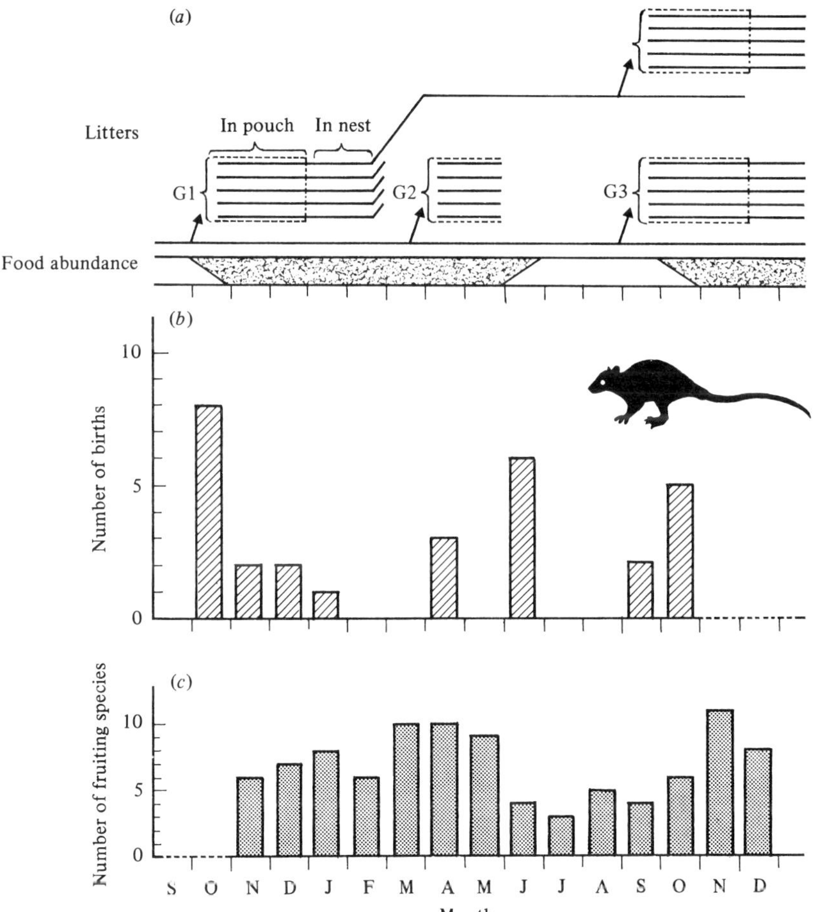

are lost before the young are independent. In Guyana *Caluromys philander* breeds twice in the year and Atramentowicz (1982) found that litters born in October–December survived to independence whereas litters born in April–June, after the former young were weaned, failed to survive pouch life (Fig. 2.3*a*). This was ascribed to the inability of the females to maintain lactation after the main fruiting season of the forest ceased in June (Fig. 2.3*c*). Young born in October–December are carried in the mother's pouch for 75–80 days and then are left in a nest for a further 30–45 days. *Caluromys philander* is an arboreal folivore which feeds for most of each night. Whereas males and non-lactating females were active for 55% of the night, females carrying pouch young more than 5 weeks old were active for 78% of the night. Later in the year, when food was less abundant, the time spent feeding by all animals increased and Atramentowicz (1982) concludes that insufficient food was then available for the additional requirements of lactation. Davis (1945) recorded information on three females from a Brazilian forest that fit the same pattern.

Philander opossum also begins to breed in February in Panama (Fleming, 1973) and Nicaragua (Biggers, 1966) and females produce 2 litters before entering anoestrus during November–January at the close of the wet season (Table 2.1). In French Guyana this species produces 3 litters (mean of 4.2 per litter) beginning in September and ceasing in May, when fruit becomes scarce (Charles-Dominique, 1983). The litter size in Panama is 4.6 (Fleming, 1973), but 6.0 in Nicaragua (Phillips & Jones, 1969). In neither *P. opossum* or *D. marsupialis* do the males show a seasonal decline in testis size or in spermatogenesis (Biggers, 1966).

The breeding season for the smaller *Marmosa robinsoni* of Panama also commences in February (Enders, 1966) but, according to Enders (1966) and Fleming (1973), only 1 litter of 10 young is produced each year. However, Fleming (1973) collected females with pouch young until September (Fig. 2.2). If weaning occurred at 70 days (Barnes & Wolf, 1971) this would suggest that 2 litters could be produced or that some females bred late. In Guyana *M. cinerea* produces litters of 6 and *M. murina* litters of 8 (Charles-Dominique, 1983).

Microbiotheriidae

Dromiciops australis in southern Chile breeds in the southern spring (October) (Collins, 1973). Females have only four teats and the litter size is accordingly small (Mann, 1958).

Caenolestidae

Information on reproduction in the Caenolestidae is restricted to three papers referring to *Caenolestes obscurus* collected from Paramos de Tame in Colombia in February–March 1911 (Osgood, 1921) and Paramos de Purace in August–September 1969 (Kirsch & Waller, 1979) and April–June 1971 (Tyndale-Biscoe, 1980*a*) (Table 2.1). Adult males with large testes and very enlarged prostate glands were obtained on all three occasions. Osgood (1921) obtained five females, one being pregnant and none lactating, Tyndale-Biscoe (1980*a*) obtained an oestrous female with four Graafian follicles on 30 April and no other adult females at all, while Kirsch & Waller (1979) collected five adult females, one probably post-partum (but without young) and four lactating. One of the latter had three teats enlarged while the others had four enlarged. Four teats is the full complement and there is no pouch (Osgood, 1921) i.e. mammary area Type 1. These observations suggest that there is a single breeding season from February to August during which a single litter is raised and that, unlike the Didelphidae, the number of ova shed is equal to or less than the number of teats.

Dasyuridae

The dasyurids are insectivorous or carnivorous marsupials that live in Australia and New Guinea. Currently, 47 living species are recognised of which 36 live in Australia and a further 12 in New Guinea (Kirsch & Calaby, 1977; Ziegler, 1977; Archer, 1982). On serological and morphological criteria (Baverstock *et al.*, 1982) the species are very uniform and the diversity observed is largely due to size, which ranges from the tiny species of *Planigale* and *Ningaui*, weighing less than 10 g to the Tasmanian devil, *Sarcophilus harrisii*, weighing 6–12 kg (Table 2.1). All are predators and most are nocturnal, feeding on insects or small or medium-sized vertebrates, depending on their size. Many aspects of the biology of these 'bright-eyed killers of the night' have been reviewed in *Carnivorous Marsupials* (Archer, 1982). The reproduction of 21 species of the Dasyuridae is now known in greater or lesser detail (Tables 2.1 and 2.2) and Russell (1982*a*) has shown that the variations observed in neonatal weight, litter size, duration of lactation and weight of young at weaning are well correlated with maternal body weight (Figs 2.4 and 2.5).

Fig. 2.4. Parental investment in the Dasyuridae: (*a*) age at weaning as a function of maternal body weight; (*b*) relative investment in litter at weaning [measured by (weight of litter at weaning/weight mother) × 100] as a function of maternal body weight. In small dasyurids investment is > 300% in several species, the highest value being 375% in *Ningaui ridei* (1), and most dasyurids have a higher level of investment for their body size than species of other families of marsupials. Symbols, (numbers) and species: ● small polyoestrous species: (1) *Ningaui ridei*, (2) *Planigale maculatus*, (3) *Sminthipsis crassicaudata*, (4) *S. macroura*, (5) *Antechinomys laniger*. ▲ monoestrous species: (6) *Antechinus bilarni*, (7) *A. flavipes*, (8) *A. stuartii*, (9) *Phascogale tapoatafa*. ■ Medium to large polyoestrous species: (10) *Dasycercus cristicauda*, (11) *Dasyuroides byrnei*, (12) *Dasyurus hallucatus*, (13) *D. viverrinus*, (14) *D. geoffroii*, (15) *D. maculatus*, (16) *Sarcophilus harrisii*, (17) *Myrmecobius fasciatus*, Redrawn from Russell (1982*a*).

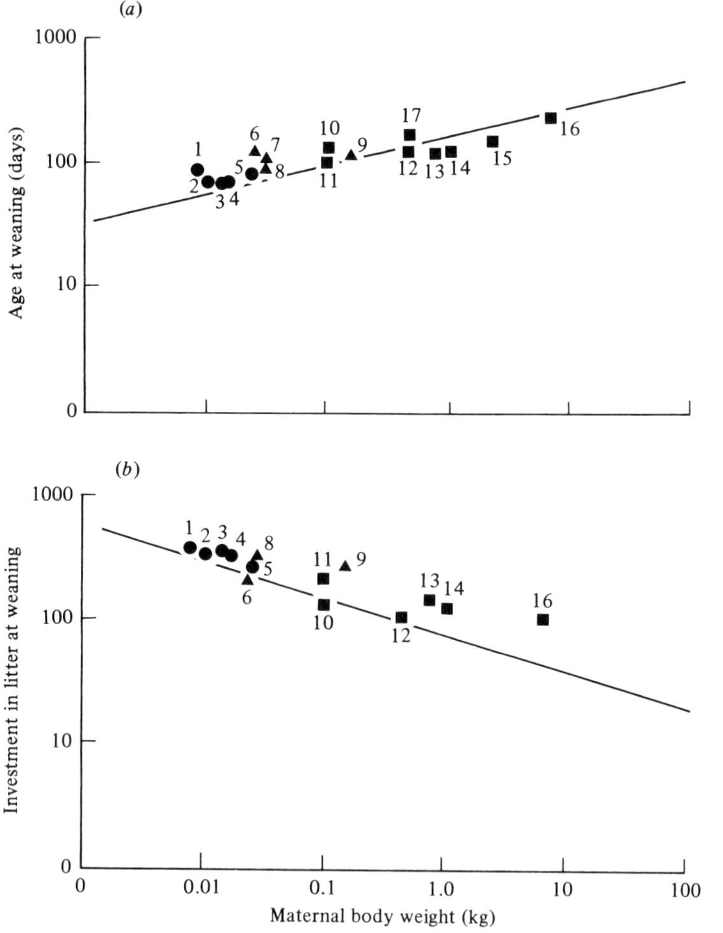

Oestrous cycle and pregnancy

There is some confusion in the literature about the condition of monoestry and polyestry in dasyurids that goes back to the first study on a dasyurid. Hill & O'Donoghue (1913) concluded that *Dasyurus viverrinus* is monoestrous because, in their experience, the females underwent only one fairly prolonged period of oestrus in a year and, if pregnancy did not ensue the female underwent very similar changes, which they termed pseudopregnancy. However, Fletcher (1985) observed that *D. viverrinus* females in captivity, which lost young prematurely, returned to oestrus and bred again and Green & Eberhard (1983) have observed the same in free-ranging females, recaptured sometime after having their first litter removed experimentally. J. C. Merchant (personal communication) has fully confirmed that the species is polyestrous and finds the oestrous cycle to be 37 days and gestation 19 days (Table 2.2). In an ecological sense *D. viverrinus* might be said to be monoestrous because females which retain

Fig. 2.5. Changes in the weight of a female *Ningaui ridei* from before her litter was born until the young were 184 days old. The period during which the young are dependent on milk, and during which they are attached and free are indicated. The dashed line is the reference weight of this female, the open square indicates the last day of calling and the closed square indicates the last day of mating. Redrawn from Fanning (1982).

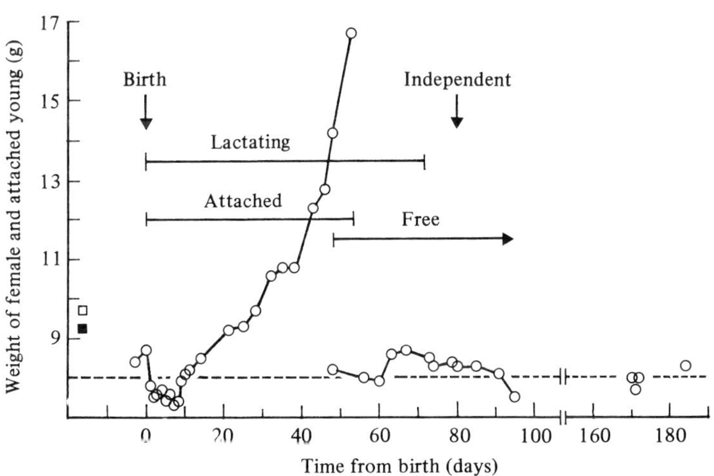

their first litter do not undergo oestrus again in that year but, in the physiological sense, the species is polyoestrous because females have the potential to undergo more than one period of oestrus in a year. Most species of dasyurid are like *D. viverrinus*; the only truly monoestrous dasyurids are species of *Antechinus* (Lee, Woolley & Braithwaite, 1982) and *Phascogale* (Kitchener, 1981; Cuttle, 1982) found in temperate Australia. Indeed, these are the only marsupials known certainly to be monoestrous and it is therefore likely that monoestry has been derived from a polyoestrous pattern, rather than the reverse.

Among the polyoestrous dasyurids, oestrous cycles range from 60 days in *Dasyuroides byrnei* (Fletcher, 1983) to 24 days in *Sminthopsis murina* (Fox & Whitford, 1982). There are several changes associated with the oestrous cycle in dasyurids that can be used to monitor it. In the smaller species there are marked changes in body weight (Woolley 1966*a*; 1973; Fanning, 1982; Fox & Whitford, 1982; Fletcher, 1983) and, since the changes are equally pronounced in non-pregnant as in pregnant females, they do not reflect changes in the gravid uterus but reflect a more general change in the animal. For *Dasyuroides byrnei* Fletcher (1983) has shown that these weight changes correlate with changes in plasma progesterone (see Fig. 6.6). During pro-oestrus in *D. viverrinus* the pouch skin becomes

Fig. 2.6. Schematic diagram of the criteria used to determine the time of ovulation of *Antechinus stuartii*. The cross-hatching shown on the pouch colour level represents the intense redness which develops in the pouch skin late in oestrus. Duration of shaded area 19.3 ± 4.4 days Redrawn from Selwood (1982*a*).

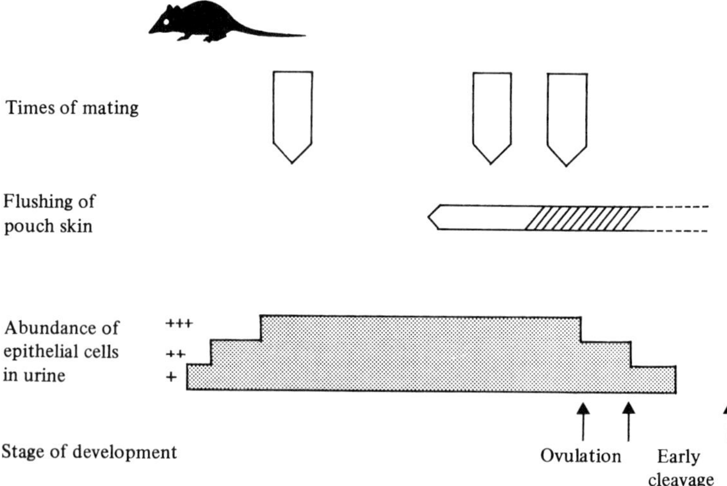

moist due to secretions of the prominent sebaceous glands therein and the vulva becomes oedematous (Hill & O'Donoghue, 1913). Similar changes have been described in other species e.g. *Planigale maculatus* (Van Dyck, 1979). As with *Didelphis*, oestrus can be detected from associated changes in the urogenital sinus, the cells from which can be sampled with a speculum or swab. In the smaller dasyurids the changes can more readily be monitored by examining the urine released by the female when handled (Godfrey, 1969*a*; Woolley, 1971*a*). Details of this technique are provided by Close (1983) and Selwood (1982*a*) (Fig. 2.6).

At the onset of breeding activity both males and females of many small dasyurids make characteristic calls (see review by Croft, 1982). Of especial interest is that females attract males to their vicinity by specific calls, and the males may respond with the same or similar call (Fig. 2.7). The frequency and intensity of calling by females is greatest during oestrus. In *Sminthopsis murina*, the mean intervals between the onset of a bout of calling was 24 days, the same duration as the interval between periods of vaginal cornification and oestrus (Fox & Whitford, 1982). In *Planigale maculatus*, *S. virginiae* and *S. murina* calling continues for several days longer when the female is not mated than when copulation has taken place (Van Dyck, 1979; Taplin, 1980).

Fig. 2.7. Sonograms of the calls uttered by *Ningaui ridei*. (*a*) the 'male call'; part of a 'female call' (*c*) the copulation call; (*d*) the triple call. Drawn from Fanning (1982).

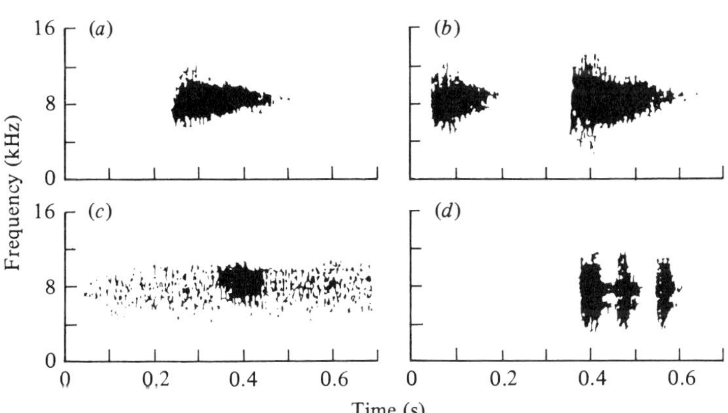

Courtship and copulation have been reviewed for 13 species of dasyurid by Croft (1982). There is an extraordinary uniformity in the behaviours, even in what might appear to be trivial features, so that one general description can suffice. In the period leading up to oestrus the female may show agonistic response to the male's approach and there is prolonged chasing by the male. In captivity these encounters may cause severe damage because the female cannot escape. At oestrus the female will stand still and allow the male to approach and investigate her mouth and genital region, and groom her flank. The male then grasps the female by the scruff of the neck with his mouth and clasps her abdomen. If this is not resisted, intromission may take place and the two remain together in coitus for 1–6 h or more. The first 1–2 h is an active phase during which the male vigorously palpates the female with fore or hind leg at the same time making pelvic thrusts. The male also rubs his chin on the nape of the neck of the female. The subsequent passive phase may last 3 h when both animals are motionless most of the time and the female curls her head under her body. The passivity is interrupted by sudden starts when both animals leap into the air and fall over onto their sides without disengaging. Eventually they disengage and the males of several species groom themselves. Oestrus lasts for 2–3 days in all species except species of *Antechinus* in which it may last 7–14 days, so that several bouts of copulation can occur.

In *Dasyurus viverrinus* (Hill & O'Donoghue, 1913), in *Antechinus stuartii* (Woolley, 1966*b*; Selwood, 1982*b*) and *Sminthopsis macroura* (Godfrey, 1969*a*) live spermatozoa are retained alive and motile in the oviduct for several days after copulation and are capable of fertilising newly ovulated eggs (see p. 264).

Ovulation is spontaneous in *D. viverrinus* and *A. stuartii* and probably in all other dasyurids but it does not occur at a fixed time in relation to oestrus, which makes it difficult to determine the true gestation accurately or to obtain precise stages of development. In *A. stuartii* Selwood (1982*a*) discovered that the onset of oestrus is associated with release of cornified cells in the urine and that the day of ovulation can be recognised by a marked decline in the numbers of these cells in the urine (Fig. 2.6). By this means she has been able accurately to determine ovulation and hence the time taken for each stage of embryo development (see p. 276, Fig. 7.11).

With these provisos the gestation periods of dasyurids range from 13 days in *Sminthopsis crassicaudata* (Godfrey & Crowcroft, 1971) and *S. murina* (Fox & Whitford, 1982) to 27 days in *A. stuartii* (Selwood, 1982*a*) and 31 days in *Dasyuroides byrnei* (Mack, 1961; Fletcher, 1983) (see Table 2.1). The variation does not bear any discernible relation to adult body size or to reproductive pattern but Sharman (1963) suggested that species with the longer pregnancies might undergo periods of embryonic diapause. This has been established for *A. stuartii* (Selwood, 1981) and will be discussed in Chapter 7.

Parturition and lactation

Birth has been observed in *Dasyuroides byrnei* (Hutson, 1976) and in *A. swainsonii* (Williams & Williams, 1982) and in both species the female stood on all four legs with the hips raised. The young travelled downwards to the teats very rapidly. Young in excess of the teats were discarded on the ground.

All dasyurids are polyovular and polytocous and the number of eggs shed and embryos developed generally exceeds by a large margin the number of teats available for the young. In *D. viverrinus* Hill & O'Donoghue (1913) recovered more than 20 eggs or embryos from each of 35 females, although the normal number of teats is 6. In *Sarcophilus harrisii*, which has 4 teats, Flynn (1922) recorded 21 eggs shed, Guiler (1970) 15 and Hughes (1982) up to 56 eggs or embryos from one female. However, Hughes also observed that the number of normal embryos among these sets did not greatly exceed the number of teats. Likewise, in *A. stuartii*, Selwood (1983) found an excess of eggs but the number of normal embryos did not exceed by much the number of teats. In this species the number of teats varies regionally (Cockburn, Lee & Martin, 1983) but Selwood (1983) found no corresponding variation in ovulation rate.

Since each young requires exclusive use of a teat for the first phase of the nursing period, there is inevitable loss of young at birth but, unlike in Didelphidae, it is usual for all available teats to be occupied initially. As a result, the initial litter size of dasyurids closely approximates the maximum for each species (see Woolley, 1973, Table 1) and teat number rather than number of eggs shed is the main determinant of fecundity in these species. Mortality has been reported to occur during pouch life in some species (Morton, 1978; Van Dyck, 1979; Begg, 1981*b*; Godsell, 1982) but not in others (Hill & O'Donoghue, 1913).

There is considerable variation in the morphology of the mammary area of the Dasyuridae and Woolley (1974) recognised four types (Fig. 2.8). Only young of species with Type 3 or 4 pouches are fully protected from the exterior for much of lactation. The smallest dasyurid, *Ningaui ridei* has Type 1 as do all species in the medium-size range from *Antechinus* (30 g) to *Dasyurus* (800 g) (Table 2.1). All species of *Planigale* except *P. subtilissima* have Type 2, as do the two largest dasyurids *Dasyurus*

Fig. 2.8. Types of mammary area or pouch in marsupials. Arrow points cranially, broken lines represent limit of pouch area, open circles enclosed teats and closed circles exposed teats. 1. The mammary area has no covering fold of skin and the teats are exposed. Marginal, usually lateral, ridges of skin develop during the breeding season. Didelphidae, Caenolestidae, Dasyuridae. 2. The mammary area is partially covered by a crescentic antero-lateral fold of skin, usually deepest anteriorly. Dasyuridae. 3. The mammary area is covered by a circular fold of skin with a central opening and all the teats are enclosed. Dasyuridae. 4. The mammary area is covered by a crescentic antero-lateral fold of skin. The teats are carried in two pockets projecting forward from the anterior margin of the skin fold. 5. The mammary area is completely covered by a fold of skin. The deep pouch so formed opens at its anterior margin. Phalangeridae, Macropodidae. 6. The mammary area is completely covered by a fold of skin. The deep pouch so formed opens at its posterior margin. Vombatidae, Phacolarctidae, Peramelidae. Types 1–4 redrawn from Woolley (1974) and classification of types from Russell (1982*a*).

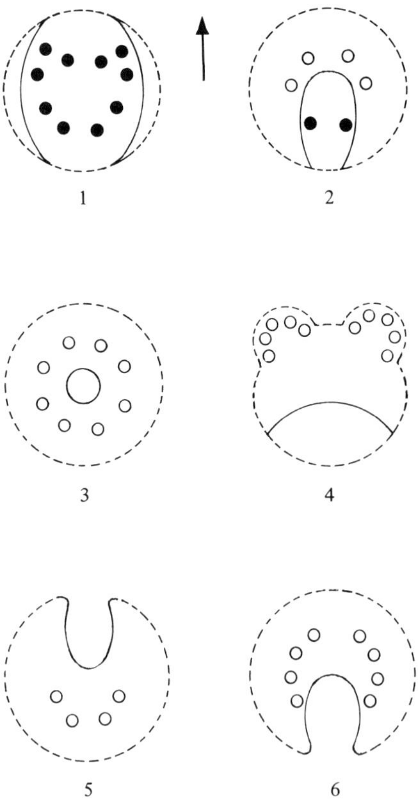

maculatus and *Sarcophilus harrisii*. All species of *Sminthopsis* have a Type 3 pouch. *Planigale subtilissima*, which Archer (1976) considered to be a subspecies of *P. ingrami* is the only dasyurid with a Type 4 pouch. However, the potoroid, *Hypsiprymnodon moschatus* also has the teats enclosed in two anterior pockets within a posterior directed pouch.

Russell (1982*a*) has recognised three patterns of maternal care in marsupials and the dasyurids all fit the first pattern, which may be related to the types of pouch accommodation available. The phase of permanent attachment to the teat is relatively short, as in didelphids, and at the end of this stage the young are left in a nest – this is clearly necessary in species with incomplete pouches (Type 1 or 2; see Fig. 2.9) – and the mother goes out to forage and returns to the nest to feed the young. They remain in the nest until their eyes open and they are clad in fur, when they venture out with the mother. At this stage, if mortality occurs, the vacated teats are sucked by the remaining litter (Merchant, Newgrain & Green, 1984), whereas at the early stage any vacant teats regress (Fig. 2.10).

The number of teats, and hence the maximum litter size that can be raised, varies from 4 in the largest dasyurid to 12 in some species of *Planigale* and *Antechinus*. Likewise, the duration of the several stages of maternal dependence and the rate of growth of the young also vary (Table 2.1). Russell (1982*a*) has shown that much of the diversity in these phenomena are highly correlated with maternal body size, as can be seen in Fig. 2.4.

Fig. 2.9. *Antechinomys spenceri* cleaning the young and pouch: (*a*) before and (*b*) after the pouch sphincter relaxes permanently. Redrawn from Happold (1972).

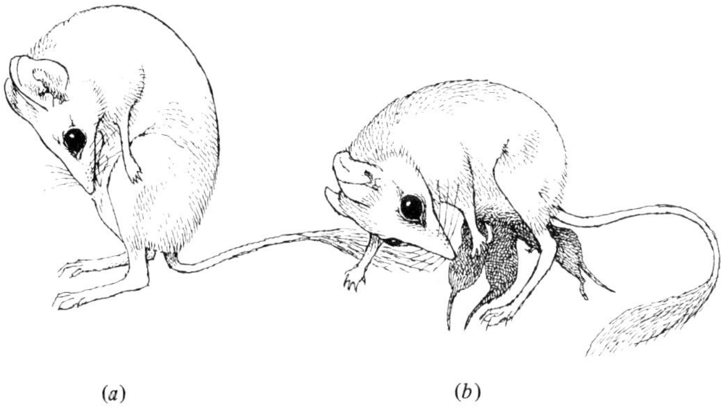

Breeding seasons

When considering the breeding strategies of dasyurids the most important of these variables are the duration of lactation, the litter size and the total weight of the litter at weaning as a proportion of maternal weight. This has been termed 'maternal investment' and, while it is a crude approximation of that, it is an indication of the constraints on reproduction for the species.

From Fig. 2.4 it can be seen that the maternal investment is highest in the smallest species, in which the total litter at weaning may weigh three or four times as much as the mother, whereas the female *S. harrisii*

Fig. 2.10. Pouch mammary area of *Phascogale tapoatafa* (Type 1) at different stages of the reproductive cycle and lactation: (*a*) immature female at pro-oestrus; (*b*) parous female at pro-oestrus; (*c*) immediately before parturition; (*d*) early lactation, with young continuously attached to the teats; (*e*) late lactation just before weaning, with young in nest. Note regression of two teats and associated mammary glands due to loss of young during early lactation. Redrawn from Cuttle (1982).

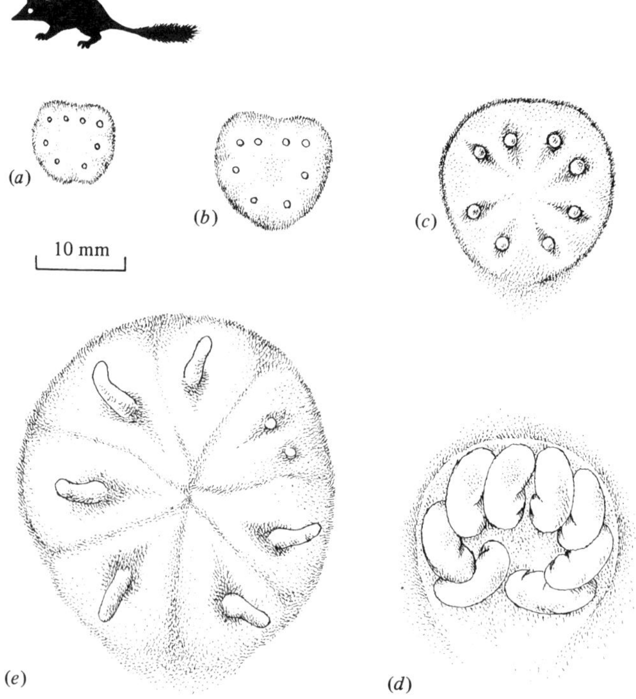

produces a small litter equal to her own weight. Furthermore, the latter species distributes that investment over 7 months, whereas the smallest species achieve it in 3 months or less (Table 2.1). Despite the wide range of habitats occupied, almost all dasyurids appear to be seasonal breeders, with the period of late lactation and pouch emergence coinciding with spring and early summer in temperate Australia and with the post-monsoonal period in northern Australia. For all except the smallest species only one litter is raised to independence in a year but, as mentioned previously, some species that are ecologically monoestrous have been found to be polyoestrous; others may also be found to be so on closer study. Only some species of *Antechinus* and *Phascogale* are strictly monoestrous. The significance of these different patterns will now be examined by considering three species for which laboratory and field data are available, and by comparing them to other species that are less well studied in this respect.

The smallest dasyurids belonging to the genera *Ningaui*, *Planigale* and *Sminthopsis* weigh between 6 and 20 g (Table 2.1). The reproduction of *Sminthopsis crassicaudata* has been studied in three field populations (Morton, 1978), in a captive outdoor breeding colony (Woolley & Watson, 1984) and in a laboratory colony (Smith *et al.*, 1978; Bennett *et al.*, 1982).

Throughout its wide range in eastern Australia the species is a strictly seasonal breeder with two peaks of birth, one in August and another in December, which coincide with a seasonal abundance of terrestrial invertebrates. The earliest births occur in the third week of July and the last young are weaned in February (Morton, 1978; Godfrey & Crowcroft, 1971).

Likewise in the laboratory colony, most litters were born during July to February (Godfrey & Crowcroft, 1971), so it is evident that the onset of breeding in both wild and captive females begins very shortly after the winter solstice and closes some time after the summer solstice. Godfrey (1969*b*) showed that females would respond to increasing photoperiod from 12L:12D to 15L:9D by returning to breeding condition, and Smith *et al.* (1978) found that it is not long day *per se* but a change from short to long day that the animals respond to. In the management of their colony, the animals are maintained on 16L:8D for most of the time but twice a year are exposed to 8L:16D for 3 weeks; oestrous cycles cease during this period and resume 20–30 d after the return to 16L:8D.

The mean number of ova shed was 14.4 ± 4.7 while the most common number of teats from all females sampled was 10 (Smith & Godfrey,

1970). The average litter size in Victoria was 7.5 (Morton, 1978) at 0–9 days, declining to 5.4 at 50–70 days.

Three other species, *Sminthopsis macroura* (Godfrey, 1969*a*), *S. murina* (Fox & Whitford, 1982), and *S. leucopus* (Woolley & Ahern, 1983) have similar patterns of reproduction, in which more than one litter is produced in a year. So do *Planigale maculatus* (Aslin, 1975; Van Dyck, 1979; Taylor, Calaby & Redhead, 1982), *P. gilesi*, *P. tenuirostris* (Andrew & Settle, 1982; Read, 1982) and *P. ingrami* (Heinsohn, 1970). *Ningaui ridei* is also polyoestrous and has a large litter and a short period of lactation (Fanning, 1982; Denny, 1982) but it is not known whether more than one litter is produced in a year. *Antechinomys laniger* is polyoestrous with cycles of 35 days. Females can undergo up to 6 cycles during the breeding season, which commences 2–4 weeks after the winter solstice and ceases 2–5 weeks after the summer solstice. Woolley (1984) concludes that photoperiod may be the proximate factor controlling its onset, as in *S. crassicaudata* and A. Valente (personal communication) has found that both males and females respond to increasing photoperiod.

The next size range of dasyurids are the species of *Antechinus*, *Parantechinus* and *Pseudantechinus*, which weigh 30–70 g (Table 2.1). Some of these species are polyoestrous and others monoestrous but none produces more than a single litter per year. The best known is *Antechinus stuartii* and the life history and ecological aspects of this species are described in detail by Lee & Cockburn (1985). The sequence of events is highly determined and in any locality very predictable from year to year.

Young males become spermatogenic, their testes enlarge (Fig. 4.8) and they become increasingly aggressive towards each other in early winter when they move greater distances (Wood, 1970). Their body weight increases to a maximum in September and they become highly territorial from then until spring when the females come into oestrus and mating occurs during a brief 2 week period (Braithwaite, 1974). Ovulation occurs spontaneously during this time and pregnancy lasts 25–31 days (Woolley, 1966*a*; Selwood, 1980) (Fig. 7.11). While the females are still pregnant the males have already begun to lose weight and they have all disappeared from the population by the time the females give birth.

More young are delivered than there are teats so generally all the teats are occupied. No true pouch is developed (Type 1, Fig. 2.8) so the attached young are exposed and some loss occurs during the first 40–50 days. After this period the young are left in a nest and suckled for a further 2–3 months. Most of the post-lactational females disappear then but up to 20% may survive and breed at the next season. The cause of the sudden and precipitate mortality of the entire adult male population before the birth of their progeny has been reviewed by Lee, Bradley & Braithwaite (1977) and Lee & Cockburn (1985). It is associated with marked increases in the free cortiscosteriod in plasma of males, with increased levels of infection by several intestinal and blood parasites and by negative nitrogen balance. These changes are induced by the combination of aggressive encounters with other males and sexual activity with females, which then predisposes the males to die from one or a combination of these factors.

This pattern, which has been termed semelparous by Braithwaite & Lee (1979), occurs in several other species of *Antechinus* from the temperate zone of Australia, but the time of onset varies between sympatric species and in the same species in different localities. In *A. swainsonii*, for instance, the time of mating varies from early May to late September being earlier at low latitudes and altitudes and later at higher latitudes and altitudes (Dickman, 1982). This species is larger than *A. stuartii* and, where the two are sympatric, the mating time of the smaller species occurs later than in the larger (Dickman, 1982). Lee *et al.* (1977) suggested that mating is so timed that the heavy burden of late lactation coincides with the spring flush of insects. In Victoria, where the ground-living species *A. swainsonii* gives birth 1 month earlier and *A. minimus* 2 months earlier than *A. stuartii*, Wainer (1976) suggested that the difference may be related to differences in feeding habits and availability of food; in *A. minimus* which feeds on larval insects, lactation coincides with the winter peak of larvae, whereas the scansorial *A. stuartii* breeds later when adult insects are abundant. While accepting this general thesis, Van Dyck (1982) points out that the abundance of insects in southern Australia is determined by increasing temperatures whereas, in northern Australia, rainfall is more important. He and Smith (1984*b*) showed that the respective timing of breeding in *A. stuartii* and *A. flavipes* in Queensland is closely related to the summer wet season.

Since most females breed once in their lifetime and since the maximum litter size is determined by the number of teats, variation in teat number is the only way for fecundity of the species to vary. Cockburn *et al.* (1983) found a high rate of teat occupancy in all four species of *Antechinus* and, while there was very little variation in teat number within any one population, there was marked variation between different populations of each species. Between regions characterised by females with 6, 8, 10 or 12 teats occur zones of overlap from only a few to 15 km wide where females with intermediate numbers of teats are found. Lowest teat numbers and litter sizes occur in coastal or insular populations and highest numbers in alpine and inland habitats and they conclude that *Antechinus* females produce litters which correspond to the maximum number of young they can nourish.

Two related species from the monsoonal climate of northern Australia, *A. bellus* and *Parantechinus bilarni*, and *P. apicalis* from south western Australia are known to be monoestrous and of these *A. bellus* is semelparous (Taylor & Horner, 1970; Calaby & Taylor, 1981), whereas males and females of *P. bilarni* breed in 2 successive years and are therefore iteroparous (Begg, 1981*a*). Limited evidence suggests that *P. apicalis* is likewise iteroparous (Woolley, 1971*b*). In contrast, *A. melanurus* and *A. naso* of the New Guinea rainforests breed throughout the year and males probably do not undergo synchronous mortality (Dwyer, 1977) but there is some doubt about their taxonomic relationship (J. H. Calaby, personal communication).

This evidence from tropical species of *Antechinus* supports the idea that semelparity is an adaptation by species living in an environment with limited but predictable resources rather than being a consequence of small size. This gains further support from two other sources. *Phascogale tapoatafa* lives in the same temperate forests as *A. stuartii* and extends into the tropics of northern Australia. Although it is considerably larger (200 g body weight), it also is semelparous (Cuttle, 1982) and – on limited evidence – so is *P. calura* (Kitchener, 1981). On the other hand *Sminthopsis murina*, which is smaller than *A. stuartii* but lives in the same region, is not. Fox (1982) considers that *S. murina* exploits a temporally transient niche in the post-fire succession at a time when *A. stuartii* has been decimated by forest fire and he concludes that polyoestry has evolved in *S. murina* as an adaptation to exploit this niche. However, as mentioned previously, polyoestry is probably the primitive condition and it is more likely to be the strict monoestry and semelparity of *Antechinus* that has arisen as an adaptation to a stable highly predictable environment.

Two rat-sized dasyurids have been studied in captivity and there is some field information on both (Table 2.1). *Dasyuroides byrnei* is polyoestrous in captivity (Woolley, 1973; Fletcher, 1983) and the males, in captivity, void spermatozoa in their urine, i.e. they are spermatorrhoeic from March to December (Woolley, 1971*a*). Gestation is 30–35 days and the young are weaned at about 120 days, so that 150 days are required to rear 1 litter to independence. Females in the wild produce 2 litters a year, in June and November.

Females of *Dasycercus cristicauda*, produce 1 litter a year but the evidence is insufficient to determine whether they are strictly monoestrous. In captivity the males are spermatorrhoeic from May to August (Woolley,

1971*a*) and copulation occurred during May–July (Michener, 1969). Oestrus lasts 5–6 days and gestation 30–44 days (Michener, 1969; Woolley, 1971*a*). There is no second oestrus and unmated females undergo pouch changes similar to those of pregnant females (Woolley, 1974). The attachment phase lasts 55 days and young are weaned at 100 days, and both males and females can breed in more than one season. The few records from the wild (summarised by Woolley, 1971*a*) are consistent with a single period of birth in June and the litters being weaned in October–December.

The several species of *Dasyurus* each produce a single litter each year, but it is now known that *D. viverrinus* is polyoestrous (Fletcher, 1985). In New South Wales *D. viverrinus*, breeds between June and August (Hill, 1910; Hill & O'Donoghue, 1913) and in Tasmania the earliest births can occur in May (Green, 1967; Godsell, 1982; Green & Eberhard, 1983). Females that fail to become pregnant at the first oestrus or that lose their young will return to oestrus respectively 37 days or 18 days later during May to September. Analysis of the specimens collected by Hill in New South Wales during 1895–1905 (Tyndale-Biscoe, 1984) show the births to have been distributed bimodally with peaks in late June and late July, which may reflect this.

There are normally six teats (O'Donoghue, 1911; Godsell, 1982) and since the number of young born exceeds this number, all become occupied after parturition. In a sample of 16 litters the mean number was 5.8 (Green, 1967). There is no pouch (i.e. Type 1, Fig. 2.8) and the young remain continuously attached to the teats for 49–56 days. They are then left in a den and are weaned at 112 days (Hill & O'Donoghue 1913, Hill & Hill, 1955). The whole period from the start of oestrus to weaning for this species is thus 120–150 days (Table 2.1). In their captive colony Merchant *et al.* (1984) observed that if young were lost during the period of attachment the associated mammary gland regressed (compare Fig. 2.9) whereas, after this stage, the gland continued to secrete milk if a young one died or was removed. At this stage the surviving members of the litter sucked from all lactating teats and their growth rates increased. For this reason indirect estimates of litter size, based on active glands of females trapped in the field, may be inaccurate.

D. hallucatus in northern Australia has essentially the same pattern except that there are 8 teats. Young first appear in early July, being left in a den in September and being weaned in November or December (Begg, 1981*b*). Mortality of young was more severe than in *D. viverrinus*, with only 30% females carrying the full number. In July the mean litter was 6.4 ± 1.1 and by September 4.4.

The largest dasyurid is *Sarcophilus harrisii*. Guiler (1970) observed

Table 2.3. *Water and sodium influxes in free-living* Dasyurus viverrinus *during early and late lactation and in non-lactating females and in males*

Month		N	Water influx (ml/kg body weight/day)	P	Sodium (mmol/kg body weight/day)	P
February	♀+♂	28	180±42		3.02±1.43	
				N.S.		< 0.001
April	♀+♂	11	185±44		5.43±1.75	
				< 0.001		< 0.001
July	♀+♂	16	284±44		8.92±1.82	
				< 0.001		< 0.001
October Non-lactating	+♂	12	211±37		5.63±0.89	
Early lactating		3	226±52		—	< 0.001
				< 0.001		
Late lactating		15	337±63		12.62±3.44	

After Green & Eberhard (1983).

matings in March and synchronous births in April after a gestation period of 31 days. As in other dasyurids the number of eggs shed far exceeds the four teats in the pouch (Hughes, 1982). The mean litter size is 2.9 (Green, 1967; Guiler, 1970) and the young remain attached for 105 days and are weaned at about 7 months in November. As in *D. viverrinus* some females give birth in July–September (Guiler, 1970), which may be due to a prior infertile cycle or to late onset of oestrus in some females.

In contrast to *Sminthopsis crassicaudata*, nothing is known about either the proximate or the ultimate factors that control breeding in the dasyurids that produce one litter per year. However, from their study on the food intake of free ranging *D. viverrinus* using [^{22}Na] and [^{3}H] isotope-dilution techniques, Green & Eberhard (1983) found that females in late lactation had turnover rates per kilogram of body weight 90% and 60% respectively higher than those of males and non-lactating females or females carrying small young (Table 2.3). From this it may be concluded that females that are in late lactation in spring will have the best chance of surviving and/or producing the largest number of young. In southern Australia and Tasmania all the dasyurids from *S. harrisii*, *D. viverrinus*, *Phascogale tapoatafa* to the several species of *Antechinus* reach this stage in spring or early summer. Because the duration of lactation varies in relation to size (Fig. 2.4*a*) the time of onset of breeding differs between species, with *S. harrisii* being the earliest in April and *A. stuartii* the latest in September (Table 2.1). If the proximate factor is photoperiod, as it appears to be for *Sminthopsis crassicaudata*, then each of these species must be responding to a different facet of the changing daylength. This has not been investigated for any of these species.

Thylacinidae

The one representative of this family, *Thylacinus cynocephalus* is now almost certainly extinct and very little is known of its reproduction. From the accounts of payments made between 1888 and 1909 on specimens presented for bounty, 'pups' and young animals were taken in all months of the year with a maximum in May to September (Guiler, 1961). The litter size was 3 or 4.

Myrmecobiidae

The sole representative of this family is the numbat, *Myrmecobius fasciatus* (Table 2.1). It has a definite breeding season, with copulation during December to April (Calaby, 1960). Friend & Whitford (1986) reported gestation to be 14 days in three females. The female has four teats but no pouch (Type 1, Fig. 2.8) and the young are born from January to April or May, and become independent by October (late spring). Friend & Burrows (1983) followed the development of one litter from first capture in March when the four young were naked (Fig. 2.11). The young were furred in July, were left in a nest from mid-August and were moving about with the mother in September.

Perameloidea

Eighteen species were recognised by Kirsch & Calaby (1977), of which sixteen belong to the Peramelidae and two to the Thylacomyidae. Seven species of Peramelidae occur in New Guinea and nine in Australia, two species being common to both lands. They are all medium-sized animals (0.5–2 kg, Fig. 6.9), nocturnal ground dwellers that occur in a

Fig. 2.11. *Myrmecobius fasciatus* from Dryandra, southwest Western Australia, (*a*) Female showing Type 1 pouch area with hairless, attached young 2 cm long and (*b*) the same female 3.5 months later with 4 attached young. Photographs by courtesy of D. A. Friend, Perth.

(*a*)

(*b*)

variety of well-vegetated habitats and their diet is insects, earthworms and plant roots. Two species, *Perameles nasuta* and *Isoodon macrourus*, have been bred in captivity (Stodart, 1966*a*, 1977; Lyne, 1976, 1982; Gemmell, 1982), which has provided material for the study of their reproduction and behaviour and there have been four field studies (Heinsohn, 1966; Close, 1977; Gordon, 1971, 1974; Stoddart & Braithwaite, 1979).

Oestrous cycle and pregnancy

All Peramelidae are polyoestrous and polytocous but their reproduction is distinguished from other marsupials by being very rapid, so that females can bear a succession of litters in one breeding season and the offspring of the first litter can be themselves breeding before the close of the season (Heinsohn, 1966). The oestrous cycle in both *P. nasuta* and *I. macrourus* is 20–1 days (Lyne, 1976), while the gestation period for both species is 12.5 days (Stodart, 1966*a*; Lyne, 1974), the shortest gestation reliably known for any mammal (Table 2.2).

Courtship behaviour and copulation has been described for two species, *Perameles gunnii* by Heinsohn (1966) and *P. nasuta* by Stodart (1966*a*) and the sequence was remarkably similar and very different from that described for dasyurids. Sexual attraction was limited to a few nights before oestrus and strong attraction only to the few hours at dusk before copulation began. It ceased at 01.00– 02.00 of the same night. In both species the male followed the female and attempted to grasp her tail but made no attempt to smell the pouch or genital region of the female (Fig. 2.12*a*). The male then mounted the female by standing up, the female raising her hindquarters at the same time (Fig. 2.12*b*). Intromission lasted 2–4 s in *P. nasuta* and 6–24 s in *P. gunnii*, but in both species these brief acts were repeated very many times over a period of 26 min – 2 h. At the peak of activity Stodart (1966*a*) recorded 13 intromissions in quick succession and Heinsohn (1966) several per minute. This pattern of copulatory behaviour resembles that of rodents, such as the golden hamster and Shaw's jird in which the frequency may exceed 100 copulations in an hour (Bourlière, 1964).

Intrauterine development has been described from cleavage stages to endoderm formation in *P. nasuta* and *I. macrourus* by Lyne & Hollis (1976, 1977*a*, *b*) and Hollis & Lyne (1977); the formation and ultrastructure of the choriovitelline and ultrastructure of the chorioallantoic placenta have been described by Padykula & Taylor (1976*b*, 1977, 1982) (see p. 319, Figs 7.20 and 7.21).

(a)

(c)

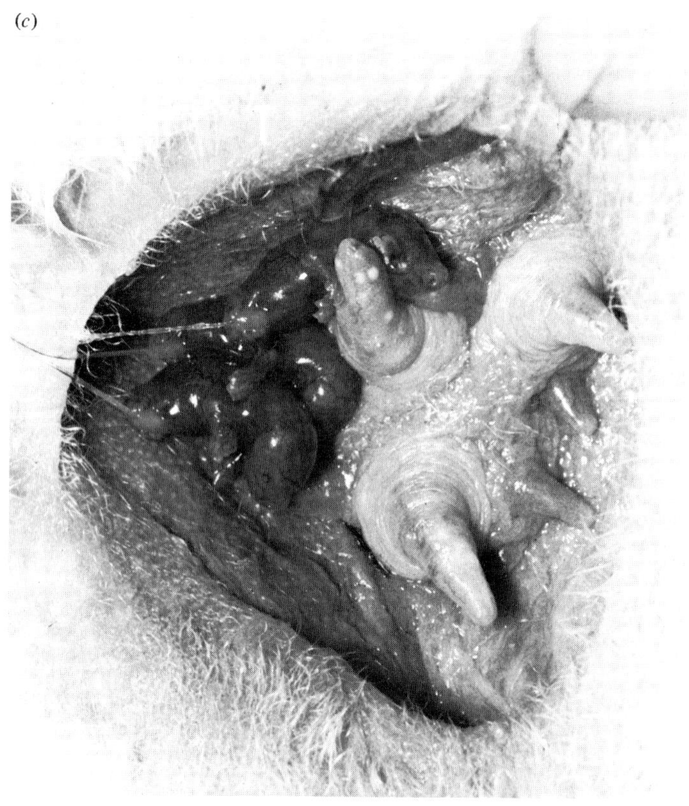

Parturition and lactation

Parturition occurs via a long median vaginal passage (see Fig. 5.5*a*) that forms each time and becomes occluded within a day or so of birth. The placentae are retained in the uterus and the young remain connected to them by long allantoic stalks until after they have become attached to the teats (Fig. 2.12*c*). The stalks become trapped in the occluded birth canal where they slowly disintegrate (Hill, 1899; Lyne & Hollis, 1982). During parturition of *P. nasuta*, observed by Lyne (1974) the female lay on one side with the leg raised and vigorously licked the vulva. The young suddenly emerged free of membranes and found their own way into the backwardly directed pouch (Type 6). Stodart (1966*a*) observed another female of this species less than 10 min after parturition; the three young were firmly attached each to a teat and still anchored by the allantoic stalk (Fig. 2.12*c*). The single young at birth is not proportionately heavier than any other marsupial (Russell, 1982*a*) but, when the total litter is considered, it is so. On morphological criteria the neonatuses are considered to be more developed (Fig. 5.5 and Sharman, 1965*c*) but there is so far no physiological or biochemical data to support this.

It has been suggested that the chorioallantoic placenta provides peramelids with a more rapid route for transport of nutrients from the mother to the fetus but, again, there is at present no experimental evidence that this is so. What is, however, indisputable is the extraordinarily rapid rate of development of the pouch young (Lyne, 1964; Russell, 1982*a*): when pouch young of the carnivorous *Dasyurus viverrinus* on the one hand and the herbivorous *Potorous tridactylus* on the other are compared with the similar-sized *Perameles gunnii* (Table 2.1), the young of *P. gunnii* release the teat and their eyes open at half the age of the other two species and

Fig. 2.12. Copulation and parturition in *Perameles nasuta*. (*a*) Mating behaviour begins with the male close following the female, whose pouch sags with the 50 day old young of the last litter. When she stops the male points his nose to her hind quarters and his lips part and curl at the corners of the mouth. Before mounting, the male paws the female's tail and anchors it to the ground. (*b*) During copulation the male stands very erect, his forelegs held out free, and the female crouches with the hindquarters slightly raised. Redrawn from Stodart (1966*a*). (*c*) The pouch 40 min after parturition contains three neonatal young still connected by umbilical cords but already attached to three of the five small teats. The three large teats were used by the previous litter until day 60, a few days before parturition. Photograph from Stodart (1966*a*), with permission.

they are weaned at 60 days compared to 135–140 days for *D. viverrinus* or 147 days for *P. tridactylus*. The metabolic rate of peramelids is no different from other marsupials (Dawson & Hulbert, 1970), so one must conclude that the quality of the milk is somehow different or the quantity produced is greater. These aspects have not so far been investigated.

In *P. nasuta* the duration of pouch life is about 50–4 days and, at the end of this period, the female returns to oestrus (Stodart, 1966*a*). The young are weaned at 61–3 days, immediately before the new litter is born. Similarly in *I. macrourus* the interval between litters is 56–8 days (Gordon 1971, 1974) and in *P. gunnii* 60 days (Heinsohn, 1966). The young become independent of their mother soon after weaning and in the case of *P. nasuta*, reach sexual maturity at 4 months (female) or 5 months (male) (Lyne, 1964). *P. gunnii* females breed at 3 months and *I. obesulus* at about the same age (Heinsohn, 1966).

Breeding season and annual productivity

While peramelids are polytocous they are more conservative than didelphids and dasyurids in the number of eggs shed. In *P. nasuta* and *I. macrourus* Lyne & Hollis (1979) observed a progressive but small decline between the mean number of corpora lutea, the mean number of embryos and the mean number of pouch young. In all species the mean number of young is always less than half the number of teats although occasional litters may exceed this. One reason for this is that the teats used by the previous litter may not be sufficiently regressed a week after weaning for the new litter to use (Fig. 2.12*c*), or may not be able to sustain full lactation. However this is not the whole reason, for litter size also varies through the season; in *P. gunnii* Heinsohn (1966) observed the successive mean litter sizes to be 2.1, 2.9 2.9 and 2.2 respectively while the same pattern (2.6–2.9, 3.6–3.8 and 2.1–2.6) was observed in *I. macrourus* near Sydney by Gordon (1971) and by Gemmell (1982) near Brisbane where the sequence was 2.8 ± 0.4, 2.8 ± 0.3, 3.3 ± 0.3, 2.0 ± 0.4.

Stoddart & Braithwaite (1979) showed a significant positive correlation ($r = 0.7$) between litter size and maternal body weight in *I. obesulus* and it may be that the changes in litter sizes observed in the other species reflects a seasonal change in maternal body weight in response to availability of invertebrate food. In *P. nasuta* Lyne (1964) found no difference in the mean litter size of 34 litters 0–24 days old and 22 litters 25–50 days old indicating negligible mortality during pouch life. However, in *I. macrourus* Gemmell (1982) found a decrease from 3.6 young per litter early in lactation to 2.1 after day 40.

The duration of the breeding season and hence the number of litters a female bandicoot can produce is correlated with latitude and presumably food availability. In Queensland at 29° 32′S *I. macrourus* breeds throughout the year and Gordon (1974) recorded one female that produced 6 litters in 13 months. However, Gemmell (1982) found more females with young in July–March than in the other months. Further south, at 37° S, breeding in this species began in July–September and the third and last litter was weaned between late February and late April (Gordon, 1971). The same pattern was found in *I. obesulus* in Victoria (38° 07′ S) by Stoddart & Braithwaite (1979), and in this species, and *P. gunnii* in Tasmania (40° 50′ S) by Heinsohn (1966). Conversely, *P. nasuta* and *I. macrourus* near Sydney breed throughout the year (Lyne, 1964) but with peaks of litters in spring (Lyne, 1982).

Gordon (1971) noticed that breeding in *I. macrourus* near Sydney began just after the females increased in weight and ceased when they lost weight and he concluded that nutritional state was the determining factor for breeding to take place. While Heinsohn (1966) came to the same conclusion for the close of breeding, both he and Stoddart & Braithwaite (1979) considered that the onset of breeding was so precise each year that photoperiod was probably the proximate signal. Gemmell (1982) also concluded that photoperiod was the predominant factor but since then Barnes & Gemmell (1984) have compared the data from the four studies just mentioned with several other environmental variables and they found that breeding, defined as percent of females lactating, was most strongly correlated with the rate of change of minimum daily temperature and only to a lesser extent with rainfall and photoperiod.

Thylacomyidae

The rabbit-eared bandicoot *Macrotis lagotis* from central Australia appears to have a much more restricted breeding season from March to May (Hulbert, 1972) but in captivity it breeds throughout the year (Johnson in Strahan, 1983). It is polyoestrous, the mean length of the cycle being 20.4 days (range 12–37 days) and gestation 14 days (range 13–16) (McCracken, 1986). There are 8 teats but the normal litter size is 2 and the young develop as rapidly as in peramelids.

Herbivorous marsupials – Diprotodonta

This diverse Australasian group includes all the exclusively herbivorous marsupials. All have a reduced dentition with a pronounced diastema and a single pair of incisors in the lower jaw. The premolars and molars show various modifications for herbivory. On other criteria as well the members of this order are considered to be related (Strahan, 1983). Their patterns of reproduction are more diverse than in either of the carnivorous groups already discussed but do, nevertheless, reflect the same constraints of size and diet. Three broad patterns can be recognised: the smallest species, which are nectarivorous and/or insectivorous are polyovular, polyoestrous and display post-partum oestrus with gestation prolonged during concurrent lactation. Included here are the pigmy possums (Burramyidae) and honey possum (Tarsipedidae). The middle-size range, display a pattern similar to that of the Didelphidae and Dasyuridae and include the arboreal folivorous Phalangeridae and Petauridae and *Phascolarctos cinereus*, and the terrestrial grazing Vombatidae. Less is known about the latter species and they may, on further study, be shown to display a distinctly different pattern. The third group comprise the rat kangaroos (Potoroidae) and the kangaroos and wallabies (Macropodidae), the largest living marsupials, almost all of which are monovular and display post-partum oestrus and delayed embryonic development but of a very different pattern from the pigmy possums.

Phalangeridae and Petauridae

These two families contain 33 species of arboreal folivores of Australia and New Guinea most of which are restricted to forest habitats (Smith & Hume, 1984). The brush-tailed possum, *Trichosurus vulpecula*, is more widespread than other species and is also well-established as an introduced species in New Zealand. The reproduction of six species has been investigated in some detail and that of *T. vulpecula* is very well known from both laboratory and field studies (Tables 2.1 and 2.2). The basic pattern of all six species is very similar; all are polyoestrous, and the gestation period occupies about two-thirds of the oestrous cycle, so that birth occurs at the end of the luteal phase, as in *Didelphis* and *Marmosa*, and oestrus and ovulation are usually suppressed during most of lactation. The females possess a large forward directed pouch, defined as Type 5 by Russell (1982*a*) (Fig. 2.8). The smaller species have four teats and usually more than one young at a time, while the larger species have two teats and are strictly monovular and monotocous. Until the eyes open, the young live permanently in the pouch but, subsequently, they may be left in a nest or den, as are the young of didelphids and dasyurids, or they may accompany the mother by riding on her back (Fig. 2.13). This is described by Russell (1982*a*) as pattern B parental care. The best known species is *T. vulpecula*, which will be considered first and to which the other species will subsequently be compared.

Trichosurus vulpecula

T. vulpecula occurs in the winter rainfall areas of Australia, and extends into drier regions along water courses, where it lives in *Eucalyptus camaldulensis*. In northern Australia it is replaced by *T. arnhemensis*, a closely related smaller species or sub-species, and in rain forest of northern Australia and New Guinea it is replaced by species of *Phalanger*. Another species, *T. caninus*, is sympatric with *T. vulpecula* in heavily timbered forests of eastern Australia. In Tasmania the body weight may reach 4 kg, while in New South Wales the average weight is 2 kg. In all populations the adult males are larger than the adult females. Similarly, the males of *T. arnhemensis* weigh 1.6 kg and females 1.3 kg, but no differences have been observed between male and female *T. caninus*, which are all in the range 2.5–4.5 kg (How in Strahan, 1983).

In the free state the adults are solitary; females occupy a fairly well

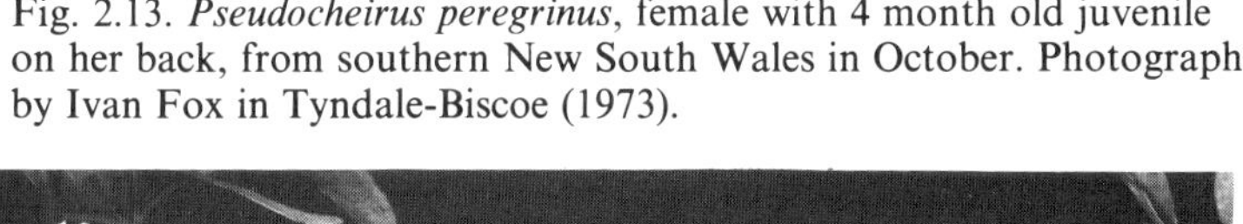

Fig. 2.13. *Pseudocheirus peregrinus*, female with 4 month old juvenile on her back, from southern New South Wales in October. Photograph by Ivan Fox in Tyndale-Biscoe (1973).

defined home-range of 1–2 ha which is marked by scent and in which there are one or more den sites (Dunnet, 1964; Winter, 1978). Males occupy larger territories which are defended from other males (Biggins & Overstreet, 1978) and are marked by rubbing the well-developed sternal gland on nearby objects. Development of the gland is stimulated by testosterone (Bolliger, 1944). According to Winter (1978) courtship lasts for 20–40 days, as it requires a considerable time for the male to develop a consort relationship with a female. Oestrus lasts less than 1 day and copulation generally occurs with only one male. The females are polyoestrous with an oestrous cycle of 26 days (Pilton & Sharman, 1962).

The onset of oestrus can be detected by changes in the cell contents of the urogenital sinus (Pilton & Sharman, 1962); during pro-oestrus the number of cornified epithelial cells increases and on the day of oestrus polymorphonuclear leucocytes appear in large numbers and persist for several days. This is followed by a thin smear of few, mostly nucleated, epithelial cells. If the female has access to a male, a plug of coagulated semen and vaginal secretions may fill the sinus for a day, being subsequently shed (Kean, Marryat & Carroll, 1964). The precise relations between behavioural oestrus, changes in the smear and ovulation have not been determined but Hughes & Rodger (1971), Shorey & Hughes (1973*b*) and Hughes & Hall (1984) state that ovulation occurs within 1 day of oestrus and they recovered unsegmented eggs in the oviduct and uterus 2 days after copulation (see Chapter 7).

The gestation period measured from copulation to birth is 17.5 days (Lyne, Pilton & Sharman, 1959; Pilton & Sharman, 1962), or about 8 days shorter than the oestrous cycle (Table 2.2). If the young is retained in the pouch, oestrus is normally suppressed until the end of lactation. On rare occasions post-partum ovulation does occur as females have been found with two young in the pouch that differ in age by 24 days. If the young are removed from the pouch experimentally, females return to oestrus 8 or 9 days later, like female *Didelphis* do.

Parturition has been observed by Lyne *et al.* (1959) and the posture was the same as seen in *Didelphis virginiana* and *Macropus rufus* (Fig. 2.23*a*) with the tail drawn forward between the legs. There was no evidence that the female aided the young in its journey to the pouch by licking the fur. The pouch is large with a forward-directed opening and contains two teats (Type 5, Fig. 2.8). At parturition both are capable of maintaining the single young but only the one to which it attaches continues to develop and secrete milk (Sharman, 1962). Sharman (1962) transferred neonates to the pouches of non-pregnant females on day 17 of the oestrous cycle, and they were successfully nourished by the foster parent to independence. This was the first experimental evidence that the endocrine conditions for lactation in the non-pregnant marsupial are fully equivalent to those in the pregnant female (see p. 348).

The growth and development of the young in the pouch have been described by Lyne & Verhagen (1957), Crawley (1973) and Winter (1978). Young first detach from the teat at 94 days and the eyes open between 100 and 110 days. At this time the young is well furred and for the first time leaves the pouch, while the mother is in the den. After 140–150 days the young no longer returns to the pouch but still feeds from the elongated teat and rides on the mother's back when she forages. Weaning takes place between 5 and 7 months and, during this time, the female may return to oestrus and produce a second offspring.

The main season of births is February–April in New South Wales (Dunnet, 1964; Smith, Brown & Frith 1969; How, 1976), Tasmania (Lyne & Verhagen, 1957) and for several parts of New Zealand (Tyndale-Biscoe, 1955; Gilmore, 1969; Crawley, 1973; Kean, 1975; Bell, 1981). Whereas all these studies suggest that the peak of breeding occurs at the autumnal equinox, Crawley (1973) and Bell (1981) found considerable variation from year to year; Bell (1981) considered dietary factors and female bodyweight to be more important than photoperiod in determining the onset of breeding, the duration of lactation and the incidence of second breeding (Fig. 2.27). Near Canberra 50% of females carried second young (Dunnet, 1964), near Brisbane 22% (Winter, 1978) and near Christchurch 20% (Gilmore, 1969) but few or none was recorded in northern New South Wales (Smith *et al.* 1969; How, 1981) or near Wellington, New Zealand (Tyndale-Biscoe, 1955; Crawley, 1973; Bell, 1981).

The males of *T. vulpecula* are spermatogenic throughout the year (Tyndale-Biscoe, 1955; Gilmore, 1969; Kean, 1975; Setchell, 1977) but in one New Zealand population the prostate glands were significantly heavier during the autumnal breeding season and, to a lesser extent, during the second spring breeding season (Fig. 4.7*b*) than at other times of the year (Gilmore, 1969). This suggests that the reproductive condition of the females may influence that of the males (see p. 248).

Other species of Phalangeridae

Trichosurus caninus has a very similar reproductive pattern to *T. vulpecula* with an oestrous cycle of 26 days and a gestation of 16 days (Table 2.2) (Smith & How, 1973; How, 1976). The breeding season is March to May but the single young is retained in the pouch for 3 months longer so there is no second peak of breeding in October. Sexual maturity is deferred to the third year (How, 1976).

Trichosurus arnhemensis breeds all year. There is high pouch survival (83%) and the young are weaned at 5.5–6.5 months when the female breeds again. Both sexes mature at 1 year (Kerle in Strahan, 1983). Very little is known about the reproduction of the tropical species of *Phalanger* except that they have four teats and *P. maculatus* and *P. orientalis* bear 1–3 young (Winter in Strahan, 1983). Births of *P. gymnotis* at Baiyer River, New Guinea, were distributed throughout the year (George, 1982).

Petauridae

The common ringtail possum, *Pseudocheirus peregrinus*, is the smallest petaurid that is known to be an exclusive folivore. It is polytocous and polyoestrous with 2 litters a year being possible. Usually 1–3 eggs are fertilised (Hughes, Thomson & Owen, 1965) but Flynn (1922) recorded 6 eggs being shed and he collected a female with a litter of 5 neonatal young (Flynn, 1928). However it is unusual for more than 2 young to survive in the pouch because the anterior pair of the 4 teats in the pouch are usually small and inverted. The young remain in the pouch continuously until 105–12 days when their eyes open and they are fully furred (Table 2.1). After permanent emergence at 4 months the young stay in a nest or ride on the mother's back (Fig. 2.13) until weaned at 6 months. The breeding season begins in May and second births occur in October in New South Wales but in Victoria a second litter is very rarely observed (Thomson & Owen, 1964). The males of this species, like those of *Petauroides volans*, show testicular regression and reduced spermatogenesis during the period of female anoestrus (Hughes *et al.*, 1965). *Ps. peregrinus* is a gregarious species and several animals may share a nest, but the larger species *Ps. herbertensis* and *Ps. archeri* are solitary, while *Ps. dahli* and *Hemibelideus lemuroides* form stable pairs. All four of the latter species have two teats and carry single young (Winter in Strahan, 1983).

The greader glider, *Petauroides* (formerly *Schoinobates*) *volans* (McKay in Strahan, 1983) is wholly restricted to the wet sclerophyll forest on which it is dependent for food and shelter (Tyndale-Biscoe & Smith, 1969; Henry, 1984). The pouch contains two teats but females are invariably monotocous (Table 2.1). Although physiologically polyoestrous, breeding is restricted to a very brief period in March–May (Smith, 1969). The males are spermatogenic only during the brief breeding season and the testes regress after May (Smith 1969; Baldwin, Temple-Smith &

Tidemann, 1974), so that females that fail to breed at the first oestrus, or those that subsequently lose their young, do not conceive again. Only about 70% of adult females breed each year, a number that is equal to the number of adult males in the population and Tyndale-Biscoe & Smith (1969) concluded that the males are monogamous, a conclusion supported by the absence of a sex difference in body size. More recently however, Henry (1984) found that about half the males in his study area were bigamous and he concluded that the mating system of the species is facultative polygyny, depending on availability of resources. The young remains in the pouch for 4 months and is thereafter carried on the mother's back or left in a den for 3 months. Weaning occurs in January and sexual maturity is attained in the second year.

Less is known about the smaller yellow bellied glider, *Petaurus australis*, which covers the same latitudinal range and forest habitat. It is a gregarious species in which groups comprise one male and one or several females and their weaned young. Females have two teats and the single young is carried in the pouch for 100 days and placed in a nest for a further 2 months. Only one young is produced each year, the breeding season in northern Queensland being November to May (Russell in Strahan, 1983; Russell (1984*b*) and in Victoria August to April (Henry & Craig, 1984).

The smaller species, *Petaurus breviceps* is gregarious and sexually dimorphic, with males larger than females and a group of up to seven animals dominated by one male. The diet consists of nectar, gum exudates and insects of the forest canopy. The general pattern of reproduction has been described by Schultze-Westrum (1969), and Smith (1971, 1973, 1979*b*) (see Table 2.2). The oestrous cycle is 29 days and gestation 15–17 days. Like *T. vulpecula* females will return to oestrus 12 days after the young are removed from the pouch. There are 4 teats but the litter size is either 1 or 2; postnatal development is rapid (111 days) and a second litter in the season is common. It is a seasonal breeder, most young being born between August and November (Suckling, 1984), pouch emergence coinciding with the spring flush of insects in the diet. Smith (1982) considers that this is the source of essential protein for lactation. Males are spermatogenic all year (Smith, 1979*b*).

Its non-volant relative, *Gymnobelideus leadbeateri*, is of similar size and has a similar social organisation. Females are polyoestrous, with an oestrous cycle of less than 30 days and gestation of less than 20 days, (Smith, 1984*a*), thus conforming to the pattern seen in *P. breviceps* and *T. vulpecula* (Table 2.2). However, the ovulation rate is greater, with up to 12 eggs shed and 12 corpora lutea formed (Strahan, 1983). The pouch contains 4 teats but the litter size, despite the high ovulation rate, is only 1 or 2 young (Table 2.1). These remain in the pouch for 80–93 days and are then left in a nest for a further 5–40 days (Smith, 1984*a*).

Burramyidae and Tarsipedidae

It is generally agreed that *Tarsipes rostratus* is not closely related to other marsupials, including the Burramyidae, but its pattern of reproduction shares features in common with the smaller members of the Burramyidae and for the purposes of this book it is convenient to consider the two groups together.

The pattern of reproduction of the largest species of the Burramyidae, *Burramys parvus*, resembles that of *T. vulpecula* more than that of its closest relatives (Dimpel & Calaby, 1972). It lives above the tree line in south eastern Australia and its pattern of very rapid reproduction, is presumably an adaptation to the brief alpine summer season (Kerle, 1984). The young are born in November after a gestation of 13–16 days and generally all 4 teats in the pouch are occupied. The young leave the pouch at 3–4 weeks, before the eyes open, they then remain in a nest until weaned at 8–9 weeks and reach adult weight at 3–4 months, before the winter. There is no evidence for post-partum oestrus, embryonic diapause or a second litter after the first is weaned, as occur in the smaller species of the Burramyidae.

Three species of the genus *Cercartetus* and *Acrobates pygmaeus* have a very different pattern of reproduction from *Burramys*. The best known is *Cercartetus concinnus* of Western Australia (Clark, 1967) (Table 2.1). It is polyoestrous and, although the gestation period and oestrous cycle length are not known, it is probable that post-partum oestrus occurs because females can be pregnant while simultaneously suckling a litter in the pouch (Bowley, 1939; Casanova, 1958; Clark 1967). In one post-partum specimen examined by Clark (1967) the ovaries contained 11 Graafian follicles and the vaginal canals were in an oestrous condition. In four lactating females Clark (1967) found that the stage of development of the embryos recovered from the uteri was positively correlated with the size of the pouch young (Fig. 2.14*a*) and she concluded that pregnancy proceeded without delay during lactation. In the living female observed by Bowley (1939) and kept isolated from adult males, the pouch young were observed to relinquish the teat and leave the pouch at 25 days and were weaned at 50 days. Very soon afterwards the female gave birth to another litter which must have been conceived at least 50 days before. Casanova (1958) made similar observations on a captive female.

Clark (1967) concluded that her observations do not support Sharman's (1963) view that there is a delay of embryonic development as in the Macropodidae, but if there is not, then *C. concinnus* has the longest gestation period known for any marsupial. The species is polyovular and polytocous, but Clark (1967) observed a progressive reduction between the mean number of corpora lutea (7.9), the number of uterine embryos (7.2) and the number of pouch young (4), which was less than the number of teats (6). Five pregnant or post-partum females were collected between 7 September and 21 January, while two other females collected in March and April had large pouch young but were not simultaneously pregnant, which may indicate the end of the breeding season in autumn.

Cercartetus nanus appears to have a similar pattern (Turner in Strahan, 1983) the gestation period is longer than 30 days, pouch life more than 40 days and the young are weaned at 50–60 days. Births occur from September to April and in that period females may produce 2 or 3 litters of 4–5 young each. Likewise in the tropical species *C. caudatus* (Atherton & Haffenden, 1982) the litter of 1–4 young leave the pouch at 45 days, but in this species there are two separate breeding seasons, January–February

Fig. 2.14. Blastocyst growth during lactation in (*a*) *Cercartetus concinnus* and (*b*) *Tarsipes rostratus* shown in relationship to size of young in the pouch. Based on data from Bowley (1939), Clark (1967) and Renfree (1980*a*). The duration of pouch life (solid line) and suckling after pouch exit (broken line) are indicated below: o, oocyte in follicle; c, cleaving eggs. Redrawn from Renfree (1981*a*).

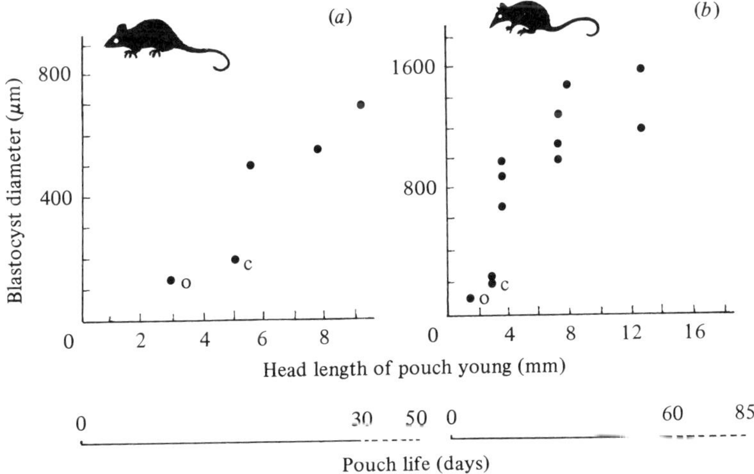

and August–November and the testes of males enlarge at each of the breeding seasons.

The feathertail glider, *Acrobates pygmaeus* has the same pattern of delayed development associated with concurrent lactation; Hill (1900*c*) observed blastodermic vesicles in the uteri of a female, which was simultaneously carrying a litter of three pouch young of unspecified size. Further evidence has been obtained by M. B. Renfree & S. Ward (unpublished observations), who recovered small blastocysts (0.22 mm) in the uteri of one recently lactating female (< 5 d) and blastocysts of 1.6 × 1.9 mm in the uteri of a lactating female carrying young aged > 70 days. This is similar to *C. concinnus* (Fig. 2.14*a*). Fleming & Frey (1984) have been able to study the species in the field because of its propensity for using telephone pole boxes as nest sites. The species displays a polygamous mating system but there is no evident size difference between the sexes. In Victoria the breeding season extends from July to January during which period the testes and epididymides of males are enlarged. There are two peaks of birth in August and November representing successive litters. Four young may be accommodated on the four teats but two is the normal litter size at weaning. The young remain in the pouch for 50 days, when they weigh 1.5 g and for a further 20 days in a nest, during which time their eyes open and they reach a weight of 3 g. They are weaned at 90–5 days when about 7 g but remain in the same social group. The life expectancy is 2–3 years.

The closely related pigmy glider from New Guinea, *Distoechurus pennatus* is currently being studied by K. Aplin (personal communication). In 20 females so far examined, the litter size was 1 and there are only 2 teats in the pouch.

Tarsipedidae

The single representative of the Tarsipedidae, *Tarsipes rostratus* is restricted to heathlands of southwestern Australia and it subsists on nectar and pollen of several species of flowering shrubs (Proteaceae) (Wooller *et al.*, 1981). *T. rostratus* is one of the smallest marsupials with adult males weighing 7–9 g and females 10–12 g (Fig. 2.15), the females being always dominant to males (Renfree, Russell & Wooller, 1984*a*). Its reproductive pattern is similar to that of *C. concinnus* (Renfree, 1980*a*; 1981*a*) (Table 2.1). There are 4 teats in a Type 5 pouch (Fig. 2.8) but the litter size is 2 or 3 (mean 2.4) and the young weigh 3–6 mg, the smallest birth weight of any mammal. In captivity, lactation lasts for about 90 days. The young first emerge from the pouch at 56–63 days and leave permanently 1 week later when fully furred. They are left in a nest while the mother forages and are weaned 3–4 weeks later. The species is polyoestrous and has a post-partum oestrus at which a pregnancy commences, but embryonic development beyond the vesicle stage, 1.2–1.7 mm diameter, ceases or is retarded during most of lactation (Fig. 2.14*b*) (Renfree, 1980*a*; 1981*a*).

While the delay in development is associated with concurrent lactation, removing the pouch young does not result in acceleration of embryonic development (Renfree *et al.* 1984*a*), which implies that the slow development

Fig. 2.15. The honey possum, *Tarsipes rostratus*. Two adult males, weighing 7 g and 8 g respectively, on an inflorescence of *Banksia menziesii*. From Renfree *et al.* (1984*a*), with permission.

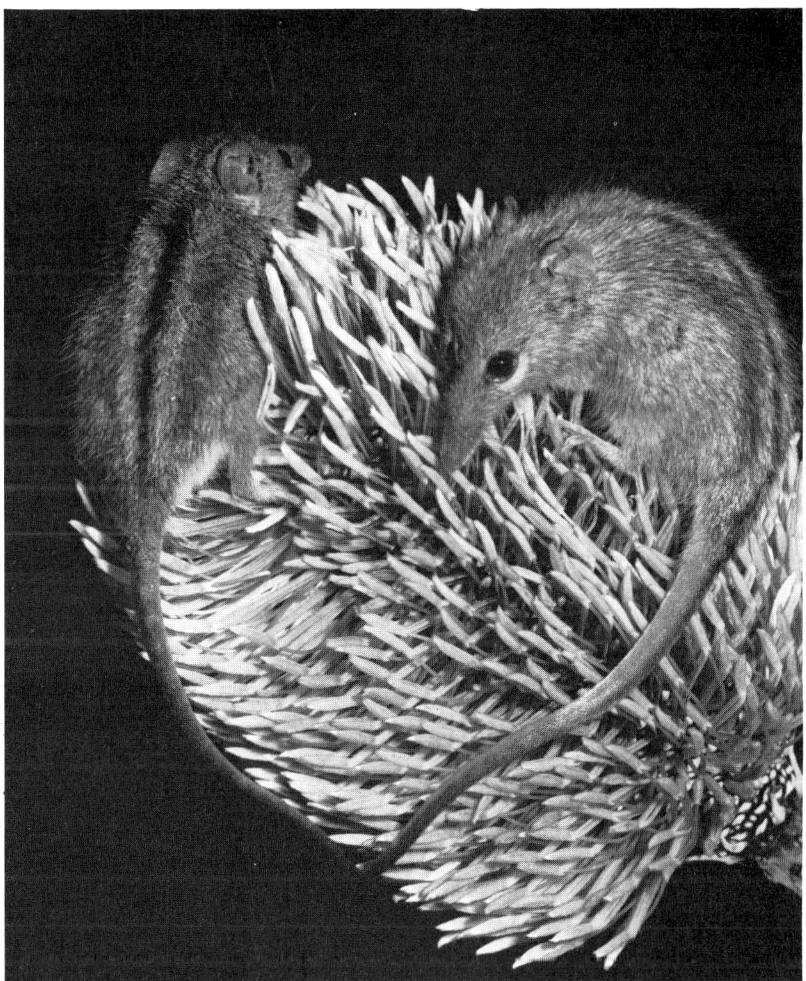

is under some other environmental control, possibly associated with the flowering periods of the plants it lives on.

Females breed at all times of the year but the proportion of females carrying pouch young is highest in January to April (51%), May to August (51%) and September to December (38%). These peaks occurred at the same time in 2 consecutive years and the interval between peaks is the time required for a female to rear a litter, so Wooller *et al.* (1981) conclude that breeding may be entrained annually by photoperiod or by increasing food supply.

Phascolarctidae

The koala, *Phascolarctos cinereus*, is the sole representative of this family and it is the largest of the arboreal herbivorous marsupials; adult females weigh 7–9.5 kg and males 9.5–13.5 kg (Martin & Lee, 1984). This sexual dimorphism is probably associated with the polygynous mating system and pronounced male dominance hierarchy. Like male *T. vulpecula*, male koalas have well-developed sternal glands, which they rub against trees during the breeding season. They also bellow loudly and exclude other males from their home-ranges (Lee & Cockburn, 1985). Their home ranges overlap those of several females with which they mate repeatedly at oestrus. Females are polyoestrous and in captivity the cycle is about 27–30 days (Smith, 1979*a*) and the gestation period 34–6 days. It is most unusual among marsupials for gestation to exceed the length of the oestrous cycle and it is possible that the oestrous cycle in the wild is longer. I. McDonald and C. Lithgow (personal communication) have been unable to detect elevated progesterone concentrations in captive females, isolated from males, but have measured peak concentrations of 10–15 ng ml^{-1} in free-ranging females. On the other hand they measured elevated oestradiol concentrations of 40 pg ml^{-1} in both captive and free-ranging females at oestrus. From these findings they conclude that ovulation in *P. cinereus* may be induced by copulation and that captive females undergo anovular cycles of shorter duration, as in *Marmosa robinsoni*, *Monodelphis domestica* and *Didelphis virginiana* (see p. 20). There are 2 teats but there are no unequivocal records of litters of more than 1 (Table 2.1).

The postnatal development is very slow; the young remains in the pouch for 5–6 months, is not fully weaned until a year old and sexual maturity is delayed until the third season. All births near Brisbane occurred between October and January and, because the young is not weaned until more than a year later, females may become so delayed that they fail to breed at all for 1 year in 3. On Kangaroo Island, births were recorded from late December to April and on French Island, Victoria from November to April (Martin & Lee, 1984) but the data are insufficient to say whether this represents a latitudinal cline as Smith (1979*a*) suggests.

Vombatidae

There are three living species of wombat but the reproduction of none has been studied systematically. From observed changes in vaginal cytology and pouch condition Peters & Rose (1979) have shown that *Vombatus ursinus* is polyoestrous with an oestrous cycle of 32–34 days. They also observed changes in the pattern of the diurnal cycle of basal body temperature, which correlated with the oestrous cycle (Fig. 2.16). In a wild population McIlroy (1973) observed females to be carrying small pouch young from May to August (Table 2.1), fully furred pouch young in September and October, when young at heel were first seen. By this age the young weighed 4 kg, compared to the adult weight of 22 kg, and continued to associate with the mother for up to a year.

The hairy-nosed wombat, *Lasiorhinus latifrons*, is monovular (Gaughwin & Wells, 1978) and the gestation period is 20–2 days (Crowcroft & Sonderlund, 1977). The latter authors describe copulatory behaviour and

Fig. 2.16. Body temperature and vaginal smear indices during the oestrous cycle of *Vombatus ursinus* in two females. T_{max}, daily maximum body temperature; BBT, daily basal body temperature; KI, karyopyknotic index; LI, leucocytic index. Redrawn from Peters & Rose (1979).

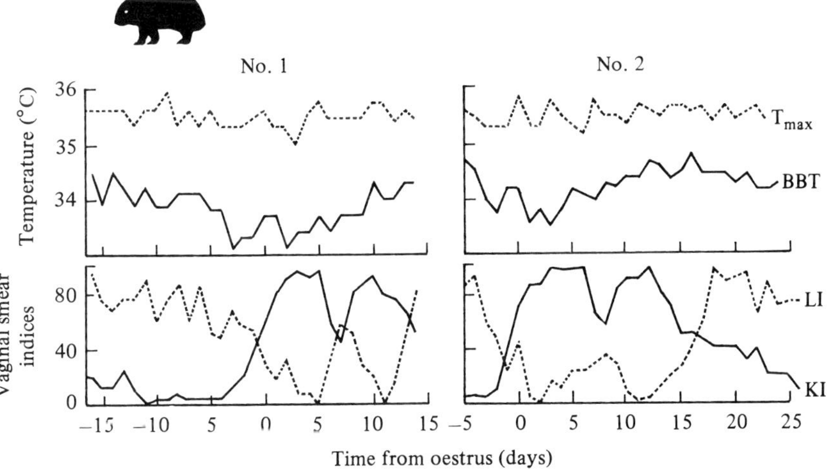

development of the young. One young first emerged from the pouch at 198 days and remained out permanently at 300 days (Table 2.1), although it was suckled until 400 days. This agrees with Gaughwin & Wells' (1978) observations of a wild population of this species in South Australia. Small young were found in the pouches of females from November to January and young of 2–3 kg were first observed outside the burrows in July. While this species is a seasonal breeder the proportion of females that ovulated or carried pouch young varied markedly between years from 0% to 70%. The maximum rates were observed in years of high rainfall and plentiful vegetation.

Macropodidae

The Macropodidae is the most distinctive and well known family of marsupials, partly because it includes the largest living species and these are crepuscular. However, the family also includes other species which are small and nocturnal. According to Kirsch & Calaby (1977) there are 56 species, distributed within two main sub-families, the Potoroinae and the Macropodinae. The Potoroinae are the small rat kangaroos and adults weigh 0.5–3 kg. They are ground living, omnivorous or herbivorous and are restricted to Australia. The Macropodinae comprise the wallabies and large kangaroos; all are herbivorous and range in size from 2 kg (*Setonix brachyurus*) to 90 kg (*Macropus rufus*) and are distributed throughout Australia and New Guinea. Three species, *Macropus r. rufogriseus*, *M. parma* and *M. eugenii* have become well-established as feral populations in New Zealand (Catt, 1977; Maynes, 1977) and breed in zoos throughout the world.

Fossil species from each sub-family have been recognised since the early Miocene (Stirton, Tedford & Woodburne, 1968), so it is reasonable to infer that features of their reproduction that are held in common are of considerable antiquity. All species of both sub-families that have been studied are polyoestrous and monovular, with the single exception of *Hypsiprymnodon moschatus* that produces twins (Johnson in Strahan, 1983). The gestation period occupies a much larger proportion (0.84 to 1.09) of the reproductive cycle than in other marsupials, which are all less than 0.6 (Table 2.2). In the majority of macropodids the next oestrus and ovulation are not suppressed by lactation and occur 1–10 days after parturition. Conception can occur at this time but, if lactation proceeds normally, the corpus luteum of post-partum oestrus and subsequent ovulation are inhibited and the embryo does not develop past the stage of a unilaminar blastocyst of less than 100 cells (Smith, 1981). This condition of ovarian quiescence and embryonic diapause is maintained for the greater part of lactation, and in two species for up to 11 months (Berger, 1966; Catt, 1977). This condition has been termed quiescence (Clark & Poole, 1967; Sharman & Berger, 1969; Tyndale-Biscoe *et al.*, 1974), to distinguish it clearly from the other phases of the normal cycle and from anoestrus. In true anoestrus, the ovaries do not contain corpora lutea or tertiary Graafian follicles, the uteri are very small and the endometrial glands are short and straight with closed lumina (Newsome, 1964*a*; Clark & Poole, 1967) (see Fig. 2.17).

In all except two species, quiescence is associated with lactation and is termed lactational quiescence. In *M. eugenii* (Fig. 2.26) and *M. rufogriseus* it also occurs regularly each year from the winter to the summer solstice (Berger, 1966; Renfree & Tyndale-Biscoe, 1973*a*; Catt, 1977) and is termed seasonal quiescence (see Figs 2.26, 9.1 and Chapter 9). Anoestrus may also occur during lactation either regularly, as in *M. fuliginosus* and *M. giganteus* (Poole & Catling, 1974) or in response to drought in *Macropus rufus* (Newsome, 1964*a*; Frith & Sharman, 1964). Seasonal anoestrus occurs in island populations of *Setonix brachyurus* (Sharman, 1955*a*; Shield, 1964). Thus female macropodids when not undergoing an oestrous cycle or pregnancy will be in one of four reproductive conditions: seasonal quiescence, lactational quiescence, seasonal anoestrus or lactational anoestrus.

Since Sharman's (1954) demonstration of embryonic diapause in *Setonix brachyurus*, the reproductive patterns of 19 species from both sub-families have been described in greater or lesser detail and much experimental work has been concerned with these phenomena, so that more is now known about reproduction in this family than in any other family of marsupials. The patterns of reproduction have been reviewed by Sharman *et al.* (1966) and Calaby & Poole (1971) and the control of reproduction by Tyndale-Biscoe *et al.* (1974). Tyndale-Biscoe (1973), Russell (1974, 1982*a*) and Newsome (1975) have reviewed macropodid reproduction in the context of ecological adaptations and Rose (1978) in relation to the evolution of the family. The phenomenon of embryonic diapause was reviewed by Sharman & Berger (1969), Tyndale-Biscoe (1973) and Renfree (1978, 1981*a*) and will be discussed in Chapters 6 and 7. Of these 19 species, all 5 potoroines and 10 of the macropodine species show the same pattern as *Setonix* (Table 2.2).

Four other macropodine species differ in that the length of gestation is several days shorter than the oestrus cycle (about 0.84) and, as in other non-macropodid marsupials, pre-ovulatory – rather than post-ovulatory –

Fig. 2.17. Uterine glands of *Macropus giganteus* in lactational quiescence and anoestrus. (*a*) Transverse section of uterine glands of non-pregnant uterus soon after parturition. Glandular epithelium of columnar cells with basal nuclei and ciliated cells in luminal epithelium. (*b*) Transverse section of anoestrous uterus, with small glands lined with low columnar epithelium and the luminal epithelium of cuboidal cells. From Clark & Poole (1967), with permission.

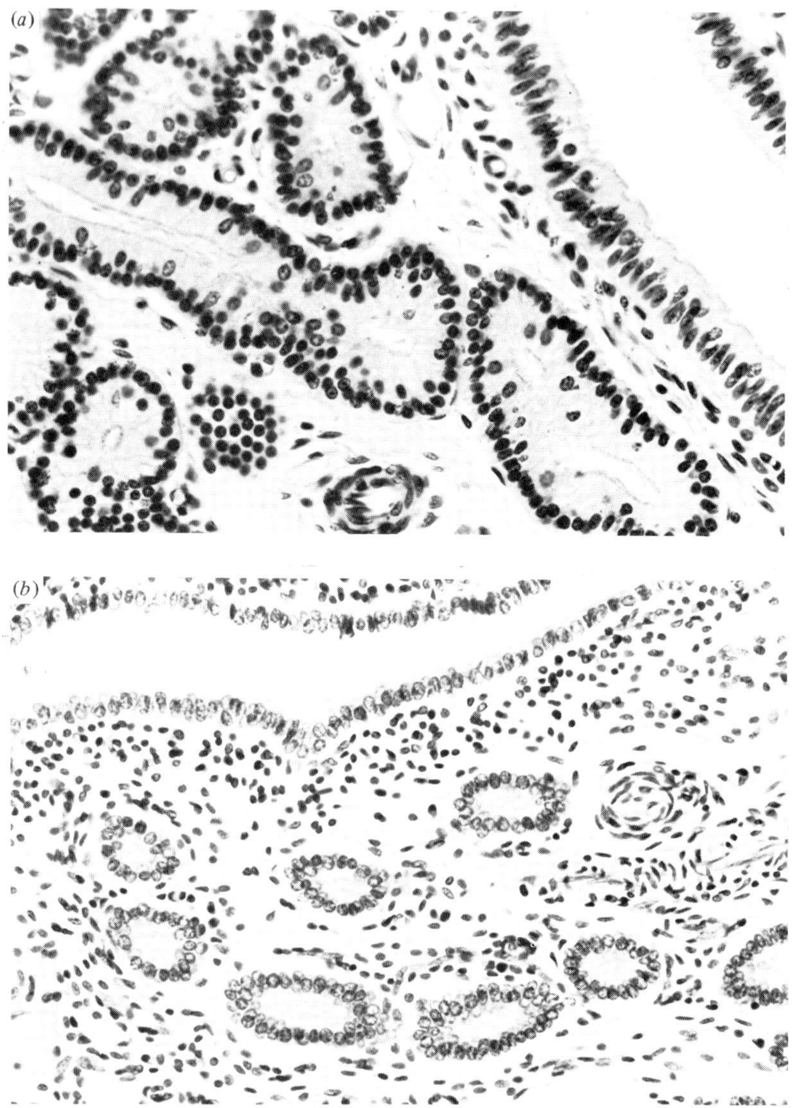

events are inhibited during lactation (Table 2.2). Nevertheless, all but one of these species can undergo quiescence and embryonic diapause in particular circumstances. Maynes (1973*a*) has argued that these four species display the primitive pattern derived from the ancestral phalangerid. Against this view is the evidence that most species of both sub-families display a common pattern that includes diapause and Maynes' hypothesis would require that this evolved twice independently since the Miocene. In our view it seems more parsimonious to suppose that the four species referred to have secondarily lost the full pattern of diapause to varying degrees, with *M. fuliginosus* being the most extreme example.

Species of macropodids occupy all climatic zones, in Australia and New Guinea from rainforest to desert and from the monsoonal tropics to the cool temperate sub-alpine zone. Where seasonal changes are regular most species have precise breeding seasons but, in areas with erratic rainfall and drought, kangaroos are opportunistic breeders (see Newsome, 1975 for review). These different patterns are achieved by slight variations of the basic macropodid reproductive cycle. So far as is known male macropodids, once they are sexually mature, are continuously spermatogenic (Setchell, 1977) although Sadleir (1965) reported some very large male *M. rufus* and *M. robustus* to have low sperm counts and Newsome (1973) reported male *M. rufus* to show reduced spermatogenesis during periods of extreme high temperatures. Even in those species in which the female shows highly seasonal breeding, such as *M. eugenii*, the males do not show a decline in spermatogenesis, testis size (Fig. 4.7*a*) or seminiferous tubule diameter (Inns, 1982). However, in this species and *M. rufogriseus* (Catt, 1977) the prostate glands of males were enlarged during the main and subsidiary periods of copulatory activity and in *M. eugenii* these periods coincide with periods of elevated testosterone in plasma (Fig. 4.7*a*). Prostate weights have not been recorded from any other species of macropodid.

Oestrous cycle and pregnancy

Female macropodids do not show marked changes during the oestrus cycle and the only reliable means for determining the cycle length is by changes in cells of the urogenital sinus, associated with changes in the pouch. These have been described for *Setonix* (Sharman, 1955*a*), *Potorous tridactylus* (Hughes, 1962*a*) *M. rufus* (Fig. 2.18) (Sharman & Calaby, 1964), *Bettongia lesueur* (Tyndale-Biscoe, 1968) and *M. giganteus* and *M. fuliginosus* (Poole & Catling, 1974). The marked change in the smear, when abundant cornified epithelial cells produce a white mass, usually occurs 1 day after behavioural oestrus and continues for several days. In some species a flush of polymorphonuclear leucocytes appears also. Some females clean out old dry secretion in the pouch 1 or 2 days before oestrus and the skin becomes vascular and moist. The ends of the teats develop small buds to which neonatal young can attach. Notwithstanding the lack of overt changes it is clear from their behaviour that males of the same species can detect changes in the female several days before oestrus, and they investigate the pouch and genital region of the female (Fig. 2.21*a*). Males of *M. rufus* (Sharman & Calaby, 1964) and *M. parma* (Maynes, 1973*a*) have also been observed to taste urine and Ganslosser (1979) and Coulson & Croft (1981) have observed males displaying flehmen behaviour like ungulates do. More subtle changes, presumably associated with the active corpus luteum, will be discussed in Chapter 6 (Fig. 6.18).

Whereas males and females of most macropodid species will associate without conflict, the presence of an oestrous female induces conflict between males (Fig. 2.19). The male of *M. rufus* stands up high on its hind feet and tail and grasps tufts of grass in its hands and rubs them on its chest (Fig. 2.20). In an established group this may be a sufficient threat posture for the dominant male to adopt, but smaller males may wrestle, striking the chest of the opponent with the hind feet. Among smaller macropodids, such as *M. eugenii* one male will jump over the other striking

Fig. 2.18. Changes in cell composition and size of smear taken from the urogenital sinus during the oestrous cycle of *Macropus rufus*. Horizontal lines, fully cornified epithelial cells; hatch, partly cornified epithelial cells; stipple, non-cornified epithelial cells; thin black line, thin smear; large dots, medium-sized smear; thick black line, thick smear. Redrawn from Sharman & Calaby (1964).

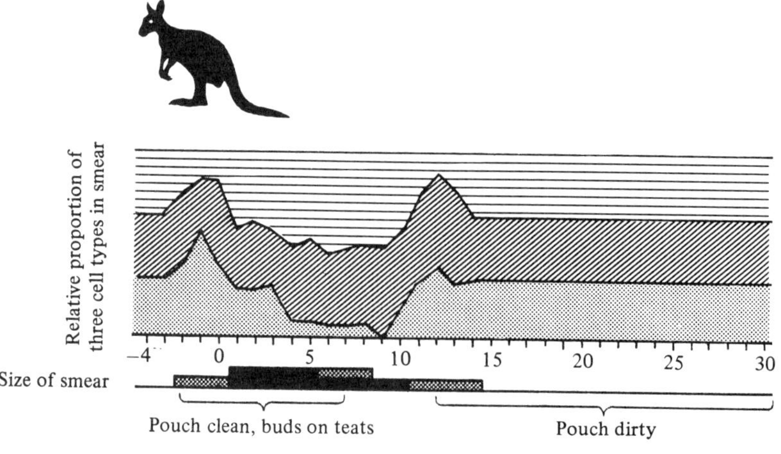

down on the opponent's back with the hind feet. In closely confined groups such conflict can lead to severe injuries or death but in larger enclosures or in wild populations a dominance hierarchy is established.

Copulatory behaviour in seven species of macropodids was reviewed by Sharman *et al.* (1966) and since then very similar patterns have been described for four other species of both sub-families (Stodart, 1966*b*; Maynes, 1973*a*; Tyndale-Biscoe & Rodger, 1978).

The female is pursued by the male who attempts to grasp her tail and investigate her pouch and genital region (Fig. 2.21*a*). In the smaller species these actions are accompanied by the male executing a sinuous lateral

Fig. 2.19. Territorial fighting in *Macropus eugenii*. Males clasp each other with their forearms, and throw their heads back as far as possible to avoid the claws of the opponent. Using the tail for balance, the powerful hind limbs rapidly kick towards the opponent's abdomen. The fights usually are of brief, but intense, duration, of up to 10–15 mins. The fight concludes with the dominant buck chasing off the loser, and frequently with a display of pawing and grabbing the ground with the forepaws, and chinning the disturbed earth. This mode of fighting is representative of all macropodids.

movement of his tail and emitting a faint clicking sound. Copulation begins when the female no longer moves away and adopts a crouching position. The male then grasps the female in front of the hips and intromission takes place to one side of the tail (Fig. 2.21*b*, *c*). The act of copulation lasts 2 min in *Potorous tridactylus*, *M. parma*, 5 min in *M. rufogriseus*, 8 min in *M.*

Fig. 2.20. Threat behaviour in male of *Macropus rufus*. There are successive positions of the head while licking and biting gular-presternal glandular area. Extra height is gained by using the tail as a prop. A brown secretion mats and stains the sternal fur. Redrawn from Sharman & Calaby (1964).

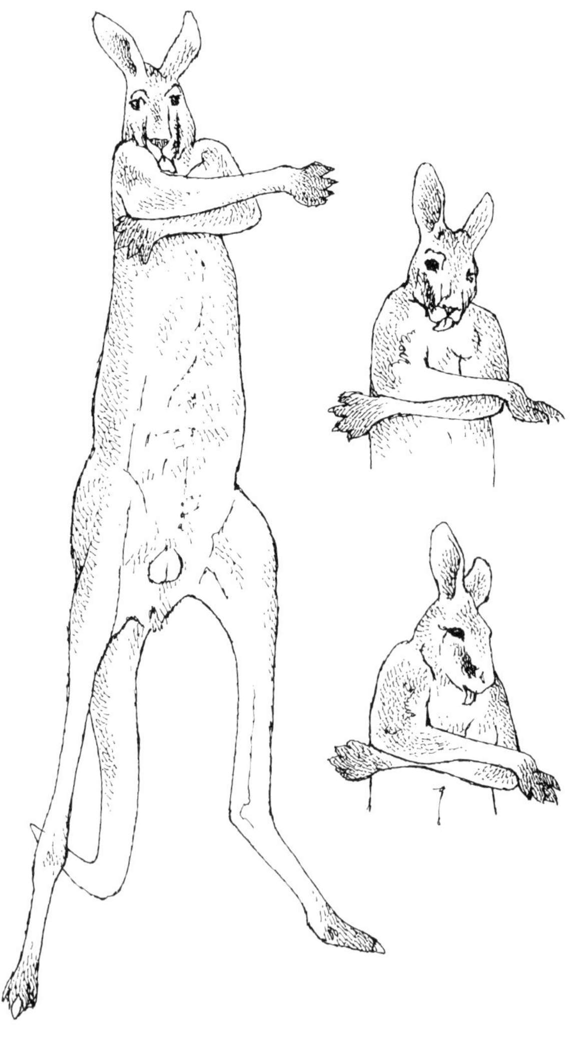

eugenii, 15–20 min in *M. rufus* and 40–50 min in *M. giganteus*. During this time, in the latter species, ejaculation occurs several times. It is unusual for a male to copulate more than once with the same female, but several other males may copulate with her subsequently during oestrus which, in *M. eugenii*, lasts for less than 12 h (Tyndale-Biscoe & Rodger, 1978). Oestrous females of *M. eugenii* may be pursued by six or eight males at once.

Since most species of the Macropodidae are not overtly territorial (Russell, 1974, 1984*a*) the very marked sexual dimorphism seen in the Macropodinae has presumably been selected for its sexual advantage (Jarman, 1983). Macropodid species of less than 5 kg body weight are not sexually dimorphic; males and females grow at the same rate to the same adult size and attain sexual maturity at about the same age. By contrast, in all species over 5 kg, males grow at a faster rate than females and attain a final size up to twice that of females (Fig. 2.22). Moreover, females attain puberty very much earlier than males of the same species (Table 2.1) and at about half their final weight and, as a result, a female at first oestrus may be one-third to one-fifth the weight of the largest males in the population. Not only is there a sexual dimorphism in weight but the forearms of males of the larger species become proportionally larger than in females (Fig. 2.20) and the two sexes may also differ in coat colour and texture. It is likely that these are phenotypic effects of testosterone since cryptorchid males display the female body form (see p. 98).

Since dominant male macropodids do not prevent subordinate males from copulating with oestrous females after they have done so, the selective pressure for large size is difficult to discern. One possibility is that the selection is between spermatozoa in the female genital tract. In Chapter 7 we discuss the evidence, from *Didelphis virginiana*, that there is only a very brief period indeed after the egg is shed when a sperm can penetrate the zona pellucida before the mucolemma begins to cover it and exclude further sperm. If this applies also to kangaroos the male whose sperm are furthest along the oviduct when ovulation occurs will have the highest probability of fertilising the egg. This would be the male that copulates first with the female and hence the largest and strongest will be at a selective advantage in fertilising the egg.

The duration of reproductive cycles in macropodids is affected by the presence of the embryo and by preceding lactation. The interval from one oestrus to the next in unmated, non-lactating females ranges from 22 days in three species of *Bettongia* to 42 days in *Potorous tridactylus* and *M. parma* and 46 days in *M. giganteus* (Table 2.2), but the most common length is

(*a*)

(*b*)

(*c*)

28–30 days. The gestation period in non-lactating females ranges from 21.5 days in *Bettongia*, to 34.5 days in *M. parma* and *M. fuliginosus* and 38 days in *P. tridactylus* and *M. giganteus*. Post-partum oestrus occurs in all species in which the length of gestation exceeds 90% of the duration of the oestrous cycle. In *M. eugenii* it occurs 8–16 h after parturition (Tyndale-Biscoe *et al.* 1983; Shaw & Renfree, 1984; Harder *et al.*, 1985) and up to 2 days post-partum in *M. rufus* (Sharman & Pilton, 1964) and *M. rufogriseus* (Merchant & Calaby, 1981).

Until 1976 it was believed that the interval from one oestrus to the next was the same in pregnant as in non-pregnant females of macropodids undergoing both normal and delayed cycles. However, in *M. agilis* Merchant (1976) observed that the interval was shorter in pregnant than in non-pregnant females (Table 2.2). As different females were involved for the two sets of data the matter could not be further resolved. He then examined the phenomenon in *M. eugenii* (Merchant, 1979), using the same group of females for both sets of observations and mating them to vasectomised males for the non-pregnant cycles. He established that there was a significant shortening of the interval in the pregnant cycle, especially in the comparison between delayed cycles (Table 6.4). Subsequently, Merchant & Calaby (1981) showed the same effect in both subspecies of *M. rufogriseus*. Merchant (1979) suggested that the effect may apply in other species such as *M. rufus* in which the oestrous cycle is 37 days (Sharman & Pilton, 1964) but post-partum oestrus occurs 35 days post-coitus in pregnant cycles (Sharman & Calaby, 1964). At present this is the only direct evidence that the marsupial fetus or placenta can affect ovarian function in the mother during pregnancy and the endocrine aspects will be discussed more fully in Chapters 6 and 7. For the minority group, where post-partum oestrus and ovulation do not occur, removal of the pouch young leads to oestrus and ovulation 8–10 days later (Table 2.2), followed by another oestrous cycle or pregnancy and there is no evidence that such an effect of the fetus occurs.

Fig. 2.21. Copulation in Macropodidae. (*a*) Pre-copulatory behaviour in *M. eugenii*. The male follows the female, grasping at her tail and hindquarters with his forepaws. When she stops, he closely inspects her pouch and vulva, sniffing and licking as shown, and tasting urine. (*b*) Mating in *M. rufus*. Male 3 years old, female 2 years old. Note difference in size of male and female. The male firmly grasps the female around the hips. Drawn from Sharman & Pilton (1964). (*c*) Mating in *Bettongia lesueur*. Initial attempt by male to mate is rebuffed by female using hind feet. Redrawn from Stodart (1966*b*).

Fig. 2.22. Sexual dimorphism in Macropodidae. Growth curves of male (solid line) and female (broken line) of (*a*) *M. giganteus*, (*b*) *M. rufus*, (*c*) *M. robustus* and (*d*) *Wallabia bicolor*, showing age at sexual maturity of females ▲. Redrawn from Jarman (1983).

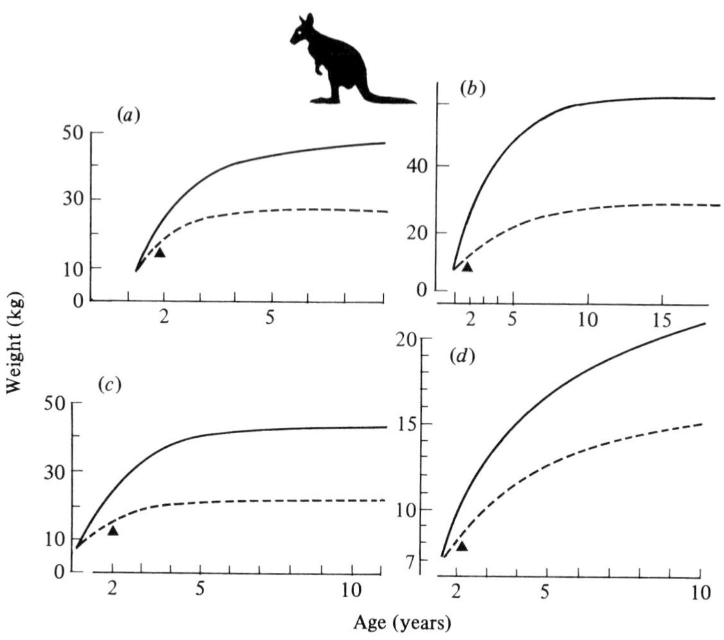

Parturition in Macropodidae

Parturition has been most fully described in *Macropus rufus* (Sharman & Pilton, 1964; Sharman & Calaby, 1964), and Sharman *et al.* (1966) reviewed previous observations of birth in seven other macropodids. These descriptions gain added significance now that the changes in several hormones at the time of parturition, post-partum oestrus and onset of lactation are known in *M. eugenii* (see Chapter 7). The following account is largely taken, edited, from Sharman *et al.* (1966), but modified where new observations on *M. eugenii* clarify some aspects.

In mid-pregnancy the pouch lining of *M. rufus* females not carrying young is covered with a brown to black scale. Twenty-four hours or more before giving birth the female begins to remove the scale by inserting its head into the pouch and licking (Fig. 2.23*a*). Pouch licking intensifies as birth approaches even though the pouch may be quite clean some hours before the birth. For birth the female adopts a characteristic position in which the tail is passed between the hind legs which are extended straight

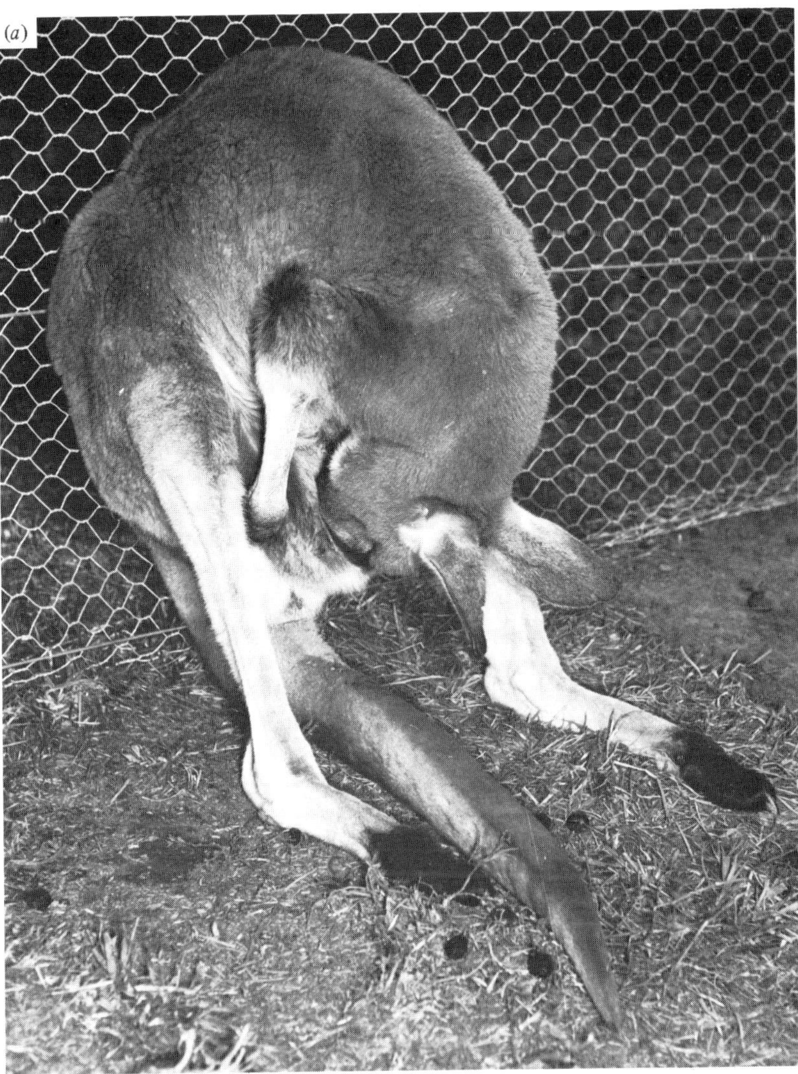

Fig. 2.23. Parturition sequence in *Macropus rufus* (*a,c,d,f,h*) and *M. eugeni* (*b,e,g,i*). (*a*) The female *M. rufus* grasps the sides of the pouch with forepaws and inserts her muzzle to clean the inside of the pouch just before giving birth. Some of the weight is still being taken by the hind legs; at later stages the legs are extended further forwards and all the weight taken on the butt of the tail. Photograph by Ederic Slater from Sharman & Calaby (1964), by permission.

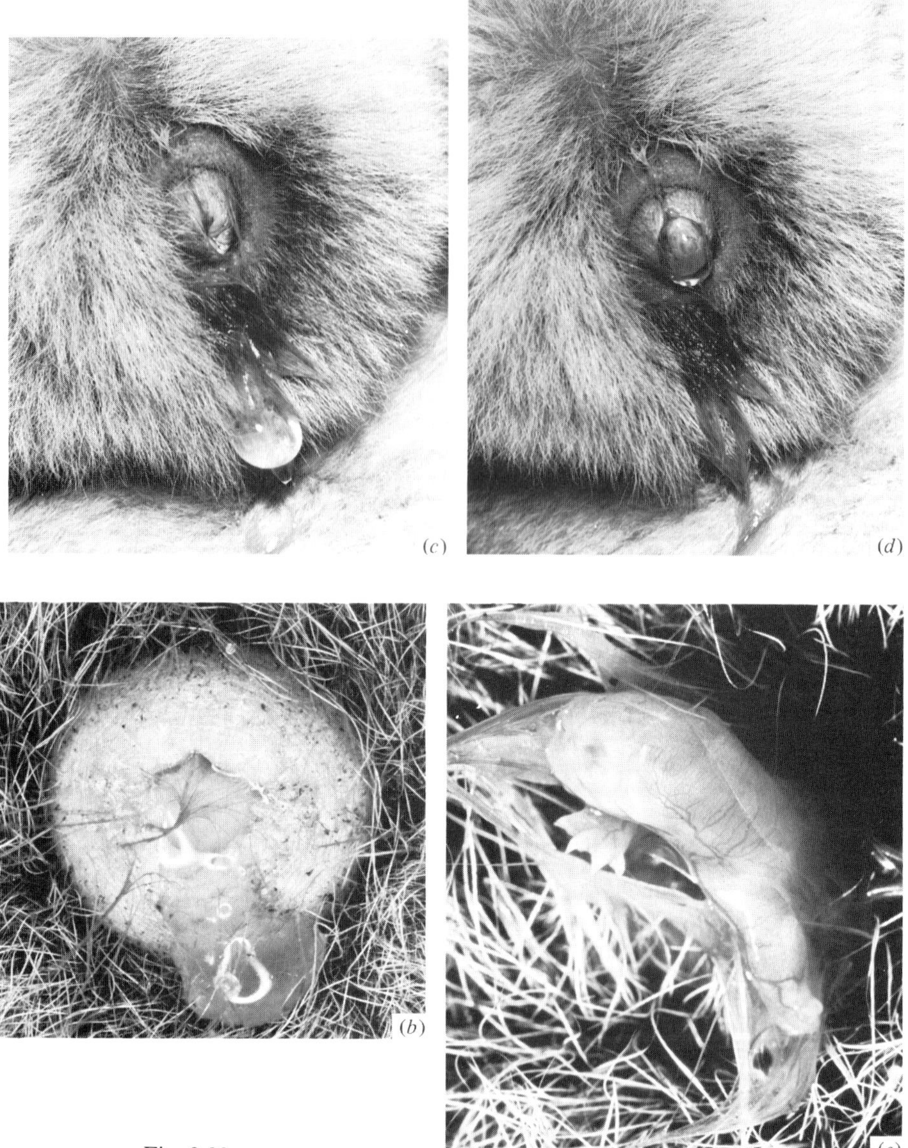

Fig. 2.23. *cont.*

(*b*) The first sign of birth is the appearance of the viscous yellow yolk sac fluid at the vulva. This is followed by (*c*) the allantoic sac, containing clear fluid leaving the vulva (above) and dropping to the ground. The yolk sac fluid which preceded it has wet the tail (below). (*d*) Head of the neonatus, enclosed in the amnion, appearing at the vulva and pointing towards the pouch. (*e*) After a brief hesitation the neonate begins to struggle free of the amnion. (*f*) The neonatus, free of the amnion, emerges from the vulva and (*g*) begins the climb to the pouch. (*h*) Neonatus free of amnion but attached by umbilical cord to yolk sac placenta which is still in the birth canal (vulva lower right).

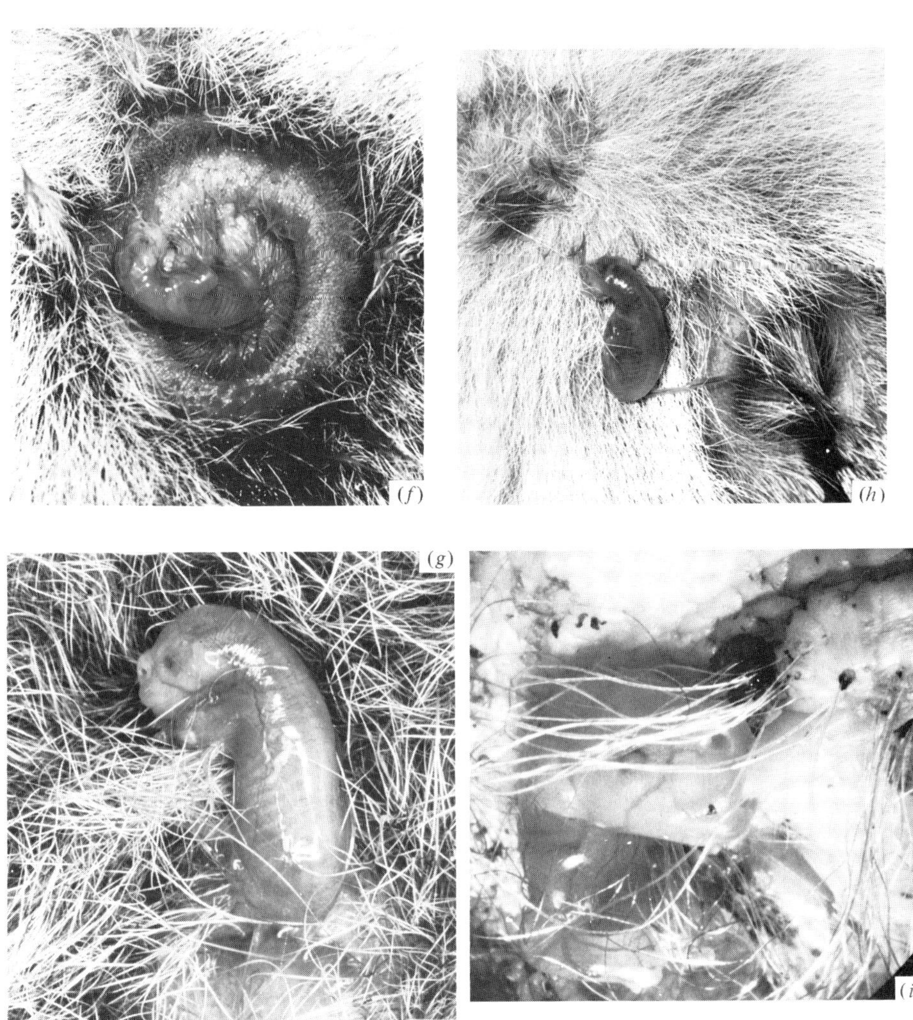

Fig. 2.23. *cont.*

(*i*) After reaching the pouch opening, the neonate changes direction and enters the pouch and attaches to one small teat. It appears to need the closeness of the ventral wall of the pouch pressed onto the dorsal wall to help it locate the teat and attach. Note the scale is not the same for the two species; the neonate of *M. rufus* weighs about 800 mg and that of *M. eugenii* about 400 mg. Photographs of *M. rufus* by Ederic Slater and of *M. eugenii* by David Parer, Mt Eliza, Melbourne, with permission.

forward (Fig. 2.23*a*). The weight is carried on the base of the tail, and the trunk is inclined forwards. The female usually, but not always, has her back supported against a tree or wall or similar rigid object. The birth position may be assumed many times beginning several hours before birth takes place. For about half an hour before birth the female licks the urogenital opening vigorously.

A minute or two before birth a few millilitres of yellow viscous yolk sac fluid run from the vulva (Fig. 2.23*b*) followed by the allantois, a nearly spherical sac about 20 mm diameter, sealed at the umbilical end, and filled with a clear fluid (Fig. 2.23*c*) with a high urea content (see Chapter 7). The young then appears, head first, enclosed in the fluid-filled amnion (Fig. 2.23*d*). After breaking free of the amnion (Fig. 2.23*e*, *f*) it crawls straight upwards from the urogenital opening to the pouch over the mother's fur (Fig. 2.23*g*) and accomplishes the journey in only a few minutes. It attaches to a teat soon after entering the pouch (Fig. 2.23*h*). The mother does not assist the young in any way either by placing it in the pouch or licking a track through the fur, and, in fact, appears to ignore it entirely. All blood and liquid escaping from the urogenital opening or left on the fur by the young is carefully licked up. The yolk sac is expelled (after the young has emerged) and is eaten by the mother. She continues to sit in the birth position for up to an hour after the birth, licking the urogenital opening and cleaning the fur, and also continues to lick around the inside of the pouch on occasions.

Other species for which there is some information on birth behaviour are *Potorous tridactylus*, *Bettongia lesueur*, *S. brachyurus*, *M. rufogriseus*, *M. giganteus* and *M. fuliginosus*. Most are known to clean the pouch prior to parturition although there is individual variation within species and a few individuals hardly clean the pouch at all. The birth position for *P. tridactylus*, *S. brachyurus*, and *M. rufogriseus* is, in general, the same as described above for *M. rufus*, except that the hind legs are not extended quite so far forwards.

However in *M. giganteus* (Poole & Pilton, 1964) and *M. fuliginosus* (Beeck, 1955) the tail was not passed forwards between the legs but was behind the animal in the normal position. The body was leaning backwards with the forepart of the trunk curved forwards; the weight was obviously being taken on the tail and heels. The cloacal eminence was thrust forwards.

It is probable that the young is normally born in the amnion but is usually difficult to see because the vulva is obscured by the head of the mother who is licking it vigorously. The birth of the young in the amnion in *M. rufus* has been seen on seven occasions, four times in normal circumstances and three times when the mother was anaesthetised. It has been observed once in *Bettongia lesueur* (Tyndale-Biscoe, 1968). It is probable that the allantoic sac is normally expelled just before the birth of the young. It has been seen on many occasions in *M. rufus* and *M. eugenii*, once in *M. giganteus* and once in *M. fuliginosus*. The sac was not seen on some occasions in *M. rufus*, probably because it ruptured before reaching the vulva.

M. rufus females have been observed eating the yolk sac and this is probably the normal fate of the membranes. In an observed birth in *M. rufogriseus* (Merchant & Calaby, 1981) the female was seen to chew something between spells of licking the vulva after the birth of the young. As only two small fragments of membranes could be found afterwards it was presumed that they had been eaten. Hughes (1962*a*) observed a birth in *P. tridactylus* and saw the female chewing what were apparently fetal membranes.

Lactation

With the sole exception of *Hypsiprymnodon moschatus* macropodids are monotocous, although all have four teats and separate mammary glands in the large forward-opening pouch (Type 5, Fig. 2.8). Two young of the same age or differing in age by the length of 1 cycle have been reported occasionally but they are invariably at an early stage of pouch life. In the captive colony at Canberra, where up to 500 *M. eugenii* have been maintained for 10 years, only one pair of twins survived to independence; in the few other natural cases, and in several experimental cases where additional young have been fostered to vacant teats, all but one have disappeared from the pouch before mid-lactation.

Pouch life (Table 2.1) ranges from 115–120 days in *Bettongia lesueur* (Tyndale-Biscoe, 1968) and *B. gaimardi* (Rose, 1978) and *Potorous tridactylus* (Hughes, 1962*a*) to 300 days in *M. giganteus* (Poole, 1975), during which two distinct phases can be recognised: a period of continuous attachment, while the young is naked, blind and poikilothermic, and a second period when the young can relinquish the teat, open its eyes and is becoming homeothermic. The change to a diet other than milk begins towards the end of pouch life but the young continues to suck from the elongated teat for a further period of 30–100 days thereafter (Fig. 2.24) (see Chapter 8). In females that mate at post-partum oestrus or later in lactation (see below) reactivation of the quiescent corpus luteum and blastocyst occurs towards the end of pouch life, presumably as the intensity of sucking declines (Sharman, 1963, 1965*a*, *b*) and the female may give birth to a young immediately or very soon after the older young has permanently vacated the pouch. When this happens the neonate attaches to one of the three vacant teats. In continuously breeding females of *M. rufus* and *M. agilis* the teat used by the penultimate, fully weaned young, is not used by the neonatus and, in this case, the female has four adjacent mammary glands each at a different stage of lactation (Figs. 8.7 and 8.11) two of which are secreting milks of entirely different composition (Griffiths, McIntosh & Leckie, 1972) (see Chapter 8).

Female macropodids reach sexual maturity before they are full grown and, in some smaller species, such as *B. lesueur* and *M. eugenii* very soon after leaving the mother's pouch and before being fully weaned (Tyndale-Biscoe, 1968; Andrewartha & Barker, 1969). Males do not reach sexual maturity so early and generally a year later than females (Table 2.1). This may be related to the longer growth phase and larger final size of males referred to earlier.

Fig. 2.24. (*a*) Female *Macropus rufus* feeding young-at-heel aged 302 days outside the pouch while a pouch young aged 67 days is suckled on another teat. (*b*) Enlargement of part of (*a*) (see Fig. 8.7 for further detail). Redrawn from Sharman & Calaby (1964).

Breeding seasons

Patterns of breeding in different macropodids have evolved by varying the time at which ovulation occurs and by varying the time at which the quiescent corpus luteum will be reactivated. Of the two phenomena, folliculogenesis and ovulation appear to be most responsive to short-term prevailing nutritional conditions, whereas the control of quiescence seems to be more conservative and controlled by long-term factors in the evolution of each species. There is, for instance, no case known in which quiescence has been induced under adverse conditions while to the contrary, the highly controlled seasonal quiescence of *M. eugenii* and *M. rufogriseus* has persisted indefinitely after the species have been transferred to other environments or to the stable conditions of captivity. These points are best illustrated by examples of the best-studied species.

Facultative breeding macropodids

The red kangaroo, *Macropus rufus*, lives in the arid and semi-arid zones of Australia where rainfall is unpredictable and in its reproduction it is very responsive to the quality and abundance of feed (Newsome, 1966, 1975). During favourable periods breeding is continuous and all adult females will be suckling a young-at-heel from one teat, while a small young will be attached to another of the four teats in the pouch and in the uterus will be a unilaminar blastocyst in diapause. If the pouch young is removed experimentally or lost the delayed corpus luteum reactivates, the pregnancy resumes and birth and post-partum oestrus occur respectively 31 and 35 days later (Sharman & Calaby, 1964; Sharman & Pilton, 1964). Normally the young makes its first excursions from the pouch at about 190 days and emerges permanently at 235 days. Pouch emergence is succeeded within a day by birth of the next young and post-partum oestrus. The young out of the pouch is weaned about 120 days later or about 600 days after conception. Despite this long period of development, under favourable conditions, a female kangaroo produces an independent young every 240 days (Fig. 2.25). Under moderately adverse conditions the growth rate of the young may be retarded (Newsome, 1965) and sexual maturity postponed. Under more severe drought the young-at-heel and older pouch young die. Death of the pouch young leads to reactivation of the quiescent corpus luteum and the diapausing blastocyst. If drought continues, a succession of young may be born in this way and each dies at an age of about 2 months (Frith & Sharman, 1964; Newsome, 1965) but, if conditions improve, the young do not die but continue to grow and the population thus reaps an immediate advantage.

If drought persists for 6 months or more, the females cease to ovulate and then enter a true anoestrus (Newsome, 1964*a*). The response to drought-breaking rainfall is almost immediate for on 2 occasions within 14 days of rainfall all the females were pro-oestrous or had recently ovulated (Newsome, 1964*b*; Sharman & Clark, 1967).

The pattern described for *Macropus rufus* has also been described for *Macropus robustus* (Ealey, 1963, 1967; Sadleir, 1965; Newsome, 1975) and for *M. agilis* (Merchant, 1976; Lincoln & Renfree, 1981*b*). However in the monsoonal climate of northern Australia there is an annual decline in breeding of *M. agilis* during the wet season (Bolton, Newsome & Merchant, 1982), analogous to the didelphids of tropical America. Two other species, from more mesic climates *M. rufogriseus banksianus* and *Petrogale penicillata*, are also continuous breeders but have not been shown to exhibit drought-induced anoestrus (Johnson, 1979; Merchant & Calaby, 1981). It is of considerable interest that in none of these species have the females been found to respond to drought by retaining a quiescent corpus luteum and embryo in diapause. Indeed Newsome (1964*a*, *b*) emphasises the point that the corpus luteum and fetus completed normal development in

Fig. 2.25. Diagram to illustrate the opportunistic breeding pattern of *Macropus rufus*, which responds to prevailing drought and rainfall. The same pattern probably holds for the euro, *Macropus robustus*. From Tyndale-Biscoe (1973).

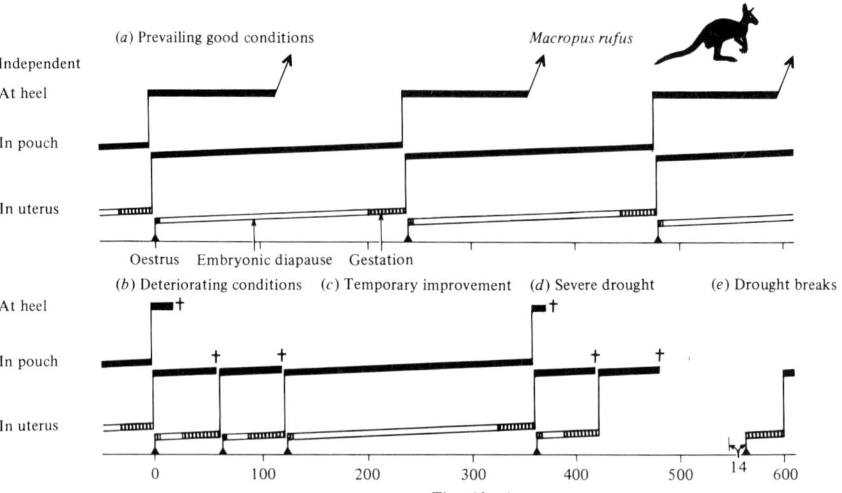

females that had already entered anoestrus and failed to undergo post-partum oestrus and ovulation. Their condition resembles that induced in *M. eugenii* after immunisation against gonadotrophin-releasing hormone (Short, Flint & Renfree, 1985) (see p. 246).

Seasonal breeding macropodids

On the mainland of Western Australia, *Setonix brachyurus* breeds continuously in the same way (Shield, 1964) as *M. rufus* but on Rottnest Island it is highly seasonal with a true anoestrus from August to January. Females come into oestrus in high summer (January–February) at the most unfavourable time of the year but the young conceived then emerge from the pouch at the most favourable time, in spring. Although most females conceive at post-partum oestrus in the early part of the year, the proportion that will produce a second offspring if the first is experimentally removed declines progressively to zero by August (Shield & Woolley, 1963), due either to failure of the corpus luteum to reactivate or of the blastocyst to survive (Wallace, 1981). This is a very different strategy from that of *M. eugenii* living on an adjacent island 10 km distant from Rottnest Island (Sharman, 1955*c*), and on Kangaroo Island (Berger, 1966), in which seasonal breeding has been achieved by prolonging quiescence.

In *M. eugenii* on Kangaroo Island, 70–80% of females give birth and undergo post-partum oestrus in late January – early February (Fig. 2.26) and the young leave the pouch in October and are weaned in November (Andrewartha & Barker, 1969). The post-partum corpus luteum remains quiescent (seasonal quiescence) and the blastocyst in diapause until after the summer solstice in December (Berger, 1966; Renfree & Tyndale-Biscoe, 1973*a*). The Tasmanian subspecies of *M. rufogriseus* (*M.r. rufogriseus*) but not the mainland subspecies, *M.r. banksianus*, has the same highly seasonal pattern, so that young emerge at the same time of year as young of *M. eugenii*.

Furthermore, young females of both *M.r. rufogriseus* and *M. eugenii*, which are weaned in October, may undergo their first oestrus at this time but then enter seasonal quiescence like the adults until after the summer solstice in December (Andrewartha & Barker, 1969; Sharman & Berger, 1969; Tyndale-Biscoe & Hawkins, 1977; Merchant & Calaby, 1981). These observations indicate that seasonal quiescence is not merely a prolongation of lactational quiescence but is a response to other factors in the environment, such as photoperiod (Hearn, 1972*a*; Sadleir & Tyndale-Biscoe, 1977; Hinds & den Ottolander, 1983). This will be discussed in Chapter 9.

Because the mainland subspecies, *M. rufogriseus banksianus*, is a continuous breeder and does not undergo seasonal quiescence it might have been expected that this condition would be lost when the island subspecies were transferred to other environments. However, *M.r. rufogriseus* has retained the same seasonal pattern since liberation in 1874 near Waimate in New Zealand (Catt, 1977) and in captivity in Canberra (Merchant & Calaby, 1981) and so has *M. eugenii* on Kawau Island near Auckland (Maynes, 1977), near Rotorua (R. M. F. S. Sadleir, personal communication) and in Canberra. Similarly, *M. eugenii* transported to the United States retained the same pattern but altered by 6 months (Berger, 1970) and *M.r. rufogriseus*, established for over 100 years at Whipsnade Park, England have done the same (Fleming, Cinderey & Hearn, 1983), with 84% of births occurring in August and September. Conversely, on Kangaroo Island and in captivity *M. eugenii* that breed in August or September out of season will give birth again after the young emerges from the pouch in February, so that two young of different ages are suckled simultaneously from different teats, as in *M. rufus* and *M.r. banksianus*.

Several other species of both sub-families that have the same basic pattern as *M. rufus* breed seasonally but it is not clear for any of them whether they display the strategy of *Setonix* on Rottnest Island, in which the corpus luteum of post-partum ovulation degenerates, the strategy of

Fig. 2.26. Annual cycle of events in the reproduction of *Macropus eugenii* on Kangaroo Island, South Australia. Females are highly synchronised, with reactivation of the diapausing blastocyst occurring around the time of the summer solstice (see also Chapter 9 and Fig. 9.1 and Table 9.2).

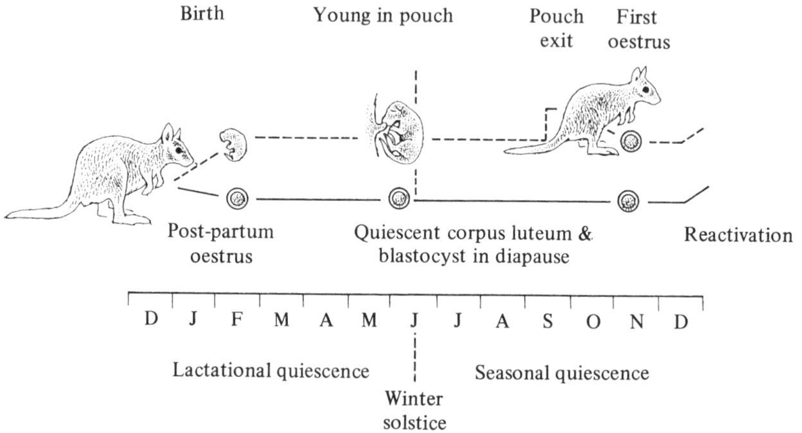

M. eugenii, with seasonal quiescence, or that of *M. rufus* in drought, in which post-partum ovulation fails to occur. In *Potorous tridactylus* in Tasmania births occur throughout the year (Guiler, 1960) but with a peak in July–September (Hughes, 1962*a*). Two successive young are normally produced, according to Heinsohn (1968), but those lost prematurely from the pouch will be replaced by reactivation of the diapausing embryo (Shaw & Rose, 1979) and this may account for the non-seasonal spread of breeding. *Aepyprymnus rufescens* in Queensland has a similar breeding pattern with births in all months of the year but with a majority in May–August (Moors, 1975; Johnson, 1978).

Bettongia lesueur appears to have a restricted breeding season on the Shark Bay islands of Western Australia but, in captivity in Canberra, females gave birth from February to September (Tyndale-Biscoe, 1968). Birth of the next young from a reactivated blastocyst can occur a few days after the pouch is vacated at 115 days, so the maximum number of young is 3 per annum and in captivity this was realised once. *Bettongia penicillata* from Western Australia and *B. gaimardi* from Tasmania show the same pattern as *B. lesueur* (Rose, 1978) (see Table 2.1). *Thylogale billardierii* in Tasmania has a shorter pouch life of 7 months and 75% of births occur in April–June (Rose & McCartney, 1982*a*).

Macropus fuliginosus and *M. giganteus* are seasonal breeders with oestrus and births occurring during September–March and lactational anoestrus from April to August (Poole, 1973; 1976). Although females may display oestrous behaviour again at the end of pregnancy, it is not associated with ovulation (Pilton, 1961). *M. giganteus* has a less precise breeding season than *M. fuliginosus*, especially in the northern part of its range in Queensland (Kirkpatrick, 1965). Lactational anoestrus is more variable from about 112 to 243 days after birth and, while lactation continues, oestrus and ovulation may take place with resulting conception (Kirkpatrick, 1965; Clark & Poole, 1967). The corpus luteum thus formed is arrested, as in other macropodids, the uteri become quiescent (Fig. 2.17*a*) and the blastocyst enters diapause (Clark & Poole, 1967). The timing and incidence of this ovulation is positively correlated with prevailing nutritional conditions: in years of drought few females ovulate during lactation but in exceptionally favourable years the incidence may be as high as 85% (T. H. Kirkpatrick, cited by Poole, 1973).

The corpus luteum resumes development shortly before the pouch is vacated at 10 months and the next young is born a few days afterwards, as in *Macropus rufus* (Kirkpatrick, 1965; Clark & Poole, 1967). These observations indicate that *M. giganteus* has the endocrine capacity for quiescence and the embryo the ability to undergo diapause as in other macropodids. *Macropus parma* and *M. parryi* (Calaby & Poole, 1971; Maynes 1973*a*, *b*) are similar to *M. giganteus* and in both species the oestrous cycle is several days longer than gestation (Table 2.2) so that post-partum ovulation is usually suppressed. However, in *M. parma* ovulation did take place post-partum in 17% of cases studied by Maynes (1973*a*), while in the rest of the cases it occurred 45 days or longer after parturition. The corpus luteum in either case became quiescent until loss of the pouch young or its emergence from the pouch.

M. fuliginosus is the only species among all the Macropodidae in which the quiescent state has not been observed to occur under any conditions (Poole, 1973; 1975; 1976; Poole & Catling, 1974). In two cases where a female ovulated during pregnancy, the pregnancy proceeded without interruption by quiescence, but the young in each case failed to survive in the pouch with the older sibling.

Conclusions

This systematic review of reproduction in species representing all families of marsupials has raised three matters which will each be discussed in later chapters. They are the reproductive cycles and their relation to gestation, the occurrence of embryonic diapause, and the factors controlling seasonal breeding. We will endeavour to draw the various pieces of evidence on each of these together here.

Reproductive cycles

The great majority of those marsupials of all families that have been studied are polyoestrous and it seems reasonable to conclude that polyoestry, not monoestry, is the basic marsupial mode as it is in eutherian mammals. Some species, such as *Dasyurus viverrinus* and *Petauroides volans*, appear to be monoestrous when wild animals only are examined. However, females are capable of ovulating more than once in a breeding season and so are polyoestrous. Some, such as *P. volans* are unable to conceive at a second ovulation but others, as *D. viverrinus* and *Sarcophilus harrisii*, are able to conceive then, if the first litter is lost. However, if the first litter survives, the duration of lactation, when oestrus is suppressed, is such that there is no further opportunity in the year for oestrus to occur and such species are ecologically monoestrous.

For other species, in which the duration of lactation is shorter, 2 or more litters may be produced in a year and these species are ecologically polyoestrous. Thus the major constraint on whether a species expresses polyoestry is the length of lactation and this, as Russell (1982*a*) has shown, is closely related to the adult body size. This applies to the smaller species among the Didelphidae (e.g. *Marmosa*), Dasyuridae (e.g. *Sminthopsis*), all Peramelidae, Petauridae (e.g. *Gymnobelideus*, *Petaurus breviceps*) and Macropodidae (e.g. *Bettongia*). Species of larger body size in all these families (except the Peramelidae) are polyoestrous but generally produce 1 litter a year; a few species, such as *Trichosurus vulpecula* and *Didelphis virginiana* and *D. marsupialis* may have a second litter but the frequency of occurrence varies between populations. True monoestry appears, on present evidence, to be restricted to species of *Antechinus* and *Phascogale* that live in temperate or sub-tropical forests of Australia, where the food supply is uneven but highly predictable through the year. As Lee & Cockburn (1985) argue, this appears to be a special adaptation of a small species to such a habitat. One other family, the Caenolestidae, may be shown to be monoestrous also but at present the evidence is insufficient.

Litter size also varies in relation to body size and diet, with the smallest species of each family having the largest number of teats, and ovulation rates that greatly exceed this number, so that teat occupancy is high and litter size large. Larger species have smaller numbers of teats and fewer eggs are shed at ovulation, so that teat occupancy is low. The extreme of this trend is expressed in the herbivorous diprotodont marsupials, where only the smallest species, e.g. *Pseudocheirus peregrinus* and *Hypsiprymnodon moschatus*, are polyovular and polytocous, and all others are invariably monovular.

Neither the length of gestation nor the length of the oestrous cycle bear any consistent relationship to body size; the longest gestation periods are those of *Cercartetus* and *Tarsipes*, which are among the smallest marsupials, whereas the shortest gestation periods occur in the Peramelidae and the Didelphidae. In the latter family the smallest and largest species have the same gestation length of 13 days. Oestrous cycles, in which a luteal phase occurs, vary from 22 days to 60 days, with most species having oestrous cycles of 25–33 days. The feature which does show a consistent pattern, however, is the ratio between gestation length and oestrous cycle (Table 2.2). For the Didelphidae, Dasyuridae, Peramelidae, Petauridae and Phalangeridae the ratio is 0.52–0.64 and, in all these species, ovulation is suppressed during lactation. Among the majority of the Macropodidae the ratio is 0.94 to 1.09 and pre- or post-partum ovulation normally occurs. For four species, however, the ratio is 0.83 and post-partum ovulation is suppressed, as in the majority of marsupials; these species have been considered to represent an intermediate condition.

For three families the evidence at present is insufficient to say what the ratio is. *Phascolarctos* appears to have a ratio of 1.25 but this may be because the present estimate of the length of the oestrous cycle is based on anovulatory cycles. For *Tarsipes* and the smaller Burramyidae there are no data on the oestrous cycle and the estimates for gestation all include periods of embryonic diapause.

Classification of reproductive patterns

The first attempt to classify the reproductive patterns of marsupials was made by Sharman *et al.* (1966) who recognised four groups. In the light of additional information Tyndale-Biscoe (1984) concluded that their Group 1, exemplified by *Trichosurus vulpecula*, represented the basic marsupial pattern and that their Group 2, exemplified by *Setonix brachyurus*, represented the only significant departure. However, the peramelid pattern is sufficiently distinct from the basic pattern to warrant separate classification, and the new information on the Burramyidae and *Tarsipes* requires a separate classification for them. Nevertheless, the pattern displayed by each of these latter three groups can be derived from the basic pattern, which we are therefore justified in concluding to be the primitive marsupial pattern. We define the four groups, which are included in Tables 2.1 and 2.2 as follows.

Group 1. Polyoestrous, polyovular marsupials, with gestation occupying less than 60% of the oestrous cycle and coinciding with the luteal phase. Monoestry in *Antechinus* is derived from this as an adaptation to special ecological constraints. Likewise the monovular condition of the folivorous species is a derived condition. Didelphidae, Dasyuridae, Petauridae, Phalangeridae.

Group 2. Polyoestrous, polyovular marsupials with ultra short gestation occupying less than the luteal phase, which is prolonged into lactation. In addition there is a well developed chorioallantoic placenta. Peramelidae, Thylacomyidae.

Group 3. Polyoestrous, monovular marsupials with gestation extending into the follicular phase and hence occupying 94%–109% of the oestrous cycle. Post-partum, rarely pre-partum, ovulation, with subsequent inhibition of the corpus luteum and embryonic diapause. An intermediate group (3i) have an extended gestation but within an extended luteal phase, so that gestation occupies 80%–88% of the long oestrous cycle. Post-partum ovulation rarely occurs, but when it does the corpus luteum is inhibited and embryonic diapause occurs. Macropodidae, Potoroidae.

Group 4. Polyoestrous, polytocous, with very prolonged pre-luteal phase and gestation, which includes a long period of embryonic diapause. It is not clear at present whether this is an obligatory diapause or is associated with concurrent lactation. Burramyidae (except *Burramys* which has the Group 1 pattern) and Tarsipedidae.

The endocrine changes for representative species from Groups 1–3 will be reviewed in Chapter 6.

Embryonic diapause

Embryonic diapause is a common feature of reproduction in the Macropodidae and, until recently, was believed to be confined among marsupials to this family. Its function, and hence selective advantage, has remained unclear. Early views (e.g. Sadleir, 1965) were that it may confer an ecological advantage in the return to breeding of desert kangaroos or that it enables more rapid replacement of lost young after predation. These reasons, however, cannot account for its widespread occurrence among seasonally breeding macropodids in temperate climates and Sharman (1965*c*) concluded that it does not have a primary selective advantage but is merely the consequence of prolonging gestation so that post-partum oestrus is not suppressed; the delay of embryonic development being a device to prevent the occupation of the pouch by a succession of young separated in age by the length of the oestrous cycle. However, this does not seem to be a satisfactory reason either since *Macropus giganteus* and *M. fuliginosus* achieved prolongation of gestation without post-partum ovulation by prolonging the oestrous cycle also.

Tyndale-Biscoe (1968, 1973, 1979) suggested that the primary selective advantage should be sought in the close relationship between the corpus luteum, uterine secretions and embryo expansion; he suggested that a period of stasis might be a normal feature of embryo development in all marsupials to ensure synchrony between embryo expansion and the secretions of the uterus needed for that expansion to take place. If this be so, the condition seen in most macropodids has come about by inhibition of the signal from the corpus luteum. Evidence from *Tarsipes*, *Cercartetus* and *Acrobates* only partly supports this view, because vesicle expansion does occur to a limited extent during diapause. On the other hand, formation of the embryo proper and organogenesis do not occur. Sharman (1963) first suggested that the relatively long gestation of *Antechinus stuartii* and *Dasyuroides byrnei* might include a period of delayed development, and this has been shown to occur in *A. stuartii* at the unilaminar blastocyst stage (Selwood, 1980, 1981) prior to expansion of the corpus luteum (Woolley, 1966*b*) and elevation of plasma concentrations of progesterone (L. A. Hinds & L. Selwood, personal communication). Notwithstanding the reason for its occurrence, the phenomenon of embryonic diapause has proved to be a very useful means for investigating the functions of the corpus luteum (Chapter 6) and the uterus in early development (Chapter 7) and the control of seasonal breeding (Chapter 9).

Cost of lactation and the timing of the breeding season

Throughout this review two points have recurred frequently: most marsupials are seasonal breeders and the duration of lactation is proportional to maternal body size. The latter point has been examined most thoroughly by Russell (1982*a*), who showed that, for 56 species of marsupial representing all major families, several other aspects of reproduction correlate to a greater or lesser degree with maternal body weight. Duration of pouch life correlates with maternal body weight and also with pattern of maternal care. Thus dasyurids that leave the litter in a nest at an early stage of development have a shorter pouch life than macropodids that carry the young to an advanced stage of development in the pouch but, within each family, the correlation with maternal size is close. Likewise the period from birth to weaning is shortest among the Peramelidae and longest among the arboreal herbivores. The weight of the whole litter at weaning correlates most closely with maternal body weight across all families but the weight of a single young at weaning does not, being much the same for all species. From this it follows that smaller species, which are invariably polytocous, make a larger investment proportionately in reproduction than do larger species (Fig. 2.4). Russell (1982*a*) expresses this parental investment as

$$\frac{\text{Weight of litter at weaning} \times 100}{\text{Maternal body weight}}$$

and shows that it varies from over 300% in the smallest dasyurids, such as *Ningaui ridei* (375%) to less than 30% in *Phascolarctos*, *Vombatus* and the largest kangaroos.

What is not clear from Russell's (1982*a*) analyses is how and when the major maternal investment is made during lactation. For the smallest species, with a short pouch life, the investment is clearly concentrated over a much shorter period than for *M. giganteus* distributing the effort over 10 months. Indirect evidence from several sources supports the idea that the major investment is made in the last third of lactation. From studies of Fleming *et al.* (1981) on *Didelphis virginiana* and Green & Eberhard (1983) on *Dasyurus viverrinus*, food intake is greatest in the second half of lactation, and in *Trichosurus vulpecula* (Smith *et al.*, 1969), *M. eugenii* (Stewart, 1984) and *M. agilis* (Lincoln & Renfree, 1981*b*) the suckled mammary gland reaches maximum size during this stage (see Chapter 8 and Fig. 8.3). For the smaller species, living in temperate habitats, birth of the litter occurs in winter or early spring, the second half of lactation with the flush of spring growth, and weaning of the young with early summer. However, for the arboreal folivores and larger dasyurids in which the young are weaned in spring or early summer the major investment in lactation must be made in winter, when it may be presumed that food resources are not abundant.

Bell (1981) analysed the reproductive performance of female *T. vulpecula* over a 5 year period and found that females that weighed more than 2.4 kg in autumn, when oestrus occurs, lost less weight in the subsequent winter

Fig. 2.27. Breeding success in *Trichosurus vulpecula* in New Zealand forests. Relationship between female body weight in the four seasons and survival of pouch young to show the importance of large size before breeding commences. Redrawn from Bell (1981).

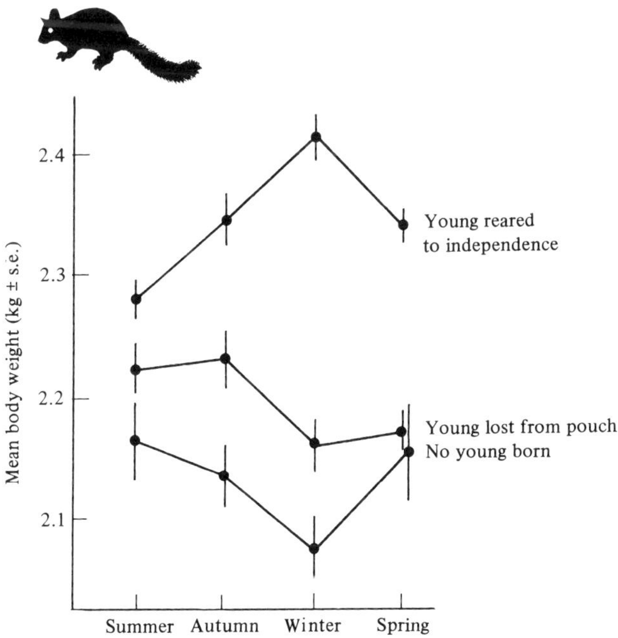

and successfully reared their young, whereas females that weighed less than 2.4 kg in autumn, continued to lose weight in winter and either lost their young or failed to breed altogether (Fig. 2.27) Furthermore, among those that reared their young successfully there was a positive correlation between the weight of the young at weaning and maternal body weight. The inference from this study, as from that of Atramentowicz (1982) on *Caluromys philander* is that, for successful reproduction in arboreal folivores, fat reserves before winter are more important than available food in winter. Since all arboreal folivores wean their young in spring, survival of the young rather than the requirements of lactation has been the factor selected for in this. Because the total duration of lactation varies in relation to maternal body weight, the mating season for each species is different. Bell's (1981) data suggest that nutritional state at that time may be an important factor in the onset of breeding, but the lack of any marked variation from year to year or in different regions suggests that photoperiod changes in autumn may determine the onset of breeding. If this is so, the change detected must vary in different species because the breeding season varies from before the autumnal equinox to after the winter solstice. Apart from *Sminthopsis crassicaudata* and *Antechinomys laniger*, however, no experimental study has been done on any of these species to investigate the control of breeding. The only species that has been thoroughly investigated so far is *M. eugenii*, and that will be reviewed in Chapter 9.

3

Sexual differentiation and development

Sex chromosomes

The chromosome number and morphology is known for 112 species of marsupials representing all families (see Sharman, 1973a; Hayman & Martin, 1974; Hayman & Rofe, 1977; Hsu & Benirschke, 1977). While the total DNA content of marsupial nuclei is approximately the same as those of eutherian species (Hayman & Martin, 1974), it is contained within fewer, larger chromosomes (Sharman, 1973*a*). The total chromosome number ranges from 2n = 10 in *Wallabia bicolor* to 2n = 32 in *Aepyprymnus rufescens*, but the distribution is bimodal around 2n = 14 and 2n = 22 (Fig. 3.1). There is lack of agreement as to which is the primitive number as species with both the modal numbers occur in South America and Australasia but there is not much doubt that 2n = 14 is the primitive karyotype for the Australian radiation from South American ancestors (Sharman, 1982). Whatever the origin may be, the large size and small number has meant that individual chromosomes, including the sex chromosomes, can usually readily be identified.

With the exception of four species, all known marsupials possess the sex chromosome pattern XY male:XX female as found in the great majority of eutherian mammals. The four exceptions have either extra X or extra Y chromosomes but their respective conditions have probably been derived from an XY:XX system by fusion of a sex chromosome with an autosome. In the macropodids *Potorous tridactylus* and *Wallabia bicolor* and in the bandicoot *Macrotis lagotis* the X chromosomes are large and metacentric and there are two Y chromosomes. At meiosis the X and the two Y chromosomes form a trivalent, from which Sharman & Barber (1952) and Sharman (1961*a*) concluded that the sex determining mechanism has been formed by fusion of the ancestral X chromosomes and a pair of acrocentric autosomes (Sharman, 1970). In males the fused chromosome is the large X, the residual autosome the new Y_2 and the original small Y the new Y_1. The validity of this hypothesis was supported by the demonstration in females, where one X chromosome would be expected to be inactive (see below), that only one arm showed delayed synthesis of DNA (Hayman & Martin, 1965*a*). In the fourth species, the macropodid *Lagorchestes conspicillatus*, both X and Y chromosomes have been translocated onto autosomes resulting in the formation of an $X_1 X_1 X_2 X_2$ female:$X_1 X_2$ Y male sex-determining mechanisms (Martin & Hayman, 1966; Hayman & Sharp, 1981). Among the monotremes *Ornithorhynchus* has an XX:XY sex-determining mechanism, whereas *Tachyglossus* has a mechanism similar to that of *Lagorchestes conspicillatus*, and *Zaglossus* most probably has also (Murtagh, 1977; Murtagh & Sharman, 1977).

Some evidence from naturally occurring intersex marsupials and cross-species hybrids suggests that again, as in eutherian mammals, the presence of the Y chromosome is male determining and, in its absence, the female sex develops. Sharman *et al.* (1970) described two intersexual *Macropus eugenii* with chromosome constitutions 15, XO and 17, XXY and Tyndale-Biscoe (1973) referred to a third which was 16, XY. The first-mentioned specimen was of female body phenotype with gonads containing testicular and ovarian elements and a female reproductive system (Fig. 3.2*d*); the

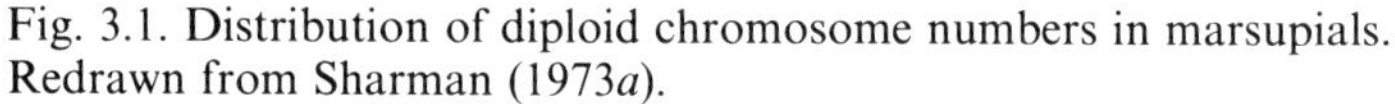
Fig. 3.1. Distribution of diploid chromosome numbers in marsupials. Redrawn from Sharman (1973*a*).

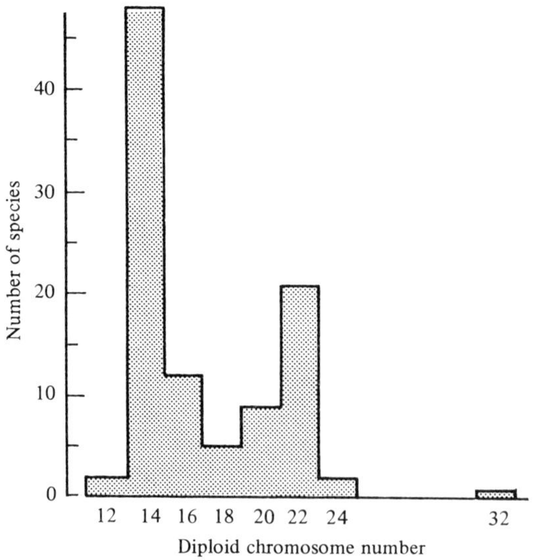

Fig. 3.2. (*a*) Comparison of reproductive systems of normal and intersexual *Macropus eugenii*. (*a*) XY male, (*b*) XXY intersex (*c*) normal XX female (*d*) XO intersex. Cross hatching, structures derived from Wolffian duct; verticle hatching, structures derived from Mullerian duct; dotted lines indicate ureters and bladder; x, ■ distribution of interstitial tissue; ●, seminiferous tubules; ⊙, Graafian follicles. Abbreviations: B, bulbo-urethral glands; C, cauda epididymis; F, fallopian tube; G, germinal epithelium; H, caput epididymis; L, lateral vagina; M, mammary gland; MV, median vagina; P, pouch; PE penis; PU, prostatic urethra; S, scrotum; T, teat; TA, tunica albuginea; U, uterus; UGS, urogential sinus, V, vas deferens. Redrawn from Sharman *et al.* (1970).

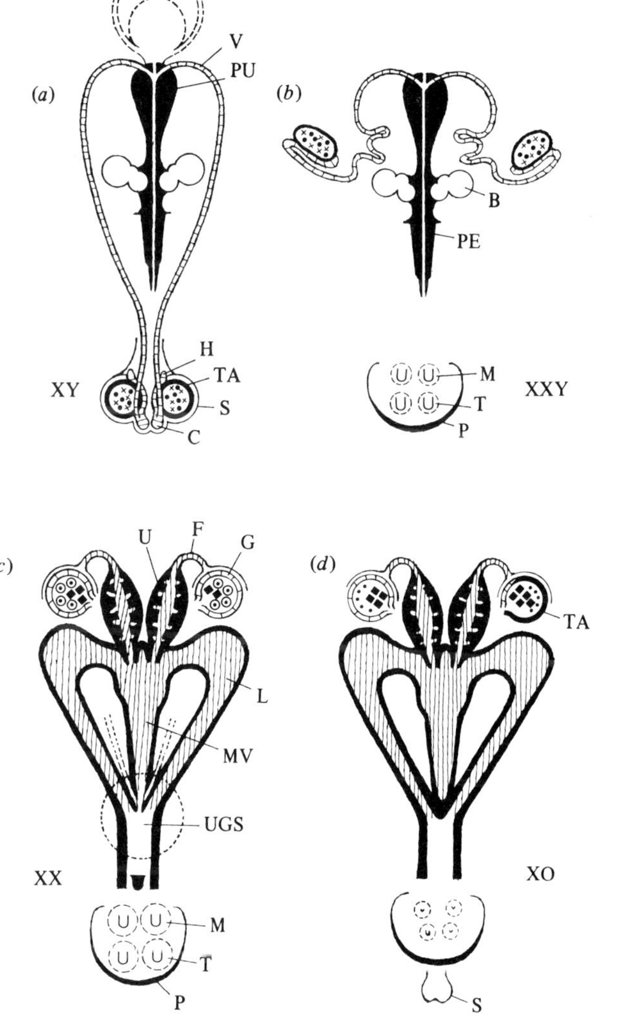

other two also had female body phenotype but contained abdominal aspermatogenic testes and male reproductive tracts (Fig. 3.2*b*). For *M. eugenii*, at least, these specimens suggest that the Y chromosome is strongly male determining. Likewise a single *M. robustus* with 16, XY chromosomes and a *Trichosurus vulpecula* with 20, XY chromosomes had abdominal testes and male reproductive tract (Sharman *et al.*, 1970).

Three other intersexual macropodids and one *Sarcophilus harrisii* (with sex-chromosome mosaicism) have been examined recently (C. Howe & G. B. Sharman, personal communication, 1984). All three macropodids had some cells with Y chromosomes and two of them were phenotypically male, whereas the third had ovary-like gonads and Mullerian ducts but possessed a small scrotum and no pouch. The specimen of *S. harrisii* was phenotypically female and its cells showed sex-chromosome constitutions of XXX, XX and XO. There are two other reports of intersex marsupials but in neither was the karyotype recorded; Gilmore (1965) described a *T. vulpecula* of similar appearance to the 20, XY specimen of Sharman *et al.* (1970) and Hartman (1920*b*) and Hartman & League (1925) described an intersexual *Didelphis virginiana* that resembled the habitus of the 15, XO *M. eugenii* of Sharman *et al.* (1970).

Smith, Hayman & Hope (1979) described the karyotype and morphology of four hybrid animals, the results of crosses between seven species of macropodid. In each hybrid the differentiation of the gonads and genital ducts were in conformity with the sex chromosomes, although in none was there any germ cell derivative. Three of the hybrids were male and in each the testes were in the scrotum, while the single female had a small pouch and four teats with associated mammary tissue. It was not possible to determine in these specimens whether primordial germ cells never reached the gonad or whether they failed to differentiate after arrival. What is clear however, is that development of the gonads as endocrine organs was sufficient to induce differentiation of the several phenotypes appropriate to the sex chromosomes.

Sex-linked inheritance and dosage compensation for X chromosomes in the Macropodidae

The small Y chromosome of mammals carries the determinants for testicular development and a few others, whereas the X chromosome carries many genes essential for life. In eutherian mammals the X chromosome is uniformly small but comprises no less than 5% of the total DNA in the chromosomes. Females therefore contain twice the dosage that males contain for the sex-linked genes but, in the development of eutherian mammals, only one X chromosome in each cell remains active. The other one replicates later than the rest of the chromosomes and remains condensed in the resting phase as a Barr body. In the extraembryonic tissues the paternally derived X chromosome is inactive (reviewed by VandeBerg, 1983*a*; VandeBerg *et al.*, 1983) but in the embryo itself, derived from the inner cell mass, the dosage compensation is apparently random with respect to the parental origin of the X chromosome, so that both alleles for each sex-linked gene are expressed but in different cells (Lyon, 1972).

The Y chromosomes of marsupials are also very small and it is likely that they carry only a few sex-linked genes, including the testis-determining genes, although even this has not been established. The X chromosomes, however, are more varied than in Eutheria and range from 2% to 20% of the total chromosomal DNA (Hayman & Martin, 1974; Hayman, Ashworth & Carrano, 1982). Hayman & Martin (1965*a*) showed that one arm of the X chromosome in their study on sex chromosomes of female *Wallabia bicolor* and *Potorous tridactylus* replicated asynchronously, as determined by the uptake of tritiated thymidine. This was complicated by the fusion of X chromosomes and autosomes but in two other species, *M. eugenii* and *M. fuliginosus*, with a normal XY:XX sex-determining mechanism Graves (1967) showed that the X chromosomes were also labelled asynchronously.

However, it was Sharman's (1971) study of hybrids between two subspecies of kangaroos *M. robustus robustus* and *M.r. erubescens* and between these and *M. rufus* that disclosed the remarkable feature of X inactivation in female kangaroos, namely that it was invariably the X chromosome derived from the male parent that showed delayed replication. He was able to demonstrate this because morphological differences between the X chromosomes of each species (Fig. 3.3) allowed them to be identified in the cells of the female hybrids. In 95% of stimulated lymphocytes examined it was the paternally derived X chromosome that was asynchronous in uptake of tritiated thymidine.

Subsequently four of the enzymes, known to occur on X chromosomes of eutherians, have been found to be coded on the X chromosomes of these kangaroos – glucose-6-phosphate dehydrogenase (G6PD), phosphoglycerate kinase-A (PGK-A) hypoxanthine-guanine-phosphoribosyl-transferase (HGPT) and α-galactosidase A (Cooper *et al.*, 1983). In the first two cases the enzymes occur as two allelic isozymes (allozymes) with different electrophoretic mobilities and so it has been possible to determine which X chromosome is active by its phenotypic expression in the tissues of the host. In hybrids between the two subspecies of *M. robustus* the expression of G6PD confirmed that the maternally derived X chromosome was invariably the one expressed in female hybrids (Richardson, Czuppan & Sharman, 1971; Johnston & Sharman, 1975). An extensive pedigree of several generations established that the paternal X chromosome would be expressed in female offspring of the hybrids, so reactivation must occur in the germ cell line. Both *M. giganteus* and *M. parryi* have two allelic polymorphisms for PGK-A (Cooper *et al.*, 1971, 1977*a*). Only one allozyme is expressed in males of these two species, and, in heterozygous females that potentially could express both, the allozyme inherited from the female parent is strongly expressed, while that from the male parent is either weakly expressed or not expressed at all.

In contrast to these results with sex-linked enzymes, two other allozymes (peptidase A and D), which are carried on the autosomal chromosomes of *M. robustus* are both expressed in heterozygote hybrids (Briscoe,

Fig. 3.3. Late replication of paternal X chromosome. The chromosomes of a hybrid female resulting from a female *Macropus rufus* and a male *M. robustus*, after exposure to autoradiographic film. The long arm of the paternally derived X chromosome (X_e) is positively labelled with respect to maternally derived X chromosome (X_r) and to autosomes. Redrawn from Sharman (1971).

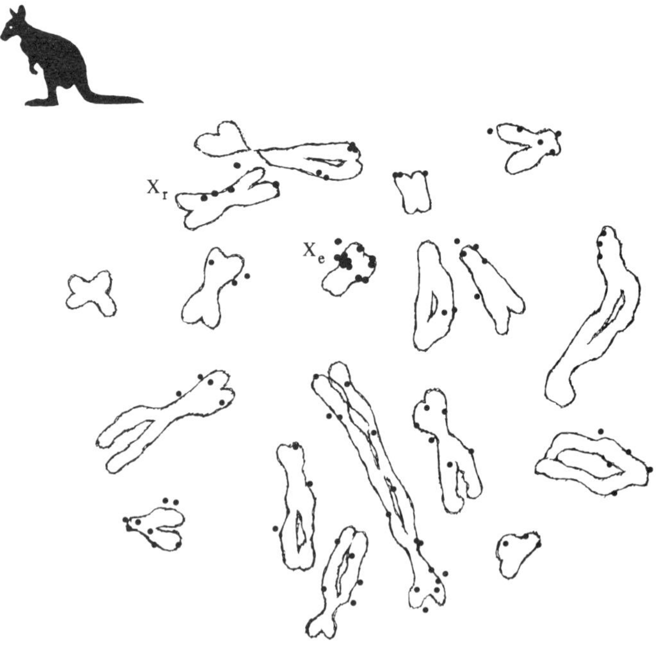

Murray & Sharman, 1981). From these and other studies (VandeBerg *et al.*, 1977; Cooper *et al.*, 1983) of several species of kangaroo it appears that paternal X inactivation is universal in female macropodids but is not complete in all tissues of the body, or for all alleles on the X chromosome. By contrast, X chromosome inactivation in the somatic tissues of all eutherian species so far investigated is normally random but, once it has occurred, it is complete for all alleles and remains so in descendant cells.

Two questions then arise: at what stage in the life cycle does the X chromosome become inactive and at what stage does it resume activity? These questions have been examined in female pouch young hybrids between the sub-species of *M. robustus* referred to before, which have different allozymes of G6PD (Johnston, Robinson & Sharman, 1976; Robinson, Johnston & Sharman, 1977). The ovaries of 8 pouch young from

Fig. 3.4. Total germ cell populations in pouch young of *M. eugenii* up to 210 days post-partum. Inset: Comparison with rat (broken line, $n \times 10^5$) and human (solid line, $n \times 10^6$) in which the peak number of germ cells occurs before birth. Redrawn from Alcorn & Robinson (1983).

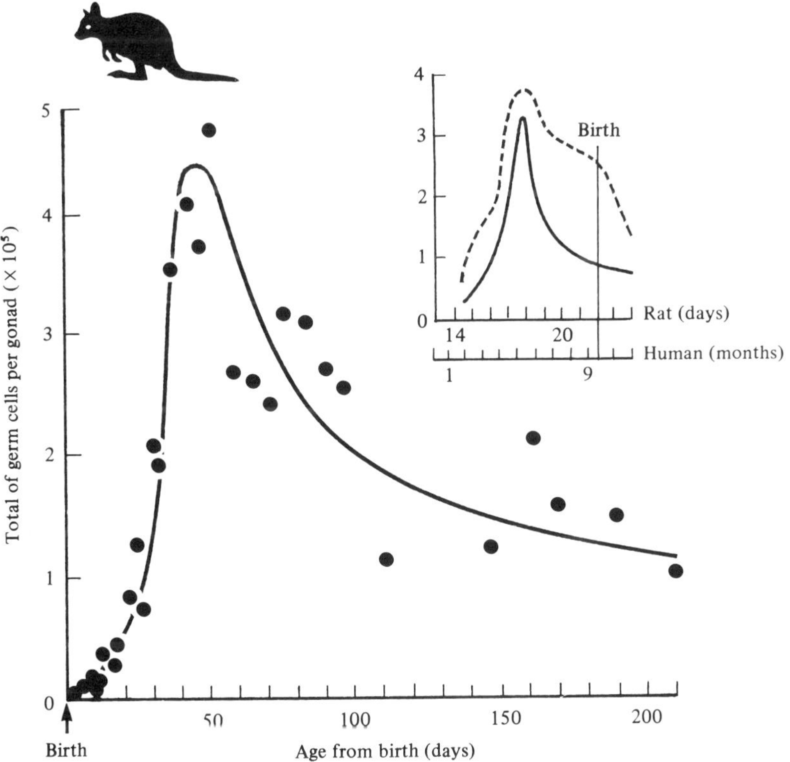

26 to 60 days old were examined. Histologically the ovaries contained very large numbers of oogonia (day 26) or primary oocytes (day 51–60) in pachytene or diplotene of meiosis. The latter group correspond in age to the stage in *M. eugenii* when the oocytes reach peak numbers (Fig. 3.4, Alcorn, 1975; Alcorn & Robinson, 1983). Specific staining of ovary slices of one pouch young showed that the oocytes had a high level of G6PD activity and, when ovary tissue was homogenised and subjected to electophoresis, only the maternally derived allozyme for G6PD was expressed in every animal.

It would appear from these results that paternal X inactivation must occur in the germ cell line from at least the oogonial stage as well as in the somatic tissues. However, VandeBerg *et al.* (1983) sound a caution in the interpretation of these results, which assume the oocytes accounted for a sufficiently high proportion of total ovarian volume and were sufficiently active to make a significant contribution to the expression of the paternally derived allele if it was active. Subsequent work (Briscoe, Robinson & Johnston, 1983) has shown that kangaroo oocytes at this stage have very low levels of G6PD activity and hence expression of the paternal alleles may have been obscured.

Inactivation may occur even earlier since Alcorn (1975) observed heterochromatin associated with the nucleolus in migrating primordial germ cells of the embryos of *M. eugenii*, which may have been the condensed X chromosome. An elegant experiment by Johnston & Robinson (1985) does not support this inference. Female *M. robustus robustus*, which have the fast allozyme for G6PD, were mated to male *M.r. erubescens* with the slow allozyme for G6PD and the resulting male and female embryos were examined at 21–25 days gestation. Homogenates of limb bud, brain, tail, amnion and allantois of male and female embryos expressed the fast (maternally derived) allozyme only, as did the yolk sac membrane of the male embryos. However, vascular and avascular yolk sac membranes (see Fig. 7.17) of the female embryos expressed both fast and slow allozymes, thereby indicating that inactivation of the paternally derived X chromosome had not occurred in these tissues. It is of especial interest that the primordial germ cells originate in the yolk sac (Alcorn, 1975) at about this time in *M. eugenii* and migrate to the genital ridge (see below). This could mean that both X chromosomes of the primordial germ cells of female embryos are active when they enter the genital ridge, and retain this state through subsequent oogenesis. While this conclusion does not support Alcorn's (1975) inference from heterochromatin in migrating primordial germ cells, it does support the reservations of VandeBerg *et al.* (1983).

As mentioned earlier, the results of pedigree data on *M. robustus* show that the paternal X chromosome is expressed in subsequent progeny so it must either remain active in the germ line or reactivate in the oocyte before the completion of the second maturation division and formation of the mature ovum, as is known to happen in Eutheria. The mechanism by which inactivation and reactivation are achieved are still unknown, but the pattern disclosed in kangaroo hybrids offers an unique and accessible system in which to investigate them.

X inactivation in other marsupials

There is growing evidence that paternal X inactivation also occurs in non-macropodid marsupials of Australia (see VandeBerg *et al.*, 1983 for review). Asynchronous labelling of one X chromosome in female tissue of *Trichosurus vulpecula* and *Antechinus flavipes* has been demonstrated by Hayman & Rofe (1977), while VandeBerg *et al.* (1979) showed that the expression of PGK-A alleles in *T. vulpecula* was in conformity with paternal X inactivation. Cooper *et al.* (1983) also showed that the expression of α-galactosidase A (AGA-A) is sex-linked in macropodids and probably in dasyurids and that in both families the dosage compensation was consistent with paternal X inactivation.

The Peramelidae possess a more extreme kind of sex-chromosome inactivation (Hayman & Martin, 1965*b*). All the species of the subfamily Peramelinae possess a normal XY:XX sex-determining mechanisms and the sex chromosomes are maintained in the germ cell line. However, in most somatic tissues of males and females, the sex chromosomes are reduced to XO, i.e. an X chromosome is eliminated in females and the Y chromosome in males. In *Isoodon obesulus*, and *I. macrourus* no somatic tissue in the adult has been found with the full chromosome number but in four other species the full complement has been found in cornea, intestinal epithelium and skin cells in culture. The stage at which the loss occurs has been sought in pouch young of *Perameles nasuta* and both species of *Isoodon* (Walton, 1971; Hayman & Martin, 1974; Close, 1979, 1984). In very young animals some cells from spleen and liver had the full complement of sex chromosomes but most were XO. They conclude that loss of the sex chromosomes does not occur synchronously in all tissues and that the transition may be of long duration and involve many cell divisions, but Close (1979, 1984) showed that loss of X chromosomes in peramelids is restricted to haematopoietic tissue and that the apparent losses reported in other tissues were probably due to the presence of macrophages within them.

Hayman & Martin (1974) suggest that the loss of an X chromosome from females and the Y chromosome from males can be derived from the pattern of X inactivation seen in the Macropodidae but, at present, there is no means of differentiating between the paternally and maternally derived X chromosomes of the Peramelidae (Close, 1984). If the inactive X chromosome replicates late, an exaggeration of this phenomenon could lead to loss of the chromosome altogether. This could be a random process operating over a number of mitotic divisions. Loss of the Y chromosome in males could be by a similar mechanism since they contend that the Y chromosome in marsupials is late in replicating compared to the autosomes. Similar patterns of loss of one X chromosome in females and the Y chromosome in males has been reported in two species of the Petauridae, *Petauroides volans* (Murray, McKay & Sharman, 1979) and *Hemibelideus lemuroides* (McKay *et al.*, 1984).

The evidence from American marsupials for X inactivation is scant and for specific paternal X inactivation almost non-existent. Chromatin bodies, which may represent the inactive X chromosome have been reported in *Didelphis virginiana* (Ohno, Kaplan & Kinoista, 1960) and *Philander opossum* (Perondini & Perondini, 1966), while Schneider (1970) using lymphocytes of *D. virginiana* found that one X chromosome took up less tritiated uridine than the other chromosomes. Also in *Didelphis virginiana* Davis & Jurgelski (1973) showed an electrophoretic variation in thyroid hormone binding globulin (TBG) which Cooper *et al.* (1977*b*) interpreted as being possibly under control of the X chromosome with paternal X inactivation. The results in which only a single allozyme was expressed in individuals of both sexes including heterozygous females, showed a striking resemblance to the patterns for G6PD and PGK-A in kangaroos. A different pattern of dosage compensation has recently been found in *Monodelphis domestica*, which, likewise, favours the maternal contribution to the genome (VandeBerg, 1983*a*). Merry, Pathak & VandeBerg, (1983) have shown that it has nucleolar organiser regions (NOR) on one pair of autosomes (No. 5) and the X chromosomes. In cultured fibroblasts from females NORs on both X chromosomes were expressed but neither of those on autosomes 5, whereas in males NORs were expressed on the single X chromosome and one or both autosomes. Thus dosage compensation for NOR activity appears to involve a balance in expression of X-linked and autosomal elements and is different from all other mammals so far described. The presence of paternal X inactivation in the American marsupials is strong evidence for its evolutionary antiquity in marsupials.

Sex differentiation

The gonads and genital ducts of marsupials, like those of other mammals are formed from three components; primordial germ cells, the coelomic epithelium that covers the genital ridge and mesodermal tissue derived from the mesonephros.

Primordial germ cells

Primordial germ cells are large, lightly staining cells with a prominent nucleus and nucleolus and they appear in the genital ridge shortly before differentiation of the gonad to testis or ovary. Until 1954 two views on the origin of the germ cells prevailed. One view was that they differentiated *in situ* from the coelomic epithelium of the genital ridge, the other was that they migrated to the genital ridge from an extragonadal site. In mice they can be recognised by their high alkaline phosphatase activity and, by this means, Chiquoine (1954) and Mintz (1957) traced their origin in the yolk sac membrane of early embryos and their migration via the dorsal mesentery to the genital ridge. Mintz also showed by genetical studies that primordial germ cells came exclusively from extragonadal sites. The prevailing view now is that all primordial germ cells originate outside the gonad and that the coelomic epithelium does not contribute any (Byskov, 1982; Ullmann, 1984). Three studies in marsupials have been made since 1954, although in none was the alkaline phosphatase reaction used to identify primordial germ cells. Ullmann (1981*a*) recognised primordial germ cells in a series of embryos of *Perameles nasuta*, *Isoodon macrourus* and *I. obesulus* (Fig. 3.5), in all of which gestation lasts 12 days (see p. 49). None was seen in a day 8 embryo but primordial germ cells were found in the endoderm and mesoderm of the yolk sac and hind gut of 9.5 day embryos and in the dorsal mesentery and dorsal region of the gonadal anlagen one day before birth. The peak of their migration occurs during the perinatal period.

A similar sequence was described in *M. eugenii* by Alcorn (1975) and Alcorn & Robinson (1983): primordial germ cells migrated from the dorsal mesentery surrounding the hind gut adjacent to its entry into the cloaca, then laterally to the posterior margins of the mesonephros, finally entering the gonadal ridges dorsally from the mesonephros (Fig. 3.6). The peak of migration occurred at about day 22 of the 27 day gestation. At day 20 they estimated there to be 500 primordial germ cells in the genital ridges and 1100 still outside, while 2 days before birth they estimated there to be 12000 cells in the gonadal ridges and only 200 outside (Fig. 3.4). This increase could have been achieved by three synchronous mitotic divisions of the primordial germ cells.

In these marsupials the origin and course of migration of primordial germ cells do not differ, except in timing, from eutherian species such as the mouse (Chiquoine, 1954), rabbit (Chrétien, 1966) or human (Mossman & Duke, 1973). Likewise primordial germ cells were recognised in the genital ridge of a 13 mm. fetus of *T. vulpecula*, that is to say about 3 days before birth, by Fraser (1919), in the neonatal *D. virginiana* by McCrady (1938) and Morgan (1943) and in the neonatal *Dasyurus viverrinus* by Ullmann (1984). These earlier authors (except Ullmann) believed that the

Fig. 3.5. Primordial germ cells in gonadal ridge of *Isoodon macrourus*. (*a–c*) Longitudinal sections through the gonadal ridge of an 11 day fetus, showing (*a*) a primordial germ cell (pgc), within the gonadal ridge in mitosis; (*b*) two primordial germ cells (pgc) of stellate shape, with pseudopod-like extensions; (*c*) a primordial germ cell in the mesenchyme, just below the gonadal rudiment. (*d*) Transverse section through the gonadal ridge (gr) of a full-term ($12\frac{1}{2}$ day) *I. macrourus* fetus showing a single large primordial germ cell and the lack of an organised epithelium on the surface of the ridge. From Ullman (1981*a*), with permission.

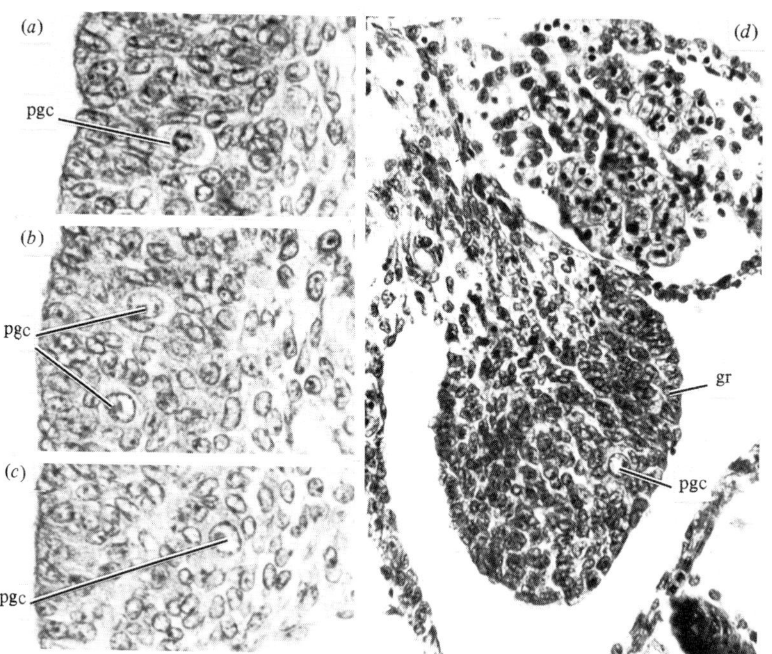

Fig. 3.6. Reconstructions in ventral and right sagittal views of the urogenital system of embryos of *M. eugenii*, showing approximate location of migrating primordial germ cells and numbers of primordial germ cells at (*a*) day 20 and (*b*) day 25 of gestation. ● approximate location of individual migrating primordial germ cells. Frequency for each antero-posterior location shown as histogram. ▨ Number of migrating primordial germ cells; □ number of primordial germ cells already in the gonadal ridge. Symbols: Ad = adrenal; Al = allantoic stalk; Cl = cloaca; G = gut; GR = gonadal ridge; Ms = mesonephros; Mt = metanephros; UGS = urogenital sinus; Ur = ureter; WD = Wolffian duct. Redrawn from Alcorn (1975).

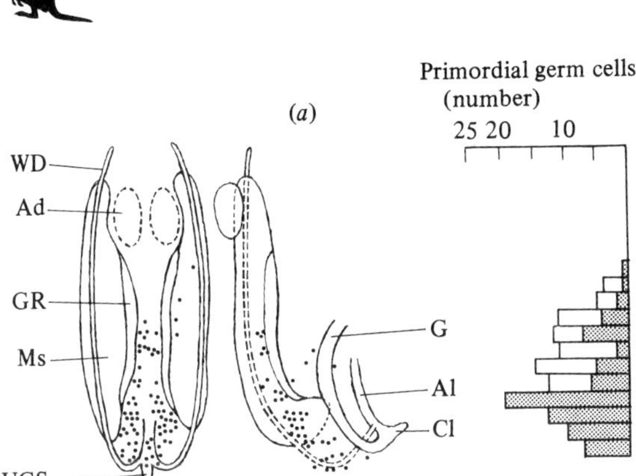

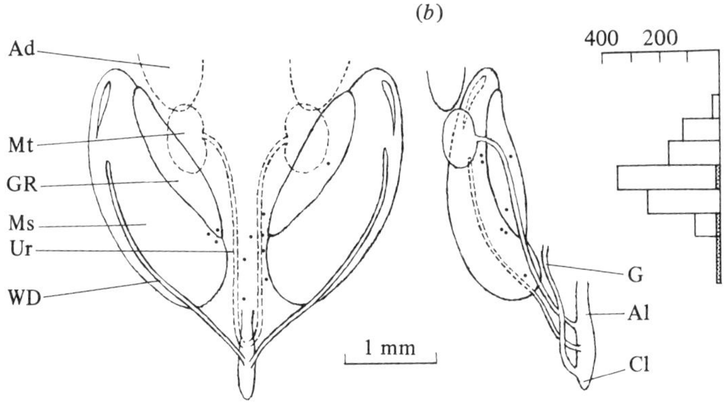

cells originated from the coelomic epithelium and they did not report seeing any in extragonadal sites. However, their observations do not conflict with the interpretations of Alcorn (1975) and Ullmann (1981*a*, 1984).

Differentiation of testis and ovary

The early development of the gonad has been described in five species of marsupial: in *Didelphis virginiana* by Moore (1939) and Morgan (1943), in *Dasyurus viverrinus* by Fraser (1919) and Ullmann (1984), in *Perameles nasuta* and *Isoodon macrourus* by Ullmann (1981*b*) and in *Macropus eugenii* by Alcorn (1975). In all these species the gonads are at the indifferent stage at birth but differentiation occurs within a few days and, as in Eutheria, the testis is recognisable before the ovary. Although the sex of the young cannot be recognised externally at birth, in at least two species, *M. eugenii* and *D. virginiana* the anlagen of the scrotum, pouch and mammary gland are histologically distinct at birth (McCrady, 1938; Burns 1939*c*; Alcorn, 1975).

The genital ridge is formed on the medial side of the mesonephros which, at birth and for some weeks after, is the functional kidney of the young marsupial (Bentley & Shield, 1962; Wilkes, 1984). As its function is progressively assumed by the metanephric kidney the mesonephros, beginning at the anterior end, regresses and shrinks (Fig. 3.7). Simultaneously, the gonad shortens and rounds up, so that it is closely attached to the regressing portion of the mesonephros. The tubules of the anterior, and now defunct, malphighian bodies coalesce into a strand of dense tissue which penetrates into the central blastema of the gonad (Fraser, 1919) (Fig. 3.8). This is the rete cord, which has been shown in the mouse to have a profound effect on the differentiation of the gonad and the development of the primordial germ cells (Byskov, 1982). In males the rete cord forms the efferent ducts that connect the seminiferous tubules to the epididymis. For a time it also becomes tubular in females and connects with the epoophoron, the homologue of the epididymis, but ultimately looses this link and its tubular structure.

Histologically the differentiation of the testis becomes apparent on day 2 or 3 after birth in *D. virginiana* (McCrady, 1938; Burns, 1939*c*), *Perameles, Isoodon* (Ullmann, 1981*b*) and *Dasyurus viverrinus* (Ullmann, 1984) and between day 3 and 7 in *M. eugenii* (Alcorn, 1975). In genetically male young, the homogeneous blastema becomes organised into a central stroma, intimately associated with the rete cords, surrounded by a zone of pale-staining cells, the medullary cords, and an outer layer of fibrous cells and coelomic epithelium.

In *Isoodon* and *Dasyurus* the middle zone consists of large chromophobic cells and the primordial germ cells are associated with these almost exclusively (Fig. 3.9*b*). By day 3 this zone has been divided up into sex cords each surrounded by small, dark staining cells from the central stroma. The cell clusters or cords so formed become hollow tubes with

Fig. 3.7. Scanning electron micrograph of the male urogenital tract of 7 day pouch young of *Didelphis virginiana*. The cranial region of the mesonephros appears shrunken, and is reduced in size as compared to the newborn. The mesonephric duct (Md) and gonad (G) show continued development. The adrenal (A), metanephros (Mt), ureter (U), urinary bladder (B), and what appears to be the gubernaculum (arrows), also are shown. From Krause *et al.* (1979*a*), with permission.

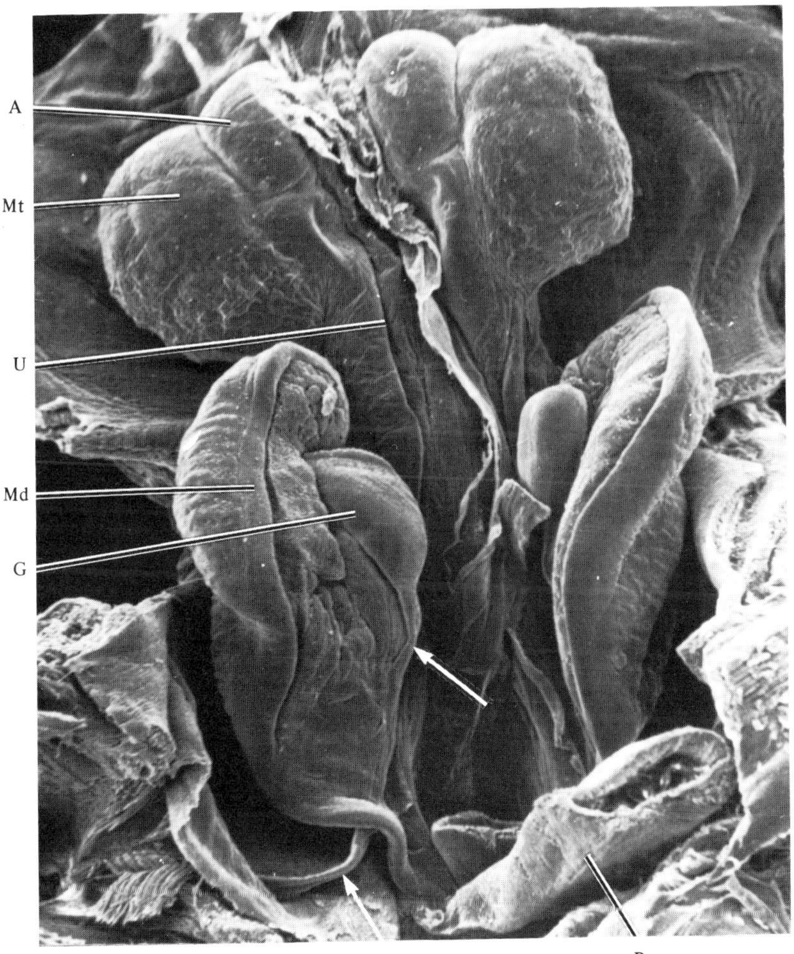

presumptive Sertoli cells arranged peripherally and primordial germ cells centrally. The central medullary region is thus devoid of sex cords, which is unusual. In most mammals, including *Didelphis* (Burns, 1961) and *M. eugenii* (Alcorn, 1975), the sex cords in genetic males occur preferentially in the central region, which led to the theory, first proposed by Witschi (1956), that the medullary component of the gonad is male determining

Fig. 3.8. Diagram of the mesonephric tubules and gonad of an embryo of *Trichosurus vulpecula*, H.L. 12.5 mm. Transverse section through the mesonephros and ovary, showing the mass of rete strands (r), which extend from the walls of the Malpighian corpuscles of the anterior mesonephros (aM), through the mesovarium (m) into the ovary (ov). The glomeruli are seen to be in progressive stages of degeneration. pM, glomeruli of posterior mesonephros; MD, Mullerian duct; WD, Wolffian duct. Redrawn from Fraser (1919).

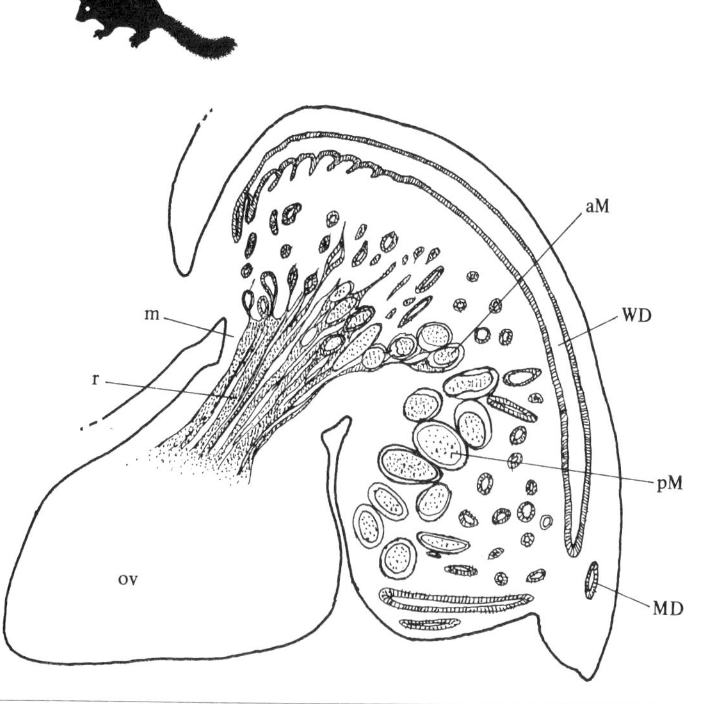

Fig. 3.9. Differentiation of testis in *Isoodon macrourus*. (*a*) Testis rudiment of 2.5 day pouch young; note well-defined Sertoli zone. (*b*) Part of (*a*), to show the Sertoli zone containing a primordial germ cell and a cellular bridge. (*c*) Testis rudiment of 3 day pouch young; the Sertoli zone has become subdivided into cords. From Ullman (1981*b*), with permission.

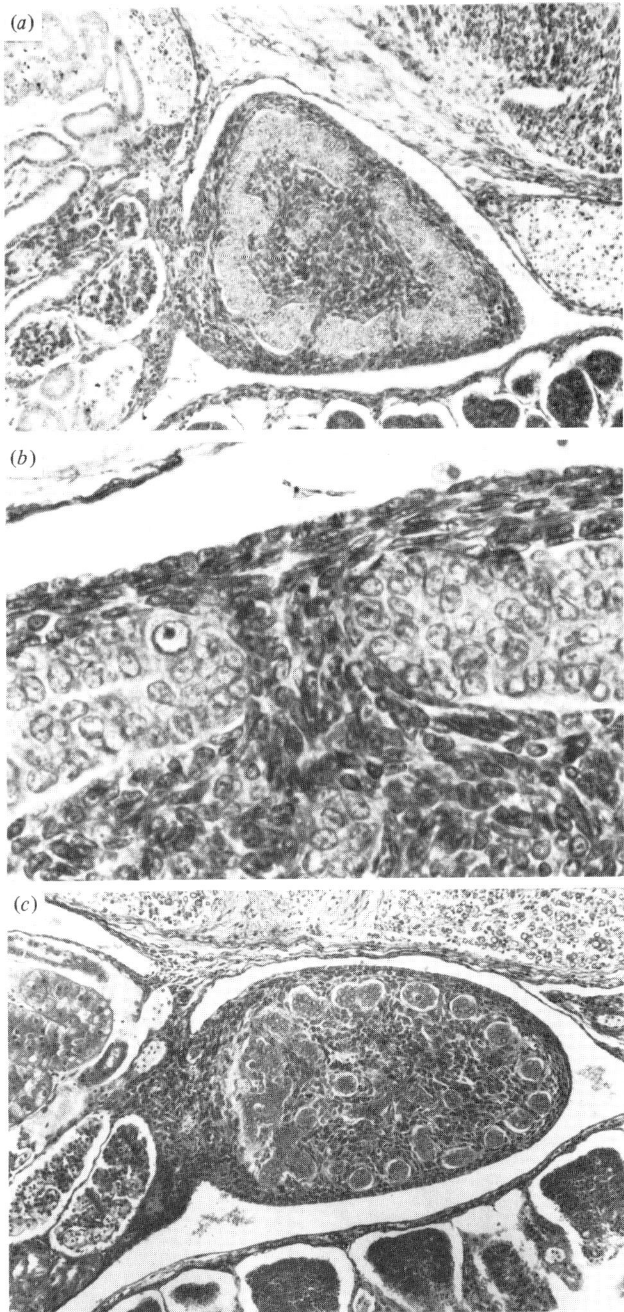

Fig. 3.9. For legend see opposite.

and sex cords in this region persist, while the cortex is female determining and sex cords in this part degenerate. The evidence from *Isoodon* and *Dasyurus* casts serious doubt on this theory. In *Dasyurus* Ullmann (1984) identified 12–14 sex cords in the developing testis (Fig. 3.10*a*) and by the age of 6 weeks these had differentiated into seminiferous tubules, which form loops whose two ends join at the hilus of the testis (see Fig. 4.5). Here the rete testis has penetrated and developed a lumen, which becomes confluent with the lumena of the seminiferous tubules (Fig. 3.10*b*, *c*). During the same period the stroma cells of the testis proliferate and some differentiate into large cells with spherical nuclei, which Ullmann (1984) considers to be the precursors of Leydig cells. This is an unusually late stage for Leydig cells to differentiate and it is likely that they have differentiated much earlier, since in *M. eugenii* testosterone has been assayed from plasma on day 1 (Catling & Vinson, 1976) and in *D. virginiana* the testes synthesised testosterone from day 10 of pouch life (George, Hodgins & Wilson, 1985). Alcorn (1975) considers that interstitial tissue, which is steroid secreting, is derived entirely from rete cords but Ullmann (1984) considers that it differentiates from stromal elements *in situ*.

Growth of the testis after the initial differentiation is more rapid than in the ovary and is due to the transformation and proliferation of germ cells into spermatogonia and growth of interstitial tissue.

In all species of marsupial so far studied the ovary can be recognised a few days later than the testis when the blastema differentiates into an inner medulla, associated with the rete cords and an outer cortex in which the germ cells are found. Alcorn (1975) recognised four stages in the development of the ovary of *M. eugenii*. The first stage of differentiation lasts from birth until day 22. During this period the mesonephros regresses and in the cortex the number of germ cells increases by mitosis. The second, proliferative, stage ends at day 50 when the increase of germ cells or oogonia reaches a maximum of 450000 per ovary (Fig. 3.4). A similar intense development of the cortex occurs during this period in *D. virginiana* (Nelsen & Swain, 1942; Morgan, 1943) and *M. robustus* (Johnston *et al.*, 1976).

In *M. eugenii* the third stage occurs between 50 and 110 days when all oogonia enter the prophase of meiosis and are transformed to primary oocytes and many become atretic. These are generally those in clusters, whereas isolated oocytes that become surrounded by follicle cells are protected from atresia and become primordial follicles. These follicles are nearest to the medulla where the ratio of follicle cells to oocytes is highest. The follicle cells are thought by Alcorn to derive from the rete cords.

Fig. 3.10. Differentiation of testis in *Dasyurus viverrinus*. (*a*) 19 day pouch young; transverse section through 12 sex cords (SC) embedded in stroma (S) and surrounded by the tunica albuginea (T). (*b*) The same to show the relationship of the rete testis (R) and sex cord. (*c*) 75 day pouch young; note the rete testis, blood vessel (BV) and mass of seminiferous tubules. From Ullman (1984), with permission.

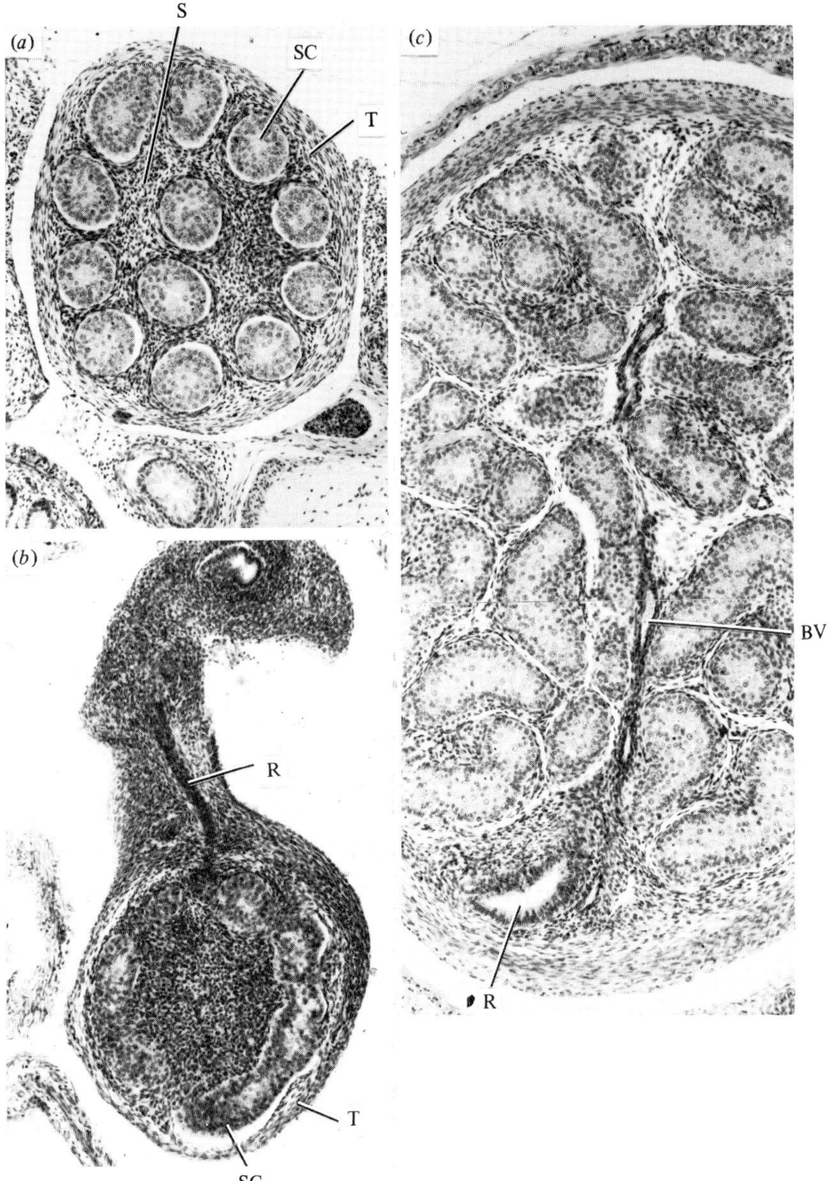

Oocytes that remain joined to others or that are too near the periphery of the ovary become atretic; this leads to the rapid decline in germ cells to less than 25% of the peak (Fig. 3.4).

From day 110 to pouch emergence at day 250 is the fourth or follicular stage, during which there is a steady growth of the ovary with formation of medium and large antral follicles. In *M. eugenii* interstitial tissue differentiates from rete cords at this stage and remains a prominent and permanent feature of the mature ovary. The structure of the primordial oocyte of *Isoodon macrourus* has been studied by light and electron microscopy (Ullmann, 1978) and the histochemistry of the oocyte of *D. virginiana* by Guraya (1968*b*). Both authors described a distinct paranuclear complex composed of RNA, protein and lipoprotein. For the later development of the oocyte and follicles see p. 195.

Differentiation of the urogenital system

The urogenital system, like the gonad, develops from the indifferent stage at birth to the distinctive male and female structures during the course of pouch life (Fig. 3.11). The best-studied species in this respect are *Didelphis virginiana* (Baxter, 1935; McCrady, 1938; Burns, 1939*c*; Chase, 1939; Moore, 1939; Rubin, 1944) and *Macropus eugenii* (Alcorn, 1975) but less-complete studies on other species such as *Macropus rufus* (Lister & Fletcher, 1881) and *Trichosurus vulpecula* (Buchanan & Fraser, 1918) are in agreement.

At the indifferent stage the mesonepheric or Wolffian duct is patent and functions as the main urinary duct conveying urine from the mesonephros to the urogenital sinus or proctodaeum (Figs 3.6, 3.7) until about day 14 after birth. Running parallel and lateral to it is the Mullerian duct, which first appears on day 3 and opens to the coelom near the genital ridge. At day 10 it has made connection posteriorly to the urogenital sinus near that of the Wolffian duct. Externally the urogenital sinus opens by the genital tubercle within a common cloacal chamber with the anus. As the metanephric kidney differentiates and becomes functional (Krause, Cutts & Leeson, 1979*b*) its duct, the ureter, begins to migrate ventrally to the neck of the urinary bladder. In so doing, it passes medial to the connection of the Wolffian (or mesonepheric duct) and Mullerian ducts with the urogenital sinus (Fig. 3.7). In eutherian mammals the ureters migrate laterally to the genital ducts. The particular course of this migration has profound implications for the future development of the genital tract, particularly of the female, because the medial position of the ureters prevents the mid-line fusion of the Mullerian ducts to form a single vagina, as occurs in eutherian mammals (Fig. 5.1). By contrast, in male eutherian mammals the Wolffian ducts, transformed to the vasa deferentia and carried with the testes into the scrotum, must loop around the ureters, while in male marsupials they connect directly with the urogenital tract (Fig. 1.1 and 4.1).

In genetically male *Didelphis* the Wolffian duct persists after day 14 but the Mullerian duct progressively disappears, beginning at its posterior connection with the urogenital sinus. The prostate first appears at day 20 as epithelial outgrowths from the lumen of the urogenital sinus into the thickened wall of that organ. Cowper's glands appear at day 5 (Rubin, 1944) and are developed from the urethral plate. By day 32 they become 3-lobed and open through paired ducts by day 52. The genital tubercle

Fig. 3.11. Genital region of *Didelphis virginiana* pouch young: (*a*) and (*b*) show the genital tubercles of a 15 day (*a*) male and (*b*) female. Bottom panel (*c*) shows day 20 male given testosterone from birth, and (*d*) a 17 day male given oestradiol benzoate from birth. Note differences in genital tubercle (gt) to penis (pe) or vulva (v), but retention of scrotum (s) and absence of pouch (p) in genetic males. Drawn from Burns (1942*b*).

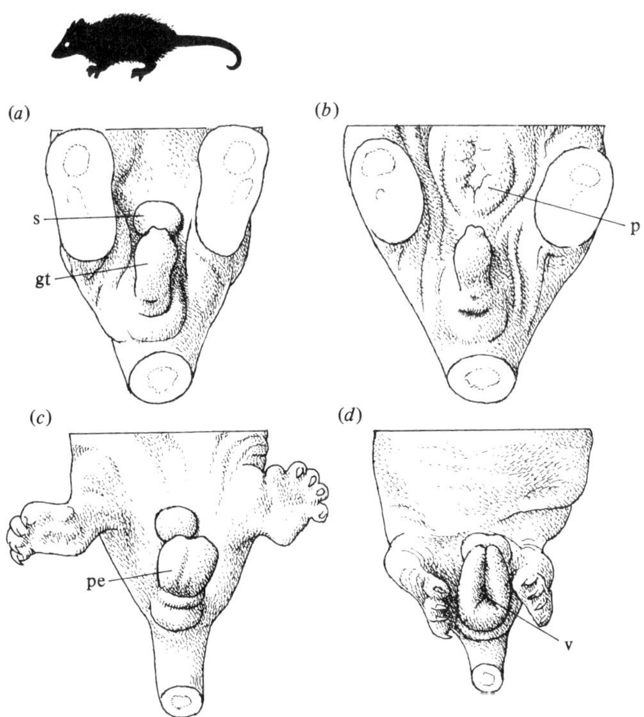

differentiates the characteristic male form with a bifid glans (see Fig. 1.1 and Chapter 4) after day 15 (Fig. 3.11*c*). The scrotum can be distinguished morphologically at day 10 but the testes do not descend into it until after day 25 (Chase, 1939).

In the genetic female, the Wolffian ducts progressively degenerate as the mesonephros ceases to function as a kidney (Krause, Cutts & Leeson, 1979*a*) but the Mullerian ducts are retained and grow (Fig. 3.13*d*). At their posterior ends they are separated by the ureters, as already mentioned, but anterior to the ureters each bends towards the mid-line (see Owen's figure of *Marmosa* in Fig. 1.2) and the mesial wall of each fuses with the other (Baxter, 1935). Anterior to this fusion the two ducts again diverge and each will subsequently give rise to an oviduct and uterus (Fig. 5.1). The unfused loops form the lateral vaginae and the fused portions the vaginal culs de sac. The further consideration of the comparative anatomy of the female tract will be left to Chapter 5. From the urethral plate are differentiated the bulbo-urethral or Bartholin's glands (homologue of Cowper's glands in males), which appear at day 5 (Rubin, 1944) but develop more slowly than Cowper's glands in the male and open by a pair of ducts after day 67. The genital tubercle differentiates the female form by day 15 (Fig. 3.11).

While the sequence and broad features of gonadal and genital tract development in marsupials are very similar to those of eutherians, the one important difference is that the events occur after birth when the young is no longer under the direct influence of maternal hormones. Burns (1939*a*) and Moore (1939) recognised that marsupials thus provide an opportunity to investigate the role of sex hormones in sexual differentiation of the gonad and the genital tract, directly. They did this by gonadectomy and by using the then newly available pure steriods, testosterone and oestradiol, at various stages of development of *Didelphis* pouch young. By present standards the doses administered were excessive and the responses accordingly difficult to interpret. However, a brief summary of their conclusions is in order because the pouch young of *Didelphis* still remains an excellent subject in which to investigate these phenomena and studies using modern techniques are just beginning (George *et al.*, 1985).

Effects of gonadectomy and gonadotrophin treatment

Moore (1941*b*; 1943) and Rubin (1943) removed gonads from male and female *D. virginiana* on day 20, when differentiation of the gonad and the genital ducts has begun, and found no effect on the subsequent development in either sex up to day 100. After that age they observed retarded development of the uterus and Bartholin's glands in females and of the prostate and Cowper's glands in males and concluded that the gonads do not secrete hormones or control development until late pouch life. This conclusion was corroborated by the findings of Moore & Morgan (1943) and Morgan (1943) that treatment of intact pouch young with gonadotrophins was without effect before day 63–70 in males and day 100 in females. However, Burns (1942*a*) was sceptical of their conclusion from the results of gonadectomy because the initial differentiation of the gonads and genital tracts already had taken place at the time of operation. In the light of later evidence in other mammals and the work of George *et al.* (1985) his reservations were well founded. George *et al.* (1985) incubated gonads of *D. virginiana* with labelled pregnenolone and testosterone and, at the indifferent stage on day 10, the gonads converted pregnenolone to progesterone and testosterone (Fig. 3.12). After differentiation, by days 19–26, only testes formed testosterone whereas, in ovaries, the label accumulated as progesterone. However, when ovaries were incubated with labelled testosterone the steriod was aromatised, presumably to oestradiol.

Effects of testosterone and oestradiol on the urogenital system

Exposure of the indifferent gonad to oestradiol, or testosterone from as early as day 3 did not affect its genotypic differentiation (Burns, 1939*a*; Moore, 1939; Moore & Morgan, 1942; Morgan, 1943) but subsequently Burns (1956) showed that if oestradiol was applied topically to neonatal young of *D. virginiana* on the day of birth and for 10 days after, the presumptive testis was altered to an 'ovotestis'. The gonad developed a female-type cortex, and germ cells within it differentiated into oocytes, while seminiferous tubules developed in the medulla. This is the only mammal in which such a transformation has been deliberately achieved and it has not been repeated by anyone else in *Didelphis*. Alcorn (1975) repeated the experiment with three neonatal young of *M. eugenii* but did not obtain an alteration of the gonad. However, the gonad of this species may be slightly more differentiated at birth than that of *Didelphis* and, as Burns (1961) points out, the more 'embryonic' the neonate was the better was the success rate in gonadal transformation.

Burns (1942*a*) reviewed the results of his own work (1939*a*, *b*, *c*, *d*) and that of Moore (1939, 1941*a*, *b*) on the effects on the genital tract and associated structures of testosterone and oestradiol applied to male and female pouch young less that 20 days old. The responses varied with the structure, the genetic sex and the age at treatment. Neither hormone had any effect on the development of the pouch in genetic females or the scrotum in genetic males even when the treatment began before the structures become morphologically visible on day 10 (Fig. 3.11*d*). Conversely, the genital tubercle in both sexes was highly responsive to the particular hormone regardless of genotype (Burns, 1939*b*): testosterone induced the characteristic male-type bifid glans in both male and female *Didelphis* prematurely early and oestradiol induced the enlarged female structure in both sexes (Fig. 3.11*d*). After normal differentiation had occurred, however, neither steriod could effect a reversal in these structures.

The responses of the Wolffian and Mullerian ducts were less clear cut (Fig. 3.13). Both hormones induced similar hypertrophy in the Mullerian

Fig. 3.12. Enzymatic differentiation in the gonad of *Didelphis virginiana*. (*a*) Progesterone and (*b*) testosterone formation from pregnenolone. Gonads from pouch young of various ages were removed and incubated with 5 μM [7–^{3}H] pregnenolone for 2 h. After the incubation pregnenolone metabolites were extracted and separated by two-dimensional thin-layer chromatography and progesterone and testosterone formation was calculated. (*c*) Aromatase activity was assessed by measuring the conversion of 0.25 μM [1 β-^{3}H] testosterone to [^3H_2O]. Each point represents the mean ± SEM. ●, ovary; ○, testis. Redrawn from George *et al.* (1985).

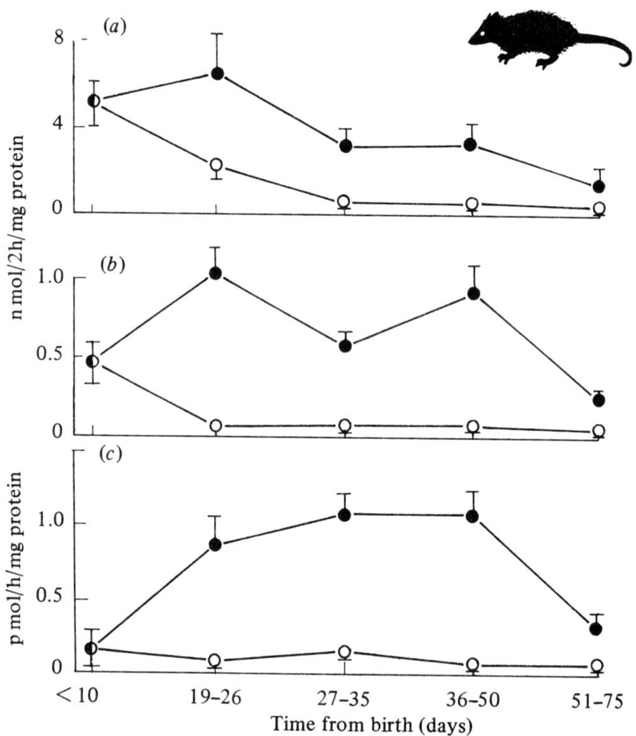

Fig. 3.13. Effects of hormone treatment on genital ducts of pouch young male and female *D. virginiana*. Diagrams illustrating the normal anatomy of (*a*) male and (*b*) female ducts and the effects of testosterone propionate (*c,d*) and oestradiol (*e,f*) at 20 days, as determined by dissection and serial sections. The ducts and related parts of one side have been separated and displayed schematically, but in proper proportion. Rete and male duct system – solid black; Mullerian duct derivatives – heavy lines; sinus cords – stipple. UGS, urogenital sinus; T, testis; OV, ovary. (*a*) normal male – note the absence of the terminal portions of the Mullerian ducts (MD); (*b*) normal female – both ducts are complete but the Wolffian duct (WD) is thinner and the efferent ducts (ED) more rudimentary than in the male. Average condition in testosterone-treated experimental (*c*) males and (*d*) females. Mullerian duct derivatives are better developed in treated females than in treated males and have differentiated into oviduct (OD), uterus (UT) and lateral vagina (LV). In males the Wolffian duct has developed as vas deferens (VD). Average condition in oestradiol-treated (*e*) males and (*f*) females. Redrawn from Burns (1939*b*).

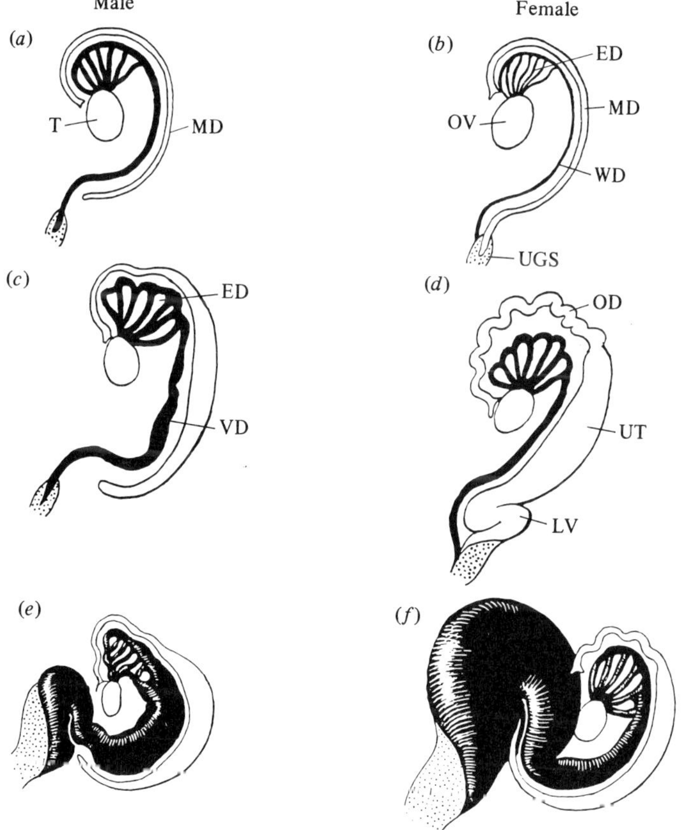

ducts of both sexes with the posterior ends of the ducts failing to maintain connection with the urogenital sinus (Burns, 1939*a*; Moore, 1939). Oestradiol also induced hypertrophy and dilation of the Wolffian ducts in both sexes. On the other hand, testosterone induced retention of the Wolffian ducts and their differentiation into vasa deferentia and epididimydes in female as well as male *Didelphis*. Only the derivations of the urogenital sinus responded to treatment in conformity with the putative roles of the steriods; testosterone induced prostatic development in males and females (Burns, 1945; Moore, 1945), whereas oestradiol inhibited this but stimulated cornification of the epithelium as in vaginal tissues (Burns, 1942*b*).

Fig. 3.14. 5α-reductase activity in urogenital tissues of *Didelphis virginiana* pouch young. (*a*) genital ducts; (*b*) urogenital sinus; (*c*) genital tubercle. 5α-reductase activity was assessed after incubating tissues with 0.25 μM [1β-^{3}H] testosterone for 1 h. Each point represents the mean ± SEM; ●, females; ○, males. Redrawn from George *et al.* (1985).

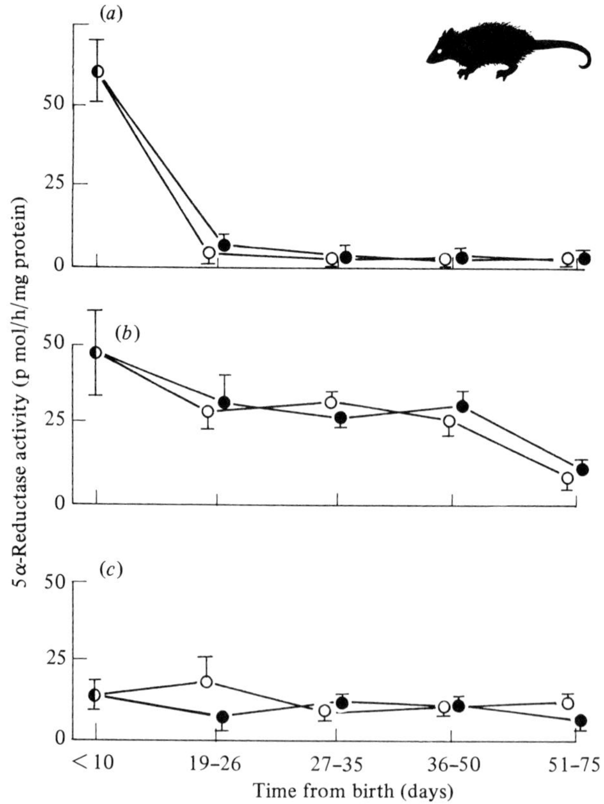

All this work antedated Jost's (1953) demonstration of the fundamental mechanism of sexual differentiation of eutherian mammals in which the female phenotype will develop in the absence of very early exposure to androgen by the Leydig cells of the developing testis. From the work of Alcorn (1975) and Ullmann (1981*b*, 1984) on other species of marsupial it is not evident that Leydig cells differentiate histologically at such an early age after birth, although it is possible that functional differentiation, as evidenced by biosynthesis of testosterone and its appearance in plasma (Catling & Vinson, 1976) may occur much earlier. If *Didelphis* resembles *M. eugenii* in this respect, essential differentiation may have taken place well before the stage at which Moore (1939) castrated *Didelphis* on day 20.

It is now recognised also that, while certain tissues of the male genital system in eutherian mammals respond to testosterone, others (such as the scrotum and genital tubercle) respond to dihydrotestosterone (Imperato-McGinley, 1983; Wilson *et al.*, 1983). George *et al.* (1985) have addressed the latter aspect by determining the occurrence and activity levels of 5α-reductase, the enzyme that is required to convert testosterone to dihydrotestosterone, in the genital tract and associated tissues of *Didelphis* pouch young. During the indifferent period, before day 10, activity was very high in mesonephric derivatives, especially the Wolffian and Mullerian ducts, but was much reduced in these tissues in older pouch young (Fig. 3.14). On the other hand, activity was consistently high through pouch life in the urogenital sinus and, to a lesser extent, in the genital tubercle, both of which tissues were shown to be responsive to steriods by Burns and Moore. Interestingly, the scrotum did not have high enzyme activity as might be expected if its differentiation is determined by exposure to androgen at an early stage.

Differentiation of pouch and scrotum

In *D. virginiana* both pouch and scrotum arise as paired bilateral folds on the ventral surface, anterior to the genital tubercle. Subsequently, the posterior ends fuse to form the scrotum in the male or the posterior lip of the pouch in the female. McCrady (1938) drew the conclusion that they are homologous structures, as are the scrotum and labia majora of eutherian mammals. However, pouch rudiments have been recorded anterior to the scrotum in the male pouch young of eight species (see Beddard, 1891; Bresslau, 1912; Pocock, 1926), and the adult males of *Thylacinus* (Beddard, 1891; Pocock, 1926) and *Chironectes* (Enders, 1966; Hunsaker, 1977) retain a well-formed pouch anterior to the scrotum, which suggests that pouch and scrotum arise from different parts of the same anlagen. It is of interest that a scrotum has not been reported to occur in any female marsupial except in the two XO intersexes referred to earlier. A further point of interest is that *m. cremaster*, which takes origin on the iliac spine inserts on the tunica vaginalis of the testis of males and, as *m. ilio-marsupialis*, in the female inserts into the substance of the mammary gland. We will return to this aspect in Chapter 8, but the question it now raises is whether differentiation to pouch or scrotum is effected by gonadal steroids, or by the genetic constitution of the cells of the anlagen itself.

The total lack of response of the anlage of pouch and scrotum of *D. virginiana* to steriod hormones in Burn's and Moore's experiments, in contrast to the genital tubercle, suggest that hormones may not effect differentiation. Subsequently Bolliger (1943) attempted to test McCrady's (1938) idea by transforming the scrotum of immature *Trichosurus vulpecula* males to a pouch by application of oestrogens, and claimed to have done this in several animals. Such a result would not have been expected from Burns' earlier work on *Didelphis* and, indeed, attempts to repeat Bolliger's result have not been successful (see Sharman, 1959). If the pouch and scrotum are determined before gonadal function has been established and are not susceptible to steriod hormones thereafter, this suggests that the differentiation of the anlagen in the first few days after birth is genetically determined and this is supported by the hybrid and intersex animals discussed earlier in this chapter.

None of the hybrid animals had functional germ cells but in all, the external anatomy was in accord with the sex chromosomes. In all except two of the intersexual marsupials referred to earlier in this Chapter (Fig. 3.2) each possessed either a pouch and mammary glands or a scrotum, and the structures were in most cases concordant with the sex-chromosome constitution. However, one specimen of *Sarcophilus harrisii* had half a pouch with two teats and mammary glands on the right side and half a scrotum on the other side (R. L. Hughes, personal communication, 1984) and its sex chromosome constitution varied between XXX, XX and XO (E. S. Robinson & G. B. Sharman, personal communication, 1984). Similarly, one of the specimens of *M. eugenii* (Fig. 3.2*d*) had a pouch with two teats on the right side, none on the left side but a small empty scrotum posteriorly. Sharman *et al.* (1970) considered the scrotum to be central, and it is so drawn in Fig. 3.2*d* but, on later re-examination of the preserved specimen, Professor Sharman (personal communication) finds that it is asymmetrically to the left. Thus, in both these unusual specimens, he interprets the condition as differentiation of the pouch/scrotum anlagen to the male condition on the left side and to the female on the right side, presumably brought about by the particular chromosomal condition of the cells of each side of the anlagen. These specimens support McCrady's (1938) hypothesis for homology of pouch and scrotum being derived from one anlagen and that differentiation is determined by the particular chromosomal constitution of the cells of the anlagen. This suggests that the Jost (1953) hypothesis may not apply in its entirety to sexual differentiation in marsupials.

Secondary sex characters

Several characters differentiate male from female marsupials after sexual maturity and it can be assumed that they are influenced by gonadal hormones. These were referred to in Chapter 2 and need only be mentioned briefly here. In a number of species there is a size dimorphism between the sexes; for *Tarsipes rostratus* adult males are smaller than adult females, whereas in many species of macropodids males are considerably larger than females and this comes about as a result of differences in growth rate at the time of puberty (Fig. 2.22). In these species, body shape also differs between the sexes with males having proportionately larger forearms and shoulders (Jarman, 1983). In *Didelphis marsupialis* and *D. albiventris* the canines of males are longer than those of females of the same age after puberty (Tyndale-Biscoe & Mackenzie, 1976). The disposition and relative development of glandular regions in the skin show a sexual difference in many species such as *Antechinus stuartii*, *Petaurus australis*, *P. breviceps* (Schultze-Westrum, 1969), *Trichosurus vulpecula* and several species of kangaroos. In *T. vulpecula* Bolliger (1944) showed development of the sternal gland was stimulated by testosterone. These and other aspects of reproductive anatomy and function of adult marsupials will be the subject of the following chapters.

4

Male anatomy and spermatogenesis

Reproduction in the male marsupial has been less well studied than reproduction in the female, perhaps because there are fewer obvious differences between marsupial and eutherian patterns in this sex. However, on closer scrutiny, many unique features emerge, including, as in the female, the anatomy of the genital tract. As Edward Tyson wrote when introducing Cowper's (1704) paper – 'the organs of generation in the male are no less surprising and remarkable than in the female; and in both they are different from any other animal that I have met with' (Fig. 1.1).

Anatomy and physiology of the male genitalia

The genital tract consists of paired testes, epididymides and vasa deferentia which open into the anterior end of the large prostate gland. A thin membraneous urethra extends from the posterior end of the prostate, and bulbo-urethral (or Cowper's) glands are present (Figs 4.1 and 4.2). There are no seminal vesicles, ampullae or coagulating glands. The only apparent exceptions to this are found in *Caenolestes obscurus* (Rodger, 1982) and in *Burramys parvus* (P. D. Temple-Smith, unpublished), which each have a convoluted secretory segment of the distal portion of the vas, which resembles the eutherian ampulla. The glans penis is cleft in some species, and the scrotum is anterior to the penis in all species (Stirling, 1891; Sweet, 1907).

The scrotum, testes and epididymides

The testes of adult marsupials are generally ellipsoid in shape, (Setchell, 1977) and, in all except two fossorial species, are permanently scrotal. In the marsupial mole (*Notoryctes typhlops*) they are permanently inguinal (Johnson in Strahan, 1983) or abdominal (Sweet, 1907), while the testes of *Lasiorhinus latifrons*, the largest fossorial mammal, are carried in a small, non-pendulous scrotum variously positioned close to the body wall but not in the inguinal canals (Brooks, Gaughwin & Mann, 1978). Like many fossorial eutherian species, which are testicond, this may be an adaptation to the burrowing habit, although other fossorial marsupials, such as *Vombatus ursinus*, *Bettongia lesueur* and *Macrotis lagotis* have more pendulous, permanently scrotal testes like other marsupials. Testicular descent is completed at about day 80 in *Didelphis virginiana* (Finkel, 1945), *Macropus robustus* (Ealey, 1967) and *M. parma* (Maynes, 1973*a*), at day 70 in *Trichosurus vulpecula* (Turnbull, Mattner & Hughes, 1981), but at about day 25 in *Perameles gunnii* (Heinsohn, 1966). The way in which the descent is achieved has been described by Hart (1909), Finkel (1945) and Turnbull *et al.* (1981).

In the Eutheria, testicular weight as a proportion of body weight varies in different species in relation to body size and mating system (Harcourt *et al.*, 1981). Table 4.1 gives these proportions for a range of marsupial

Fig. 4.1. Male reproductive tracts of (*a*) *Didelphis virginiana*, ventral view; (*b*) *Antechinus stuartii*, dorsal view; (*c*) *Perameles nasuta*, ventral view; (*d*) *Trichosurus vulpecula* dorsal view; (e) *Macropus eugenii*, ventral view. Bl, bladder; Cp, Cowper's gland; Cr, crus penis; G, glans; MU, membraneous urethra; P. penis, Pr, prostate; U, ureter; UB, urethral bulb; VD, vas deferens. Redrawn from Rodger & Hughes (1973).

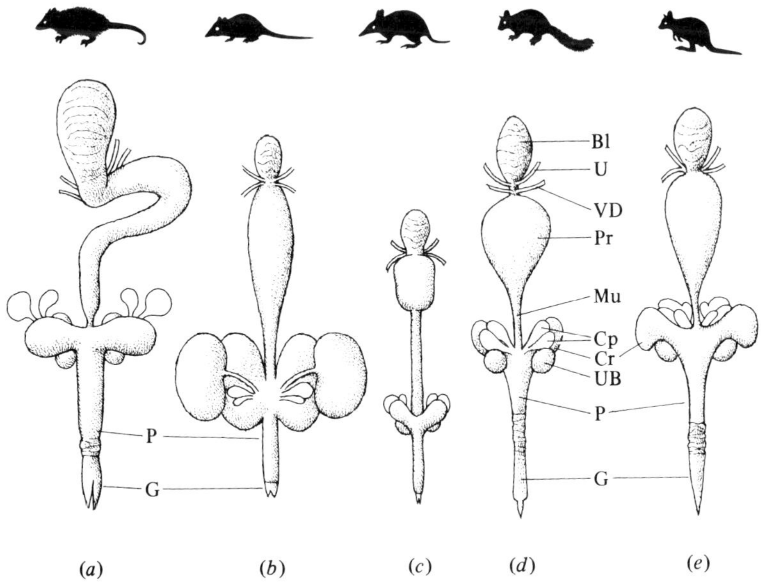

Fig. 4.2. Photograph of ventral view of male reproductive tract of *Macropus eugenii*, to show the pigmented tunica on the left testis, and right testis (T) removed from tunica to show the efferent duct (ED), epididymis (Ep) and rete mirabile (Rm). See Fig. 4.1 for other abbreviations.

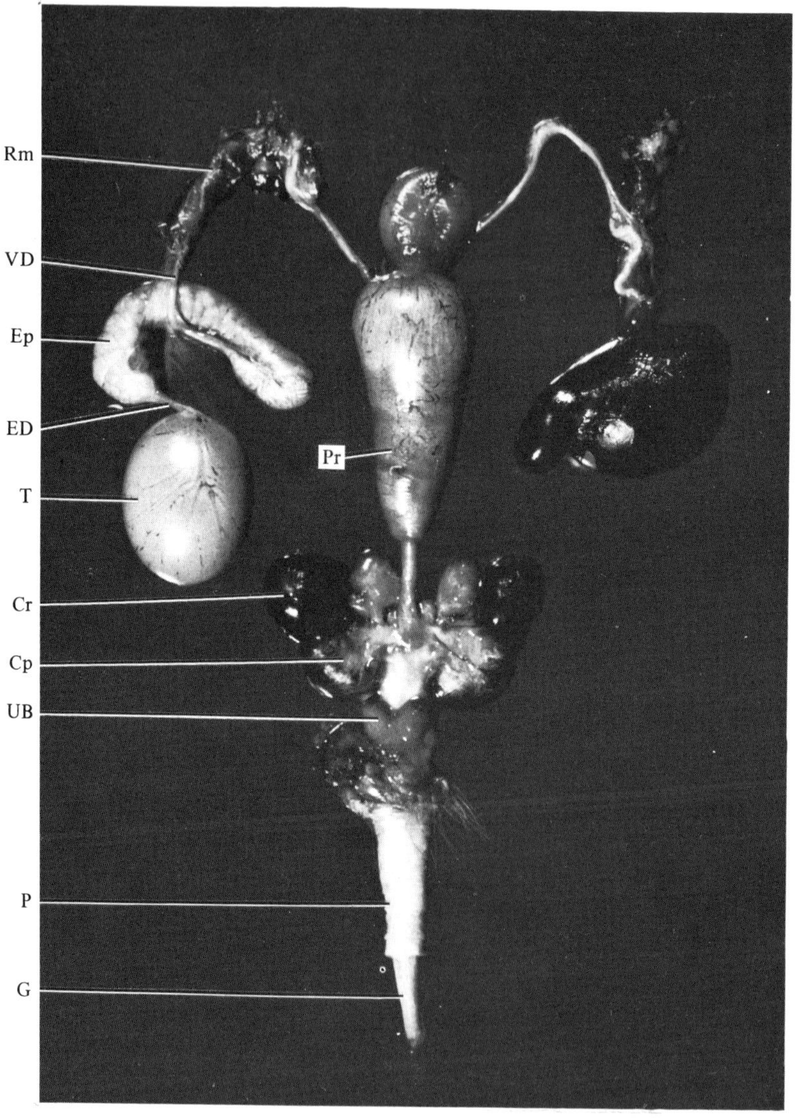

species (R. V. Short, unpublished observations). In most it is less than 0.5% but in one of the most diminutive marsupials, the honey possum *Tarsipes rostratus*, it is a remarkable 4–5% of the body weight.

The tunica vaginalis of the scrotum was first noted by Finkel (1945) to be highly pigmented with melanin in American marsupials, and it has since been found to be so in several other species (Table 4.2, Fig. 4.2 and Fig. 4.3). Its function is unclear: Biggers (1966) suggested that it has a role in regulating the temperature of the testis by acting as a black body radiator but its distribution among different species does not seem to correlate with climatic conditions as this might predict. For instance it is pigmented in *Didelphis virginiana* and in three other tropical didelphids (*D. albiventris*, *Philander opossum* and *Marmosa murina*) as well as in several small macropodids (Setchell, 1977), including *Macropus eugenii* (Fig. 4.2), and

Table 4.1. *Body weights and combined testes weights in some male marsupials*[a]

	n	Body weight (g)	Combined testes (g)	Body weight (%)
Dasyuridae				
Dasyurus hallucatus	7	450 ± 84.1	1.35 ± 0.21	0.30
Pseudantechinus macdonnellensis	8	27.1 ± 5.4	0.11 ± 0.06	0.41
Dasykaluta rosamondae	6	35.3 ± 8.7	0.126 ± 0.56	0.36
Burramyidae				
Cercartetus concinnus	5	14.3 ± 2.4	0.062 ± 0.013	0.44
Acrobates pygmaeus	4	12.3 ± 2.2	0.178 ± 0.08	1.45
Tarsipedidae				
Tarsipes rostratus	24	8.9 ± 1.9	0.365 ± 0.064	4.12
Peramelidae				
Isoodon obesulus	1	1155	4.4	0.38
Phalangeridae				
Trichosurus vulpecula	15	3350 ± 410	8.26 ± 1.13	0.25
Macropodidae				
Macropus eugenii	17	5850 ± 640	31.0 ± 4.4	0.53
M. agilis	6	$11\,400 \pm 2400$	25.64 ± 6.9	0.23
M. rufogriseus rufogriseus	9	$18\,500 \pm 3000$	54.57 ± 9.47	0.30
M. giganteus	9	$40\,720 \pm 994$	42.02 ± 10.62	0.10
M. fuliginosus	10	$34\,150 \pm 944$	51.62 ± 10.08	0.15

[a] Weights given as mean ± SD.
R. V. Short, unpublished data.

in *Trichosurus vulpecula* and *Tarsipes rostratus* (Fig. 4.3*b*). It is unpigmented in *Macropus rufus* (Setchell, 1977) and *Lasiorhinus latifrons* (Brooks *et al.*, 1978), both inhabitants of hot dry climates and in *Caenolestes obscurus*, an inhabitant of a cold moist environment. In this last species it is completely unpigmented and each testis and epididymis is easily visible through the thin membrane (Rodger, 1982). Likewise *Sarcophilus harrisii*, *Burramys parvus* and *Acrobates pygmaeus*, which inhabit cool environments, have unpigmented tunicas (Guiler & Heddle, 1970; Temple-Smith, 1984*b*).

Nevertheless, as Rodger (1982) points out, the role of tunic pigments in regulation of scrotal temperature cannot be entirely ruled out, since the distribution of the pigment in the tunica vaginalis, particularly of the prominent cauda epididymis (Fig. 4.3*a*), may be of greater significance than the overall presence or absence of pigment, because thermoregulation in the epididymis, where spermatozoa are stored, may be far more critical than that in the testis (Bedford, 1978). The distribution of scrotal hair varies between species and it too may influence temperature (Temple-Smith, 1984*b*). There may be some correlation with environmental factors, such as temperature and humidity, but no systematic studies have been made of this.

Testicular temperature is maintained several degrees lower than body temperature (Table 4.3) by a counter-current exchange of heat in the spermatic cord, and the primary loss of heat is from the scrotal skin, assisted by the evaporation of sweat or moisture from licking. Local heating of the scrotum, or returning the testes and epididymides to the abdominal cavity experimentally, can cause disruption of spermatogenesis

Table 4.2. *Distribution of pigmentation of the tunica vaginalis amongst male marsupials (see text for sources)*

Pigmented	Unpigmented
Didelphis virginiana	*Caenolestes obscurus*
D. albiventris	*Burramys parvus*
Philander opossum	*Acrobates pygmaeus*
Marmosa murina	*Lasiorhinus latifrons*
Tarsipes rostratus	*Sarcophilus harrisii*
Cercartetus lepidus	*Macropus rufus*
C. nanus	
Trichosurus vulpecula	
Macropus giganteus	
M. fuliginosus	
M. eugenii	

(Setchell & Thorburn, 1969, 1971; Setchell, 1977), as in eutherians, and spermatogenesis has been reported to be impaired in *Macropus rufus* during hot weather (Newsome, 1973). In *T. vulpecula*, testicular temperature is maintained 3–6 °C below body temperature in ambient temperatures of 21–40 °C and this differential was increased to 14 °C when the animals were able to lick their scrotums (Table 4.3).

The neck of the marsupial scrotum is very narrow, a consequence perhaps of its prepenial position, and each of the two spermatic cords contains a testicular artery, testicular vein, lymph ducts, and vas deferens. There is also a well developed cremaster muscle, which is capable of pulling the scrotum well up against the body, but the narrow neck prevents the testes moving up into the body cavity as occurs in the rat, rabbit and other Eutheria which have an 'open' inguinal canal. The testicular artery and vein divide close to or in the spermatic cord to form a complex rete mirabile (Fig. 4.4) of intermingled parallel branches of blood and lymphatic vessels (Harrison, 1949; Barnett & Brazenor, 1958; Heddle & Guiler, 1970; Lee & O'Shea, 1977; Setchell, 1977). By contrast, in the Eutheria, the testicular artery remains undivided but highly convoluted in the pampiniform plexus. Setchell & Waites (1969) have demonstrated that the vascular rete mirabile of *M. eugenii* attenuates the pulse from arterial blood, with only a small reduction in mean blood pressure and provides a counter-current heat exchange, whereby the arterial blood is pre-cooled by returning venous blood which has lost heat from the superficial veins on the testis close to the scrotal skin (Setchell & Thorburn, 1969). Testicular blood flow appears to be of the same order of magnitude as in eutherians (Setchell, 1977), and it increases if the testicular temperature is raised experimentally by heating the scrotum but not in experimental cryptorchids (Setchell & Thorburn, 1969, 1971).

The development of the seminiferous tubules in the testis and their relation to the rete testis and Wolffian duct were discussed in Chapter 3 (p. 112) and spermatogenesis and the cell cycle are dealt with in detail later in this Chapter (p. 156). The convoluted seminiferous tubules join to the rete testis by the tubuli recti (Fig. 4.5*b*), which are short straight tubules continuous with the seminiferous tubules, as in eutherians, and are lined with Sertoli-like cells (Orsi, Ferreira & de Melo, 1979; Rodger, 1982). In adult macropodids and *T. vulpecula* the rete testis is a horseshoe-shaped area on the surface of the testis (Lee & O'Shea, 1977; Setchell, 1977). It lies under the veins on one surface and under the arteries on the other. However, in American marsupials (Rodger, 1982; Woolley, 1986) and dasyurids (Woolley, 1975) it is centrally placed deep within the testis. In

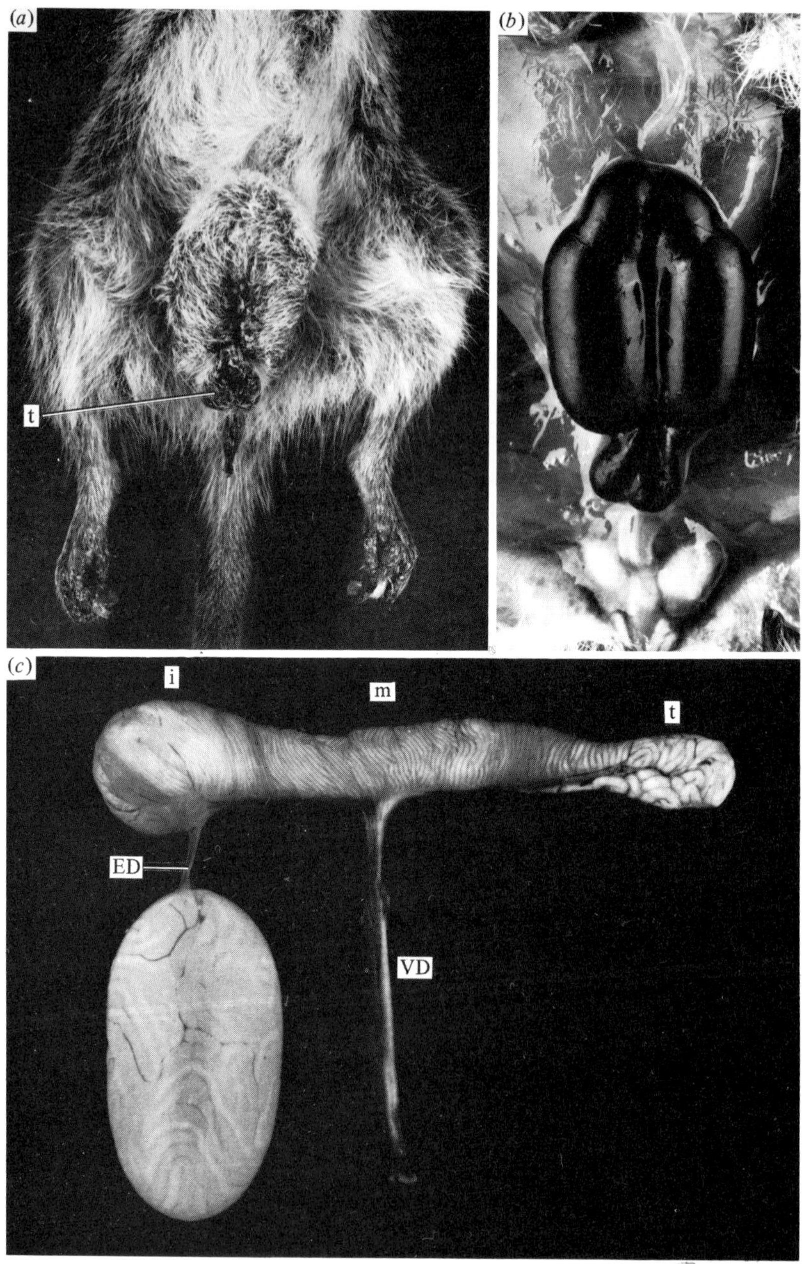

Fig. 4.3. Male genitalia of *Tarsipes rostratus*. (*a*) Intact scrotum with bare skin overlying storage region of the cauda epididymis (t); (*b*) with the skin removed to show the densely pigmented tunica albuginea; (*c*) the testis and epididymis dissected out to show the efferent duct

the latter there are only 1–4 seminiferous tubules opening into the rete (Fig. 4.5*a*), and these have a wider diameter (0.4–0.5 mm, see Fig. 4.8) and are shorter (< 1 m) than in other marsupials or in eutherians (Woolley, 1975).

The efferent ducts, which connect the rete testis to the single duct of the epididymis (Fig. 4.3*c*), are comparatively long, and are joined to the two parts of the rete near the hilus of the testis where the blood vessels enter. Under the electron microscope, the ciliated epithelium appears to have an absorptive capacity (Ladman, 1967), which supports other evidence (see next section). In the Didelphidae, Dasyuridae and Peramelidae, there is a relatively long single efferent duct (Figs 1.1 and 4.5), which together with the associated blood vessels links the epididymis to one pole of the testis by the membraneous mesorchium (Hughes, 1965; Martan, Hruban & Slesers, 1967; Noqueira, Godinho & Cardoso, 1977), whereas in the Phalangeridae, Phascolarctidae and Macropodidae (Fig. 4.2), it is linked by a tract of 12–15 efferent ducts (Hughes, 1965; Setchell, 1970*a*).

The epididymis is similar to that of eutherians, except that it can be separated further from the testes (Figs 4.2 and 4.3*c*). The head of the epididymis is not fused with the testis in any of the 18 species (Families Macropodidae, Dasyuridae, Phascolarctidae, Peramelidae and Phalangeridae) examined by Hughes (1965) nor in *Tarsipes rostratus* (Cummins, Temple-Smith & Renfree, 1986).

The epididymis of marsupials, like that of eutherians can be divided morphologically into three main regions (Fig. 4.3*c*), which broadly reflect the different functions of fluid absorption (initial segment or caput epididymis), sperm maturation (middle segment or corpus epididymis) and sperm storage (terminal segment or cauda epididymis). The cytology of the epithelia of these regions have been described in *Didelphis albiventris* (Orsi *et al.*, 1981; De Melo *et al.*, 1982), *T. vulpecula* (Cummins, 1981; Temple-Smith, 1984*a*), *M. eugenii* (Jones, Hinds & Tyndale-Biscoe, 1984) and *Tarsipes rostratus* (Cummins *et al.*, 1986). In the initial segment the epithelial cells are tall, have long stereocilia and pinocytotic vesicles, indicative of fluid resorption, and the density of sperm progressively increases along the duct, as fluid is removed. In the middle segment the duct is wider and the lumen is packed with immature sperm. During

Fig. 4.3. *cont.*
(ED), the caput epididymis or initial segment (i), corpus epididymis or middle segment (m) and cauda epididymis or terminal segment (t), and vas deferens (VD). Derived from Cummins, Temple-Smith & Renfree (1986).

Table 4.3. *Temperature differential between testis and the body under various conditions in* Macropus eugenii *and* Trichosurus vulpecula

Species	Conditions of measurement	Temperature (°C)		
		Body	Testis	Difference
Macropus eugenii	Conscious	36.7	32.3	4.4
	Anaesthetised	36.5	31.1	5.4
	Anaesthetised:			
	scrotal testis	35.5	31.5	4.0
	cryptorchid testis		35.4	0.1
Trichosurus vulpecula	Conscious, ambient temp. 21 °C	36.1	33.0	3.1
	Conscious, ambient temp. 40 °C:			
	scrotal licking prevented	39.8	34.2	5.6
	after scrotal licking		25.9	13.9
	Conscious:			
	scrotal testis	35.4	29.1	6.3
	cryptorchid testis		35.4	0
	Anaesthetised	35.0	32.2	2.8
	Anaesthetised	33.9	31.6	2.1

Data from Setchell & Thorburn (1969), Setchell & Waites (1969) and F. N. Carrick (unpublished data) cited by Setchell (1977).

passage through this segment cytoplasmic droplets are shed from the spermatozoa and these are engulfed and phagocytosed by specialised principal cells of the epithelium (Fig. 4.19), which are able to differentiate between the droplets and sperm (Temple-Smith, 1984*b*). In the terminal segment the duct becomes considerably wider, the epithelium lower and the investing muscular coat thicker. This part of the epididymis extends

Fig. 4.4. Testicular and associated blood vessels of *Trichosurus vulpecula*. Arteries injected with white latex, veins filled with blood. (*a*) Blood vessels in situ to show relation to organs. (*b*) Testicular artery, latex filled, after maceration to show rete mirabile. Note that the artery becomes a single vessel before entering the testis. B, bladder; P, prostate; E, epididymis; EB, branches to epididymis; IC, inguinal canal; PV, caudal vena cava; RM, rete mirabile of testicular artery (TA) in spermatic cord; T, testis; TV, testicular vein. From Lee & O'Shea (1977), with permission.

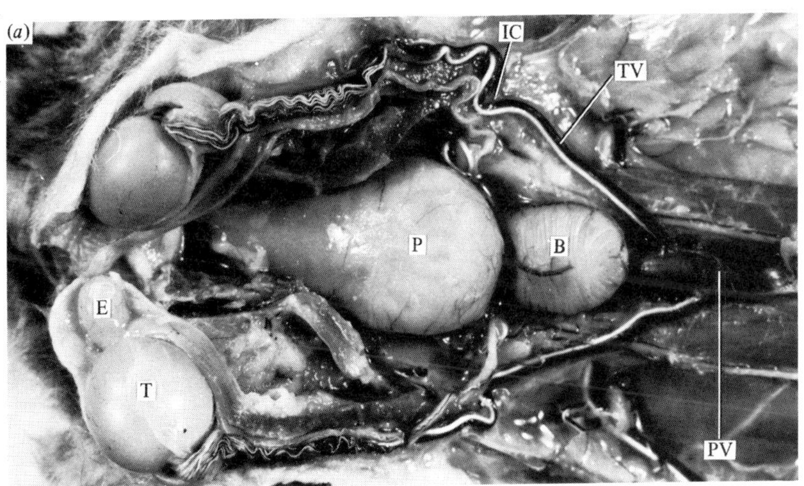

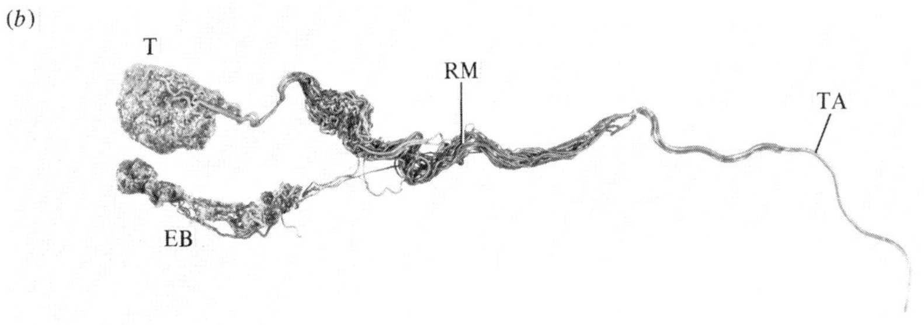

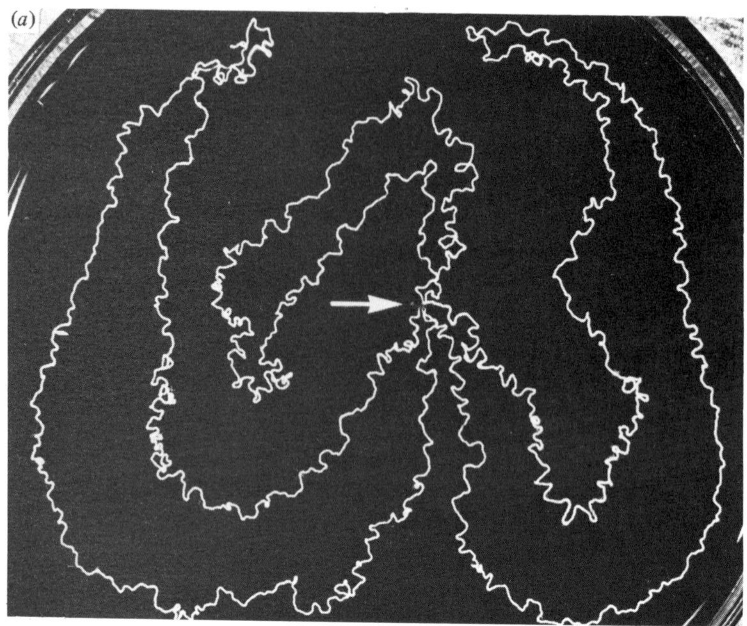

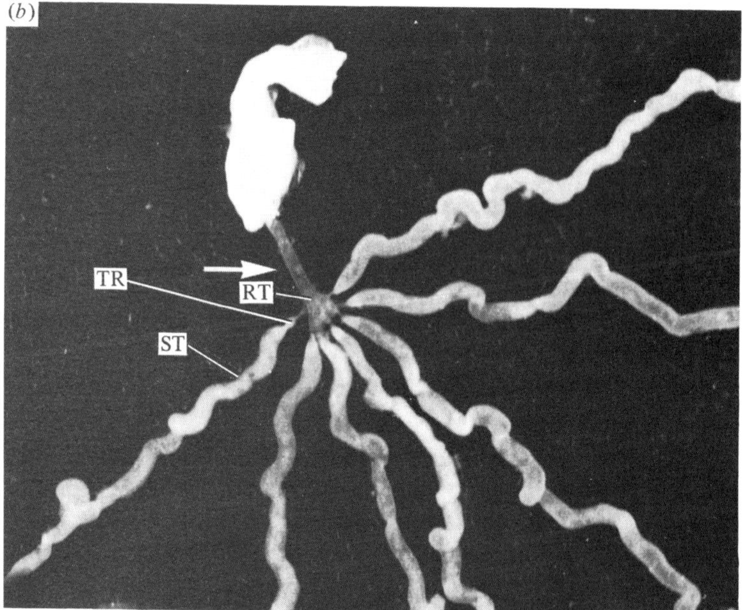

Fig. 4.5. Seminiferous tubules attached to rete testis in *Dasyuroides byrnei*. (*a*) Teased-out testis showing the three complete seminiferous tubules which formed the testis. Each tubule was a single loop opening

beyond the limit of the testis and forms a conspicuous projection of the distal end of the scrotum (Fig. 4.3*a*).

The ductus or vas deferens in most marsupials is a relatively simple muscular tube connecting the cauda epididymis to the prostatic urethra but, in *Caenolestes obscurus*, the distal end is enlarged and convoluted to form an ampulla-like structure which appears to be secretory (Rodger, 1982). Spermatorrhea occurs in marsupials, as in other mammals, even when the animal is not sexually active (Biggers, 1966; Rodger & White, 1976*b*). In *T. vulpecula*, spermatozoa are largely immotile in the urine, except in the breeding season when they are strongly motile (Bolliger, 1940).

Testicular secretions

Rete testis fluid

The spermatozoa leave the testis as a dilute suspension in a fluid secreted by the testis (Setchell, 1974). In *M. eugenii*, rete testis fluid collected from the efferent ducts contains 8×10^7 spermatozoa ml^{-1} and has a flow rate of about 12.5 $\mu g\ g^{-1}$ testes per hour, a rate intermediate between rams and rats (Setchell 1974, 1977). The fluid has no detectable inositol, almost no glucose or protein, but relatively high amino acid concentrations (Setchell, 1970*b*). As in other mammals, 80–90% of this fluid is reabsorbed by the efferent ducts before it enters the initial segment of the epididymis.

Testicular endocrinology

Fawcett, Neaves & Flores (1973) recognised three types of testicular interstitial organisation in mammals, and they included *Didelphis virginiana* in Type 3, characterised by very abundant Leydig cells with only small lymphatic spaces. To this Type may be added all other Didelphidae, *Caenolestes obscurus* (Rodger, 1982) and the Peramelidae (Setchell, 1977), whereas the Phalangeridae and Macropodidae conform to Type 2, characterised by a predominance of loose connective tissue in the intertubular

Fig. 4.5. *cont.*
at both ends into the rete testis at arrow. (*b*) Connections of the four seminiferous tubules (ST) with the rete testis (RT) in another *D. byrnei* (one of the connections was broken and cannot be seen). Note the simple rete, the single efferent duct (arrow) and the narrow tubuli recti (TR), connecting the seminiferous tubules to the rete. From Woolley (1975), with permission.

spaces, with Leydig cells occupying only 5% of the total volume of the testis (Figs 4.15 and 4.17).

The ultrastructural characteristics of Leydig cells of *D. virginiana* and *T. vulpecula* have been described by Christensen & Fawcett (1961) and Temple-Smith (1984*b*). Both species contain abundant agranular endoplasmic reticulum, mitochondria and lipid inclusions, which are characteristic of Leydig cells, and it may be presumed that – as in Eutheria – these cells are the source of androgen production in marsupials. Testis slices from *D. virginiana* were able to synthesise androstenedione and testosterone from precursors such as acetate, cholesterol and pregnenolone (Cook *et al.*, 1974).

The main androgen secreted by all male marsupials so far studied is testosterone (Figs 4.6, 4.7 and 4.8), although 5 α-dihydrotestosterone has been detected in some species as well. Whereas testosterone occurs in high

Fig. 4.6. Response to castration in *Macropus eugenii*. Concentrations (mean ± s.e.m.) of (*a*) Follicle stimulating hormone; (*b*) Luteinising hormone and (*c*) testosterone in the plasma of four males. Redrawn from Catling & Sutherland (1980).

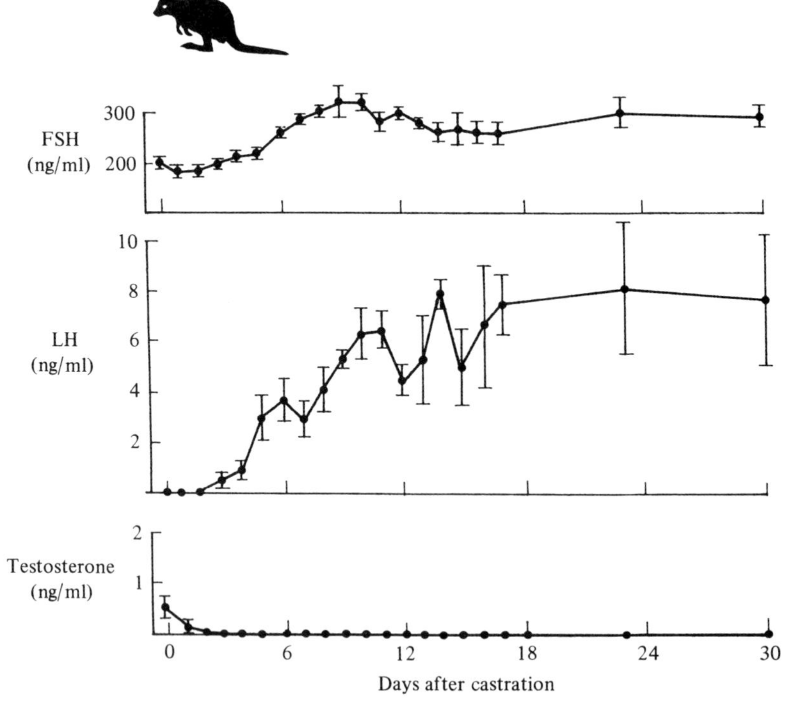

concentration in plasma from the spermatic vein in *T. vulpecula*, *Macropus robustus*, *M. rufus* and *Perameles nasuta*, much less was measured in *Pseudocheirus peregrinus* (Table 4.4). However, this last was a single specimen obtained out of the breeding season when the testes are regressed. Testosterone concentrations measured in peripheral plasma of males of several species is around the middle of the range recorded for Eutheria (Table 4.5). Androgens have been measured in several dasyurids, notably *Antechinus stuartii* (Fig. 4.8*a*) and *A. minimus* (Fig. 4.8*b*), and testosterone and dihydrotestosterone have been identified as the major androgens (McDonald et al., 1981).

In *T. vulpecula*, and in several other species (Table 4.5), peripheral plasma contains significant amounts of 5α-dihydrotestosterone (DHT) as well as testosterone (Cook, McDonald & Gibson, 1978; Allen & Bradshaw, 1980; Curlewis & Stone, 1985a). The concentration of DHT in *T. vulpecula* is about 10 times human levels, and there is a high specific activity of 5α-reductase in epididymal and prostatic tissues (Cook *et al.*, 1974; Curlewis & Stone, 1985*b*), the enzyme involved in its conversion to testosterone.

The rate of testosterone secretion is comparable to that of eutherians and is apparently under similar control. The testis of *T. vulpecula* secretes 2.3 ng min^{-1} (g testis)$^{-1}$ (Table 4.4) and the release of testosterone is apparently not pulsatile (Curlewis & Stone, 1985*a*). In this species and in five macropodids studied by Lincoln (1978) the stress of sampling caused a drop in testosterone concentration. Allen & Bradshaw (1980) reported a diurnal variation in testosterone concentrations in *T. vulpecula*, with highest values in the morning, but Curlewis & Stone (1985*a*) consider that this result may be due to the stress of collection, since the samples were collected every 3 h, commencing at 12.00, and were followed by a decline in peripheral androgen levels. Then between 18.00 and 09.00 the following day, androgen levels increased and corticosteroids decreased.

Synthesis of testosterone by testis of *D. virginiana* was stimulated *in vitro* by the addition of luteinising hormone (LH) to the medium, or *in vivo* by injecting the animal with LH 3 days before the time of the incubation (Cook *et al.*, 1974). Likewise, testosterone release was induced *in vitro* by the injection Gn-RH in five macropodid species (Lincoln, 1978). Macropodid marsupial testes have distinct receptors for follicle-stimulating hormone (FSH) and LH, which exhibit very similar properties to FSH and LH testicular receptors of the rat (Stewart, Sutherland & Tyndale-Biscoe, 1981). Gonadotrophins were undetectable in the plasma after hypophysectomy (Hearn, 1975*a*) and both testes and seminiferous tubules had shrunk

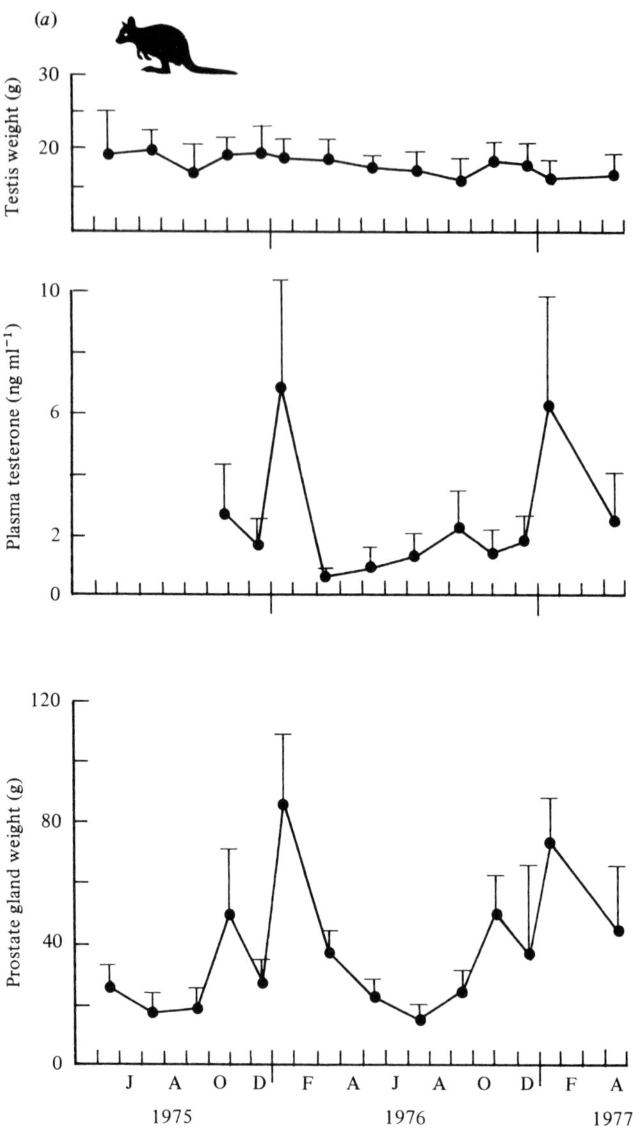

Fig. 4.7. Seasonal changes in prostrate weight of male marsupials. (*a*) Mean ± SD values for testis weight, plasma testosterone concentration and prostate gland weight in *Macropus eugenii* during 2 years. Note peaks of testosterone concentration and prostate weight during January–February, the time of maximum number of oestrous females, and minor peaks in October when young females enter first oestrus (see Fig. 2.26). Redrawn from Inns (1982). (*b*) Mean monthly weights of testis (broken line), epididymis (dotted line) and prostate (solid line)

within 1 week of the operation and spermatozoa were almost completely absent from the epididymis within 60 days after hypophysectomy; the prostate and other accessory glands also showed marked atrophy (Hearn, 1975*a*). These results indicate that male marsupials require gonadotrophins to stimulate and maintain testicular function and androgen-dependent structures. Conversely, castration of *M. eugenii* (Catling & Sutherland, 1980) caused a rapid fall in testosterone to undetectable levels in 2 days and a significant increase in plasma LH and FSH levels to a maximum respectively 17 and 9 days later (Fig. 4.6). Likewise in *Dasyuroides byrnei*, Fletcher (1983) observed that testosterone fell to undetectable levels in 5 days and LH reached a maximum 10 days after castration. Thus the evidence available so far shows that male marsupials have a negative-

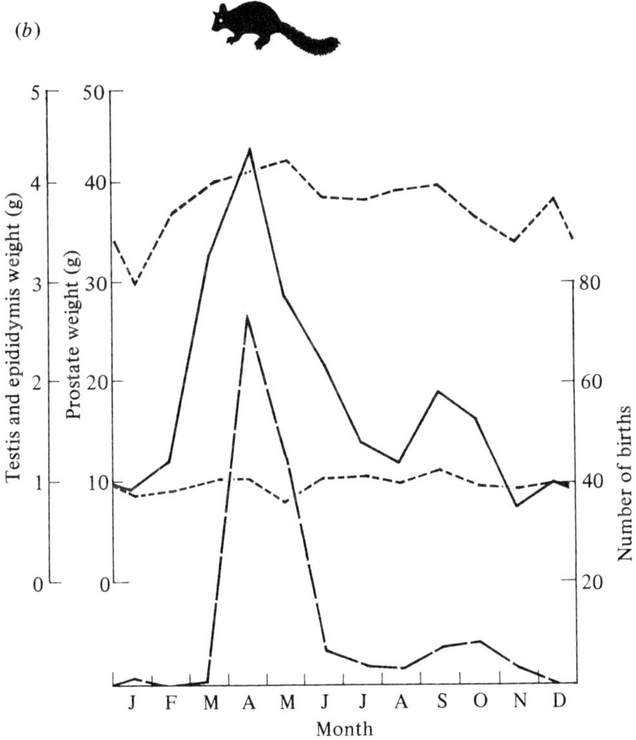

of *Trichosurus vulpecula* males compared to the number of births recorded each month for the same period. Note the marked increase in prostate weight during the major and minor breeding seasons. Redrawn from Gilmore (1969).

feedback loop between the gonad and the pituitary, like other mammals and female marsupials (Evans, Tyndale-Biscoe & Sutherland, 1980).

As is to be expected, castration affects the organs of reproduction. In *D. virginiana* it caused a decrease in the size of the prostate and Cowper's glands (Chase, 1939) and in *T. vulpecula* and *M. eugenii* epididymal size was also found to be reduced (Curlewis & Stone, 1985*b*; R. C. Jones, personal communication). This interfered with sperm maturation and it

Fig. 4.8. (*a*) Mean ± SEM of testicular weight (broken lines), diameters of the seminiferous tubules (solid line) and plasma androgens of *A. stuartii* from February through to August. (*b*) The same in *Antechinus minimus* and associated changes in the weights of Cowper's gland, epididymis and prostate gland. Differences are significant ($p < 0.001$) between May and July for Cowper's gland and prostate, and between April and July for the epididymis. Note that the changes occur about 1 month earlier in *A. stuartii* than in *A. minimus* in accordance with the earlier onset of oestrus in the former species. (*a*) Redrawn from Kerr & Hedger (1983); (*b*) Redrawn from Wilson & Bourne (1984).

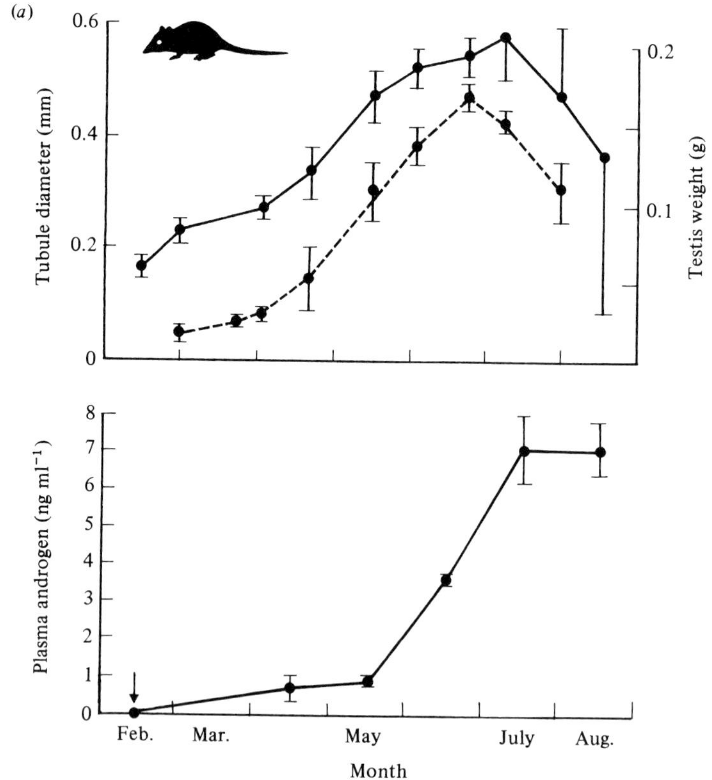

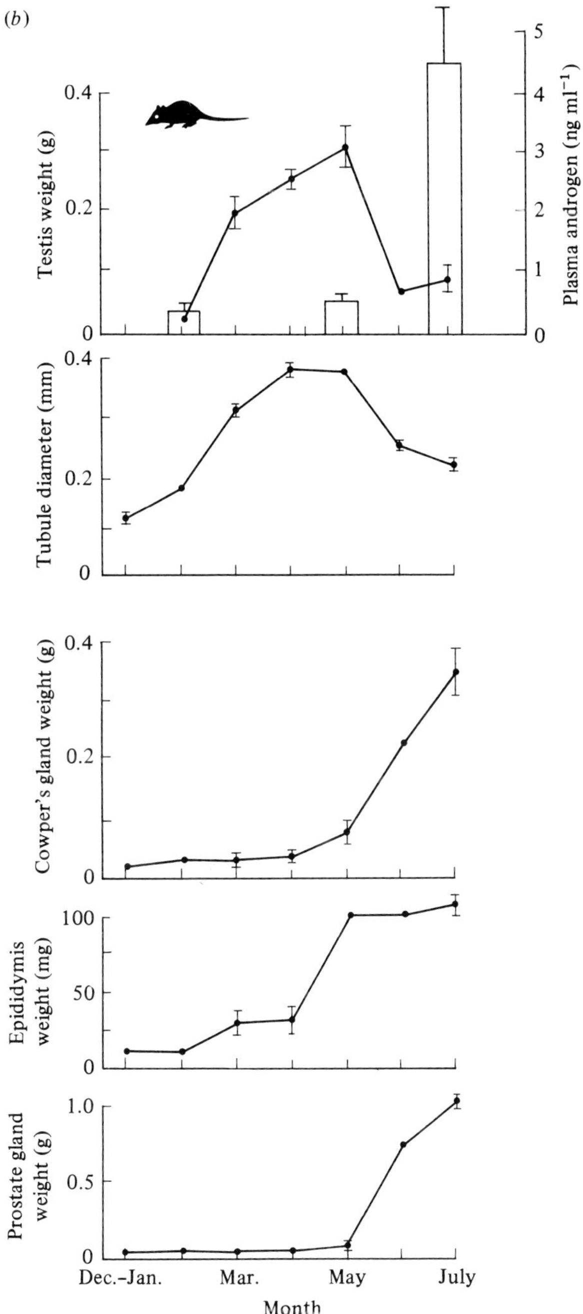
(b)
Testis weight (g)
Plasma androgen (ng ml^{-1})
Tubule diameter (mm)
Cowper's gland weight (g)
Epididymis weight (mg)
Prostate gland weight (g)
Dec.-Jan.
Mar.
May
July
Month

Table 4.4. *Testosterone production* in vivo *by the testes of some Australian marsupials*

Species	Body weight (kg)	Weight of both testes (g)	Testosterone concentration in plasma from the internal spermatic vein (ng ml^{-1})	Testosterone secretion rate: (ng min^{-1} per g testes)	(mg day^{-1})
Perameles nasuta	1	4	13	0.1	0.001
Trichosurus vulpecula	3	8	100	2.3	0.025
Pseudocheirus peregrinus	1.5	8	1	—	—
Macropus robustus	30	40	98	6.8	0.4
Macropus rufus	50	40	83	2.6	0.15
Ovis aries	70	500	60	6.0	4

Compiled by Setchell (1977) from data in Carrick & Cox (1977) for marsupials and Setchell, Waites & Lindner (1965) for sheep.

Table 4.5. *Concentrations (ng ml^{-1}) of testosterone (T) and 5α-dihydrotestosterone (5αDHT), or total immunoreactive androgens (A) in the peripheral plasma of various male marsupials during non-breeding (NB) and breeding (B) periods*

	A		T		5αDHT		
	NB	B	NB	B	NB	B	Source
Dasyuridae							
A. minimus	< 1.0	> 5.0					Wilson & Bourne 1984
A. swainsonii	0.6–1.7	4.8 ± 0.2		3.2 ± 0.3		0.6 ± 0.08	McDonald *et al.* 1981
A. flavipes	0.6–1.7	12.4 ± 1.1		12.1 + 1.1		1.2 ± 0.2	McDonald *et al.* 1981
S. crassicaudata		4.0 ± 0.9[a]	2.6 ± 0.6[a]		1.2 ± 0.2[a]		McDonald *et al.* 1981
S. crassicaudata	1.32 ± 0.7	7.7 ± 0.7					McDonald *et al.* 1981
Dasyuroides byrnei			< 1.8	3.0–4.0			Fletcher 1983
Phalangeridae							
T. vulpecula			0–9.7[b]		0.2 ± 0.09	1.4 ± 0.5	Curlewis & Stone 1985*a*, *b*
T. vulpecula			0.2–9.5[a]		0.6 ± 0.1[a]		Allen & Bradshaw 1980
Macropodidae							
M. eugenii	3.2 ± 0.4	10.1 ± 2.1					Catling & Sutherland 1980
M. eugenii			1–3	6.0			Inns 1982
M. eugenii			2.0–5.0[a]				Lincoln 1978
M. rufus			4.0–5.0[a]				Lincoln 1978
M. rufogriseus			3.0[a]				Lincoln 1978
M. giganteus			2.0–9.0[a]				Lincoln 1978
Thylogale thetis			2.0–9.0[a]				Lincoln 1978

[a] Time of year of sampling unspecified.
[b] No consistent changes between breeding and non-breeding periods.

seems that it is the initial and middle segments of the epididymis, where sperm maturation mainly occurs, that are especially susceptible to androgen depletion (Cummins, 1977); after treatment with androgen antagonists like diethylstilbestrol there is regression of the epididymal duct and 50–60% of the sperm in the initial and terminal segments are found to be dead or fragmented (Cummins, 1981). Nevertheless, no seasonal change in epididymal weight, corresponding to plasma testosterone concentrations, has been observed in *T. vulpecula* (Gilmore, 1969; Curlewis & Stone, 1985*a*, *b*), *M. eugenii* (Inns, 1982) or *Dasyurus viverrinus* (Fletcher, 1985).

In *T. vulpecula* a marked seasonal fluctuation in prostate size has been described by Gilmore (1969), which correlates closely with the peaks of births (Fig. 4.7*b*). It is probable that this is controlled by testicular androgen but the precise means is not clear. In captive animals a marked relationship between prostate weight and plasma concentrations of testosterone, 5α-dihydrotestosterone (5α-DHT) and androstanediol was reported by Cook *et al.* (1978) and in five males sampled through the year Gemmell (1986) found the concentration of testosterone to vary from a peak of 38.6 ng ml^{-1} at the start of the breeding season in March to a nadir of 0.1 ng ml^{-1} at the end in September. Using pneumopertinography McFarlane, Carrick & Brown (1986) showed that the prostate volume of individual males increased from 7–20 ml in the non-breeding season to 40–60 ml in the breeding season, thus confirming Gilmore's (1969) conclusions from post-mortem samples. However, Curlewis & Stone (1985*b*) were unable to find any correlation between testosterone concentration and prostate weight but did find a significant seasonal increase in accumulation of 5α-DHT in the prostate, although its source remains uncertain (Curlewis & Stone, 1985*a*). They conclude that 5α-DHT is probably the active androgen in the prostate and that the gland's growth is due to changes in concentration of specific receptors for androgen coupled with the raised concentration in the tissue.

The prostate and Cowper's glands are probably under androgen control in other species of marsupial since, in most seasonal breeders, the maximum enlargement of these glands coincides with peak androgen concentrations. Males of *Macropus eugenii* are not strictly seasonal in themselves and show no change in testis weights (Fig. 4.7*a*) but do show significant increases in the size of the prostate and Cowper's glands and in peripheral testosterone during the main and subsidiary periods of the year when females are in oestrus (Inns, 1982) (Fig. 4.7*a*). In captivity, only males that were run with females during the breeding season showed increased concentrations of LH and testosterone (Fig. 6.17) (Catling &

Sutherland, 1980) and the same effect was observed by Fletcher (1983) in *Dasyuroides byrnei*. Preliminary studies suggest that mature male *T. vulpecula* may respond in this way also (C. A. Horn, unpublished results; Curlewis & Stone, 1985*b*).

In several species of *Antechinus* androgens show a marked seasonal increase from undetectable to peaks of 7 ng ml^{-1} for *A. stuartii* (Fig. 4.8*a*) and 5 ng ml^{-1} in *A. minimus* (Fig. 4.8*b*) and maximum enlargement of the prostate and Cowper's glands in June and July coincides with the peak androgen concentrations. Androgens have been shown to play a major role in the marked stress effects that are associated with reproduction and male mortality in these species (Lee *et al.* 1977; Bradley, McDonald & Lee, 1980; Lee & Cockburn, 1985).

Penis morphology

The structure of the penis of marsupials was first described in *Didelphis virginiana* (Cowper, 1704), and subsequently in *Potorous* (Owen, 1868), *Phascolarctos cinereus* (Young, 1879), *Tarsipes rostratus* (Rotenburg, 1928) and more recently in many others (reviewed by Woolley & Webb, 1977; Woolley, 1982). The penis is withdrawn into the body in an S-shaped curve when not erect. A preputial sac is formed by an invagination of skin surrounding the terminal free portion of the penis (Fig. 4.1). The opening of the preputial sac is directed to the posterior.

The glans penis of marsupials is variously shaped; in the majority of species it is bifurcate while in the Macropodidae it tapers to a single point. Intermediate forms are found in *Phascolarctos* and *Vombatus* (Biggers, 1966). In *Tarsipes rostratus* there does not appear to be a distinct glans penis, but the two corpora cavernosa and the corpus spongiosum continue side by side to the apex (Rotenberg, 1928). In *Macrotis lagotis* the urethra as well as the glans penis is cleft (Owen, 1868). The glans penis of all South American marsupials, *Philander opossum*, *Didelphis marsupialis*, *Marmosa sp.* and *Caluromys sp.*, is cleft (Biggers, 1966) and, as in other species with a bifid penis, the urethra continues as a groove along the inner aspect of each half of the split and may terminate at the tips (*Marmosa*, *Caluromys*) or at some distance from the tips (*Didelphis*, *Philander*). The urethral grooves in most species form functional ducts along the inner surfaces of the split glans penis, and it was originally suggested that these served to direct the spermatozoa into each lateral vaginal canal (Cowper, 1704; Biggers, 1966). However, since macropodids have a single, uncleft penis and yet the sperm still travel up the two lateral vaginae (Tyndale-Biscoe and Rodger, 1978), this explanation does not seem to be sufficient.

Several species of the Dasyuridae have an accessory erectile body on the penis derived from the corpora cavernosa (Woolley & Webb, 1977; Woolley, 1982) (Fig. 4.9). The appendage is found only in three genera, *Dasyurus*, *Antechinus* and *Myoictis*. Amongst the species of *Antechinus*, Woolley (1982) has shown that two major groups can be recognised on the basis of the form of the penis tip, the length of the urethral grooves and the presence of a median lobe which projects from the dorsal surface of the penis at the base of the cleft. Group 1 species have long urethral grooves and a bifid penis, and a dorsal lobe. Group 2 species have short urethral grooves, a single tip, and have no dorsal lobe. Although Woolley has used these criteria to recognise four or possibly five genera amongst the 12 species, the functional significance of the structural diversity for reproduction remains unclear.

Fig. 4.9. External anatomy of the penis of representative species of *Antechinus* found in Australia. lm = levator muscle; tlm = tendon of levator muscles; rm = insertion of rectractor muscle; p = level of attachment of preputial skin; dl = dorsal lobe; ug = urethral groove; acc = accessory structure. Redrawn from Woolley (1982).

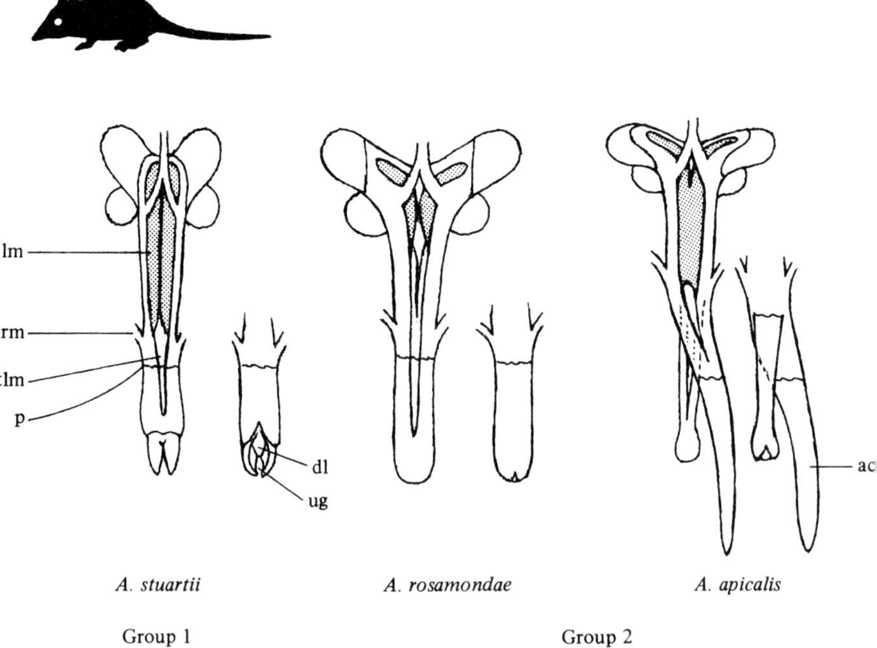

Accessory gland structure

The prostate and Cowper's glands are the only accessory sex glands in the reproductive tract of male marsupials, and the prostate is proportionately larger than that of eutherians (Setchell, 1977). Whilst the prostate is a lobed structure in most eutherians, it is a single carrot- or heart-shaped structure in most marsupials (Rodger & Hughes, 1973). The carrot-shaped prostate occurs in the Macropodidae, Dasyuridae, Thylacinidae, Tarsipedidae, Petauridae, Phalangeridae, Notoryctidae and in *Lasiorhinus latifrons* and the heart-shaped prostate occurs in the Peramelidae and Phascolarctidae (Figs 4.1 and 4.10a). The prostate gland of *Caenolestes obscurus* is a relatively large organ, similar in appearance to the bulbous heart-shaped peramelid prostate but segmented along its length as seen in American didelphids and other Australian marsupials (Rodger, 1982; Rodger & Hughes, 1973). The prostate gland is divided into distinct segments and all of these are different in their microanatomy and in their secretory products (Rodger & White, 1980) (Fig. 4.11). Since marsupials lack other accessory glands, such as seminal vesicles and ampullary glands, this segmentation probably arose as a result of specialisation of regions of the prostate to produce different constituents of the seminal plasma (Rodger & Hughes, 1973).

The marsupial prostate is entirely disseminate, with all the glandular tissue lying between the urethral lumen and the outer urethral muscle. In *T. vulpecula*, the prostate during the breeding season becomes the largest organ in the body cavity except the liver (Bolliger, 1946; Gilmore, 1969). As mentioned in the last section, this hypertrophy is controlled by androgens, and Cook *et al.* (1978) showed that the arterial supply to the prostate arises from a dense vascular network surrounding the prostatic urethra. They suggested that there may be direct transfer of androgens present in the fluid contents of the urethra to the prostatic parenchyma via this vascular system.

In the Macropodidae, Phalangeridae, Dasyuridae and Didelphidae, the prostate gland is divided into regions along its length (Fig. 4.10*a*). In the Macropodidae, and in *Trichosurus*, and *Didelphis*, there are three segments, designated either as Anterior, Central or Posterior (Chase, 1939; Hruban *et al.*, 1965; Rodger & Hughes, 1973). In *Pseudocheirus peregrinus* and the Dasyuridae, segmentation also occurs along the length of glandular tubules in one region of the prostate (Rodger & Hughes, 1973). In the Peramelidae the prostatic segmentation is quite different and the prostate is made up

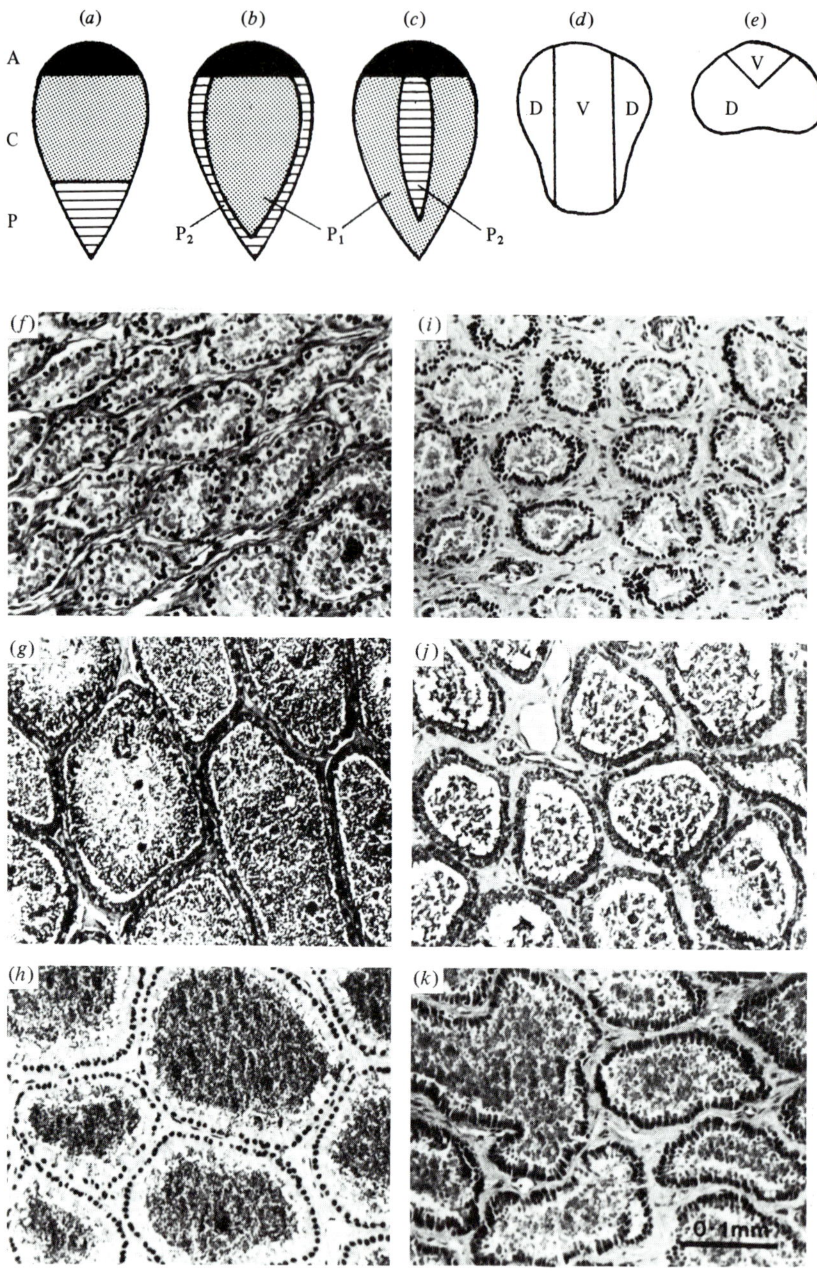

Fig. 4.10. Diagram of the segmentation of the prostate in (*a*) macropodids and *Trichosurus vulpecula;* (*b*) dasyurids; (*c*) *Pseudocheirus peregrinus;* (*d*) peramelids, ventral view; (*e*) peramelids,

of two parts. The dorsal segment forms the bulk of the gland and into it is set a wedge of tissue, the ventral segment (Rodger & Hughes, 1973).

The numerous single branched tubular glands, supported by varying amounts of connective tissue and lined by a single layer of columnar cells make up the prostate and the volume of intertubular space is a characteristic of different gland segments (Fig. 4.10*b*). The height of the columnar epithelium differs between segments, species and season of the year. The tubules end blindly beneath the urethral muscle or are coiled at their distal extremity, but do not form acini or other enlargements. In cross-section they appear as radially arranged straight tubes, running towards the urethral lumen to which they are joined often by non-secretory collecting ducts (Chase, 1939; Rodger & Hughes, 1973). The position of these various segments, and the histochemical characteristics of their secretions, suggest that there is a homology of the anterior prostate segment in *Trichosurus* and *Macropus* species (Carrodus & Bolliger, 1939), whilst the peramelid prostate is quite distinct (Rodger & Hughes, 1973).

The number of Cowper's glands varies from one to three pairs in different species of marsupial (for detailed references see Rodger & Hughes, 1973; Brooks *et al.*, 1978). They are bulbous structures joined to the urethra by ducts (Figs 1.1, 4.1 and 4.2), which are surrounded by smooth muscle. Each gland is surround by striated muscle and is of a branched, tubular type, with the tubules greatly expanded with wide lumina lined by columnar, occasionally pseudostratified secretory epithelium and supported by connective tissue. The tubules of Cowper's glands are of much greater diameter than those of the prostate and appear to produce a mucus secretion.

Prostatic secretions

Prostatic secretions of eutherian mammals contribute carbohydrates to the seminal plasma, which are important for the metabolism of the spermatozoa (Mann, 1964). They are characterised by high concentra-

Fig. 4.10. *cont.*
transverse section. (*a–c*) In frontal section, with shading to indicate the suggested homologies. Segments: A, anterior; C, central; D, dorsal; P, posterior; P_1, posterior 1; P_2, posterior 2; V, ventral. Redrawn from Rodger & Hughes (1973). Changes in the histology of the prostate gland of *Macropus eugenii* (type a) in the breeding season (*f–h*) and the non-breeding season (*i–k*); (*f*) and (*i*) Anterior segment; (*g*) and (*j*) central segment; (*h*) and (*k*) posterior segment. From Inns (1982), with permission.

tions of fructose, and also include citric acid, ergothioneine and inositol. By contrast, marsupial prostatic secretions, apart from those of *Lasiorhinus latifrons*, lack fructose (Fig. 4.11), and the Macropodidae, Phalangeridae and Peramelidae have instead *N*-acetylglucosamine as the major free sugar of the prostate gland tissue and seminal plasma (Rodger & White, 1976*a*). The central segment of the prostates of *T. vulpecula* and *Macropus* is the source of the majority of seminal *N*-acetylglucosamine, while in the Peramelidae both glandular regions contain appreciable amounts. In the Dasyuridae and Didelphidae there is neither fructose nor *N*-acetylglucosamine, and glucose is only a minor constituent, whereas glycogen concentrations are generally high (Martan & Allen, 1965; Rodger & White,

Fig. 4.11. The prostatic sugars of 10 species representing six families of marsupial. The figures given indicate the level in the glandular segment with the highest concentration of each particular sugar in milligrams of sugar per gram of wet tissue: Ng, *N*-acetylglucosamine; Gu, glucose; Fr, fructose; G, glycogen. Data from Rodger & White (1980).

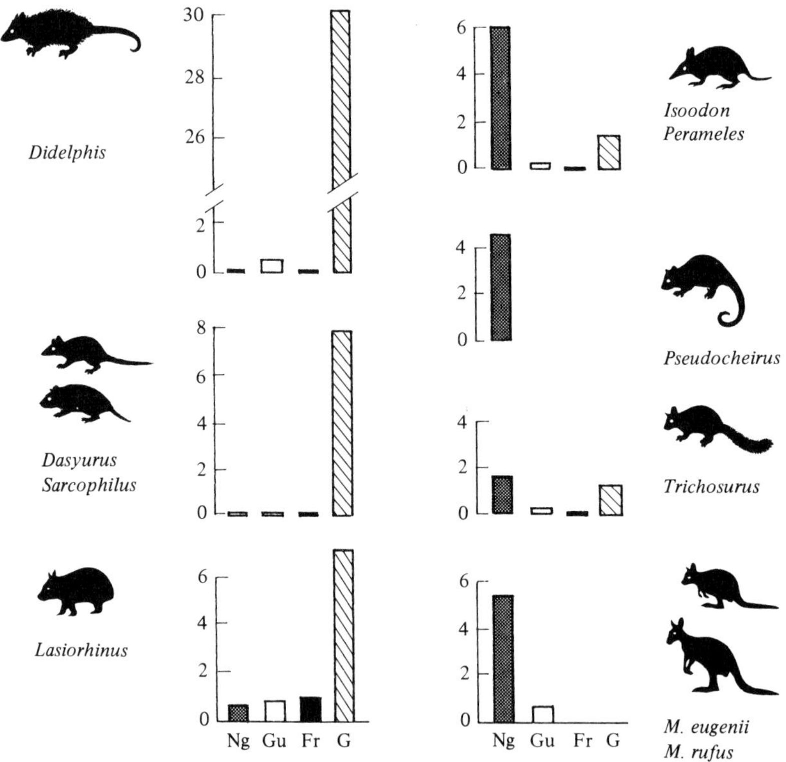

1980), especially in the posterior segment in *Didelphis*. In *Lasiorhinus*, the anterior, central and posterior segments contain sorbitol, fructose and glycogen respectively; the central and posterior also contain some glucose and *N*-acetylhexosamine (Brooks *et al.*, 1978). There appears therefore to be at least three prostatic carbohydrate patterns in marsupials (Rodger & White, 1980) (Fig. 4.11). Hypophysectomy reduces the concentration of *N*-acetylglucosamine in *M. eugenii*, but does not affect glucose or anthrone-reactive carbohydrate levels in any segment (Rodger & White, 1976*a*).

The Macropodidae are the only marsupials that form a true copulatory plug, and the semen coagulates shortly after ejaculation independently of the female tract or its secretions (Rodger & White, 1975; Rodger, 1978). Since transport of sperm up the lateral vaginae is quite rapid, semen coagulation may be restricted to the Macropodidae because of the retention in this group of a patent median vagina (Rodger, 1978).

Spermatozoa

Sperm morphology

Sperm structure and function has been better studied than most other aspects of the male marsupial. Detailed studies have been made on spermiogenesis in *Perameles nasuta* (Cleland & Rothschild, 1959; Sapsford & Rae, 1969; Sapsford, Rae & Cleland, 1967, 1969*a*, *b*, 1970) and descriptions of various aspects of spermiogenesis or mature sperm structure on *Trichosurus vulpecula* (Olson, 1975; Harding, Carrick & Shorey, 1975, 1976*a*, b), *Didelphis virginiana* (Holstein, 1965; Rattner, 1972; Olson *et al.* 1977), *Caluromys philander* (Phillips, 1970), and *Marmosa mitis* (Rattner, 1972). In addition aspects of epididymal sperm maturation have been examined in *Trichosurus vulpecula* (Harding *et al.*, 1975, 1976*a*, *b;* Cummins, 1976, 1977; Temple-Smith & Bedford, 1976), in *Tarsipes rostratus* (Harding *et al.*, 1981; Cummins *et al.*, 1986); in 13 species of dasyurid (Harding *et al.*, 1982*b*); in *Caluromys philander* (Olson & Hamilton, 1976) and in *Didelphis virginiana* (Olson *et al.*, 1977; Olson, 1980; Temple-Smith & Bedford, 1980).

Marsupial sperm display a number of unusual features when compared with those of eutherians. However, the only reliable distinguishing characteristic of marsupial sperm appears to be the location and form of the acrosome (Harding *et al.*, 1979). In monotreme and eutherian sperm the acrosome forms a cap which surrounds the proximal region of the nucleus and projects beyond its rostral tip, whereas in all marsupials so far studied the acrosome covers only one surface of the flattened nucleus (Fig. 4.12*a*).

The nucleus is flattened perpendicular to the long axis of the flagellum

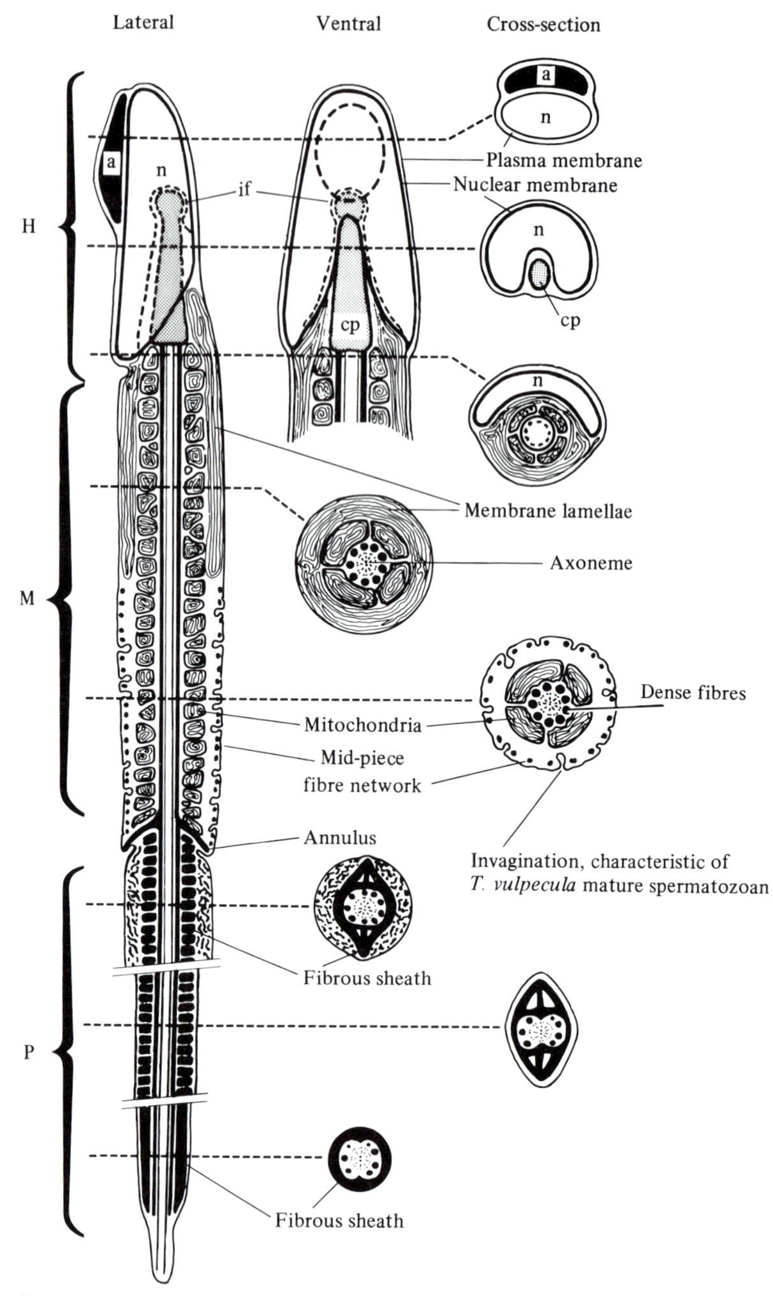

Fig. 4.12. (*a*) Generalized diagram of the spermatozoon of *Trichosurus vulpecula* in lateral, ventral (head, neck and anterior midpiece only) and cross-sectional views to show the distinctive structural features of

during spermiogenesis, but this is not a diagnostic feature of marsupial sperm because *Phascolarctos cinereus* and the Vombatidae have a different morphology (Hughes, 1965; Harding, 1977; Harding *et al.*, 1979).

Spermatozoon morphology within the Macropodidae, Dasyuridae, Phascolarctidae and Peramelidae is relatively homogeneous (Fig. 4.12) but, within the Phalangeridae, gross morphology is more diverse (Hughes,

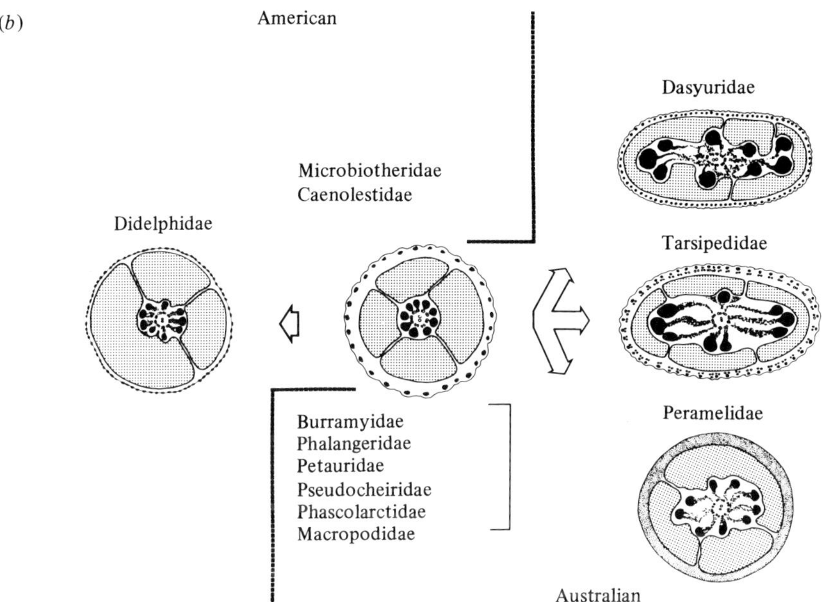

bilateral symmetry, nuclear shape, and neck insertion. H, head; M, midpiece; P, principal piece; T, tail; a, acrosome; if, implantation fossa; n, nucleus; cp, connecting piece of the neck. From Temple-Smith & Grant (1986), with permission. (*b*) Comparison of the morphology and distribution of midpiece organelles shown in transverse sections through the distal midpiece of spermatozoa from representatives of various families of American and Australian marsupials. This evolutionary scheme suggests that simple radial symmetry, circular midpiece section and rounded dense fibres immediately adjacent to the outer doublet microtubules of the axoneme, as seen in the Microbiotheridae, Caenolestidae, the Phalangeroidea (except the Tarsipedidae) and the Macropodidae, represent the primitive ancestral condition in marsupials. The minor radial displacement and structural asymmetry of dense fibres in the Didelphidae, and the more extreme displacement of dense fibres and asymmetry of tail structures in the Peramelidae, Dasyuridae and Tarsipedidae probably represent derived structural specialisations superimposed on the symmetrical ancestral form. From Temple-Smith & Grant (1986), with permission.

1965; Harding *et al.*, 1982*b*). The head of all the Australian species is cone-shaped, but open on one side toward the base, and the mid-piece is inserted into this cleft (Hughes, 1965) (Fig. 4.12*a*), so that the nucleus overlies the proximal end to varying degrees. All have thick tails. In the Dasyuridae the head is long and slender, but in the Macropodidae and Phalangeridae it is short. The peramelid sperm has a broad head for most of its length (Fig. 4.13) and in the Vombatidae its long tip is curved back towards the insertion of the mid-piece. *Tarsipes*, one of the smallest marsupials has the longest spermatozoon (360 μm) described for any mammal (Fig. 4.13). The Dasyuridae also have large spermatozoa (Harding *et al.*, 1982*b*) for example *Dasyurus viverrinus*, in which the length is 232 μm (Fletcher, 1985).

Amongst the American marsupials, groupings of morphological types also occur (Fig. 4.12*b*). Biggers & DeLamater (1965) recognise three

Fig. 4.13. Whole spermatozoon of (*a*) *Isoodon macrourus*, (*b*) *Tarsipes rostratus* 360 μm long and (*c*) head of latter. Courtesy of Dr J. M. Cummins.

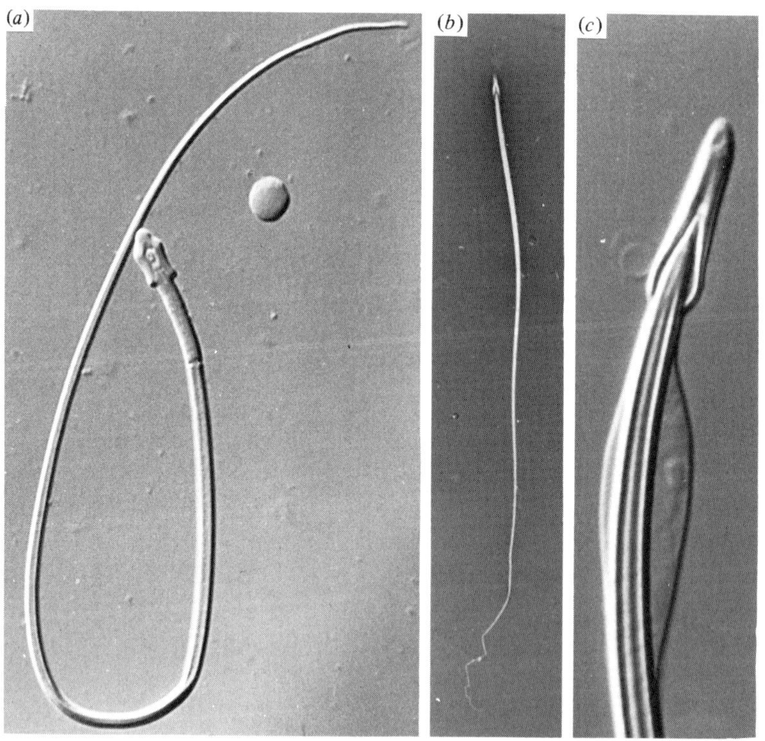

Fig. 4.14. Diagrammatic representation of the morphology and orientation of sperm pairs in the three extant families of American marsupials in lateral and abacrosomal view. Shaded portions represent the acrosomes and the dotted outline on each abacrosomal diagram corresponds to the peripheral limits of the acrosome. From Temple-Smith & Grant (1986).

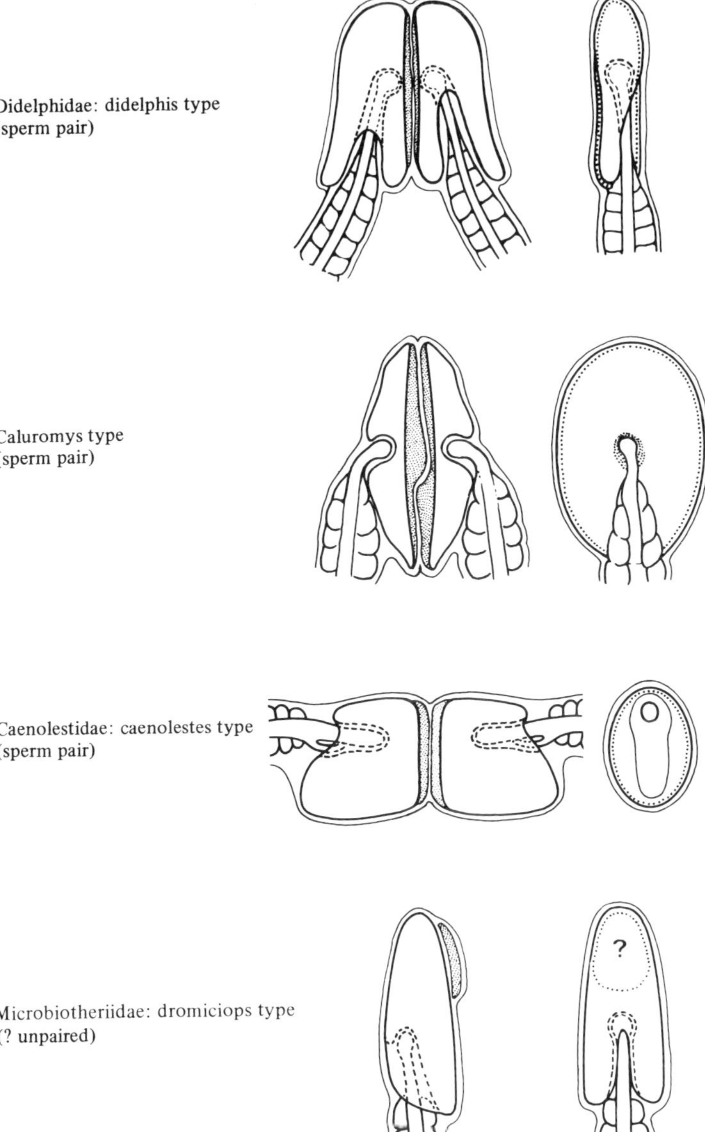

distinct morphological types (Fig. 4.14), the 'didelphis' type, which *Monodelphis*, *Philander*, *Metachirus*, *Chironectes* and *Marmosa* share, the 'caluromys' and the 'caenolestes' type. The didelphis type is characterised by a flattened, hook-shaped head with one thickened limb, and one longer and tapered (Biggers, 1966). The acrosome lies on the anterior part of the thicker limb. The mid-piece is attached by a fine filament to the base of the cleft separating the two limbs of the head. The caluromys type has a saucer-shaped head, with the acrosome lying in the centre of the concave side, and the insertion of the mid-piece into the centre of the convex side. *Caenolestes* sperm heads have a cleft on one side into which the mid-piece is inserted. *Dromiciops australis* may represent a fourth type in which conjugation does not occur, but more evidence of this is required. Marsupial spermatozoa appear to be especially fragile and the nucleus will decondense spontaneously if unfixed suspensions are smeared and air dried, unlike eutherian spermatozoa which are highly resistant to breakdown (Cummins, 1980).

Spermatogenesis

The seminiferous tubule epithelium of marsupials consists of the germinal cell layers and supporting Sertoli cells as in the Eutheria (Figs 4.15 and 4.17). The diploid spermatogonia undergo a series of mitotic divisions, eventually producing the primary spermatocytes which enter the long meiotic prophase. Meiotic divisions of the spermatocytes produces the haploid spermatids, which eventually become transformed from the round, undifferentiated cell to the complex spermatozoon, which is liberated into the tubule lumen (Setchell, 1977).

The Sertoli cells extend from the wall of the tubule to the lumen, and spermatogonia are found between the wall and the Sertoli cells. The spermatocytes and spermatogonia are either sandwiched between adjacent pairs of Sertoli cells or are completely or partially embedded in individual Sertoli cells (Sapsford *et al.*, 1967, 1969*a*).

Spermatogonia of *M. rufogriseus* appear to have at least four synchronised divisions (Setchell & Carrick, 1973). The meiotic divisions to form the spermatids are very similar to the process as described in the Eutheria, and it is of interest to note that it was in the course of such studies on *Didelphis virginiana* that Painter (1922) demonstrated the existence of the Y chromosome in mammals (Setchell, 1977). Spermiogenesis is even more complex in marsupials than in Eutheria because of the transformation from testicular spermatozoa, which are released from the germinal epithelium with the head set at right angles to the tail, to the mature sperm in which the long axis of the head is parallel to the tail, having rotated about the point of attachment (Fig. 4.16). This process occurs during passage through the epididymis (Hughes, 1965) and will be discussed in more detail below.

From extensive ultrastructural studies on spermiogenesis in *Perameles nasuta* the process has been divided into six stages comprising the early spermatid, nuclear protrusion, nuclear flattening and condensation, nuclear rotation, early post-rotation, and late post-rotation (Sapsford *et al.*, 1967, 1969*a*, 1970; Sapsford & Rae, 1969).

The following is derived from a concise summary of this work given by Setchell (1977). The middle piece and the principal piece of the sperm tail develop from the longitudinal centriole near the Golgi complex which itself forms first the acrosomal vacuole and then the acrosome when it attaches to the nucleus. The condensation of the acrosome occurs comparatively late in marsupials. The development of the tail filaments continues after migration of two centrioles to the pole of the nucleus opposite to the acrosome.

The perinuclear spermatid cytoplasm next moves away from the spermatid nucleus towards the tubular lumen. The caudal sheath or manchette develops and, as it comes to lie immediately beneath the plasma membrane of the spermatid, it probably contributes to the moulding of the shape of the anterior extremity of the spermatid and possibly also the nucleus.

The annulus at the base of the flagellum is prominent in marsupials and it moves down the flagellum to allow the mitochondria to assume a helical arrangement around the mid-piece. The nuclear membrane thickens, the nuclear volume is reduced (condensation) and flattened, at which time the thickening disappears. The nucleus is now flattened in a plane at right angles to the axis of the tail (Fig. 4.16).

The nucleus rotates about the point of attachment to the tail, possibly caused by invasions by Sertoli cell cytoplasm and, once complete, the invasion is less marked. At the beginning of the late post-rotational stage, a second wave of invasion of the spermatid by tongues of Sertoli cell cytoplasm begins, but are less intrusive and occur more distally. The spermatid nucleus then is moved towards the lumen of the seminiferous tubule through its own cytoplasm by a specialised tongue of Sertoli cell cytoplasm which is applied to the acrosome. Surplus spermatid cytoplasm is retained by the invading Sertoli cell branches as the residual body. When the spermatozoa are shed into the lumen, the residual bodies move within the Sertoli cell to the wall of the tubule, become more dense and gradually break up.

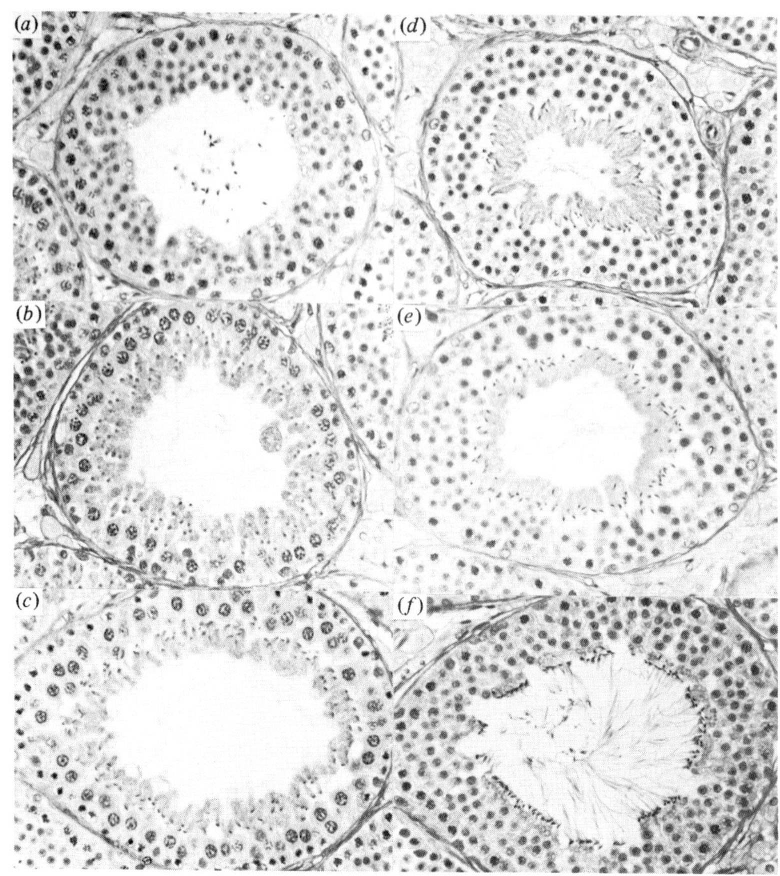

Fig. 4.15. The pre-meiotic and post-meiotic stages of the spermatogenic cycle in *Macropus eugenii*. (*a*) Stage 1: the lumen of the tubule is lined with round spermatids; outside these are pachytene spermatocytes and a new generation of pre-leptotene spermatocytes; these are the last cells to take up [^{3}H]thymidine; all sections contain Sertoli cells and spermatogonia near the tubule wall. (*b*) Stage 2: the spermatid nuclei have condensed and the pachytene spermatocytes outside them have entered diplotene, with large nuclei and prominent chromosomes; the outer spermatocytes still take up [^{3}H]thymidine early in this stage so the transition from pre-leptotene to leptotene must occur during this time. (*c*) Stage 3: the nuclei of the spermatids are now elongated and bifid and their cytoplasm is eccentrically placed; the inner spermatocytes are still diplotene but the outer spermatocytes have entered zygotene. (*d*) Stages 5 and 6: these two stages are very difficult to separate as the distinction depends on the granularity of the spermatid nucleus; the old spermatids are grouped in bundles with their nuclei arranged radially; the new spermatids lie outside them and then the pachytene spermatocytes. (*e*) Stage 7: this

The duration of the spermatogenic cycle is about 16 days in *M. eugenii*, *M. rufogriseus* and *T. vulpecula* (Setchell & Carrick, 1973), towards the upper end of the reported range for eutherians. The times taken for the remaining stages are also similar to those of Eutheria (Table 4.6). Spermiogenesis in the dasyurids follows this general pattern with a few unusual structural features (Harding *et al.*, 1982*b*). An apparently unique pattern of spermatogenesis has recently been described in *Antechinus stuartii* (Kerr & Hedger, 1983) and probably also occurs in the other semelparous dasyurids (see p. 43), such as *A. minimus* and *A. swainsonii* (Wakefield & Warneke, 1963; Wilson & Bourne, 1984), *A. flavipes* (Woolley, 1966*a*; Inns, 1976), *A. bellus* and *Dasyurus hallucatus* (Taylor & Horner, 1970). In other marsupials, as in other mammals, from four to six types of germ cell occur in a single association (reviewed in Setchell, 1977), but in species of *Antechinus* only a simple arrangement of germ cells occurs, with commonly only two or three different types in a single germ cell association (Woolley, 1966*b*, 1975; Taylor & Horner, 1970). This appears to be caused by a spontaneous arrest of spermatogenesis (Kerr & Hedger, 1983), due to a selective arrest of spermatogonial divisions (Fig. 4.17). However, later phases of spermatogenesis continue and the support given by Sertoli cells to these germ cells allows them to complete meiosis and spermiogenesis to yield spermatozoa. This phenomenon has not previously been described in any mammal and, as yet, the controlling factors are unknown. As mentioned previously, plasma androgens are low during spermatogenesis in this species, and no androgen-binding protein could be detected at any time, so Sertoli cell function could only be assessed on morphological grounds but they appeared to be normal throughout (Kerr & Hedger, 1983). These authors therefore suggest that the inability of the testes to maintain spermatogenesis results from intrinsic changes to testicular function exerted at the level of the spermatogonial population.

Other types of testicular changes at the biochemical level occur during spermatogenesis, for example, *Petauroides volans*, which has an annual cycle of spermatogenic activity between December and July (Smith, 1969), also shows a cyclic expression of lactic acid dehydrogenase (LDH)

Fig. 4.15. *cont.*
stage is similar to the previous two but the spermatozoa are evenly arranged around the luminal edge of the epithelium, instead of being grouped in bundles (*f*) Stage 8: in this stage, the nuclei of the spermatozoa are set parallel to the wall of the tubule, not radially as in earlier stages; the other cell types are unchanged. From Setchell (1977), with permission.

Fig. 4.16. Sequential changes in the maturation of the spermatozoon of *Trichosurus vulpecula* during epididymal transport. These figures illustrate the nuclear reorientation, consolidation of the acrosome, transformation of the mitochondrial cristae, modification and loss of the cytoplasmic droplet and the differentiation of the anterior and posterior segments of the mid-piece. (*a*, *b*) Immature spermatozoa from the initial segment or caput epididymis: acrosomal consolidation and nuclear reorientation have begun, the outer layer of cytoplasm around the droplet has disappeared and, concomitantly, a thick perimitochondrial sleeve of cytoplasm has appeared in the posterior mid-piece. There are now fibrous bands beneath the plasma membrane, the mitochondrial cristae have become condensed, a fissure has developed between the droplet and the mitochondiral sheath and membraneous scrolls are visible around the neck. (*c*) Maturing spermatozoon from the middle segment or corpus epididymis. The droplet has been drawn into an eccentric position, and aggregates of membrane now infiltrate the fissure, along the anterior mid-piece. (*d*) Newly mature spermatozoon from the terminal segment or cauda epididymis. The nucleus now lies parallel to the long axis of the tail, the vesicles of excess plasma membrane around the acrosome have been sloughed, the cytoplasmic droplet has disappeared, the anterior segment of the mid-piece is characterised by profuse stacks of membrane, and plasma membrane invaginations have developed between the fibrous bands in the posterior segment. Compare this last stage with the mature spermatozoon in Fig. 4.12*a*. From Temple-Smith & Bedford (1976), with permission.

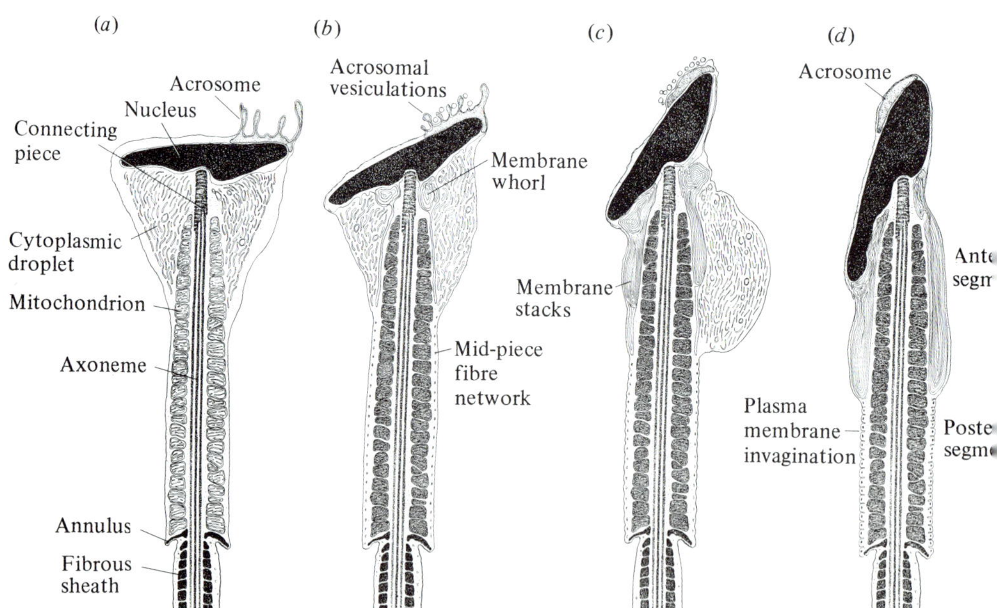

Table 4.6. *The duration (in days) of various stages of spermatogenesis in three marsupials compared to the eutherian range*

Species	*Macropus eugenii*	*M. rufogriseus*	*Trichorurus vulpecula*	Eutheria (range)
Spermatogenic cycle	16	17	15	8–16
Meiotic prophase	21	24	21	12–23
Spermiogenesis	25	25	22	13–23
Pre-leptotene to release of spermatozoa	48	51	45	24–48
Pre-leptotene to first labelled spermatozoa in urine	61		56	
Minimum epididymal transit	13		11	7–25

Compiled by Setchell (1977) from Setchell & Carrick (1973) for marsupial species and from Courot *et al.* (1970) and Hamilton (1972) for eutherian comparisons.

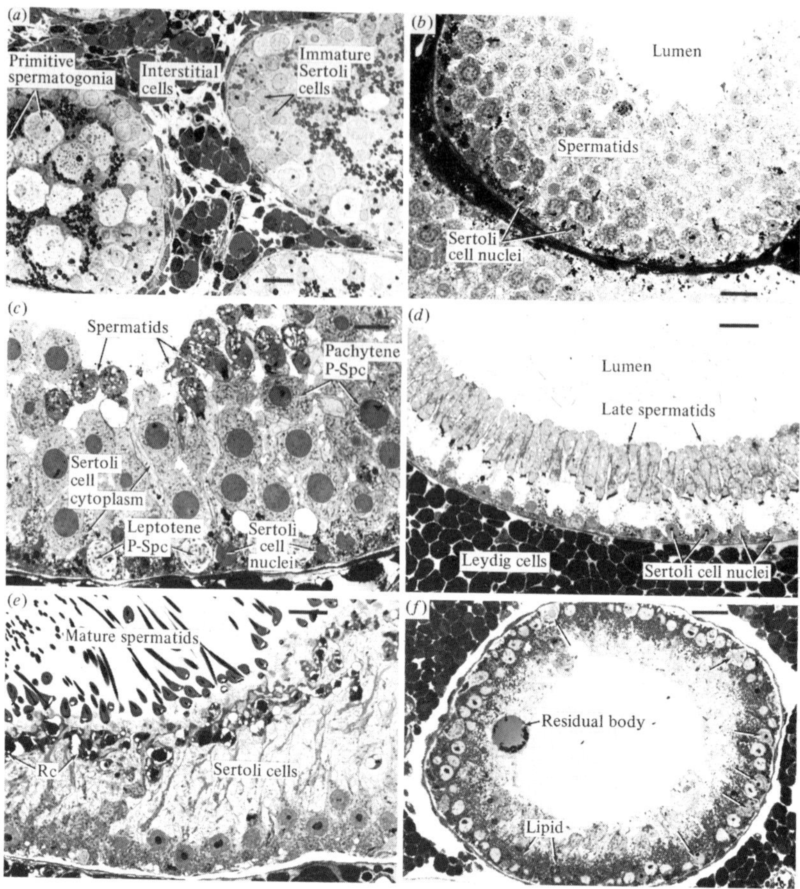

Fig. 4.17. Stages in the spermatogenic cycle of *Antechinus stuartii*. (*a*) Early February: the seminiferous cords contain immature Sertoli cells and primitive spermatogonia. A variety of interstitial cells occupy the intertubular tissue. Bar, 20 μm. (*b*) May: the seminiferous epithelium exhibits maximum depth due to abundant numbers of spermatids. Pachytene primary spermatocytes (arrows) are identified by their elevation away from the basal lamina and presence of sex vesicle in the nucleus. Bar, 10 μm. (*c*) May: the process of spermiogenesis results in the transformation of early round spermatids into elongating spermatids exhibiting condensed nuclei. Columns of Sertoli cell cytoplasm extend from the basal lamina to the lumen of the seminiferous tubules. P. Spc, primary spermatocyte. Bar, 20 μm. (*d*) Early July: a single row of spermatids is seen with their heads associated with Sertoli cell cytoplasmic processes. Other types of germs cells are rarely observed. The intertubular tissue contains many closely packed Leydig cells. Bar, 10 μm. (*e*) July: release of mature spermatids is illustrated leaving behind lobes of residual cytoplasm (Rc) at the

isozyme genes, some of which are testis-specific (Baldwin *et al.*, 1974). Similar testis-specific LDH isozymes have been reported in representatives of all the major families (Holmes, Cooper & Vandeberg, 1973; Renfree & Fox, 1975), but not in the monotremes (Baldwin & Temple-Smith, 1973).

Spermatozoa take 11–13 days to pass through the epididymis (Table 4.6) and, during this time, pronounced morphological changes occur in the acrosome and mid-piece unlike anything seen in eutherian mammals. It is during epididymal transit that the conjugation or pairing of sperm seen in the American marsupials (Fig. 4.14) also occurs (Biggers & Creed, 1962; Biggers & DeLamater, 1965) and both these aspects of sperm maturation will now be considered.

Post-testicular maturation of the spermatozoa

T. vulpecula will serve as an example of the changes in sperm morphology that occur in the epididymis (Fig. 4.16). The sperm is released from the seminiferous epithelium with the nucleus nearly at right angles to the neck and, during maturation, it becomes streamlined with the long axis of the nucleus in line with the flagellum (Cummins, 1976, 1981; Temple-Smith & Bedford, 1976). This streamlining coincides with contraction of the cytoplasmic droplet around the neck, with loss of its internal complexity and is accompanied by pitting and vesiculation of the plasma membrane. The droplet is sloughed directly from the anterior mid-piece (Temple-Smith & Bedford, 1976). At the same time as this is occurring, the acrosome changes from a bowl-shaped structure on the anterior dorsal third of the nucleus into an inconspicuous button-shaped structure; this contraction is accompanied by the budding off of vesicles of plasma membrane (Temple-Smith & Bedford, 1976; Cummins 1981; Harding *et al.*, 1983).

The mid-piece also changes dramatically; in the immature spermatozoon there is a space between the mitochondrial sheath and the plasmalemma which overlies it; this contains a network of membraneous cisternae and is linked anteriorly with similar endoplasmic-reticulum-like structures in

Fig. 4.17. *cont.*
apical extensions of the Sertoli cell cytoplasm. Bar, 20 μm. (*f*) August: the removal of spermatozoa from the seminiferous tubules is accompanied by shrinkage of the tubules, which now contain large residual bodies. Sertoli cells are identified by pale nuclei containing a large single nucleolus. Basally situated spermatogonia are illustrated (arrows) and exhibit a pale cytoplasm. Bar, 10 μm. From Kerr & Hedger (1983) with permission.

the cytoplasmic droplet (Cummins, 1981). During maturation the cisternae are replaced by a spiral fibrous sheath underlying the plasma membrane of the posterior two-thirds of the mid-piece. The fibrous network is helically arranged with the gyres of the sheath running counter to the gyres of the mitochondrial spiral, and are interspersed with caveolae-like invaginations of the plasma membrane. The sheath may be strengthened by the formation of cross-linking between protein-bound thiols (Temple-Smith & Bedford, 1976; Cummins, 1981). In addition to these alterations, modification of the sperm surface over the head and tail occurs during epididymal transit (Temple-Smith & Bedford, 1976) and also an increase in potential for motility – spermatozoa released from the testis and the

Fig. 4.18. Changing motility of spermatozoa in successive segments of the epididymis of *Trichosurus vulpecula*. Mean motility and range shown for five points. Degree of motility was assessed on a scale from 1 (occasional feeble twitching) to 6 (widespread highly active motility). After Cummins (1976).

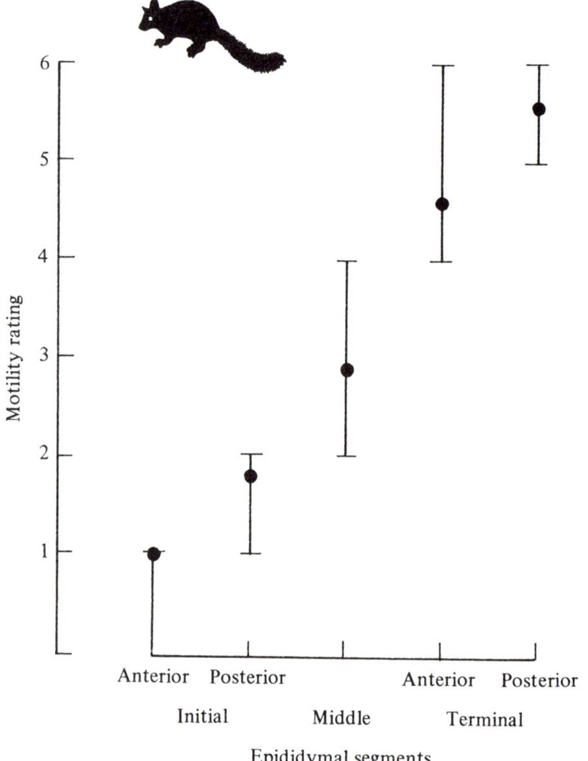

initial segment of the epididymis are immotile or have only uncoordinated movements (Fig. 4.18), whereas many of those released from the intermediate segment and most from the terminal segment have an energetic, vigorous beat with active forward progression (Cummins, 1976; Temple-Smith & Bedford, 1976; Jones *et al.* 1984).

The full range of morphological changes seen in maturation of the sperm of *T. vulpecula* do not occur in all marsupials (Harding *et al.*, 1979). Development of the helical mid-piece fibre network is common, but invagination of the mid-piece plasma membrane and the pronounced acrosomal modifications as seen in *T. vulpecula* occurs only amongst members of the Phalangeridae and Petauridae but not the Macropodidae. In *Didelphis virginiana* epididymal maturation is, for the most part, comparable to *T. vulpecula*, though there is no acrosomal modification (Temple-Smith & Bedford, 1980). In the Dasyuridae, Peramelidae and in *Tarsipes rostratus* the intraflagella structure sets these species apart from other marsupials (Harding *et al.*, 1981; Harding *et al.*, 1982*b*), but care should be taken in implying phylogenetic relationships on the basis of these characters alone. *Phascolarctos cinereus* has some unusual features in sperm and Sertoli cell structure, which include: an uneven condensation

Fig. 4.19. The sequence of events associated with the selective removal and degradation of detached cytoplasmic droplets of the spermatozoa by the specialized phagocytic principal cells in the epididymis of *Trichosurus vulpecula*. From Temple-Smith (1984*a*), with permission.

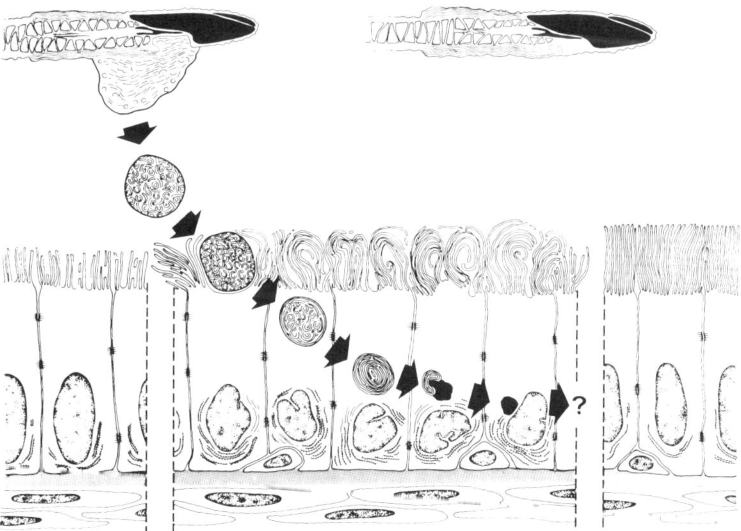

Fig. 4.20. Longitudinal section of the heads of two mature spermatozoa from the cauda epididymis of *Didelphis virginiana* to show the apposition of the acrosomal faces during conjugation. From Temple-Smith & Bedford (1980), with permission.

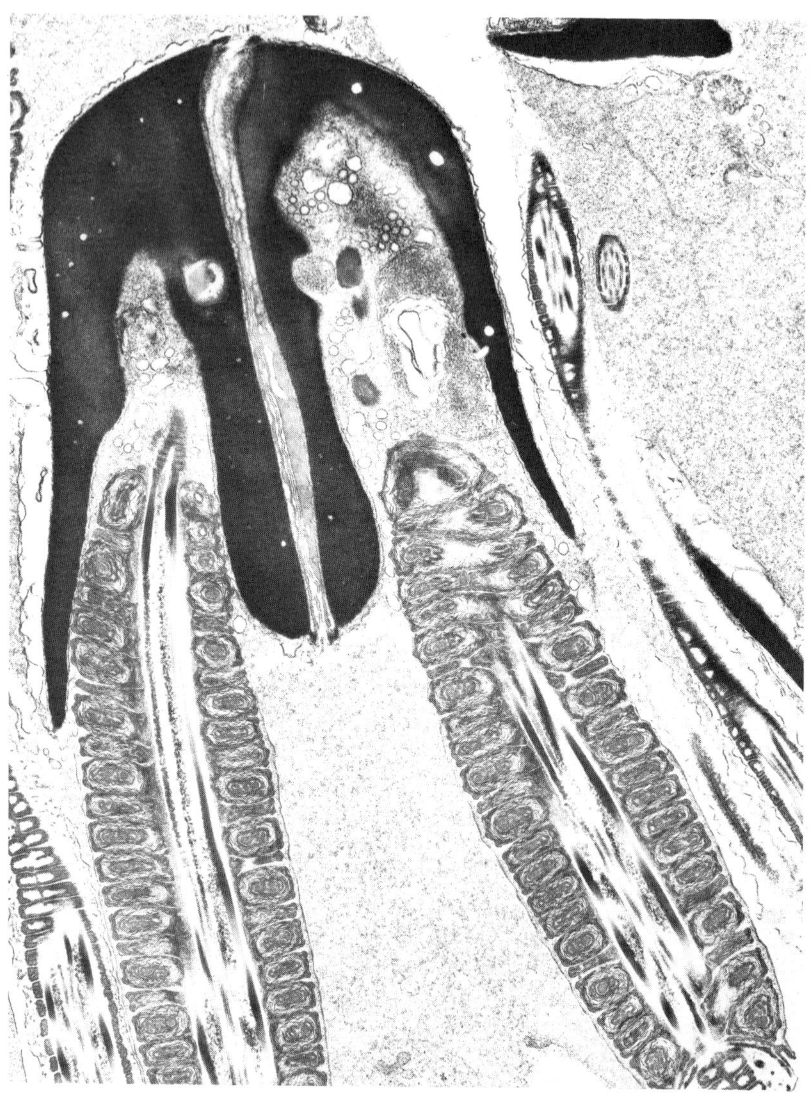

of the chromatin during nuclear development; a different mode of nuclear flattening relative to the flagellum which results in a more 'eutherian-like' neck insertion of the sperm; features of early acrosome development; organisation and internal structure of the mid-piece mitochondria and the presence of crystalloid inclusions in the basal region of the Sertoli cell (Harding *et al.*, 1982*a*, 1983).

In *M. eugenii* sperm maturation occurs in the initial and middle segments of the epididymis, and there is phagocytosis of the cytoplasmic droplet by principal cells of a specialised region of the middle segment, suggesting that the epithelium can distinguish sperm from droplets (Jones *et al.*, 1984). This also occurs in *T. vulpecula* (Temple-Smith, 1984*a*, *b*) (Fig. 4.19). The principal cells in this region have microfolds rather than stereocilia.

Fig. 4.21. Graph showing the percentage (mean and individual estimates) of paired spermatozoa in different segments of the epididymis of *Didelphis virginiana*. Conjugation is first established as spermatozoa pass through segments 4 and 5 (corpus epididymis) and is the predominant condition of spermatozoa in the cauda epididymis and the vas deferens. From Temple-Smith & Bedford (1980).

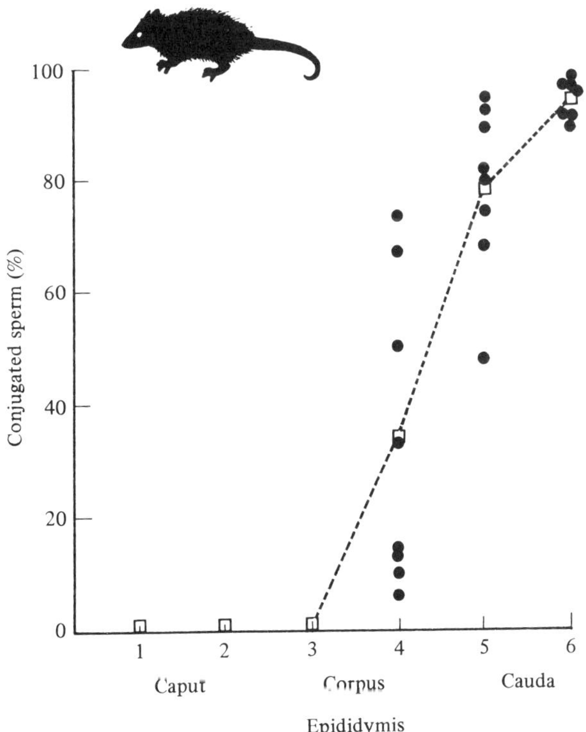

Sperm pairing. The conjugation of the spermatozoa into pairs (Fig. 4.20) occurs in all American marsupials, with the possible exception of *Dromiciops australis*, but is unknown in any other species of vertebrate, let alone mammal. The phenomenon was first reported by Selenka (1887) in the semen recovered from female *Didelphis virginiana* soon after coitus. Von Korph (1902) found that they left the testis unpaired and became coupled in the epididymis by intimate association of the acrosomes (Fig. 4.20) and this has been confirmed by later workers (Biggers, 1966). This pairing has since been described in at least eight genera of American marsupials (Biggers, 1966) and at an ultrastructural level in *Caluromys derbianus* (Phillips, 1970; Olson & Hamilton, 1976) and *D. virginiana* (Krause & Cutts, 1979; Olson, 1980; Temple-Smith & Bedford, 1980). Conjugation is established by a close association of the plasma membrane overlying the acrosomes of apposed spermatozoa, together with peripheral junctions, which effectively seal off the acrosomes from the exterior. In *Didelphis* most sperm pair as they pass through the intermediate segment to the terminal segment of the epididymis (Temple-Smith & Bedford, 1980, Fig. 4.21). Granular material is seen between the apposed acrosomal surfaces in the early phase of pairing.

The biological significance of this pairing is unclear. Sperm pairs separate in the oviduct, and the unpairing is associated with the appearance of membrane vesicles within the matrix of the intact acrosome (Rodger & Bedford, 1982*b*). Phillips (1970) suggested that, as the flagella of the paired spermatozoa beat in a coordinated alternating manner, they may have some mutual influence on each other; Rodger & Bedford (1982*b*) have observed, however, that a single spermatozoon of a pair may be immotile and its partner be normally active. Phillips (1970) further suggests that the junction surrounding the acrosome which probably serves to hold the cells together may also protect the acrosome in some way during passage through the male and female tracts. The acrosome of *D. virginiana* contains at least four hydrolytic enzymes seen in eutherian sperm (acrosin, arylsulphatase, hyaluronidase, and *N*-acetylhexosaminidase) (Rodger & Young, 1981) and, since the zona pellucida surrounding the ovum appears most vulnerable to enzymic digestion (Rodger & Bedford, 1982*b*) the pairing may prevent their untimely release before fertilisation. However, as Rodger & Bedford (1982*b*) point out, if this pairing is so important for normal sperm function and is a consequence of the shape of the acrosome, why has it not evolved in any of the Australian marsupials?

Bedford, Rodger & Breed (1984) view sperm conjugation as one aspect of efficient sperm transport to the vicinity of the egg, the other being the special crypts in which many sperm are housed in the oviducts of didelphid and dasyurid species (Fig. 4.22). They suggest that both aspects increase the survivorship of sperm in the female tract and so allow a much lower level of sperm production to be maintained without compromising fertility. In support of their hypothesis they provide evidence for *Didelphis virginiana* that the number of sperm in the epididymides is very low compared to other species and the male inseminates the female with a relatively low number of sperm (about 13×10^6), compared to rabbits (150×10^6) or cattle and sheep (10^9). However, the proportion of these that reach the site of fertilisation is very much higher in *D. virginiana* (5%) compared to about 0.01% in the rabbit.

To test this hypothesis, Bedford *et al.* (1984) examined several other species of marsupial from Australia, which do not show conjugation, and *Monodelphis domestica* and *Caenolestes obscurus*, which do (Table 4.7). The later two species were similar to *D. virginiana*, whereas representatives of the Macropodidae, Phalangeridae, Vombatidae and Peramelidae conformed more nearly to the eutherian species in having larger numbers of spermatozoa in the epididymides. One of these, *M. eugenii*, also has been

Fig. 4.22. A section of the oviducal isthmus of *Didelphis virginiana* showing detail of a sperm-containing crypt below the main epithelium. The epithelium of the main duct is formed by ciliated cells (large arrows) with large oval adluminal nuclei, and mucoid secreting cells (small arrows) with darkly stained cytoplasm (PAS) and basally located nuclei. The crypts are lined by cuboidal cells, lacking cilia and mucoid. From Rodger & Bedford (1982*a*), with permission.

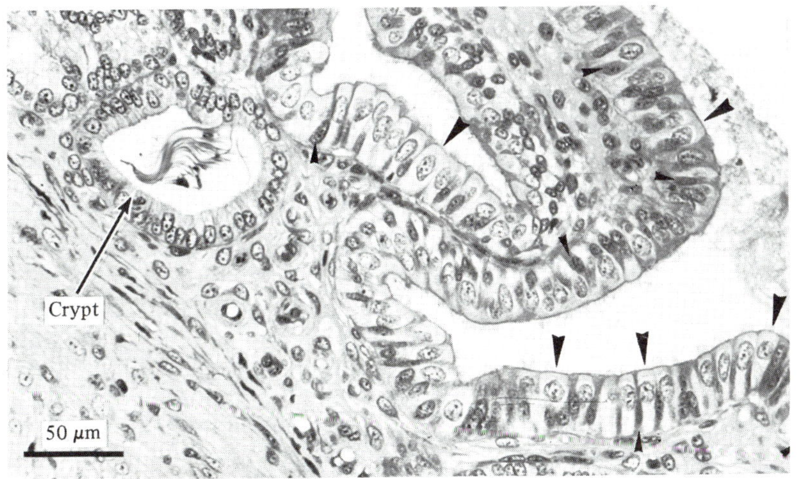

shown to have higher numbers of sperm deposited in the female tract and the survival is similar to that in the rabbit (Tyndale-Biscoe & Rodger, 1978). However, in males of seven species of the Dasyuridae the sperm numbers in the epididymides were low, as in *D. virginiana* and, while none of these species show sperm pairing, the oviducts of the females contain

Table 4.7. *Number of spermatozoa (millions) per epididymis in American marsupials with conjugated sperm and Australian marsupials with separated sperm*

Species[a]	Total sperm per epididymis	Total sperm per gram of testis	Body mass (kg)
American			
Didelphidae			
Didelphis virginiana (4)†	18.6±2.4	12.1±1.1	3.7
Monodelphis domestica (4)†	2.1±0.01	70.0±3.2	0.7
Caenolestidae			
Caenolestes obscurus (3)	0.9±1.1	—	0.04
Australian			
Dasyuridae			
Dasyurus viverrinus (1)	2.0	1.1	1.5
Antechinus flavipes (1)	1.8	5.1	0.07
A. swainsonii (1)	3.6	33	0.15
Dasyuroides byrnei (2)	0.9, 2.5	3.2, 6.8	0.12
Sminthopsis macroura (1)	0.24	3.0	0.03
S. crassicaudata (3)	0.7, 0.3, 0.4	10, 3.8, 6.7	0.02
S. virginiae (1)	0.15	1.2	0.03
Peramelidae			
Isoodon obesulus (1)	48	25	0.79
I. macrourus (2)	70, 97	29, 33	1.9
Phalangeridae			
Trichosurus vulpecula (3)	146, 250, 130	44, 74, 36	1.5
Vombatidae			
Lasiorhinus latifrons (2)	73, 218	10, 27	29.0
Macropodidae			
Macropus fuliginosus (2)	840, 1224	38, 35	35.0
M. rufus (3)	406, 202, 174	18, 11	45.0
M. rufogriseus (1)	800	42	13.0
Wallabia bicolor (1)	354	44	32.0

[a] Numbers in brackets are number of specimens; daggers indicate that values are mean ± standard error of the mean.
Data from Bedford *et al.* (1984).

crypts (Fig. 4.22) in which the sperm are stored for up to 10 days (see p. 265). At present the hypothesis of Bedford *et al.* (1984) seems to be the most promising to account for the phenomenon and further reference will be made to it when we discuss fertilisation in Chapter 7.

Conclusions

We may conclude this brief review by reiterating that, in most respects, the male marsupial resembles the eutherian. The unusual features that distinguish the marsupial and invite further research are related to the spermatozoon itself; the unique manner in which spermatogenesis is stopped in *Antechinus*, the unusual manner of formation and rotation of the head piece, and the transformations of the sperm that take place in the epididymis, especially the conjugation of sperm pairs. The great development of the prostate gland and the different energy substrates that it produces in different species of marsupial and compared to Eutheria are secondary but interesting aspects that merit further investigation.

5

The female urogenital tract and oogenesis

The unusual nature of the anatomy of the female marsupial was first recognised by Tyson (1698), when he described the paired vascular 'cornua uteri' of *Didelphis virginiana*, each leading into a vaginal cul de sac and then to separate lateral vaginae. The paired uteri are invariably separate and open by separate cervices (Fig. 5.1) unlike in many species of Eutheria, which have a single chambered uterus and a single cervix. The lateral vaginae open posteriorly into a urogenital sinus, which also receives the urethra.

Anatomy of the urogenital tract

The oviduct

Study of the oviduct in marsupials has been neglected, possibly because the eggs pass through it so rapidly. Nevertheless, as in other mammals, it is a complex structure with a transport and secretory function. The oviducts vary considerably between species in length and in the degree of convolution but the transport of sperm and eggs must involve muscular contractions. The musculature consists of inner circular and outer longitudinal layers which allow peristaltic movement (Hartman, 1924).

The mucosal lining of the mammalian oviduct consists of an intricately folded, simple ciliated columnar epithelium which is interspersed with mucus-secreting cells (Nalbandov, 1969). A similar arrangement of cells occurs in the few marsupials examined (Hughes, 1974; Arnold & Shorey, 1985; Armati-Gulson & Lowe, 1985; Figs 4.22 and 5.2). Four regions of the mammalian oviduct can be distinguished according to the structure of the epithelium (Nilsson & Reinius, 1969), and the suggested classification of these authors is largely followed here.

The pre-ampulla includes both the fimbria and the infundibulum which pass into the ampulla, where fertilisation occurs in most mammals. The delicate and membraneous fimbria surrounds the ovary and is variable in structure amongst the marsupials observed. The ampulla passes into the end of the convoluted isthmus and this, in turn, connects to the uterus through the junctura containing the uterotubal junction, which prevents fluid backflow from uterus to oviduct in *Macropus eugenii* and other macropodids (Tyndale-Biscoe, 1968; Renfree & Tyndale-Biscoe, 1978). Andersen (1928) observed in *Didelphis virginiana* that the isthmus is tortuous with the loops held firmly in place by connective tissue, and it has a wide lumen that narrows down at the uterotubal junction which is not guarded by any special uterine folds. In *Trichosurus vulpecula* the lower end of the tube is about two-thirds the uterine width, but narrows abruptly

Fig. 5.1. Ventral view of (*a*) the female reproductive tract of *Macropus eugenii* and (*b*) in labelled outline. The right ureter is just visible passing between the median and right lateral vagina; the left ureter was lost in dissection. The tract was dissected from an oestrous, parturient female; the post-partum uterus is on the left side of the animal, ipsilateral to the corpus luteum. A large Graafian follicle is present in the right ovary. b, bladder; c, cervix; CL, corpus luteum; f, fimbrium; Gf, Graafian follicle; lv, lateral vagina; mv, median vagina; ov, ovary; od, oviduct; u, ureter; UGS, urogenital sinus; ut, uterus. From Renfree (1983), with permission.

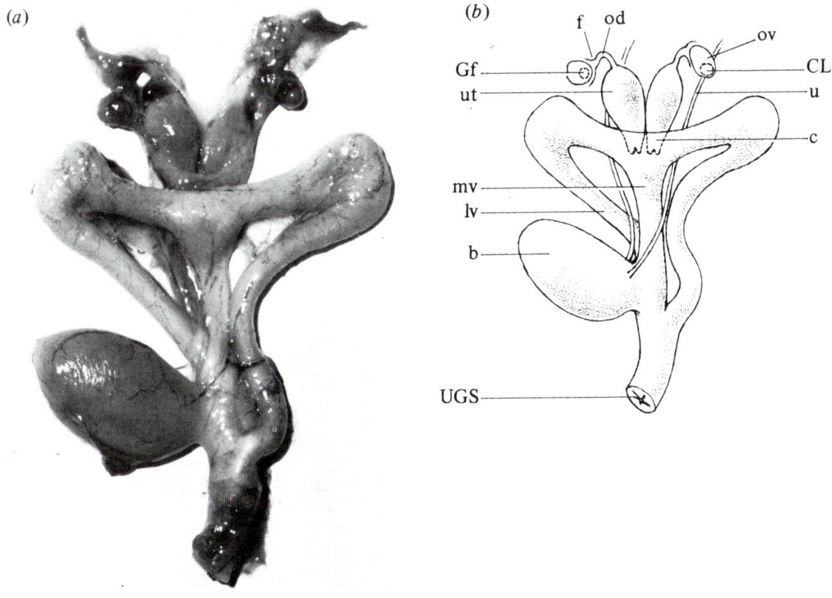

at the uterotubal junction, which is marked by a sharp constriction, and the mucosa at this point forms a circular ridge protruding down into the uterus (Andersen, 1928). Nalbandov (1969) concludes that a uterotubal barrier in the form of a sphincter or folds, arranged in such a way as to prevent the passage of fluid from the uterus, is generally present in polytocous mammals and absent in monotocous species. From the above evidence in a few species this does not appear to hold for marsupials but more species need to be examined to say whether these findings are general.

During pro-oestrus and oestrus the oviducal epithelium secretes mucopolysaccharide which entraps the spermatozoa and becomes deposited outside the zona pellucida of the newly ovulated ovum as the mucoid coat (Hughes, 1974). The shell membrane is largely produced in the uterus (see Chapter 7) but Hartman (1916) and Hughes (1974) suggest that some deposition of shell material may occur at the uterotubal junction or, perhaps more strictly, at the junctura. In *Didelphis virginiana* the last few millimeters of the oviduct have numerous mucosal glands and Andersen (1928) supports Hartman's (1916) conclusion that these secrete the first layers of the shell membrane, as they do in monotremes (Hill, 1933). This is a matter that will need to be resolved with the electron microscope.

Fig. 5.2. Scanning electron micrograph of the lining of the oviduct of a lactating *Pseudocheirus peregrinus*. c. cilia; mv, microvilli. From Armati-Gulson & Lowe (1985), with permission.

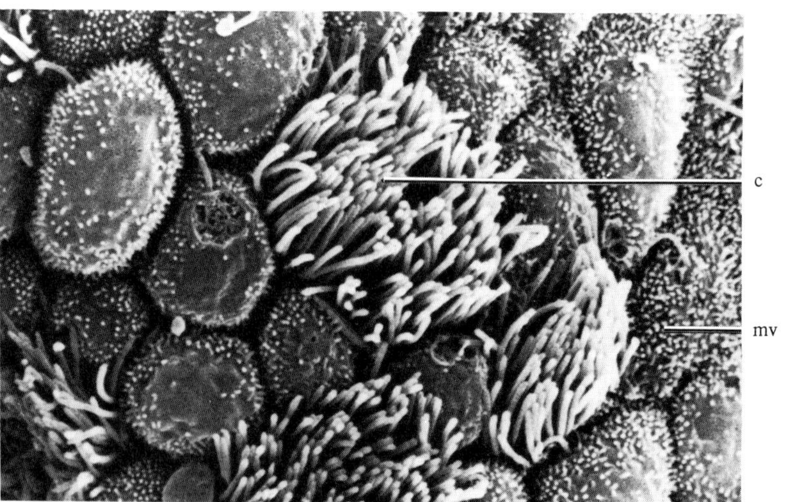

The uterus

The paired but separate uteri, sometimes referred to erroneously as uterine 'horns', consist of an inner, glandular endometrium and an outer, muscular myometrium. In monovular species the uteri are well-defined fusiform bodies, with their posterior extremities lying close together (Figs 5.1 and 5.3*a*, *b*) on each side of the median line and loosely joined by connective tissue (Pearson, 1945). In polyovular species each uterus is usually deflected laterally at the anterior ends into a comma-shaped structure which increases the size of the uteri (Fig. 5.3*d*). The endometrium consists of a stroma and convoluted glandular extensions of the central uterine lumen, which are capable of producing a copious secretion (see below). Capillaries are numerous in the connective tissue. The parallel uterine necks, up to 1 cm or so in length, open into the vaginal culs de sac by two separate cervices.

The myometrium consists of an outer longitudinal and inner circular layer of muscle, with some oblique muscle fibres (Baxter, 1935). In *Didelphis virginiana* the myometrium is a polarised structure with the myocytes arranged in a similar way to that which occurs in the human deferent duct (Lierse, 1965). There is a primary muscle system which is limited to the uterine wall, distinguishable from a branching secondary muscular system. The secondary muscle system is not spread over the body of the uterus as in the human. The muscle fibres run from the outer to the inner region of the uterus in a tube-like structure and in a circular direction (Fig. 5.4). This arrangement is not seen in the uteri of the Eutheria. Lierse (1965) concludes that this relatively simple arrangement reflects the short gestation and 'minimal' growth of the embryo that occurs in the uterus of *D. virginiana* before birth and so does not require a gross distension of the uterine wall. This point has not been further investigated, but the myometrial activity through pregnancy and parturition has been (Shaw, 1983*a*, *b*) and is discussed in Chapter 7 (p. 334).

The vaginal complex and birth canal

The vaginal canals are the most variable part of the anatomy of the female tract (Figs. 5.3 and 5.5) and much of the early work concentrated on their gross anatomy. This was reviewed first by Lister & Fletcher (1881) and Fletcher (1881, 1883), and subsequently by Nelsen & Maxwell (1942), Pearson (1945, 1946), Eckstein & Zuckerman (1956), and Barbour (1977). The vaginal apparatus consists of two lateral vaginae, each connecting the uterus of the same side to the urogenital sinus (Fig. 5.6) The anterior end

Fig. 5.3. For legend see opposite.

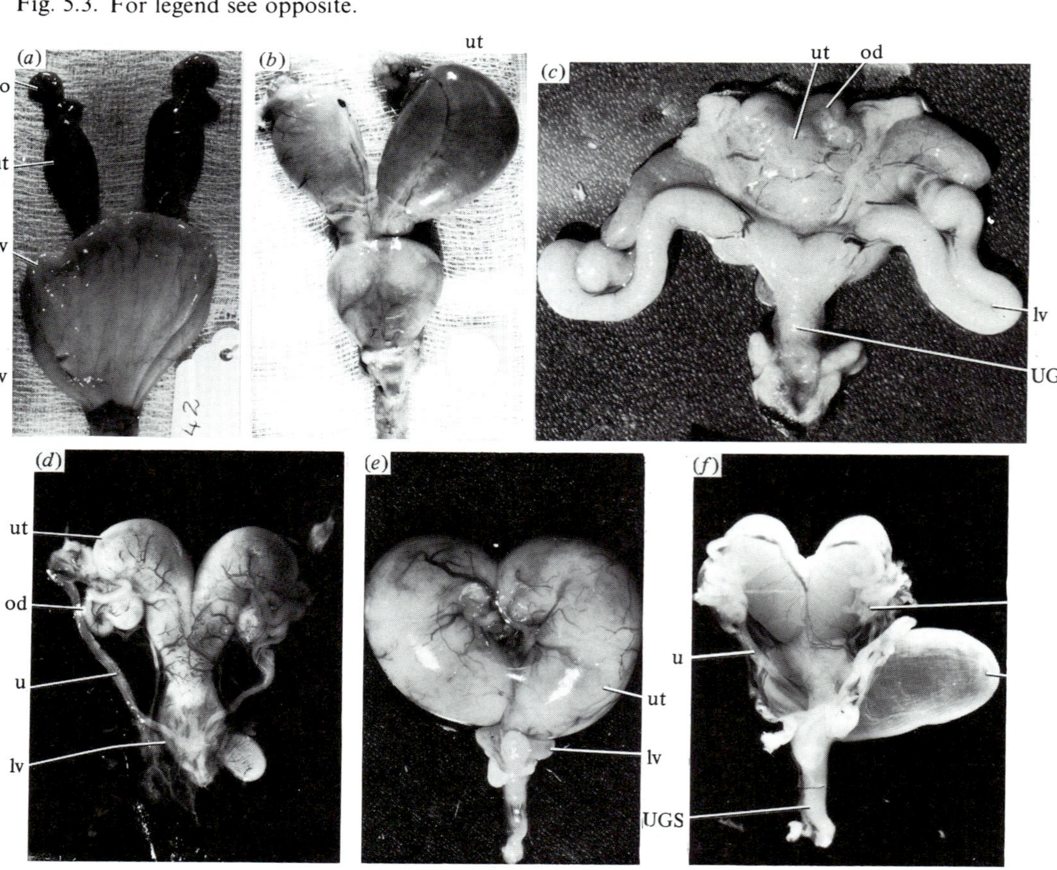

of each lateral vagina, separated only by a common median septum, form the vaginal culs de sac, into which the uterine cervices open. In the Macropodidae the septum also extends anteriorly between the cervices so as to form a pair of small anterior vaginal culs de sac (Fig. 5.6) (Tyndale-Biscoe, 1965, 1968).

Between the posterior margins of the two vaginal culs de sac and the anterior end of the urogenital sinus is a mass of fibrous connective tissue, the urogenital strand, which supports the posterior part of the lateral vaginae, the two ureters and the urethra (Figs 5.5 and 5.6). It is in this urogenital strand that a direct connection known as the pseudo-vaginal canal or birth canal is formed at parturition. Its discovery by Home (1795) and subsequent confirmation by Lister & Fletcher (1881) was described in Chapter 1. It is now well established that in most species of the Macropodinae, and in *Tarsipes rostratus* (Fig. 5.5*b*, *c*) the canal remains open after the first parturition and becomes lined with cuboidal epithelium like that of the culs de sac, merging posteriorly with the stratified squamous epithelium of the urogenital sinus. In the majority of marsupials, however, the canal is not permanent and the pseudo-vaginal canal forms in the connective tissue strand between the ureters prior to each parturition (Fig. 5.5*a*, *d*–*g*). The connective tissue cleft closes rapidly after passage of the young and the epithelia of the urogenital sinus and of the vaginal culs de sac re-form so that in most species no indication of the pseudo-vaginal canal remains (Tyndale-Biscoe, 1966). However, in the Peramelidae, remnants of the umbilical cords are trapped in the urogenital strand where they are recognisable for some time after parturition (Hill, 1899, 1900*a*; Lyne & Hollis, 1982). This pattern of regular opening and closure of the passage associated with parturition occurs in the Didelphidae, Caenolestidae, Dasyuridae, Peramelidae, most species of Phalangeridae, all Potoroinae, and in *Lagostrophus fasciatus* and *Macropus giganteus* among the Macropodinae.

Although Flynn (1923) and Pearson (1946) concluded that the ancestral

Fig. 5.3. Photographs of representative female genital tracts to illustrate the variation between species and between reproductive states. (*a*) *Trichosurus vulpecula* in oestrus. Ventral view. Enlarged median vagina. (*b*) *T. vulpecula* in late pregnancy. Ventral view. Enlarged left uterus. (*c*) *Caenolestes obscurus* in oestrus. Dorsal view Convoluted lateral vaginae. (*d*) *Tarsipes rostratus* in post-partum oestrus. Dorsal view. (*e*) *Tarsipes* in late pregnancy. Ventral view. Both uteri enlarged. (*f*) *Acrobates pygmaeus* in lactational quiescence. Ventral view. Abbreviations as in Fig. 5.1.

route for birth was via the lateral vagina, as evidenced by a single entrapped fetus in a specimen of *Potorous tridactylus*, evidence from Shaw & Rose (1979) for this species supports the conclusions of Hill & Fraser (1925) and Tyndale-Biscoe (1968) that 'the median mode of birth will ultimately be found to hold good for the whole of the Marsupialia'.

The septum separating the right and left vaginal culs de sac is present throughout life and the culs de sac small and just covering the cervices which project through them, in *Antechinus* (Fig. 5.5*e*) and *Marmosa* (Pearson & de Bavay, 1951; Hill & Fraser, 1925). These marsupials, and others that give birth to the smallest young (Dasyuridae and Didelphidae) have the smallest vaginal culs de sac, the most complete septa, and the greatest distance between the cervices and the urogenital sinus (Fig. 5.5). The exception here is *Tarsipes rostratus*, which gives birth to the smallest

Fig. 5.4. Diagram of the muscle fibre tracts in the uterus of *Didelphis virginiana*. The muscle fibres run from the exterior to the interior of the uterus in tube-like structures in a circular direction. The angle of ascent is about 45°. Redrawn from Lierse (1965).

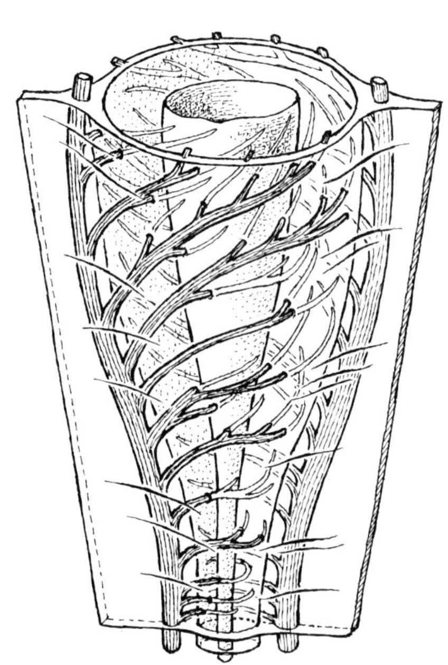

Fig. 5.5. Drawing of newborn young and diagrammatic representations of the vaginal apparatus in parous females of various marsupials. Newborn young drawn to scale but the illustrations of the vaginal apparatus are not to scale. X – positions of openings of uteri into vagina; ● – positions of ureters; ○ – position of opening of urethra into urogenital sinus. Median vaginal septum and its remnants shown as in post-parturient females; degree of development of the median vaginal canal indicated by amount of distance between openings of uteri and position of urethral opening it occupies; vertical hatching indicates extent of pseudovaginal canal which opens at each parturition. *a*, *Perameles nasuta; b*, *Tarsipes rostratus; c*, *Macropus rufus; d*, *Dasyurus viverrinus; e*, *Antechinus stuartii; f*, *Trichosurus vulpecula; g*, *Pseudocheirus peregrinus*. Redrawn from Sharman (1965*c*).

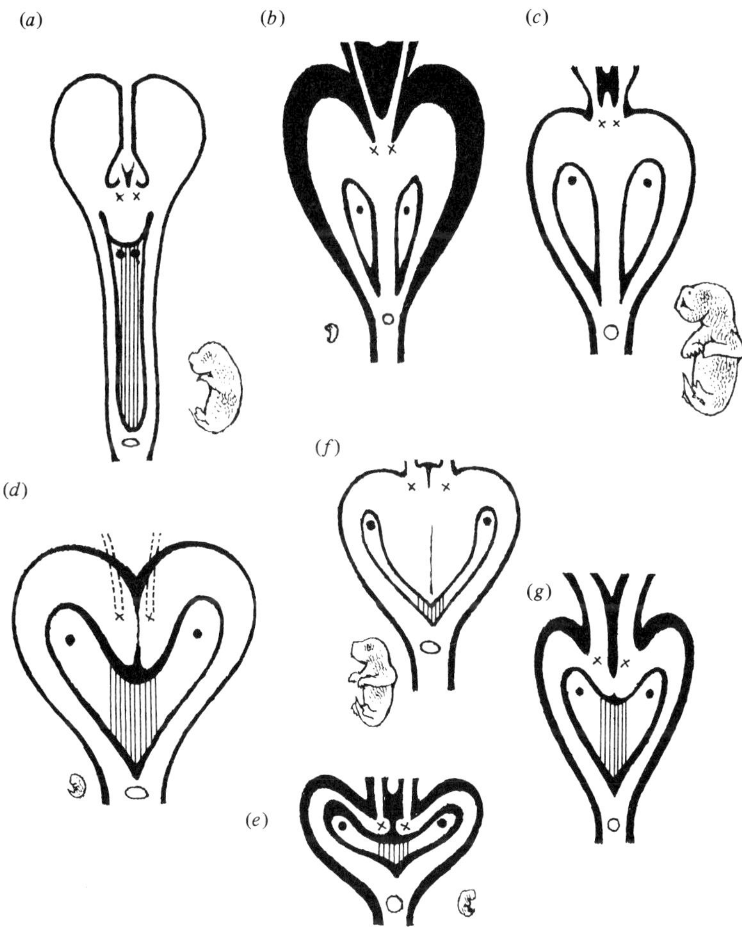

Fig. 5.6. Urogenital system of a parturient *Bettongia lesueur*. (*a*) sagittal section; (*b*) section through mid-horizontal plane; (*c*) representative cross-sections, traced from direct projection, from those used to make the reconstructions in (*a*) and (*b*). avc, anterior vaginal cul de sac; ave, anterior vaginal expansion; ov, ovary; pvs, posterior vaginal sinus; ugs, urogenital sinus; ur, urethra; wd, Wolffian duct; other abbreviations as in Fig. 5.1. From Tyndale-Biscoe (1968).

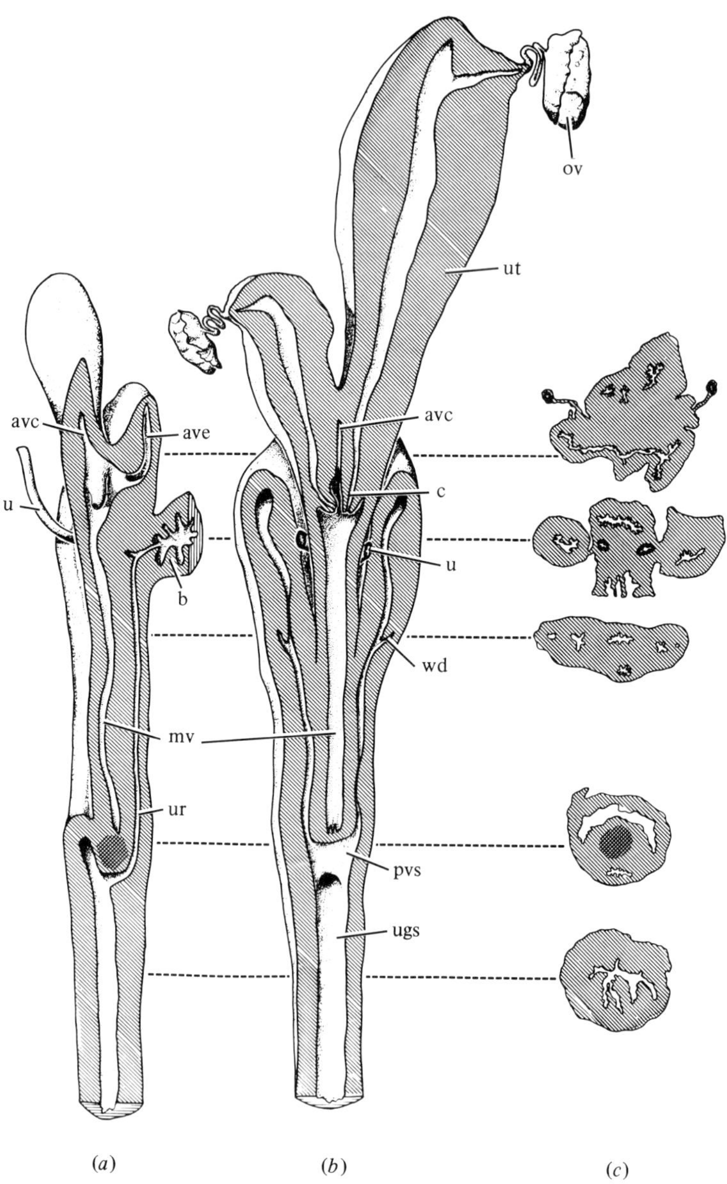

mammalian young yet recorded, but has a permanently open birth canal (de Bavay, 1951; Renfree, 1980*a*). Conversely, the species that have the largest neonates have large culs de sac, incomplete (or absent) median septa, and a short birth canal, which in the Macropodidae may remain permanently open after the first birth (Fig 5.5) (see Sharman, 1965*c*).

The shape of the lateral vaginae varies in the several families of marsupials. In *Caenolestes obscurus* they are especially long and convoluted (Fig. 5.3*c*), whereas in *Trichosurus vulpecula* and *Tarsipes rostratus* (Fig. 5.3*a*, *d*) they are short and straight. In some species, the anterior parts may be enlarged as seminal receptacles. When separate they are termed anterior vaginal expansions (Macropodinae, Fig. 5.1) and when the pair become fused as a single median chamber it is called the vaginal caecum (Peramelidae, Potoroinae) (Fig. 5.6).

Histology and functional aspects of the uterus and vaginal canals

The functional aspects of the urogenital tract and the changes associated with pregnancy, parturition, and the oestrous cycle have received more attention in recent years. In all marsupials distinct and well-marked changes occur in the genital tract during the follicular or proliferative phase and during the luteal phase of the reproductive cycle.

The proliferative phase

The proliferative phase spans the pro-oestrous period of follicular growth and maturation, oestrus and the first few days after ovulation. This phase is characterised by growth of the whole genital tract. In the cells of the uterine epithelium and glands mitotic figures are numerous (Hartman, 1923*a*; Sharman, 1955*a*; Pilton & Sharman, 1962; Hughes *et al.*, 1965; Clark & Poole, 1967; Shorey & Hughes, 1973*a*). In polyovular species, like *Didelphis virginiana*, no differences can be seen between the two uteri or between pregnant and non-pregnant animals (Fleming & Harder, 1981*a*) but in the monovular *Trichosorus vulpecula* von der Borch (1963) observed that at oestrus the frequency of mitoses was greater in the endometrial cells of the uterus ipsilateral to the Graafian follicle than in those of the contralateral uterus, which was suggestive of local effects of the ovarian hormones, presumably oestrogens, on the endometrium. The hormones were not transferred down the oviduct, as ligation did not affect the difference, and she suggested that the follicular influence might be conveyed by a direct vascular connection. Such a route has since been described and will be discussed below.

During the follicular phase, the vaginal complex and the urogenital sinus

(*a*)

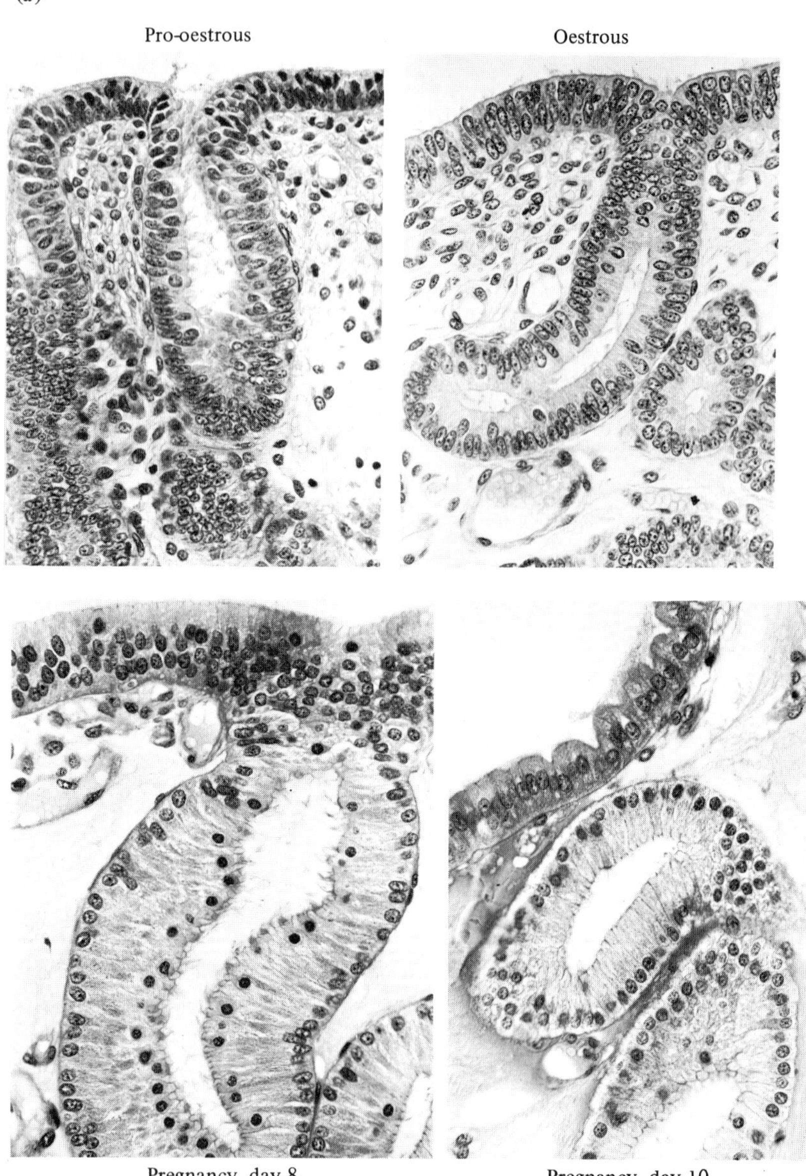

Fig. 5.7.(*a*) Endometrium of *Didelphis virginiana* during pro-oestrus, oestrus and late pregnancy. Note the lower height of the glandular epithelium at pro-oestrus compared to oestrus, and the superficial small round nuclei of the ciliated cells in the glandular epithelium on day 8, which are absent by day 10 of pregnancy. Photographs courtesy

also enlarge by hyperplasia and hypertrophy (Figs 5.3*a*, 5.7*b*), reaching a maximum at oestrus when the total weight may be several times as great as in the quiescent or anoestrous state. In the Didelphidae and Macropodinae it is the lateral vaginal canals which enlarge (Fig. 5.1), but the median vagina in *Trichosurus vulpecula* (Fig. 5.3) (Pilton & Sharman, 1962). In the Peramelidae (Lyne, 1976), Potoroinae (Fig. 5.6) (Tyndale-Biscoe, 1968) and *Lagostrophus fasciatus* (Tyndale-Biscoe, 1965), it is the vaginal caecum which enlarges. These changes have been induced by treatment with oestradiol in *D. virginiana* (Risman, 1947), *T. vulpecula* (Bolliger, 1946; Khin Aye Than & McDonald, 1976) and *M. eugenii* (Evans *et al.*, 1980; Renfree, Wallace & Young, 1982).

The mucosal lining of cuboidal epithelium of the median vagina and vaginal caecum is highly secretory at oestrus (Risman, 1947; Hughes & Rodger, 1971) whereas the lateral vaginae and urogenital sinus are lined with squamous epithelium (Barbour, 1977), which becomes cornified and is sloughed in clumps into the lumina (Fig. 5.7*b*). After copulation the semen passes up the lateral vaginae to the vaginal culs de sac, where it mixes with the mucus and coagulates. Like the oviducts, the lateral vaginae are

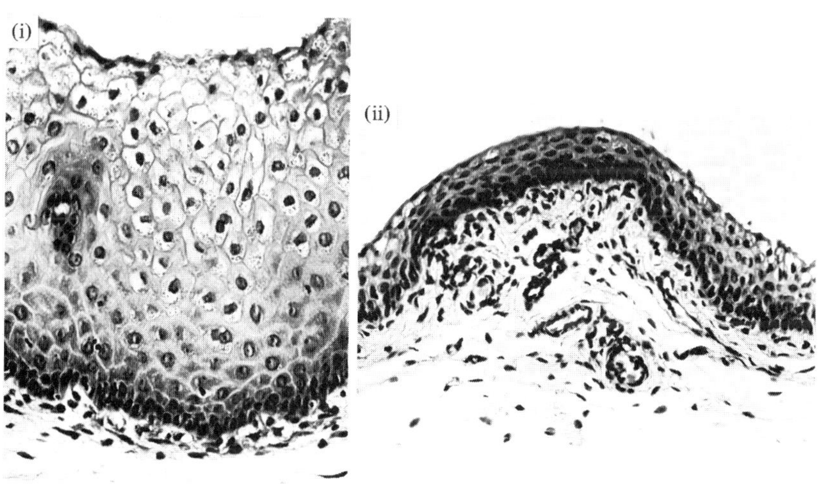

Fig. 5.7. *cont.*
of Professor H. A. Padykula and Dr. M. Taylor. (*b*) Lateral vagina of *Macropus eugenii* to show (i) the proliferation of the epithelium associated with follicular development and oestrus compared with (ii) an animal at the same day in which follicular growth had been prevented by immunisation with antigonadotrophin releasing hormone. From Short, Flint & Renfree (1985), with permission.

highly motile and undergo active peristalsis at oestrus (Hartman, 1924), which may facilitate this transport and mixing. In *M. eugenii* several males copulate with each female during the 12 h oestrus period and the vaginal complex can contain up to 100 ml of coagulum (Tyndale-Biscoe & Rodger, 1978). The coagulated mass in the urogenital sinus disappears within a day in *M. eugenii*, other macropodids (Sharman & Calaby, 1964) and in *T. vulpecula*, but the part remaining in the median vagina is retained for several days and is only slowly resorbed. Indeed, in some instances, a hard remnant may be found some months after the last mating.

The luteal phase

After oestrus the entire vaginal complex decreases in size (Fig. 5.3*b*, *e*) whilst the uteri become highly secretory and more vascular and oedematous. All authors cited in the last section agree that mitotic activity ceases completely within 4–6 days after oestrus, so that the marked and progressive enlargement of the uteri during the luteal phase must be due to hypertrophy of pre-existing cells, and oedema. For *Trichosurus vulpecula*, these conclusions from histology have been confirmed by analysis of the ratios of RNA:DNA, DNA:tissue and DNA:protein, sampled at oestrus and days 5, 9 and 13 post-oestrus. Curlewis & Stone (1986) conclude from these analyses that there are two phases of growth with the transition occurring between days 5 and 9. During the second phase, cellular hypertrophy and increased metabolic activity occur and this coincides with increasing levels of progesterone in circulation (Fig. 6.7). Since oestradiol levels are very low after oestrus (Curlewis, Axelson & Stone, 1985) they conclude that the hypertrophy is due largely to progesterone.

In this monovular species von der Borch (1963) observed that the hypertrophy of the two uteri was unequal, and at days 8–10 the uterus ipsilateral to the growing corpus luteum was significantly heavier than the contralateral uterus. She concluded that this was caused by the greater hyperplasia in the ipsilateral uterus at oestrus referred to above, but McDonald & Waring (1982) suggest that other factors could be involved, such as a difference in steroid hormone receptors, or a local effect of progesterone from the adjacent corpus luteum. Curlewis & Stone (1986) have confirmed von der Borch's (1963) observation but they support the view that differences in receptor concentration may be the major factor, as will be discussed below. Renfree (1973*a*) also reported differences in weight between the two uteri of *M. eugenii*, which is also monovular, with the heavier one being that associated with the corpus luteum and the same factors may be involved.

Fig. 5.8. Electron micrograph of the endometrial secretory cells of *Trichosurus vulpecula* at day 8 of the oestrous cycle showing a basal gland releasing its secretion into the gland lumen. Courtesy of Dr. C. D. Shorey, Sydney.

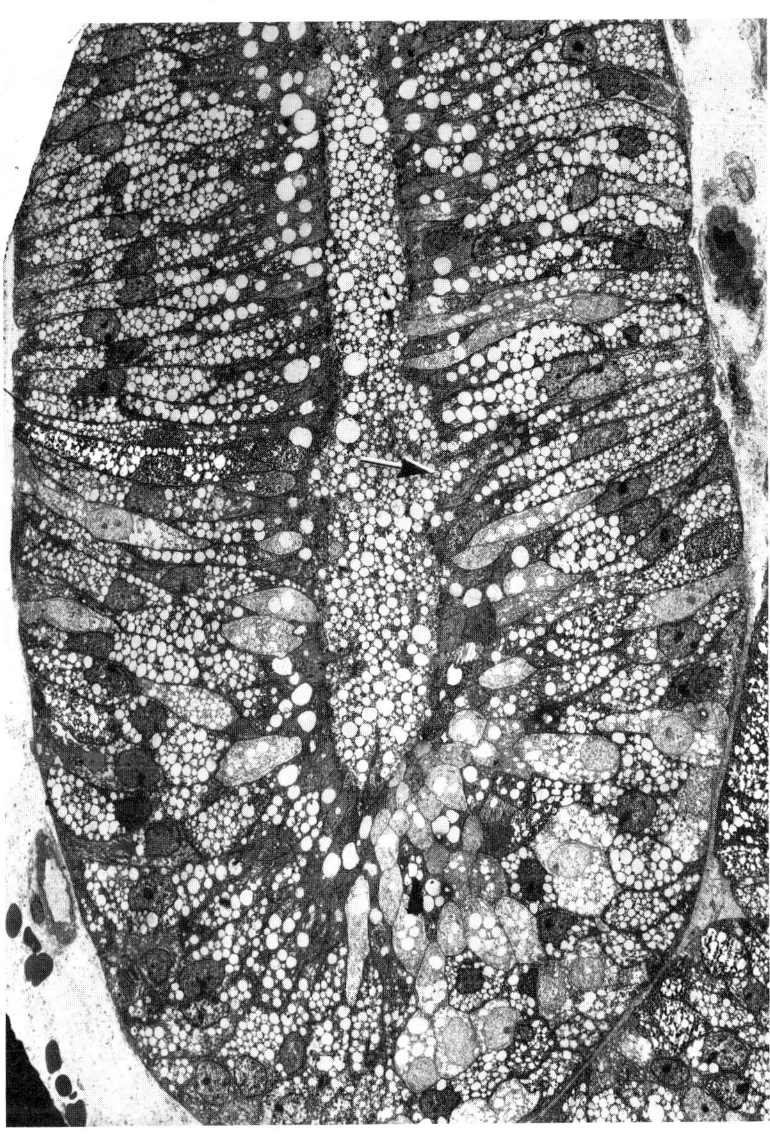

The uterine hypertrophy is associated with marked histological changes in all species. In the uterine glands and epithelia the cuboidal cells with large central nuclei become transformed to tall columnar cells with small basally situated nuclei and chromophobic cytoplasm (Fig. 5.7*a*; Sharman, 1959; Clark & Poole, 1967). This is termed the 'luteal' phase of the uterus (Sharman, 1959; Pilton & Sharman, 1962), because it is associated with an enlarged corpus luteum and can be induced by progesterone (Tyndale-Biscoe, 1963*b*; Renfree & Tyndale-Biscoe, 1973*a*). The cytoplasm contains abundant granular endoplasmic reticulum and secretory vesicles which are poured into the uterine lumen by apocrine secretion (Fig. 5.8) and provide nutrient material during expansion of the free-floating vesicle. In polyovular species, such as *Didelphis virginiana*, both uteri produce the abundant secretion (Renfree, 1975; Fleming & Harder, 1981*b*) but, in some monovular species, only the gravid uterus maintains this secretory activity (Renfree & Tyndale-Biscoe, 1973*a*; Wallace, 1981), which suggests that the fetus or placenta may provide a stimulus. This is discussed further in Chapter 7 (p. 328).

The post-luteal phase

At the end of the luteal phase the glands of the endometrium and their epithelial cells regress, and the sub-epithelial stroma shrinks. There is as yet no general agreement on the fate of the cells in different species. Hill & O'Donoghue (1913) attributed the changes to a process of shedding of glandular elements in *Dasyurus viverrinus*. Tyndale-Biscoe (1955) came to a similar conclusion for the epithelium of *T. vulpecula* and Shorey & Hughes (1972) showed that a new ciliated epithelium and basement membrane was reformed from stromal cells (Fig. 5.9*a*) being completed during the subsequent follicular phase. Lactation blocks the regeneration process, in which case the endometrium becomes anoestrous in appearance.

In *Didelphis virginiana* the glands consist of low columnar or cuboidal cells at the end of the luteal phase and the lumina are filled with degenerating cells, which disappear after a few days (Hartman, 1923a). More recently, Padykula & Taylor (1976*a*) have examined this process with the electron microscope and conclude that the degenerating cells originate as macrophages in the stroma, having there engulfed extracellular stromal ground substance (Fig. 5.9*b*). The degenerating cells and investing fibroblast cells penetrate the glandular lumen and are carried from there to the uterine lumen by the newly differentiated ciliary lining of the glands. They conclude that the interpretation of Shorey & Hughes (1972) for *T.*

vulpecula does not apply to *D. virginiana*, and that the integrity of the glandular epithelium is maintained throughout. The macrophages move by a cellular mechanism which removes large amounts of extracellular material without disruption of the integrity of the endometrium.

Flynn (1930) described an extrusion of lymph into the glandular lumina of *Bettongia gaimardi* at the end of the luteal phase, which was less profound than in the other species, and suggested it may be important as histiotrophe for the embryo when present. Similarly the post-luteal changes in *Setonix brachyurus* (Sharman, 1955*a*; Tyndale-Biscoe, 1963*a*), *M. eugenii* (Renfree

Fig. 5.9. Post-luteal uterine gland regeneration. (*a*) Low-power electron micrograph showing a double uterine gland of *Trichosurus vulpecula* on about day 22. Cilia (arrowed) protrude into the secondary lumen between the two glands from the outer gland cells (OG). The necrotic cells of the inner gland (IG) enclose the primary lumen (PL) into which cilia protrude. SC, stromal cells. From Shorey & Hughes (1972), with permission. (*b*) Gland and stroma of *Didelphis virginiana* on day 2 post-partum or about day 15 post-oestrus. The glandular epithelium has a less regular appearance than during pregnancy. Ciliated cells (C) occur among non-ciliated cells. Macrophages (M) occur in the stroma, in apposition to the basal glandular surface, and in an intra-epithelial position. Cells occur in the glandular lumen and mitoses in the epithelia. Stromal cells are numerous and consist of fibroblasts, monocytes, lymphocytes, and plasma cells in addition to the macrophages. From Padykula & Taylor (1976*a*) by permission.

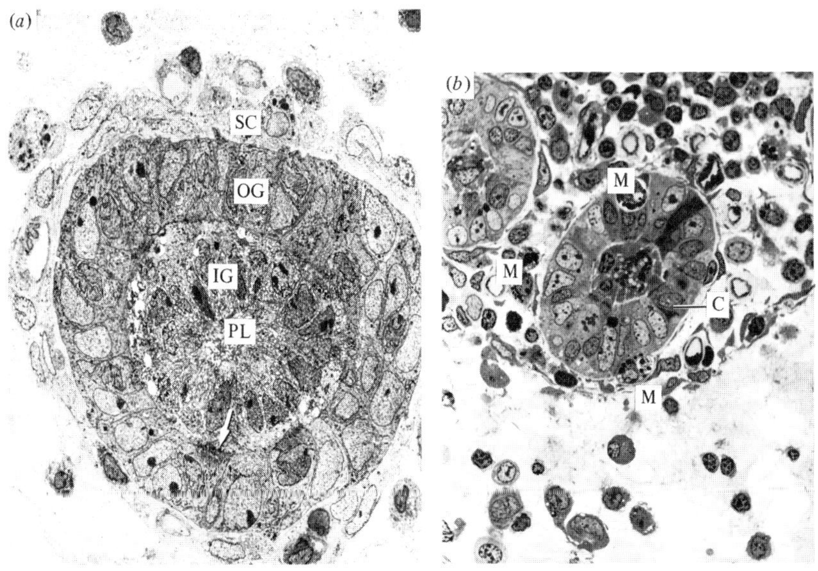

& Tyndale-Biscoe, 1973*a*) and *Potorous tridactylus* (Shaw & Rose, 1979) are not so dramatic and the epithelia are not replaced. This may be a reflection of less endometrial development in macropodids in the luteal phase and less secretion, as in *M. eugenii* (Renfree, 1972*b*, 1973*a*). In *M. giganteus* the gland cells degenerate in the post-luteal phase, and the small irregular nuclei are no longer uniformly basal, and the cell membranes adjacent to the glandular lumen are ruptured (Clark & Poole, 1967). In macropodids which are in lactational or seasonal quiescence the uteri remain small but not flaccid (Fig. 2.17). The uterine glands are numerous and irregularly scattered, with connective tissue between them, their cells are columnar with round basal nuclei, no mitosis is observed, and the lining of the uterine lumen has both ciliated and non-ciliated cells (Clark & Poole, 1967; Sharman & Berger, 1969).

From the foregoing it is clear that the several parts of the female urogenital tract respond differently to the changing levels of steroid hormones in circulation during the oestrous cycle and pregnancy. Two factors could bring this about: differences in the concentration of specific steroid receptors in different tissues; and local patterns of circulation, such as that invoked by von der Borch (1963). Both of these aspects have received some attention recently but, in both instances, it is too early to draw firm ideas about their roles. We therefore present the evidence available in the anticipation that both are areas that are likely to develop in the near future.

Steroid receptors in the urogenital tract

Three studies have been published on steroid receptors in the urogenital tract of *Trichosurus vulpecula* (Young & McDonald, 1982; Curlewis & Stone, 1986) and one on progesterone receptors in the uterus of *Setonix brachyurus* (Owen, Cake & Bradshaw, 1982). In *T. vulpecula* receptors with low capacity/high affinity for oestradiol have been described by Young & McDonald (1982) from cytosolic and nuclear fractions of endometrial cells and vaginal epithelia. The concentrations of each type varied between the two tissues and with the reproductive state. Both types were low in anoestrous and lactating females and in ovariectomised animals but were at highest levels in oestrous females and in ovariectomised females pre-injected with oestradiol. In animals at the luteal phase, cytosolic receptor concentration remained high in endometrial tissue but not in the vaginal epithelia.

Curlewis & Stone (1986) confirmed these findings for oestradiol cytosolic receptors and also described specific progesterone receptors in

Table 5.1. *The effects of oestradiol administration on organ weight, RNA:DNA, protein:DNA and DNA:g tissue ratio, total DNA and oestradiol and progesterone-cytosol-receptor levels in uterus of ovariectomized* Trichosurus vulpecula[a]

Organ	Treatment	Wet weight (g)	Progesterone receptor [pmol (mg DNA)$^{-1}$]	Oestradiol receptor [pmol (mg DNA)$^{-1}$]
Oviduct	C	0.04 ± 0.005	ND	ND
	E	0.09 ± 0.001**	ND	ND
Uterus	C	0.27 ± 0.037	0.15 ± 0.047[b]	2.82 ± 0.433[b]
	E	1.20 ± 0.085***	2.22 ± 0.089***	8.61 ± 0.447***
Vaginal cul de sac	C	0.90 ± 0.118	0.10 ± 0.039	2.44 ± 0.828
	E	4.83 ± 0.986**	0.71 ± 0.161**	4.95 ± 1.658
Lateral vagina	C	0.26 ± 0.038	0.15 ± 0.028[b]	4.48 ± 0.776[b]
	E	0.81 ± 0.166*	0.66 ± 0.141**	4.02 ± 0.721
Urogenital sinus	C	1.52 ± 0.120	0.14 ± 0.25	2.40 ± 0.564
	E	3.46 ± 0.377**	0.37 ± 0.068*	2.54 ± 0.420

[a] Animals were treated with either peanut oil (C) or 5 μg oestradiol in peanut oil (E) for 3 days and killed on the fourth day. Results are means ± S.E.M. for 5 animals. Difference from the mean within each parameter and organ was tested using Student's 't' test.
*, $0.05 > p > 0.01$; **, $0.01 > p > 0.001$; ***, $0.001 > p$.
[b] One lost sample.
From Curlewis & Stone (1986).

cytosol from endometrium (Table 5.1). Concentrations of the latter receptors were maximal at oestrus, lower at day 5 and much reduced at day 13. In ovariectomised females, progesterone receptors were induced by treatment with oestradiol, but suppressed by progesterone. They suggest that the unilateral effect of von der Borch (1963) could be in part due to increased synthesis of progesterone receptors in the ipsilateral uterus at oestrus, which then allowed greater uptake of progesterone in the uterus subsequently, with consequent greater hypertrophy. Owen *et al.* (1982) also observed that oestradiol induced progesterone-receptor synthesis in cytosol of endometrium in *S. brachyurus*. Conversely, progesterone plus oestradiol depressed it, but stimulated progesterone-receptor concentration in the nuclear fraction. They interpreted this as being due to translocation of the bound complex into the nucleus.

Thus it could be that the high levels of oestradiol found at oestrus in *Macropus eugenii* (Figs. 6.15 and 9.3, Flint & Renfree, 1982; Shaw & Renfree, 1984; Harder *et al.*, 1984), *T. vulpecula* (Curlewis *et al.*, 1985) and *Didelphis virginiana* (Fig. 6.3, Harder & Fleming, 1981) may have the effect of inducing progesterone-cytosol-receptor synthesis at this time and the receptors are then available when the subsequent rise in progesterone from the corpus luteum occurs.

In *Setonix brachyurus* there is a difference in the concentration of progesterone receptors between gravid and non-gravid uteri, and the levels peak in uterine tissue about a day before the early peripheral plasma peak of progesterone (Fig. 6.13, Owen, 1984). Preliminary results from *M. eugenii* show a similar trend (M. B. Renfree, unpublished results), and since there is also a peak at day 5–6 in peripheral oestradiol which coincides with the progesterone peak (Fig. 6.15) it is possible that the progesterone receptor is induced by oestradiol in these two macropodid species, as it is in *T. vulpecula*. It would be of interest to know the fate of oestradiol-induced progesterone receptors in the quiescent uteri of macropodids during the 9–11 month diapause.

Vascular anatomy of the urogenital tract

In several eutherian species (ewe, cow, sow and guinea pig) the uterine and ovarian vasculature forms a complex association of closely apposed ovarian arteries and uterine veins (Del Campo & Ginther, 1972, 1973). It is thought that this allows prostaglandin $F_{2\alpha}$, produced by the endometrium, to pass directly into the ovarian artery and induce luteolysis in the ipsilateral ovary (McCracken, Baird & Goding, 1971). Very few marsupials have been studied in this regard but, as we have seen, unilateral effects do occur in marsupial genital tracts, so it is of interest to know the possible routes by which such ovarian and uterine hormones might circulate.

The only detailed studies so far are those of Lee & O'Shea (1977), who described the origin, distribution and structure of the blood vessels of the female reproductive tract of *Trichosurus vulpecula*, and also gave some details of *Vombatus ursinus* and five macropodids, *M. giganteus*, *M. eugenii*, *M. agilis*, *M. rufus* and *Thylogale billardierri*. Towers, Shaw & Renfree (1986) described the vascular anatomy of *M. eugenii*, *Setonix brachyurus* and *M. agilis*.

In all these species the urogenital tract is supplied and drained by four paired sets of arteries and veins: the ovarian, cranial urogenital, caudal urogenital, and internal pudendal (Table 5.2).

In *T. vulpecula*, the ovarian arteries arise caudal to the renal arteries, either separately or by a common trunk. The artery runs along the surface of the corresponding vein for most of its length, and is closely apposed to it. Branches are given off to the ipsilateral ovary, oviduct, ureter and uterus (Fig. 5.10). The vessel branches several times near the ovary and these are interwoven with branches of the ovarian vein. An additional, usually single branch to the ureter arises at the same level in some cases, but more often arises independently, closer to the aorta at about the level that the ureter crosses dorsally over the ovarian vessels.

Parallel to the arteries are the larger ovarian veins (Fig. 5.11). Branches from the ovarian venous plexus join with branches from the oviduct, uterus and ureter to form the ovarian vein which enters the caudal vena cava.

The cranial urogenital artery and vein branch from the internal iliac artery and vein to supply the caudal segments of the uteri, the lateral and

Table 5.2. *Names of main blood vessels of the female urogenital tract*

Names proposed by Lee and O'Shea (1977) for marsupial urogenital vasculature	Names of closest corresponding structure given by Schaller *et al.* (1973)
Ovarian artery and vein	Ovarian artery and vein (*A. ovarica, V. ovarica*)
Cranial urogenital artery and vein	Uterine artery and vein (*A., V. uterina*)
Caudal urogenital artery and vein	Vaginal artery and vein (*A., V. vaginalis*)
Internal pudendal artery and vein	Internal pudendal artery and vein (*A., V. pudenda interna*)

median vaginae and ureters. Anastomoses occur with branches from the corresponding ovarian artery or vein supplying or draining the uterus. The caudal urogenital arteries and veins serve the urogenital sinus, with anastomoses occurring between branches of the cranial and caudal urogenital arteries on each side. The internal iliac urogenital arteries and veins continue as the internal pudendal vessels which supply the most caudal regions of the urogenital tract. In *M. eugenii*, the mammary glands receive arterial supply from branches of the external iliac arteries, arising 3–4 cm from the aortic bifurcation.

All the species of macropodid studied by Lee & O'Shea (1977) and Towers *et al.* (1986) have a similar blood vascular arrangement, and were generally similar to that of *Trichosurus vulpecula* with minor variations.

Unlike in *T. vulpecula*, in which they branch from a single tract, the ovarian arteries of *M. eugenii* always arise independently from the aorta, caudal to the renal arteries, with the left ovarian artery usually just cranial to the right (Fig. 5.11) (Towers *et al.*, 1986) as in most eutherian mammals.

Fig. 5.10. Latex cast of right ovarian artery and its branches in *Trichosurus vulpecula*. o, ovarian network; u, network at cranial tip of uterus; oa, ovarian artery; ob, much divided ovarian branches of ovarian artery; fb, branch to Fallopian tube; ub, branch to uterus. From Lee & O'Shea (1977), with permission.

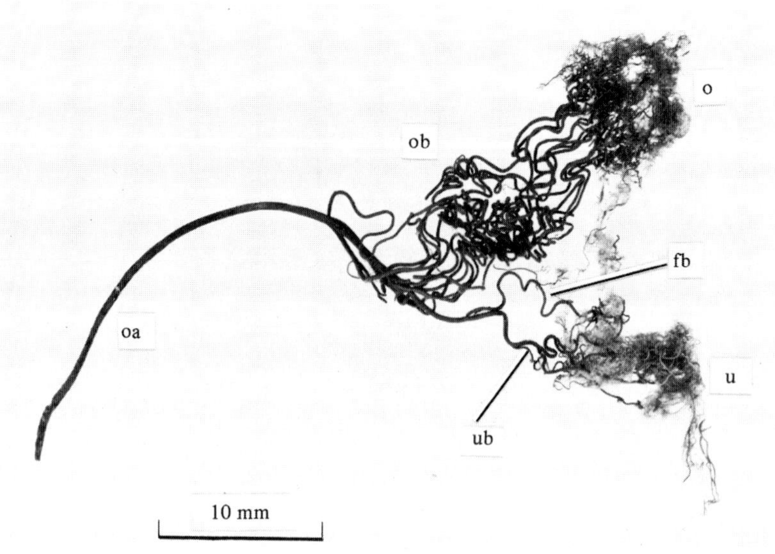

All arterial supply to the remainder of the urogenital tract is derived from the internal iliac arteries (Towers *et al.*, 1986). Likewise, the cranial and caudal urogenital vessels of at least one specimen of *Macropus rufus*, *Macropus giganteus* and *Vombatus ursinus* (Lee & O'Shea, 1977), and all specimens of *Setonix brachyurus* and *Macropus agilis* examined by Towers *et al.* (1986) arose by common trunks from the internal iliac vessels.

Fig. 5.11. Semi-diagrammatic representation of the blood vascular system of the urogenital system of *Macropus eugenii*. Right side of the figure (left side of animal) complete. The anterior lateral vaginae have been deflected posteriorly towards the bladder, as shown in the small diagram bottom left. A, arteries; V, veins. From Towers *et al.* 1986, with permission.

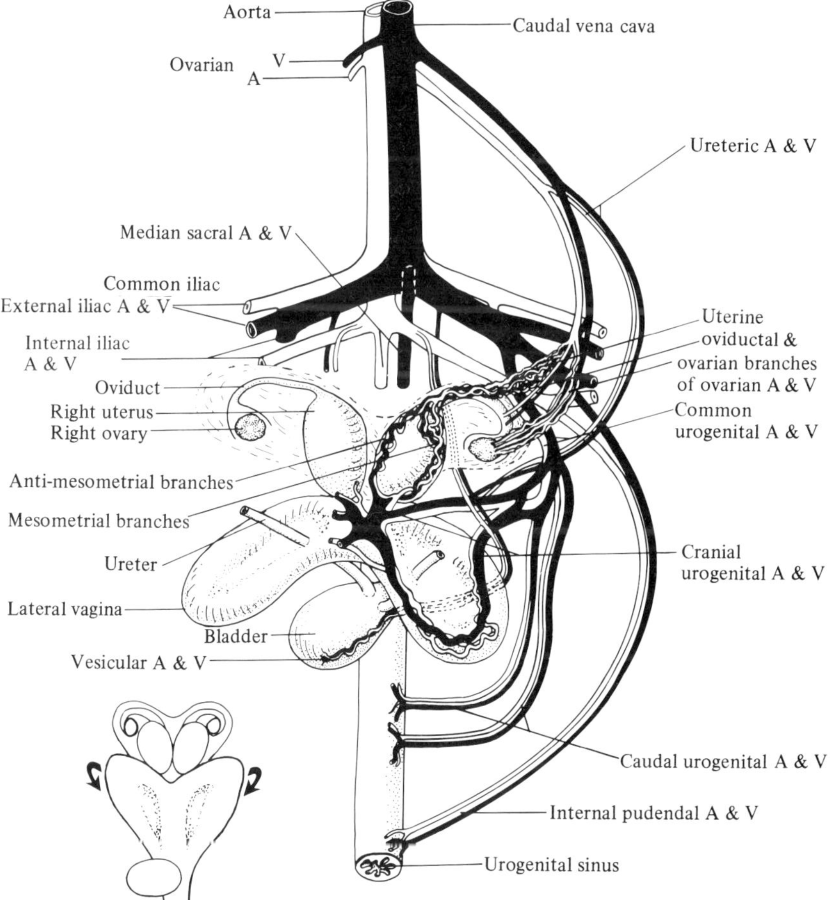

All the macropodids have ovarian veins and arteries which run in close relationship to one another, giving off branches to the oviduct and uterus which anastomose with branches of the cranial urogenital vessels, and in most supply between 7 and 12 branches to the ovary. As in *T. vulpecula*, the first few millimeters of the many branches running to the ovary are almost straight, but then become markedly tortuous and have many U- and S-shaped bends and regions of spiralling. These arteries branch infrequently and show occasional anastomoses and are intimately apposed to the branches of the venous plexus draining the ovary. In *Setonix brachyurus* the ovarian branch of the ovarian artery does not branch as extensively as in *M. eugenii* and *M. agilis* (Towers *et al.*, 1986) and *V. ursinus* has fewer (4–5) ovarian branches of the ovarian artery (Lee & O'Shea, 1977).

In *M. eugenii*, the uterine branch of the ovarian artery is, like the ovarian branch, very convoluted and with many U- and S-shaped bends. On reaching the uterotubal junction, the major uterine branches run along the mesometrial and antimesometrial borders of the uterus in a spiral pattern and branches from these run around the surface of each uterus. In *T. vulpecula* the ovarian veins closely correspond to the ovarian arteries, and the branch of the ovarian vein draining the ovary is, like the artery, formed by a fusion of a plexiform group of small veins draining the ovary, several veins from the cranial end of the uterine horn, and many small veins draining the oviduct (Lee & O'Shea, 1977). A similar arrangement is observed in the macropodids. The largest uterine branches of the ovarian vein run along the mesometrial and antimesometrial sides of each uterus, the mesometrial being the larger. These anastomose with the cranial urogenital vein and also between circular branches.

The ramification of the ovarian artery in *M. eugenii* and *M. agilis* almost completely covers the uterine branch of the larger ovarian vein, so that an anatomical basis exists for a direct transfer of substances between the two vessels, and therefore between the uterus and ovary. A part of the uterine venous effluent passes to the caudal vena cava via the ovarian veins, and arrangement of the ovarian vessels and the branches of the cranial urogenital vessels supplying the uterus suggest that the ovarian artery contributes to the uterine blood supply (Lee & O'Shea, 1977). The local effects described by von der Borch (1963) for *T. vulpecula*, by Renfree (1973*a*) for *M. eugenii* and by Shaw & Rose (1979) for *Potorous tridactylus* could be ascribed to a local hormonal effect of the ovary on the uterus via a veno-arterial transfer from ovarian venous blood to ovarian arterial blood thence via the uterine branches of these (ovarian) vessels to the ipsilateral uterus (Lee & O'Shea, 1977).

At the anterior ends of the separate uteri there are no across-the-mid-line anastomoses, so a localised effect on the ipsilateral uterus could be effected, but across-the-mid-line anastomoses do occur at the first possible place where the uteri or vaginae unite. Lee & O'Shea (1977) suggest that the ovarian arterio-venous intermingling may be a specialisation for counter-current exchange. In the sheep, such a counter-current transfer mechanism has been demonstrated for the passage of progesterone between ovary and uterus (Einer-Jensen & McCracken 1981).

Evidence for the physiological significance of this vascular arrangement comes from endocrinological studies in *M. eugenii* (Towers *et al.*, 1986). Blood collected from the uterine branch of the ovarian vein had a higher concentration of progesterone than from the peripheral circulation, and progesterone concentration was higher on the side corresponding to the active corpus luteum. Harder *et al.* (1984) have shown that there was a significantly higher concentration of oestradiol 17β in the ovarian vein draining the ovary with the Graafian follicle and the ipsilateral uterus than from the other ovarian vein or the peripheral circulation, thereby showing that the follicle is the main source of oestrogen in the peripheral plasma. As the site of collection was anterior to the branching of the ovarian branch of the ovarian artery it is not possible to say whether any of this oestrogen reached the uterine branch of the ovarian vein. Nevertheless, if this were so, the uterus ipsilateral to a new corpus luteum could be preferentially exposed to oestrogen during the pro-oestrous phases, which could then induce hyperplasia and steroid-receptor synthesis, and to progesterone after ovulation, which could induce hypertrophy and be reflected as an unilateral difference in endometrial or uterine weights in the luteal phase.

While these results suggest that there may be physiological significance in the intimate structural relationships between the ovarian arteries and veins in monovular species, further studies need to be made of the precise routes by which substances can be transferred from ovary to uterus, and uterus to ovary.

Oocyte and follicular growth and development

The differentiation of the ovary during pouch life was discussed in Chapter 3 and the final stage of ovulation and formation of the corpus luteum will be considered in Chapter 6. Here we will consider the growth of the oocyte and its associated follicle in the prepubertal and mature ovary. The main studies are those of Morgan (1943) on *Didelphis virginiana*, Alcorn (1975) and Panyaniti, Carpenter & Tyndale-Biscoe

(1985) on *Macropus eugenii* and Hughes *et al.* (1965) on *Pseudocheirus peregrinus*. Lintern-Moore *et al.* (1976) analysed the relationship of oocyte and follicular growth in nine species of marsupial and Lintern-Moore & Moore (1977) compared their findings in marsupials with eutherians and monotremes. In eutherian mammals, growth of the oocyte and its associated follicle is biphasic (Brambell, 1956). During the first phase there is no antrum and the diameter of the oocyte and the diameter of the follicle increase synchronously, the latter by an increment of layers of the membrana granulosa. During the second phase, which coincides with antrum formation, only the follicle grows with increase in cell layers and by expansion of the antrum.

Folliculogenesis

Pedersen & Peters (1968) have subdivided each phase of folliculogenesis into several follicle types, each one characterised by the size of the oocyte and follicle, by the number of layers of the membrana granulosa and by the presence and extent of the antrum. The development of the follicle in *M. eugenii* conforms to a similar pattern (Fig. 5.12).

The final size reached by the oocyte at the end of phase 1 is very similar for all eutherians examined (83 ± 14 μm, Table 5.3) but the final diameter of the pre-ovulatory follicle at the end of phase 2 bears a close relationship to the body weight of the adult female of the respective species. Alcorn (1975) and Lintern-Moore *et al.* (1976) found that, in marsupials, oocyte and follicular growth also conformed to a biphasic pattern and that the maximum diameter of the oocyte was similar in all species studied (135 ± 16 μm, Table 5.3). The main difference to eutherian mammals was that the diameters of the follicle, oocyte and oocyte nucleus were about 1.6 times larger at the completion of phase 1 (see Table 5.3).

More surprising, however, is the conclusion of Lintern-Moore & Moore (1977) that oocyte growth in the monotreme, represented by *Ornithorhynchus*, also conforms to a biphasic pattern. Whereas the monotreme oocyte, just prior to ovulation, is many times larger than that of other mammals and completely fills the follicle so that no antrum is formed, the oocyte nuclear and nucleolar diameters cease to increase after the follicle attains a diameter of 380 μm (Table 5.3). These similarities assume more significance since these same authors (Moore *et al.* 1974; Moore & Lintern-Moore, 1978) have shown that nuclear and nucleolar size in the growing oocyte of the mouse are linearly and positively correlated with RNA synthesis and endogenous RNA polymerase activity. In contrast to marsupials and eutherians, phase 2 in monotremes involves growth of the oocyte as well as of the follicle which surrounds it (the correlation being 0.92). Although most of the increase in the follicle is due to an expansion of the oocyte, the follicle cells do secrete a small amount of fluid which Flynn & Hill (1939) considered to be homologous with follicular fluid of other mammals and to set the monotremes apart from Sauropsida whose follicles do not secrete fluid.

Whereas the several dimensions of the oocytes are similar in all the marsupials examined regardless of the adult body size, as in Eutheria, the final sizes of prė-ovulatory follicles do vary at the end of phase 2 between species and are positively correlated with adult body size (Table 5.4).

Fig. 5.12. The mean nuclear, oocyte and follicle diameter of follicle types occurring during development in *Macropus eugenii*, based on the classification of follicle types described by Pedersen & Peters (1968). Redrawn from Alcorn (1975).

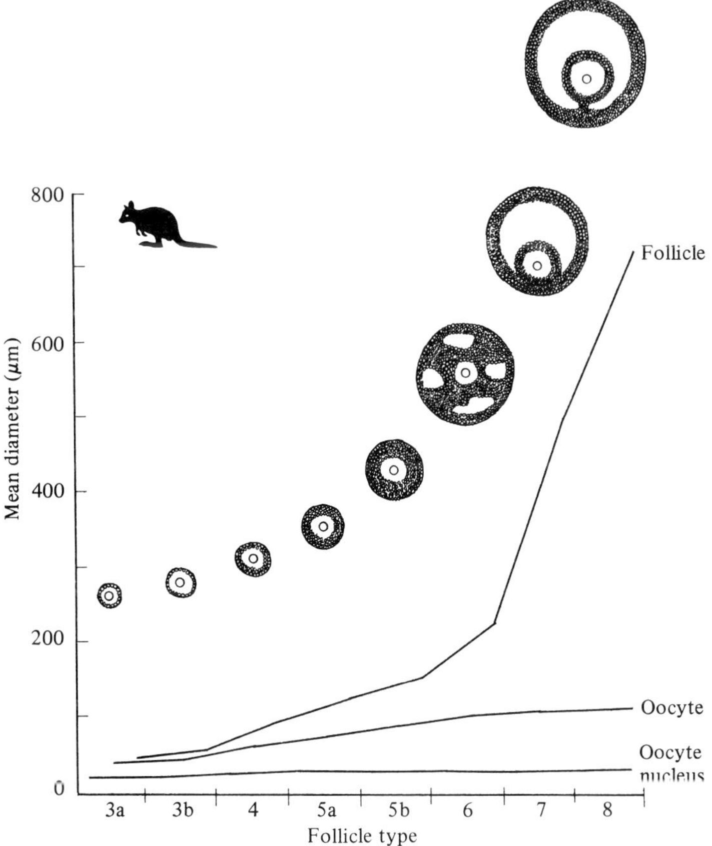

Throughout follicular growth, the oocyte remains in prophase of the first maturation division. Completion of the first division with extrusion of the first polar body (illustrated in Fig. 7.1) has been seen to occur in the mature Graafian follicle in *Didelphis virginiana* (Hartman, 1919; McCrady, 1938), in *Dasyurus viverrinus* (Hill, 1910; Hill & O'Donoghue, 1913), in *Pseudocheirus peregrinus* (Hughes *et al.*, 1965), in *T. vulpecula* (Hughes & Hall, 1984), in *Sminthopsis macroura* (Godfrey, 1969*a*) and *Antechinus stuartii* (Selwood, 1982*b*). It probably occurs at this stage in other species also (Selwood, 1982*b*). Hill (1910) observed that the second maturation division had also commenced in the follicles of one animal but Hartman (1919) and McCrady (1938) state that this division is not completed in *D. virginiana* until after ovulation and then only if a sperm enters the egg (see Rodger & Bedford, 1982*b*). Among eutherian mammals, the timing of the maturation divisions in respect to ovulation and fertilisation varies between species (Austin, 1961) but, for marsupials, insufficient species have been examined to say whether a similar variation applies.

The development of follicles in pre-pubertal females and the role of the pituitary in this has only been examined in *M. eugenii* (Alcorn, 1975; Hearn, 1972*a*, 1975*b*; Panyaniti *et al.*, 1985). During the pre-pubertal period some follicles develop antra and attain to type 5 or 6 (see Fig. 5.12) before becoming atretic, few reach type 7 and no type 8 follicles are found. Nevertheless, the biphasic pattern of growth is displayed, the slopes and inflection being the same as in the adult so that it can be concluded that failure to develop past type 6 is due to absence of pituitary stimulus.

In the post-pubertal ovary, type 7 and 8 follicles occur and the latter may include several pre-ovulatory Graafian follicles as well as atretic follicles of all types. Panyaniti *et al.* (1985) argue that the pattern of oocyte and follicular growth is best described by a parabola rather than by two intersecting regressions and they show that the same curve expresses the

Table 5.3. *Oocyte and follicle dimensions in mammals*

Sub-class	Number of species	Maximum diameter (μm) of			Follicle at end of phase 1
		Oocyte	Nucleus	Nucleolus	
Monotremata	1	3500	97	14	380
Metatheria	10	135 ± 16	40 ± 3	8 ± 1	227 ± 32
Eutheria	13	83 ± 14	26 ± 6	7 ± 1	141 ± 36

From Lintern-Moore & Moore (1977).

Table 5.4. *Oocyte diameter and follicle diameter in marsupials in relation to adult female size*

Species	Adult weight (g)	Oocyte diameter (μm)	Follicle diameter at phase 1 (μm)	Pre-ovulatory follicle diameter (μm)	Reference
Sminthopsis macroura	16–35	250	300	474	Godfrey (1969*a*)
Antechinus stuartii	30	128	194	611	Lintern-Moore *et al.* (1976); Selwood (1982*b*)
Cercartetus concinnus	30	130	—	420	Clark (1967)
Marmosa robinsoni	40	—	—	826	Godfrey (1975)
Pseudocheirus peregrinus	700	—	—	3900	Hughes *et al.* (1965)
Potorous tridactylus	1000	158	—	3000	Hughes (1962*a*)
Petauroides volans	1500	114	224	3500	Bancroft (1973)
Didelphis marsupialis	2000	124	208	—	Lintern-Moore *et al.* (1976)
Trichosurus vulpecula	2500	170	220	4900	Hughes & Rodger (1971); Lintern-Moore *et al.* (1976)
Setonix brachyurus	2750	125	—	3000	Waring *et al.* (1955)
Macropus eugenii	4500	120	195	4200	Lintern-Moore *et al.* (1976); Tyndale-Biscoe & Rodger (1978)
Macropus giganteus	22000	133	257	6000	Clark & Poole (1967); Lintern-Moore *et al.* (1976)

data from prepubertal ovaries and from the ovaries of hypophysectomised adults as well as form ovaries of mature adults.

In Eutheria the completion of phase 2 is dependent on pituitary gonadotrophin and this is probably true for marsupials. In *M. eugenii*, the only species examined, development of follicles past phase 1 was blocked after hypophysectomy and 60 days after the operation there were no follicles greater than 0.5 mm in diameter in either ovary (Hearn, 1972*a*) and the ovaries were reduced to half the initial weight (Hearn, 1975*b*). Similarly, in females passively immunised against gonadotrophin-releasing hormone, the growth of follicles was prevented (Short *et al.*, 1985). Panyaniti *et al.* (1985) examined this further by comparing the abundance of normal and atretic follicles of type 3b to type 8 in ovaries of adult females 21 days after hypophysectomy or sham hypophysectomy. No difference could be detected in the number of follicles of each type or the frequency of atresia at each stage between the two groups except for type 8 follicles: these were absent from the ovaries of the hypophysectomised animals and they concluded that only the final stages of follicular growth are dependent on gonadotrophins.

No other species has been hypophysectomised to investigate this aspect but, in species that undergo true anoestrus, such as *Didelphis virginiana* (Hartman, 1923*a*; Martinez-Esteve, 1942), *Pseudocheirus peregrinus* (Hughes *et al.*, 1965), *Petauroides volans* (Smith, 1969), *Setonix branchyurus* (Sharman, 1955*a*) and *Trichosurus vulpecula* (Tyndale-Biscoe, 1955), the ovaries become small and contain only small follicles less than 0.2 mm in diameter, resembling the ovaries of hypophysectomised *M. eugenii*.

The number of follicles that ovulate varies widely in marsupials (see Table 2.2). In *Didelphis virginiana* Rafferty-Machlis & Hartman (1953) recorded a maximum of 60, with 16 as normal, and similarly in *D. marsupialis* (Hill, 1918) recorded more than 20 eggs shed at one ovulation. The very high numbers of eggs shed at ovulation in *D. virginiana* may be associated with another unusual finding in this species. Hartman (1926) reported that polyovular follicles occur in 'astounding numbers' and polynuclear ova also occur by hundreds in the ovaries of some animals. Godfrey (1975) also recorded 12–27 follicles or corpora lutea in *Marmosa robinsoni*. In all the Dasyuridae examined, ovulation rates are high: in *Dasyurus viverrinus* (Hill, 1910), *Sarcophilus harrisii* (Flynn, 1922; Guiler, 1970; Hughes, 1982), *Sminthopsis macroura* (Godfrey, 1969*a*), *Antechinus stuartii* (Woolley, 1966*b*) and *Dasyuroides byrnei* (Woolley, 1971*a*).

In the didelphids *Dromiciops* and *Caluromys*, and in the Peramelidae, Burramyidae and *Tarsipes rostratus* on the other hand, the ovulation rate is low and the number of corpora lutea generally equal the number of young in the pouch (Clark, 1967; Lyne & Hollis, 1979; Renfree *et al.*, 1986). The herbivorous monotocous species are almost invariably monovular and, on circumstantial evidence, it has generally been supposed that ovulation occurs alternately from the ovaries in successive cycles. After ablation of one ovary, however, female *M. eugenii* ovulated in successive cycles from the remaining ovary (C. H. Tyndale-Biscoe & L. A. Hinds, unpublished results).

Follicular atresia and the origin of interstitial tissue

Differentiation of the mature follicle involves an increase in the number of layers of cells in the membrana granulosa (Fig. 5.12) and differentiation of theca interna and externa. The theca interna is inconspicuous in *Dasyurus viverrinus* (Sandes, 1903) but is well developed in other species (Sharman, 1959).

The development of the theca folliculi in several marsupials was described and compared by O'Donoghue (1916) and for *D. virginiana* by Martinez-Esteve (1942). The fate of the theca in follicles that do not ovulate varies in different marsupials. The commonest fate is for the cells of the membrana granulosa to swell and die and for their nuclei to become pycnotic (Martinez-Esteve, 1942). The antrum disappears and the theca interna shrinks.

However, in *Antechinus stuartii*, two types of follicle develop at oestrus (Woolley, 1966*b*). In a minority the granulosa tissue predominates in the usual manner and these follicles are destined to ovulate but, in the majority, the membrana granulosa remains a single layer, while the theca interna become greatly enlarged and the theca cells become luteinised. The follicle does not rupture and the oocyte and surrounding granulosa cells remain enclosed within it (Woolley, 1966*b*). This formation has been described in the water shrew (Brambell, 1956) but is so far not known from any other marsupial even from the closely related species *Antechinus minimus* (Wilson, 1986). Martinez-Esteve (1942) states that after dissolution of the membrana granulosa in *D. virginiana* the theca cells remain as a conspicuous body, which he termed interstitial tissue. This is different from the origin of interstitial tissue in *M. eugenii* which, according to Alcorn (1975), is derived from the rete cords. It is also at variance with the conclusions of O'Donoghue (1916) who found interstitial tissue in the ovaries of 10 species of phalangeroid marsupials and absent from the ovaries of six dasyurid and peramelid species. In *M. eugenii* it is a prominent feature of all ovaries, and while it has the appearance of a steroid-secreting tissue it does not change in any regular way in relation to the reproductive state of the female (Renfree & Tyndale-Biscoe, 1973*a*) nor is it affected by hypophysectomy (Panyaniti *et al.*, 1985). It does, however, contain some of the enzymes of steroid metabolism, namely 3β-hydroxysteroid dehydrogenase-4, $5'$-isomerase and 17β-oxidoreductase but no aromatase or 5α-reductase (Renfree *et al.*, 1984*b*).

6

Ovarian function and control

The growth and development of ovarian follicles has been discussed in Chapter 5. In this Chapter we review the evidence for hormonal secretion and the several functions of the ovary in marsupials, as well as the evidence for endocrine control of the ovary. Because of its profound importance in the regulation of the oestrous cycle and pregnancy of marsupials, consideration of the functions and control of the corpus luteum forms a large part of this Chapter.

Serious study of the corpus luteum of marsupials dates from Sandes' (1903) paper on *Dasyurus viverrinus* and O'Donoghue's (1912, 1914, 1916) papers on this and nine other species. Sandes concluded that the corpus luteum is probably a gland of internal secretion with the functions of influencing the genital organs and of stopping further ovulation during pregnancy. O'Donoghue (1912) extended these conclusions to the corpus luteum of the non-pregnant female by showing that in its growth and development and in the related changes in the genital tract it was wholly equivalent to the corpus luteum of pregnancy, and this was subsequently shown to hold for almost all other marsupials (Sharman, 1970). During the past decade, the application of hormone-assay and surgical techniques has fully confirmed and extended Sandes' conclusion as to the importance of the corpus luteum in marsupial reproduction. There is now evidence that it is involved in the regulation of follicular growth and ovulation; in oestrous behaviour and, possibly, in male response; in embryo development and endometrial growth and secretion; and in the preparation of the genital tract for parturition and of the mammary gland for lactation. Notwithstanding this the evidence is gathering that, after its formation the marsupial corpus luteum (unlike its eutherian counterpart) is autonomous, depending neither on the pituitary for luteotrophic support nor subject to luteolysis at the close of its active phase.

Because very much more is currently known about ovarian function in *Macropus eugenii* than in any other marsupial, much of the discussion will be concerned with this species. However, sufficient is known about several other species to indicate that there are considerable differences between families, as discussed in Chapter 2, so that generalisations based on *M. eugenii* must be made with caution. This chapter divides into four main parts. We begin by reviewing what is known about the control of oestrus and ovulation in *M. eugenii* and compare this to what is known for three other species.

Formation, growth and endocrine secretions of the corpus luteum is then reviewed for three of the four types of reproduction defined in Chapter 2, including a review of the macropodid corpus luteum associated with quiescence and embryonic diapause. This is followed by a review of the interactions of the corpus luteum and ovary; the chapter closes with a review of the endocrine control of the corpus luteum and of its eventual demise.

Oestrus and ovulation

In all adequately studied species of marsupial, ovulation is spontaneous. *Phascolarctos cinereus* may, on further study, be shown to be an exception; some other species, when in captivity, fail to ovulate for reasons that are not presently known. Oestrus precedes ovulation by 1 or 2 days in most species, exceptions being *Antechinus stuartii*, *Dasyuroides byrnei* and *Dasyurus viverrinus*, in which the interval may be up to 10 days (Woolley, 1966b; Selwood, 1980), 4–6 days (Fletcher, 1983) and 5 days (Hill & O'Donoghue, 1913) respectively.

The short-term profiles of several hormones at post-partum oestrus and ovulation are now known in some detail for *M. eugenii*, as well as some information on the control of the sequence of events from late pregnancy through parturition, oestrus and ovulation (Fig. 6.1). Progesterone, secreted by the corpus luteum of pregnancy, remains elevated until less than 8 h before parturition, when it falls rapidly to basal levels (Hinds & Tyndale-Biscoe, 1982*a;* Tyndale-Biscoe *et al.*, 1983; Shaw & Renfree, 1984; Harder *et al.* 1985). This is preceded by a brief but substantial pulse of prolactin 8 h before or at parturition. This pulse does not occur at the end of the non-pregnant oestrous cycle, when a similar decline in progesterone occurs (Fig. 7.27). The significance of this will be considered, under the control of parturition later (p. 337).

Oestradiol is the major oestrogen (Shaw & Renfree, 1984; Renfree *et al.*, 1984*b*) and levels begin to rise 1 day before birth and reach a peak of 22 pg ml^{-1} about 8–12 h after the fall in progesterone and parturition. Copulation occurs at this time only in those females that have a level of oestradiol greater than 13 pg ml^{-1} (Shaw & Renfree, 1984; Harder *et al.*, 1985). The origin of the oestradiol peak is the single Graafian follicle growing in the ovary opposite the one carrying the corpus luteum of pregnancy; this was shown by a 2- to 4-fold increase in the venous outflow

Fig. 6.1. Summary of events at parturition, post-partum oestrus and ovulation in *Macropus eugenii*. (*a*) Growth of the Graafian follicle, ovulation and formation of the corpus luteum (CL): an, antrum; mg, membrana granulosa; o, oocyte; t, theca folliculi. (*b*) Changes in the plasma concentrations of progesterone, oestradiol and luteinising hormone related to hours before and after birth and progesterone fall. (*c*) Mean intervals between birth, oestrus, oestradiol peak, LH peak, minor progesterone peak and ovulation. Derived from data in Tyndale-Biscoe *et al.* (1983), Harder *et al.* (1984; 1985) and Shaw & Renfree (1984).

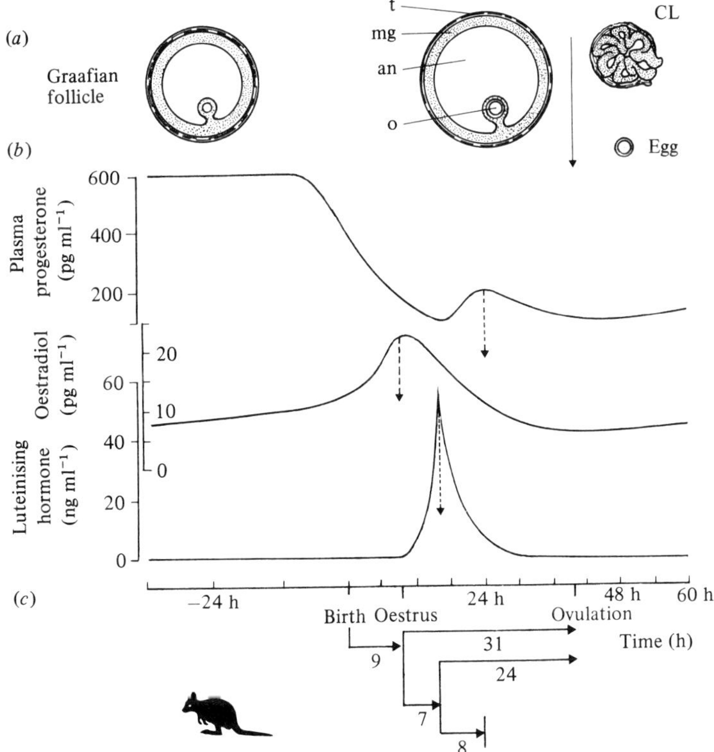

of oestradiol from this ovary (Harder *et al.*, 1984) and by abolishing the pulse in animals from which the ovary bearing the follicle was removed four days before expected oestrus (Shaw & Renfree, 1984; Harder *et al.*, 1985).

Of the two pituitary gonadotrophins, no changes have been observed in follicle stimulating hormone (FSH) (Evans *et al.*, 1980) but a marked pulse of LH occurs about 16 h after the progesterone fall and parturition (Tyndale-Biscoe *et al.*, 1983) or 8 h after oestrus (Sutherland, Evans & Tyndale-Biscoe, 1980) and the peak of oestradiol (Harder *et al.*, 1985). It is very probable that the surge of oestradiol initiates the LH pulse, since ovariectomised females given a single injection of oestradiol, which elevated peripheral concentrations to a similar extent, induced a similar LH pulse within 24 h (Horn, Fletcher & Carpenter, 1985). Oestradiol levels in both the ovarian vein draining the Graafian follicle and in the peripheral circulation decline shortly before the LH pulse (Harder *et al.*, 1984, 1985) and have reached basal levels 12 h before ovulation. A transient pulse of progesterone occurs in some animals at the same time as, or 8 h after, the LH pulse (Tyndale-Biscoe *et al.*, 1983). Although there is no experimental evidence as yet, these sequential changes in oestradiol and progesterone could result from the LH surge inhibiting aromatase activity and so redirecting steroid synthesis in the follicle from oestradiol to progesterone. Such an effect of LH has been demonstrated in the pre-ovulatory ewe (Goodman *et al.*, 1981). The interval from oestrus (mating) to ovulation varies from 24 h to 48 h (Tyndale-Biscoe & Rodger, 1978; Sutherland *et al.*, 1980) but the interval from the LH pulse to ovulation is very close to 24 h (Harder *et al.*, 1985). Thus *M. eugenii* is similar in this respect to the sheep, monkey and human.

The respective roles of the pituitary, corpus luteum and Graafian follicle in controlling this sequence of events has been investigated by surgical removal, and by immunisation against gonadotrophin releasing hormone (GnRH). Female *M. eugenii* hypophysectomised during late pregnancy failed to give birth, failed to show oestrus and failed to show a pulse of LH or to ovulate (Hearn, 1974), whereas females immunised against GnRH, which presumably prevented the secretion of FSH and LH, gave birth at the normal time but failed to show oestrus or to ovulate (Short *et al.*, 1985). None of these animals had enlarged Graafian follicles but their corpora lutea of pregnancy were of normal size and appearance.

Although these experiments confirm that the pituitary plays an essential role in the maturation of the follicle prior to the LH pulse, no evidence for this can be detected in the peripheral circulation of the intact animal before parturition. Removal of the Graafian follicle 3 days before expected birth abolished the rise in oestradiol, as mentioned previously, but these animals also failed to show oestrus, an LH pulse, or ovulation up to 4 days after the normal time (Harder *et al.*, 1985). However, those examined 9 days after expected ovulation, that is to say 12 days after removal of the Graafian follicle, had grown new follicles, which had ovulated, but none of those animals had mated.

The role of the declining corpus luteum in oestrus and ovulation is less clear cut. As mentioned in Chapter 2, the interval between one oestrus and the next is significantly longer in the non-pregnant cycle than in the pregnant cycle and the difference is associated with a slow decline of the corpus luteum and plasma progesterone. Tyndale-Biscoe *et al.* (1983) concluded that the decline in progesterone is the factor that sets the timing of the subsequent events and that in pregnancy some signal from the fetus, possibly involving the prolactin pulse, accelerates the demise of the corpus luteum and hence the earlier onset of oestrus. This idea was tested by removing the corpus luteum (Harder *et al.*, 1985), which had been shown to cause progesterone in circulation to decline to basal level in 1 day (Findlay, Ward & Renfree, 1983). When the corpus luteum was removed less than 3 days before expected oestrus it did not advance or affect the rise in oestradiol, oestrus or ovulation (Harder *et al.*, 1985). However, if it was removed more than 6 days before expected oestrus animals did not show oestrus, even though the Graafian follicle developed, and ovulation occurred (Evans *et al.*, 1980; Stewart, 1984). This suggests that an active corpus luteum must be present shortly before the oestradiol rise in order to elicit oestrous behaviour but is not involved in the timing of the oestradiol pulse or ovulation. On the other hand, when the corpus luteum had been missing and progesterone levels low for more than a few days, oestrus will not occur even when a Graafian follicle is present and ovulates.

This role of the corpus luteum may be restricted to *M. eugenii* since *M. rufus*, which responded to rains breaking a long drought, mated and ovulated in less than 14 days (Frith & Sharman, 1964; Newsome, 1964*b*; Sharman & Clark, 1967). In the latter species, all had been anoestrous beforehand so none would have had an active corpus luteum to provide the postulated pro-oestrous stimulus. Likewise, in all species that have the Type 1 pattern, the corpus luteum has regressed and progesterone declined to basal levels many days before the next oestrus occurs (see below).

To conclude, the sequence of events in *M. eugenii* (Fig. 6.1) involves first a pituitary stimulation of follicular growth in the ovary contralateral to the corpus luteum, which leads to rising levels of oestradiol from the follicle. This, possibly combined with declining progesterone and the resultant change in the oestrogen : progesterone ratio, elicits oestrous behaviour and, by positive feedback on the pituitary, produces the LH surge. The LH surge redirects follicular steroid synthesis from oestradiol to progesterone and sets in train the events that lead to ovulation 24 h later.

Comparative data from other species are sparse. In *Didelphis virginiana* Harder & Fleming (1981) measured peak values of oestradiol of 24 pg ml^{-1} on the 4 days preceding oestrus, at the time when the ovaries contained Graafian follicles, and they measured low values during the luteal phase of the oestrous cycle and pregnancy (Fig. 6.3*a*). Similarly, Curlewis *et al.* (1985) were able to detect oestradiol in peripheral plasma of *Trichosurus vulpecula* only on the day of oestrus or pro-oestrus. They identified several steroids in the venous outflow from the ovary of this species and found oestradiol to be the major oestrogen and oestrone a minor component. In this species C. A. Horn (unpublished results) detected a transient LH pulse only on the day of oestrus. As ovulation occurs about 1 day after oestrus (Hughes & Rodger, 1971), the sequence is probably similar to that in *M. eugenii*, except that the decline of progesterone occurs very much earlier (Fig. 6.7).

An unusual pattern of LH secretion has been observed in *Dasyuroides byrnei* (Fletcher, 1983). As mentioned earlier, ovulation in this species occurs 4–6 days after oestrus but a large pulse of LH was only detected 12 days before oestrus, or some 16–18 days before ovulation (Fig. 6.6). Because of the small size of the species, blood samples could only be taken at 3–day intervals, so it is possible that a later pre-ovulatory pulse of LH was not detected, but this does not help to understand the function of the very large pro-oestrous pulse. Fletcher (1983) offers the suggestion that it may induce the pro-oestrous rise in progesterone (Fig. 6.6) by redirecting steroid biosynthesis in the follicles or interstitial tissue, and that this may be important in the induction of oestrus, as we have suggested may obtain in *M. eugenii*. As will be apparent later (p. 214) its larger relative *Dasyurus viverrinus* has a similar cycle (Fig. 6.5) and this matter could be resolved using this species.

Formation and development of the corpus luteum

At ovulation the wall of the Graafian follicle ruptures on the outer surface and the ovum is expelled into the fimbria without any cells of the membrana granulosa adhering to it (Figs 6.1 and 7.5*b*). Some granulosa cells and extravasated blood cells extrude through the point of rupture and form a conspicuous pink ovulation tip or *bouchon epithelial* (Fig. 6.2*a*).

Fig. 6.2. Corpus luteum formation in *Macropus eugenii*. (*a*) Section through the ovulation point (↓) of a newly ruptured follicle, 49 h after oestrus or about 20 h after ovulation. (*b*) Section of a corpus luteum (CL) on day 6 post-oestrus. Note the centrally placed interstitial tissue (IT) in both ovaries. Bar, 1.0 mm.

(*a*)

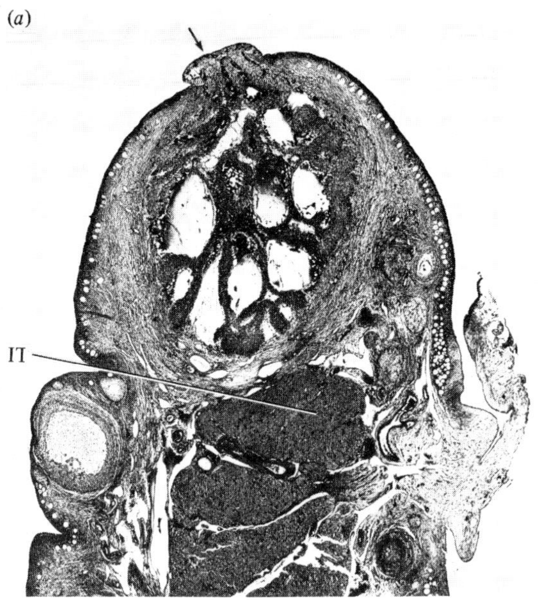

(*b*)

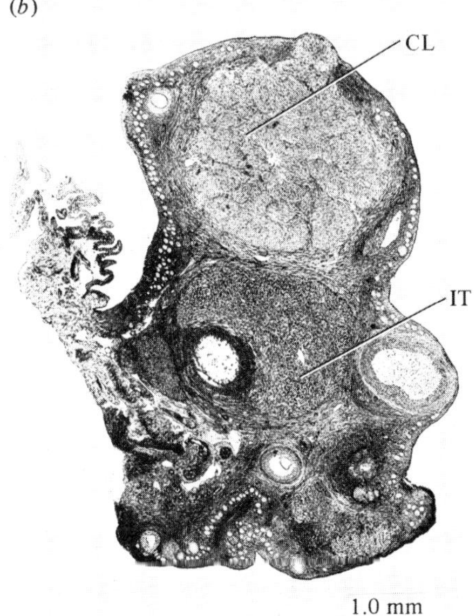

While this is small in most species in *T. vulpecula* a large quantity of membrana granulosa may become everted at this time. In most species the basement membrane supporting the granulosa cells is penetrated by elements of the theca interna which then fill the central cavity with connective tissue and blood capillaries and thus provide a matrix which the granulosa cells proceed to fill as they expand. This pattern occurs in *D. viverrinus* (Sandes, 1903), *Perameles nasuta*, *Isoodon obesulus*, *Macropus rufogriseus* and *Phascolarctos cinereus* (O'Donoghue, 1914, 1916), *I. macrourus* (Lyne & Hollis, 1979) and *M. giganteus* (Clark & Poole, 1967). In *P. cinereus* the central cavity remains throughout pregnancy and O'Donoghue (1916) suggested that this may be a factor in the common occurrence of cystic ovaries in this species. In other species the basement membrane is not penetrated until several days after ovulation, so that the central cavity has been obliterated by hypertrophied granulosa cells before the capillary network is established in it. This pattern occurs in *Didelphis albiventris*, *Trichosurus vulpecula* (O'Donoghue, 1916), *Antechinus stuartii* (Woolley, 1966*b*) and *Macropus eugenii* (C. H. Tyndale-Biscoe, unpublished observations; Fig. 6.2).

Although cells of the theca interna can readily be identified before ovulation it is difficult to recognise them afterwards, so that it is not clear whether they contribute to the luteal cell population or not. O'Donoghue (1914, 1916) came to the firm conclusion that thecal cells do not become luteinised in any species of marsupial and Martinez-Esteve (1942), Guraya (1968*a*) and Godfrey (1969*a*) agreed with him in respect to *D. virginiana* and *Sminthopsis macroura*. However, Sharman (1955*a*) said they probably did contribute to the corpus luteum of *Setonix brachyurus*.

If thecal cells do contribute to the luteal cell population, as they do in the pig, it might be supposed that they could be distinguished with the electron microscope. However, only one type of luteal cell was described in the corpora lutea of *T. vulpecula* (Shorey & Hughes, 1973*b*), *Perameles nasuta*, *Isoodon macrourus* (Hollis & Lyne, 1980) and *M. rufogriseus* (Walker, Gemmell & Hughes, 1983).

While all authors agree that mitoses occur in thecal cells during the first few days after ovulation, there is disagreement as to whether granulosa cells undergo cell division after ovulation. This is probably a consequence of the difficulty of identifying thecal cells that may have become luteinised. All agree that the corpus luteum is devoid of mitoses by 4 days after ovulation and that subsequent growth of the corpus luteum to its full size results from hypertrophy of luteal cells. This is in accord with eutherian species.

Involution of the corpus luteum in the non-pregnant female is recognised by degeneration and shrinkage of luteal cells, pycnosis of luteal cell nuclei and a preponderance of connective tissue. In some species, such as *Didelphis virginiana*, the corpora lutea have almost disappeared by the beginning of the next cycle (Hartman, 1923*a*). In *M. giganteus* the corpus albicans lasts for 150 days (Clark & Poole, 1967) while those of *T. vulpecula* and *M. eugenii* are recognisable for less than a year. All the corpora albicantia of *Trichosurus caninus*, however, are retained as large structures, probably for life (Smith & How, 1973) and the corpora albicantia of *Phascolarctos cinereus* may also survive for several years (O'Donoghue, 1916).

From structural and endocrine studies on 13 species, representing the major families of marsupials, 3 broad patterns of corpus luteum growth and decline can be recognised. These coincide with 3 of the 4 types of reproduction referred to in Chapter 2. In the majority of marsupials, represented here by *Didelphis virginiana*, *Dasyurus viverrinus*, *Dasyuroides byrnei*, *Antechinus stuartii* and *Trichosurus vulpecula*, the luteal phase occupies no more than 60% of the oestrous cycle and is followed by a follicular phase leading to the next oestrus and ovulation. Pregnancy is accommodated within the luteal phase, and parturition coincides with corpus luteum regression. If lactation follows, the subsequent follicular phase is suppressed and the corpus luteum of pregnancy slowly disappears or remains as a corpus albicans.

In the second pattern, seen in *Perameles nasuta* and *Isoodon macrourus*, gestation is equally short but parturition occurs during the peak of the luteal phase. If lactation follows, the corpora lutea remain large and the follicular phase is suppressed for most of lactation.

In the third pattern, seen in the majority of the Macropodidae, the luteal phase lasts for more than 90% of the cycle and the follicular phase is not suppressed. Gestation occupies all of the extended luteal phase and is followed by post-partum oestrus and ovulation. If lactation follows, the corpus albicans associated with pregnancy declines in the same way as in the non-pregnant female, but the new corpus luteum formed at post-partum ovulation is held in a state of quiescence, which may persist for the whole of pouch occupancy and, in two species, for some months after.

Insufficient is known about the corpora lutea of the Burramyidae or *Tarsipes rostratus* so the subsequent discussion will consider each of the three other patterns separately, and then compare them with respect to the endocrine functions and control of their corpora lutea.

Type 1: short gestation, short luteal phase

The luteal cells of the corpus luteum of *Didelphis virginiana* are derived wholly from the membrana granulosa and, since mitoses were rarely seen, Martinez-Esteve (1942) concluded that the chief growth of the gland is due to hypertrophy of these pre-formed cells. As in *D. albiventris* (O'Donoghue, 1916), blood vessels and connective tissue trabeculae from the theca interna form the skeleton of the gland. Growth of the gland is unusually rapid: according to Hartman (1923*a*) and Martinez-Esteve (1942) full size is attained by day 3 post oestrus but, according to Fleming & Harder (1983), not until day 7 (Table 6.1) The latter authors found no significant difference between the growth of corpora lutea in pregnant and non-pregnant animals. Decline of the corpora lutea begins after day 8 and by day 12, just prior to parturition, the glands have become shrunken, lost vascularity and are infiltrated with leucocytes. By day 20 they are reduced to yellow specks by which time, in non-lactating females, a new crop of Graafian follicles dominates the ovary. Oestrus recurs on about day 28. These changes in size of the corpora lutea are closely reflected in levels of progesterone in peripheral plasma (Fig. 6.3*b*). Cook & Nalbandov (1968) indentified progesterone by recrystallisation as the main steroid synthesised by luteal tissue of *D. virginiana in vitro*. Under these conditions the tissue converted cholesterol and pregnenolone to progesterone but could not utilise acetate (Cook *et al*., 1977*b*).

Because of the relatively high levels of progesterone measured in plasma Cook *et al*. (1977*b*) tested the pooled plasma of 13 animals for corticosteroid binding globulin (CBG) and sex hormone binding globulin (SHBG) using steady-state gel electrophoresis. This showed that the plasma contains CBG but lacks SHBG, a conclusion later confirmed for a wide variety of marsupials by Sernia, Bradley & McDonald (1979).

Table 6.1. *Dimensions of corpora lutea and ovaries in* Didelphis virginiana

	Day 3	Day 7	Day 11
Ovarian weight (mg)	394 ± 20	620 ± 28	513 ± 31
Corpora lutea diameter (mm)	1.43 ± 0.02	1.86 + 0.03	1.80 ± 0.02
Calculated volume (μl)	1.5	3.4	3.1
Number of corpora lutea	30.6 ± 2.4	32.6 ± 2.2	29.7 ± 2.0
Total luteal tissue (μl)	45.9	109.8	90.7

From Fleming & Harder (1983).

Harder & Fleming (1981) determined levels of progesterone in peripheral plasma throughout pregnancy and the oestrous cycle (Fig. 6.3*b*) and showed that, as with the growth of the corpora lutea, there was no significant difference that could be ascribed to pregnancy but the profiles clearly reflected the growth and decline of the corpora lutea.

Whereas progesterone is being largely secreted by the corpora lutea, the profile of plasma oestradiol-17β, obtained by Harder & Fleming (1981) (Fig. 6.3*a*), suggests that it is probably secreted by the pre-ovulatory follicles or other ovarian tissue; maximum levels of more than 22 pg ml^{-1}

Fig. 6.3. Concentrations (mean ± s.e.m.) of oestradiol and progesterone in the peripheral circulation of *Didelphis virginiana* during the oestrous cycle (●—●) and gestation (○--○). Redrawn from Harder & Fleming (1981).

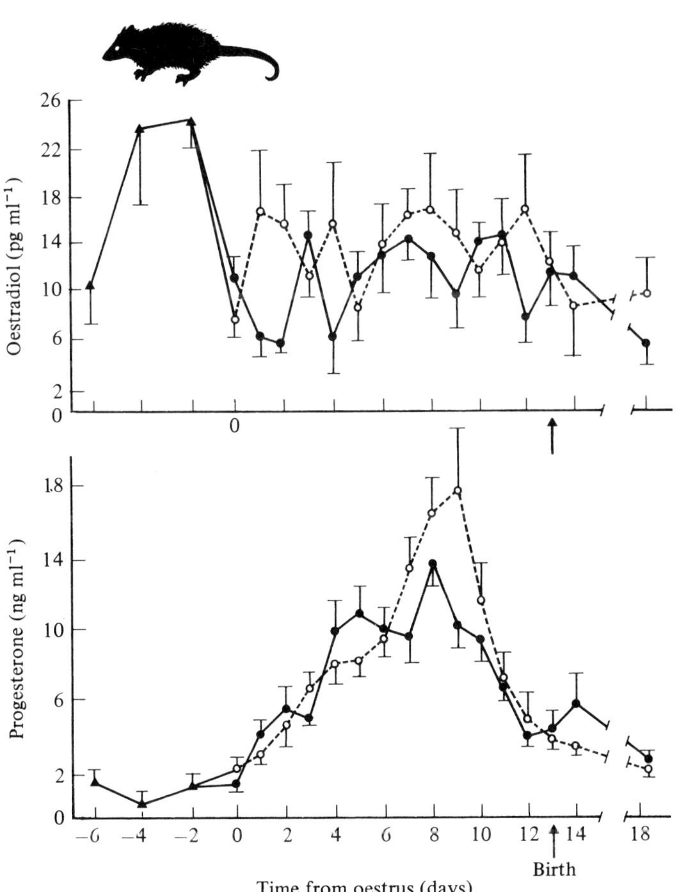

occurred during the 4 days preceding oestrus and declined to less than 6 pg ml^{-1} during the period of corpus luteum growth and elevated plasma progesterone. As with progesterone, no significant differences could be detected between levels in pregnant and non-pregnant animals. However a significant difference in the ratio of oestrogen: progesterone for the two states was found on day 12, which may be associated with the onset of parturition.

Changes in the weight of the endometrium (Renfree, 1975) show a similar pattern to the profile of plasma progesterone (Fig. 6.4), suggesting that its size and function is controlled by the corpora lutea. In pregnant *D. virginiana* expansion of the blastocysts from less than 0.25 mm on day 4 to 5 mm on day 8 (Hartman, 1928) occurs during the period of increasing output of progesterone by the corpora lutea and increasing weight of the endometrium. Because of the large number of corpora lutea (*c.* 30) it is not possible to test their role directly, except by total ovariectomy. While bilateral ovariectomy caused the endometrium to collapse (Hartman, 1925*b*) the contained embryos continued to develop and grow for several days after ovariectomy as early as day 2, and continued to full term after ovariectomy on day 6 (Hartman, 1925*b*; Renfree, 1974*a*). However, parturition was never successful, even after ovariectomy as late as day 11. In a single female that had haemorrhagic corpora lutea but otherwise normal ovaries, Hartman (1927) observed endometrial collapse and embryo death similar to that seen after bilateral ovariectomy and he concluded that this was further evidence for the paramount role of the corpora lutea in the maintenance of pregnancy in this species.

Three species of the Dasyuridae display a somewhat different Type 1 reproductive cycle. *Dasyurus viverrinus* was the object of much research at the turn of the century and, as a result, there is a detailed record of the histological changes that take place in the corpus luteum (Sandes, 1903; O'Donoghue, 1912), uterus (Hill & O'Donoghue, 1913) and embryo (Hill, 1900*b*; 1910), throughout gestation and of associated changes in the mammary gland and pouch (O'Donoghue, 1911). Until very recently, however, these studies could not be properly related in time, or to hormonal changes. Part of the difficulty in obtaining timed stages is due to the variable duration of oestrus and the variable interval between oestrus and ovulation already referred to. Now that the gestation period is known to be 19 days and the oestrous cycle in unmated females to be 37 days (Fletcher, 1985; J. C. Merchant, unpublished results), it is possible to obtain much more information about intrauterine development from the early data in Hill (1910), Hill & O'Donoghue (1913), and O'Donoghue

(1911) and from the unpublished records of the J. P. Hill collection at the Hubrecht Laboratory, Utrecht. Because it is so scattered all the timed cases have been collated in Table 6.2. This can be related to plasma progesterone profiles though pregnancy, determined by Hinds (1983) (Fig. 6.5). As in *Didelphis* the progesterone profiles through pregnancy cannot be distinguished from those of the oestrous cycle.

The pro-oestrous period, which last for 4–10 days, in *Dasyurus* can be recognised by the increasing moistness of the pouch region and the swelling of the vulva, both of which reach maximum development at oestrus (Sandes 1903; Hill & O'Donoghue, 1913). Oestrus usually lasts for 1 day, less often for 3 days, but live, active spermatozoa can be found in clumps in the oviducts of mated females for up to 14 days *post coitus* (*p.c.*) (Hill & O'Donoghue, 1913). Ovulation occurs 4–7 days after oestrus according to 12 records reported by Hill (1910) and Hill & O'Donoghue (1913) (Table 6.2). Three females killed on day 5 *p.c.* had not ovulated but the oocytes were completing their maturation divisions in the follicles, the first polar body was extruded and, in one animal, the spindle for the second division was established. Five females with unsegmated uterine eggs, containing the male and female pronuclei, were obtained on days, 4, 5, 6, 7, and 8, 2-cell

Fig. 6.4. The weight of endometrium of *Didelphis virginiana* during lactation (lact.), pregnancy (●) and the luteal phase of the oestrous cycle (○). Redrawn from Renfree (1975).

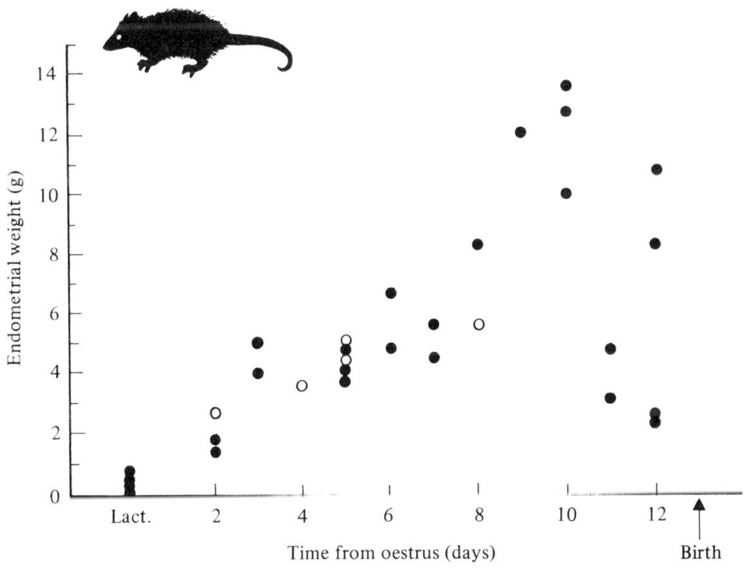

Table 6.2. *Summary of all embryos of known age of* Dasyurus viverrinus *in J. P. Hill Collection.*

Days Post-coitum	Stage of development									Reference
	Oocyte	1-cell	2-cell	4-cell	Blastocyst	Bilaminar vesicle	Primitive streak	Fetus	Neonate	
4			2							1, 2, 3
5	3		1	1						1, 2, 4
6		1	1							1, 2, 4
7		1	1							1, 2
8		1							1	1, 2, 4
9										
10										
11				1	1					1, 4
12										
13						1				4
14										
15					1					1
16									1	2, 4
17						1	1	1		4
18	1									3
19								2		4

Data from (1) Hill, 1910; (2) Hill & O'Donoghue, 1913; (3) O'Donoghue, 1912; (4) Marsupial catalogue, Hubrecht Embryological Institute.

eggs were also recovered from the same female on day 4 *p.c.* and from the uteri of three others on days 5, 6, and 7. Four-cell eggs were recovered from three females on days 5, 11 and 18 *p.c.* respectively. Sandes (1903) described the formation of the corpus luteum in some of these animals and O'Donoghue (1912) described its formation in females that had either not been mated or from which unfertilised eggs were recovered. He could find no differences from Sandes' account.

Sandes (1903) states that formation of the corpus luteum from the ruptured follicle takes 3 days but no evidence is given for this nor measurments of dated corpora lutea. Nevertheless this may be near to the truth when the level of progesterone in plasma is considered (Fig. 6.5); the concentration is less than 2 ng ml^{-1} until day 10 *p.c.* and then rises to maximum levels of 8–10 ng ml^{-1} by day 13 *p.c.* If ovulation occurs at day 5 or 6, the formation of the corpora lutea would be complete by day 8–9 and the growth phase would coincide with the rise in peripheral progesterone. This is consonant with Sandes' (1903) description of the luteal cells becoming filled with granules and drops of secretion and much enlarged. The period of high progesterone concentration from days 13–19 is the period when expansion of the blastocysts occurs and embryogenesis proceeds to full term (Table 6.2).

The decline in progesterone is equally rapid, falling to less than 1 ng ml^{-1} by day 21 *p.c.* and remaining low until 2 days before the next oestrus, when there is a minor but significant pro-oestrous peak. Presumably the source

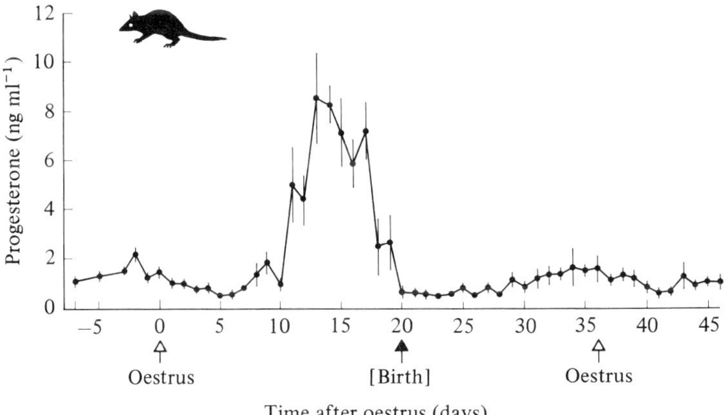

Fig. 6.5. Concentration (mean ± s.e.m.) of plasma progesterone in *Dasyurus viverrinus* through the oestrous cycle. The time that birth would occur in a pregnant animal is shown. From Hinds (1983), with permission.

of this is the new crop of developing follicles. However, the rapid decline in progesterone at day 21, reported by Hinds (1983), does not correspond to the histological appearance of the corpora lutea. O'Donoghue (1912) found fully grown corpora lutea in non-pregnant females, 20, 22 and 25 days *p.c.*, while Sandes (1903) and O'Donoghue (1911), state that the corpora lutea remain in the same condition as at the peak of growth for 7–8 weeks of lactation. A similar anomaly between declining secretory activity and persistent large corpora lutea occurs in the Peramelidae (see below). Regression in *Da. viverrinus* is marked, as in *Didelphis virginiana*, by shrinkage and leucocytic infiltration. The pattern and levels of peripheral progesterone in *Da. viverrinus* resemble those of *D. virginiana* except for the initial post-oestrous period, which can be accounted for by the delay in ovulation. The maximum of 10 ng ml^{-1} is somewhat less than *D. virginiana* corresponding, perhaps, to the slightly smaller number of corpora lutea produced.

The plasma progesterone profiles in pregnant and non-pregnant cycles of *Dasyuroides byrnei* (Fletcher, 1983) (Fig. 6.6*a*) are not different from each other and resemble the profiles of *Da. viverrinus* already described (Fig. 6.5). During the pro-oestrous period progesterone remains at less than 2 ng ml^{-1} except for a small peak of short duration 4–5 days before oestrus, which, as mentioned earlier, is preceded by a pulse of LH. Ovulation occurs 4–6 days after oestrus and progesterone reaches 4 ng ml^{-1} about this time. However, the main peak of progesterone begins about day 15, reaching 12 ng ml^{-1} at day 25. A steep decline on day 30–31 coincides with parturition in pregnant females or in non-pregnant females with 'pseudo-birth', when body weight drops and vaginal bleeding occurs. In the absence of lactation, progesterone remains low until the pro-oestrous peak is repeated just before the next oestrus at day 60.

Hormone profiles have also been determined in *Antechinus stuartii* (L. A. Hinds, personal communication) and the relationships between ovarian events, especially corpus luteum size, and the course of gestation can be examined. Females are monoestrous and polyovular and, at the start of the breeding season, each ovary contains up to 6 Graafian follicles (Selwood, 1983) and a very much larger number of follicles in which the thecal cells have proliferated and become luteinised (Woolley, 1966*b*). In these the oocyte is surrounded by a single layer of membrana granulosa cells and there is no antrum. As in *Da. viverrinus*, active spermatozoa are retained in crypts in the oviducts for up to 10 days (Woolley, 1966*b*) and ovulation may occur several days after oestrus. As mentioned in Chapter 2, ovulation coincides with a marked decline in the cell contents of the urine

Fig. 6.6. Summary of changes in a group of six *Dasyuroides byrnei* which underwent an oestrous cycle (broken line) followed by a pregnancy (solid line). (*a*) Concentration (mean ± s.e.m.) of plasma progesterone to show sharp drop before birth, or vaginal bleeding in non-pregnant cycle. (*b*) Change in body weight during oestrous cycle and pregnancy, calculated as per cent change from the value taken 25 days prior to the first oestrus of the year. (*c*) All values were synchronised to day 2 and correlated with the urine cell cycle, mating and birth. Redrawn from Fletcher (1983), with permission.

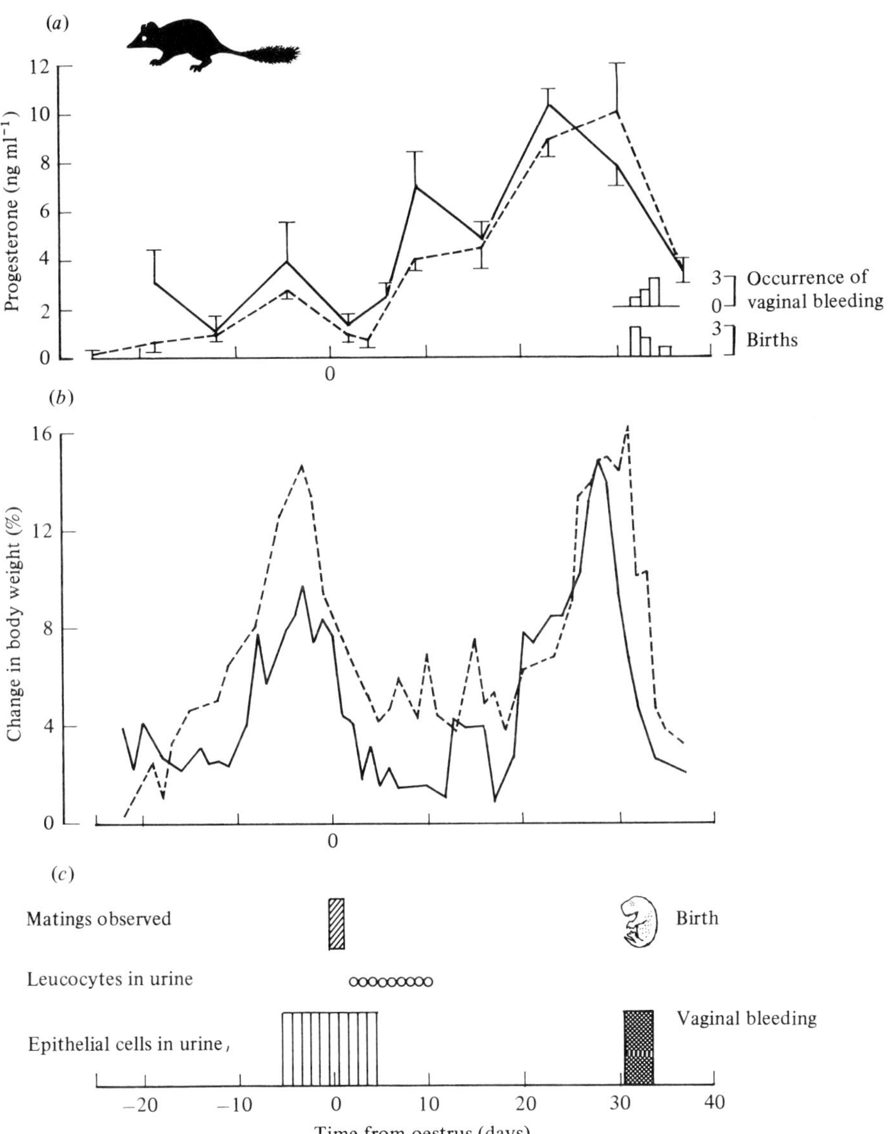

(Selwood, 1980). Much of the variation in gestation (27.2 ± 1.83 days, Selwood, 1983) is due to a slow rate of early cleavage and a period of developmental stasis before blastocyst expansion begins on day 16 (Selwood, 1981), whereas the period that follows is less variable (Fig. 7.11). During the first part of pregnancy the corpora lutea have a diameter of less than 0.6 mm and are composed of small luteal cells, and they grow to 0.9 mm diameter by hypertrophy at the time blastocyst expansion begins on day 16. Woolley (1966*b*) suggested that blastocyst expansion might depend directly or indirectly on stimulation by secretions of the enlarging corpora lutea, an idea now supported by measurements of progesterone in peripheral plasma (L. A. Hinds, unpublished observations). In females collected on days 0–12 the concentrations were less than 5 ng ml^{-1} and rose to more than 10 ng ml^{-1} from day 13.

Trichosurus vulpecula is the only Type 1 species studied, which is monovular and, as in other monovular species, ovulation takes place in alternate ovaries in successive cycles (O'Donoghue, 1916; Pilton & Sharman, 1962). This distinguishes monovular marsupials from monovular eutherian species in which ovulation is not so regular. At ovulation the follicle collapses completely, obliterating the antrum, and drawing thecal strands in with the folds. At the same time, granulosa cells can be everted as a mushroom-like outgrowth. O'Donoghue (1916) saw no mitoses in either component but Pilton & Sharman (1962) reported mitoses in the thecal elements during the first 3 days. Maximum size of 4.0 to 4.5 mm diameter is reached 7–10 days after oestrus (Shorey & Hughes, 1975) as a result of hypertrophy of luteal cells (Pilton & Sharman, 1962) and no difference in size or rate of growth was detected between pregnant and non-pregnant animals. From day 17 to 19 histological signs of degeneration are seen – luteal cells shrinking, their nuclei irregular in shape and the connective tissue more prominent.

These morphological changes in the corpus luteum are precisely reflected in the levels of progesterone in the peripheral circulation as first determined by Thorburn, Cox & Shorey (1971) and Shorey & Hughes (1973*a*) in single samples from a mixed group of 4 pregnant and 14 non-pregnant females. Hinds (1983) and Curlewis *et al.* (1985) have confirmed their findings by sequential sampling through the oestrous cycle and pregnancy and have shown that the profiles are not significantly different (Fig. 6.7*a* and *b*). This amply confirms Pilton and Sharman's (1962) prediction, based on their measurements of total pregnanediol excreted in the urine on days 10–16 in pregnant (8.2 μg) and non-pregnant (9.9 μg) females and males (3.9 μg). Until day 8 the levels of progesterone in peripheral plasma are less than

1 ng ml^{-1} but rise rapidly thereafter to a peak of more than 8 ng ml^{-1} on days 12–14. The subsequent decline is equally rapid beginning on day 16 and reaching the basal level by day 19–20. In pregnant females parturition occurred 16.5 or 17 days *p.c.*, immediately after the start of progesterone decline.

The secretory function of the corpus luteum through the oestrous cycle was examined by Thorburn *et al.*, (1971) who measured progesterone in plasma collected from the ovarian vein draining the ovary bearing the

Fig. 6.7. Progesterone concentrations (mean ± s.e.m.) in *Trichosurus vulpecula* during (*a*) the oestrous cycle and (*b*) pregnancy. Occurrence of birth in 3 females indicated by histogram. Redrawn from Hinds (1983), with permission.

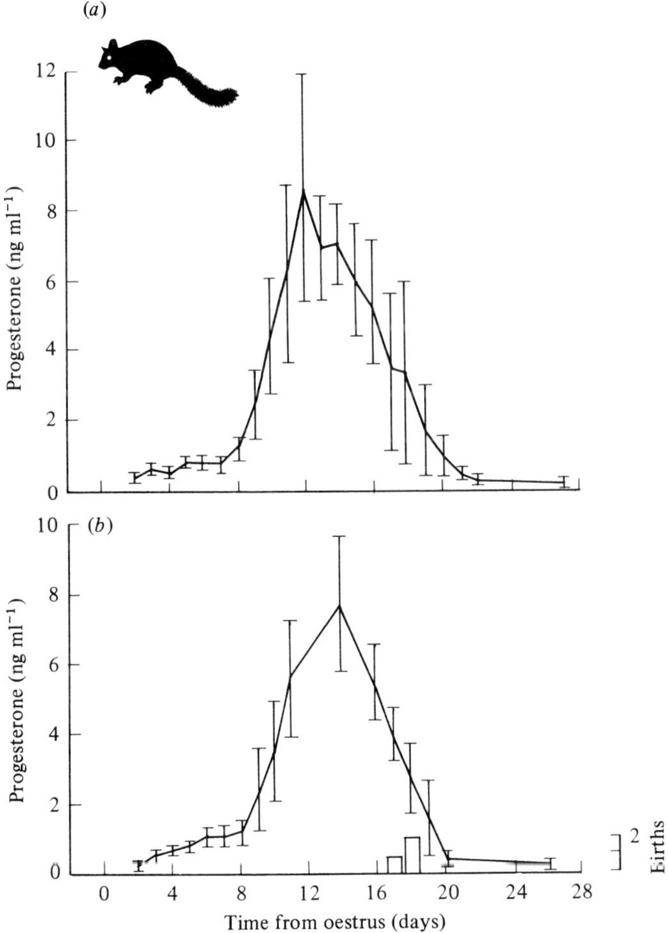

corpus luteum, using a competitive protein binding (CPB) assay. Shorey & Hughes (1973*b*) correlated this with changes in the fine structure of the luteal cells from the same animals. For the first 7 days after ovulation, the luteal cells underwent rapid intracellular differentiation with proliferation of vesicular elements of the agranular endoplasmic reticulum, and formation of closely associated lipid droplets; there were relatively few mitochondria. During this period the concentration of progesterone (Fig. 6.8*a*) and its secretion rate (Fig. 6.8*b*), as measured in the ovarian vein (called utero-ovarian vein by Shorey & Hughes, 1973*b*) draining the ovary bearing the corpus luteum, was low and hardly distinguishable from anoestrus. During the luteal phase, which begins on day 8 the luteal cells were

Fig. 6.8. (*a*) Concentrations and (*b*) secretion rate of progesterone in the ovarian vein draining the active corpus luteum in *Trichosurus vulpecula*, during the oestrous cycle. Redrawn from Shorey & Hughes (1973*a*).

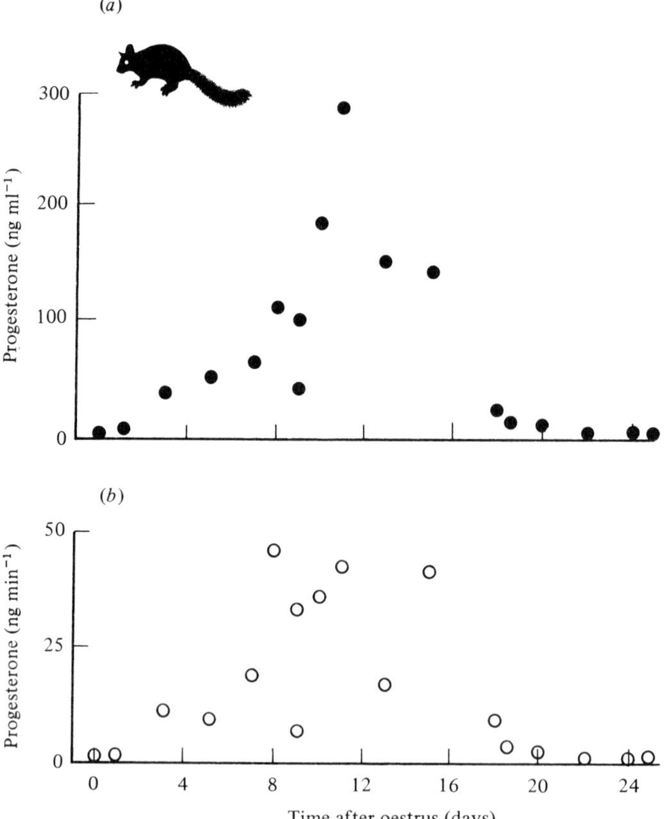

characterised by intense development of agranular endoplasmic reticulum arranged in whorls, a decrease in lipid and an increase in mitochondria. Membrane-bound, electron-dense vesicles measuring 150 nm also appeared in association with, and probably formed by, the Golgi complex. These changes coincided with a very marked increase in the concentration of progesterone in the ovarian vein plasma, which rose to a peak on day 11, and in the secretion rate. Ablation of the corpus luteum in four females on day 7 reduced the progesterone concentration in the ovarian vein from 62.7 ng ml^{-1} to 18.5 ng ml^{-1} after 30 min and the secretion rate from 18.62 ng min^{-1} to 1.10 ng min^{-1} (Shorey & Hughes, 1975), thus clearly implicating the corpus luteum as the major source of progesterone.

Relaxin activity, as measured by the mouse pubic symphysis bioassay, was significantly elevated in corpora lutea collected from *T. vulpecula* on days 9 and 15 of the oestrous cycle compared to days 4 and 19 (Tyndale-Biscoe, 1969), which correlates with the appearance of the membrane-bound vesicles.

Involution of the luteal cells had begun on day 16 with the appearance of prominent hyaline bodies derived from mitochondria and irregular-sized, densely-stained lipid droplets. Progesterone concentration in the ovarian vein plasma and the ovarian secretion rate had declined to basal levels by day 20 (Shorey & Hughes, 1973*b*). Four pregnant females were included in this study and the values for progesterone were the same as for the non-pregnant females and no differences could be detected in the structure of the corpora lutea.

Thorburn *et al.* (1971) also reported the secretion of an oestrogen-like substance at a high rate (12 ng h^{-1}) just before oestrus, and at low levels 1 day after oestrus. More recently Curlewis *et al.* (1985) have confirmed that this steroid is oestradiol by gas chromatography/mass spectrometry. Oestradiol concentration was 2 ng ml^{-1} at proestrus and 0.1 ng ml^{-1} at day 13. From its occurrence in these two studies it seems unlikely that oestradiol is secreted by the corpus luteum and more probable that it comes from the Graafian follicle before ovulation, as in *D. virginiana* and *M. eugenii*.

The effects of ablation of the corpus luteum or of the ovary bearing the corpus luteum on the oestrous cycle and pregnancy have been investigated by Shorey & Hughes (1975) and Sharman (1965*a*). Removal of the corpus luteum on day 2 or 4 prevented the development of the luteal phase in the endometrium and the animals returned to oestrus 8 or 9 days later. Removal on day 8 did not prevent luteal development but full growth was affected, while removal on day 12 was without apparent effect. Removal of the corpus luteum on day 7 and 10 was followed by death of the embryos but after lutectomy on day 11 two out of four animals gave birth on day 17 (Sharman, 1965*a*). It is now evident that the rise in peripheral progesterone that begins on day 8 is important for the full development of the secretory endometrium, which, in turn, supports rapid expansion of the embryo vesicle.

In all five species that display the Type 1 pattern, the profile of peripheral progesterone has common features; there is an early post-oestrous period of low concentration followed by a rise to peak concentration of 8–15 ng ml^{-1}, which coincides with growth of the corpora lutea and expansion of the blastocysts in pregnant animals. A rapid decline in progesterone follows several days before the next oestrus and ovulation. No difference is evident in any of these species in the pattern or concentration of progesterone between pregnant and non-pregnant cycles until parturition, which coincides with the decline in progesterone. Thereafter lactation prevents the subsequent oestrus and ovulation. In the two species examined, oestradiol 17β is the main oestrogen and reaches peak concentration at or before oestrus, coincident with the presence of Graafian follicles in one ovary.

Type 2: short gestation, prolonged luteal phase

The pattern of corpus luteum growth and secretion in *Perameles nasuta* and *Isoodon macrourus* (Fig. 6.9) during pregnancy resembles the Type 1 pattern already described but the subsequent fate of the corpora lutea at, and after, parturition is very different indeed; for they persist as large, and apparently functional glands, for two-thirds of the 60-day lactation.

Formation of the corpus luteum was described for *P. nasuta* and *I. obesulus* by O'Donoghue (1914) and for *P. nasuta* and *I. macrourus* by Lyne & Hollis (1979). At ovulation the follicle wall does not collapse but much blood enters the antrum, presumably from ruptured thecal capillaries. By day 1, thecal elements have penetrated the basement membrane and the layers of granulosa cells, where they begin to fill the antrum with connective tissue. During days 4–6 the granulosa cells undergo hypertrophy so that the whole gland is uniformly filled with luteal cells and a central core of connective tissue. Because of the manner of formation there is very little change in diameter from the Graafian follicles (1.7 mm) to the fully formed corpus luteum (Fig. 6.10*a*) but the volume of luteal tissue more than doubles (Fig. 6.10*b*) as it comes to fill the antrum (Lyne & Hollis, 1979). After parturition and until day 45 of lactation the corpora lutea retain the same size and the luteal cells, as judged with the light microscope, remain of the same size and appearance (Hughes, 1962*b*; Lyne & Hollis, 1979). After day 45 the decline in size is rapid and is caused by marked shrinkage of luteal cells and their nuclei and a proportionate increase in connective tissue, much as has been described for the regression of the corpora lutea of *D. virginiana* at day 13 and *T. vulpecula* at day 17. Coincident with regression of the corpora lutea, a new crop of follicles is growing and ovulation may occur by day 45.

The ultrastructure of the corpora lutea of *Isoodon macrourus* was described by Gemmell (1979) and for this species and *P. nasuta* by Hollis & Lyne (1980). The luteal cells were shown to have the structures common to steroid-secreting cells and little change could be detected between corpora lutea in pregnancy and those in lactation, prior to regression. However, small membrane-bound, electron-dense granules of 0.2 μm, similar to those described for *T. vulpecula* (Shorey & Hughes, 1973*b*) and *Philander opossum* (Enders, 1973), were present in both species during pregnancy, but more abundant in *P. nasuta*, in which species they were also found by Hollis & Lyne (1980) in lactation. Gemmell (1979) reported them to be present in pregnancy and lactation in *I. macrourus* during those

Fig. 6.9. Adult short-nosed bandicoot, *Isoodon macrourus*. Photograph by Ederic Slater.

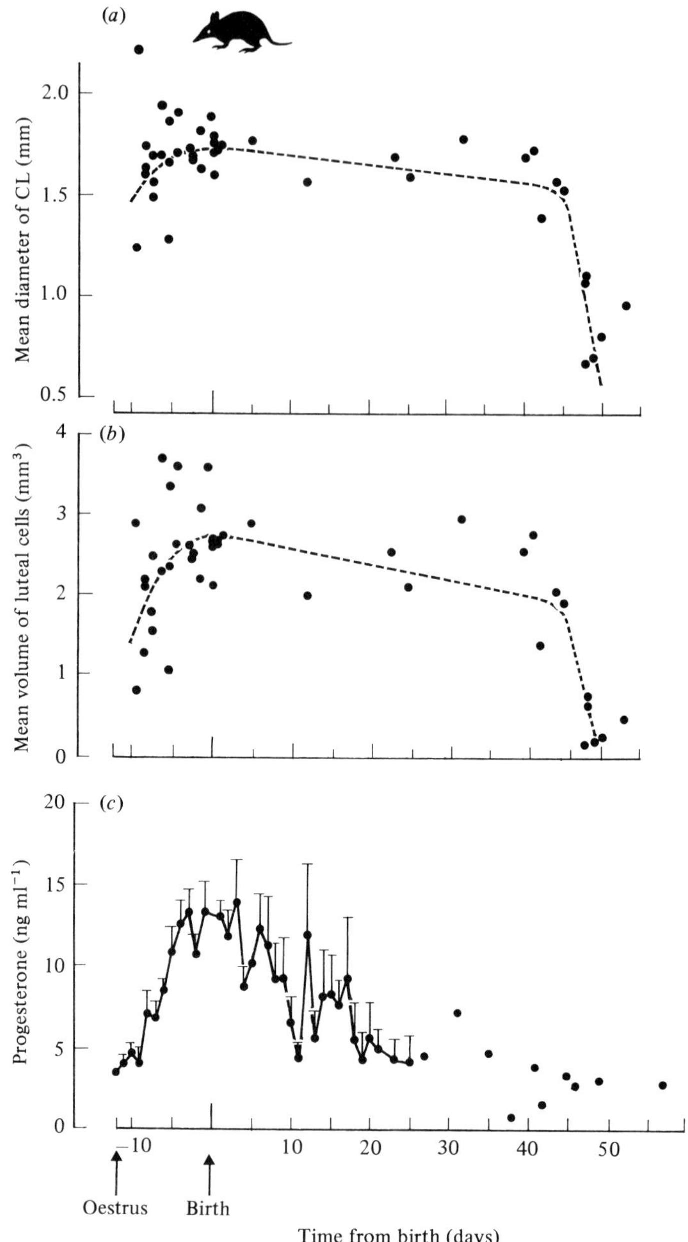

Fig. 6.10. (*a*) Diameter of the corpus luteum (CL) and (*b*) mean volume of the constituent luteal tissue in *Isoodon macrourus* during pregnancy (between arrows) and the subsequent lactation. During the period 45–53 days of lactation, ovulation may recur. The curve was drawn

times when he observed plasma progesterone to be elevated. Similar structures have been shown to be associated with storage of relaxin in the pig (Belt *et al.*, 1971) but Gemmell, Stacey & Thorburn (1974) hold that they are involved in the storage and controlled release of progesterone from luteal cells. Evidence from *T. vulpecula* and *Isoodon macrourus* (Gemmell, 1979) is consistent with both interpretations.

Gemmell (1979, 1981, 1984) has reported on the profile of plasma progesterone of *I. macrourus* through pregnancy and lactation and Gemmell, Jenkin & Thorburn (1980) measured progesterone and prostaglandin levels at more frequent intervals around the time of parturition. To date, the profile through the oestrous cycle has not been reported. The following discussion is based on the cumulative set of values for progesterone, as reported in Gemmell (1981, 1984).

The basal level of progesterone in females is about 3–5 ng ml^{-1}. Although it falls below this level immediately after ovariectomy, it returns to this level in 10 days, so Gemmell (1979, 1984) concluded that the adrenal cortex may compensate for the loss of ovarian progesterone.

For the first 3 days of the 12 day pregnancy, plasma progesterone is less than 5 ng ml^{-1}, but it then rises rapidly to reach a maximum of about 13 ng ml^{-1} on day 9 (Fig. 6.10*c*). This level is maintained for the remaining 3 days of pregnancy and the first 4 days post-partum. Gemmell *et al.* (1980) could detect no change in progesterone through the peripartum period, even when samples were taken at intervals of a few hours. However, from day 11 of pregnancy to day 4 of lactation they observed a significant elevation in the metabolite of prostaglandin $F_{2\alpha}$ (PGFM). From day 4 of lactation, progesterone declined steadily until the basal level of 3–5 ng ml^{-1} was reached on about day 19 (Gemmell, 1979). This level was maintained until growth of the next set of corpora lutea formed at ovulation after day 45. In his later paper Gemmell (1984) found that progesterone remained elevated after day 19 when the female suckled a larger litter (3) than when she suckled one. In females with larger litters, removal of the young on days 25–31 was followed by a rapid decline in progesterone. Both these results suggest that sucking exerts a luteotrophic stimulus on the corpora lutea.

Fig. 6.10. *cont.*

freehand. Redrawn from Lyne & Hollis (1979). (*c*) Concentrations (mean ± s.e.m.) of plasma progesterone in *Isoodon macrourus* during pregnancy and lactation. Individual samples after day 25 are shown. Redrawn from Gemmell (1981).

The profile of progesterone before parturition matches very closely the formation of the corpora lutea during the first 3 days and the subsequent increase in volume of luteal tissue, referred to above (Fig. 6.10*b*), but the decline in progesterone levels during the first half of lactation is not matched by a corresponding decline in volume of luteal tissue. Thus, if the corpora lutea of lactation are active endocrine glands, as Hughes (1962*b*) and Lyne & Hollis (1979) concluded from the histological appearance, progesterone is not the principal hormone after day 11 of lactation. Nevertheless, the results of Gemmell's (1981) experiments suggest that the corpora lutea may have two endocrine roles in lactation, independent of progesterone secretion (Table 6.3). Hughes (1962*b*) observed that, in females from which young were removed after the first week of lactation, the corpora lutea declined prematurely and oestrus occurred about 10 days later. Close (1977) confirmed this but noticed also that the interval to next oestrus was about 20 days rather than 10 if the young were removed earlier than day 7.

Since progesterone remains elevated for the first 6 days it could be inferred that the corpora lutea are responsible for this longer delay to ovulation and Gemmell (1981) investigated this by removing either pouch young, corpora lutea, or both at once from females in early lactation and by measuring plasma progesterone thereafter (Table 6.3).

In females from which the pouch young were removed by day 6 the decline in progesterone was the same as in females with pouch young, and ovulation occurred at about day 21–24, confirming Close's (1977) findings. However, in 2 animals from which the corpora lutea and the pouch young were removed on day 2 post-partum, progesterone fell to undetectable levels by day 3 and ovulation occurred between day 10 and 15, that is to say, 5–10 days earlier than in intact females. The same rapid decline in progesterone occurred in two females lutectomised on days 1 and 12, which retained their young, but in these two ovulation had not occurred when last examined on day 27 and 41 respectively. Thus the corpora lutea do exert an inhibition on folliculogenesis during the first 10 days but the main agents of follicular inhibition during this period and throughout lactation are the young in the pouch. Since the period of corpus luteum inhibition coincides with the period of elevated progesterone, it might be concluded that progesterone is the agent of this inhibition, although it is clearly not involved in the inhibition exerted by the pouch young. The mean length of the oestrous cycle in *I. macrourus* is 20 days with a range of 9–34 days (Lyne, 1976) and this corresponds approximately to the duration of pregnancy and the subsequent period of corpus luteum inhibition of

Table 6.3. *Effects of removing the corpora lutea of pregnancy (CLX) and/or removing the pouch young (RPY) on peripheral progesterone concentration and time of next ovulation in* Isoodon macrourus

Treatment	Number of animals	Progesterone decline (days post-partum)	Ovulation (days post-partum)	Lactation duration (days)
Intact	—	20	45–50	60
RPY before day 6	7	16	17–26	—
RPY after day 6	14	—	[> 16][a]	—
RPY on day 24	4	26–31	> 38	—
RPY on day 30	3	31	> 40	—
RPY + CLX on day 2	2	2	10	—
CLX days 1, 8, 12, 13, 14, 20, 26, 38	9	2	> 41	< 7 (7 females), 27, 41

[a] 5–10 days after RPY.

Data from Close (1977) and Gemmell (1981, 1984).

ovulation. This may indicate a similarity between the pregnant and non-pregnant cycles of *Isoodon*, which is masked by the intervention of lactation. Unlike in other marsupials, however, the corpora lutea of pregnancy do not decline in size during lactation, although progesterone levels do. This suggests that they may have another role unconnected with the inhibition of ovulation.

In all, Gemmell (1981) lutectomised 9 animals during lactation but only 2 retained their young. The 7 other animals lutectomised on days 1, 8, 13, 14, 20, 26 and 38, lost their young after 1 day (3) or after 7 days (4). This may have been due to the stress of surgery, or it may indicate that the corpora lutea have a role in the maintenance of lactation in *Isoodon macrourus* and, when larger litters are maintained, their capacity to secrete progesterone is prolonged by the stronger sucking stimulus exerted. This would be different from all other marsupials so far examined but then, lactation in the Peramelidae is also suspected of being very different from lactation in other marsupials (Russell, 1982*a*; Green, 1984). The results of Gemmell's (1981, 1984) experiments have further emphasised the unusual nature of the corpora lutea of the Peramelidae among marsupials and stimulate our curiosity about them.

Type 3: long gestation, delayed luteal phase

The growth and development of the corpus luteum, and its secretory activity and functions in reproduction are better known for the Macropodidae than for any other family of marsupials. This is because of the special nature of macropodid reproduction, in which a period of quiescence of variable duration is often interposed between formation of the corpus luteum and completion of its growth (see Chapter 2). This phenomenon has provided unrivalled opportunities for controlled experiments in several species, so that comparisons between species have enabled the fundamental aspects to be distinguished from the particular. The species that has received the most attention is *Macropus eugenii*. Findings in this species can be compared with findings in *Setonix brachyurus*, *M. rufogriseus* (referred to in early papers as *M. ruficollis*) and less complete observations on several other species.

Almost all the work on the macropodid corpus luteum has been done on the gland reactivated after lactational quiescence by removing the pouch young (RPY). This is because of the convenience of being able to synchronise groups of animals and to programme events to take place at precise times, rather than being dependent on the vagaries of individual animals. Hence little is known about the initial formation of the corpus luteum and the subsequent events in the cycle uninterrupted by quiescence, but it is necessary to review what is known as a prelude to considering the far more extensive work on the events that take place after removing the pouch young.

O'Donoghue (1914) described a Graafian follicle of 4.5 mm diameter and a well-formed corpus luteum of 7 mm in *M. rufogriseus*. The formation resembled that of *Perameles* and *Da. viverrinus* with thecal elements breaking through the membrana granulosa and establishing a connective tissue network in the antral cavity. No mitosis was seen in the granulosa cells of the Graafian follicle or the newly forming corpus luteum. A similar formation was described for *Setonix brachyurus* by Sharman (1955*a*) but he observed mitoses in the luteal cells up to 3 days after ovulation. As he considered that some luteal cells developed from the theca interna, the mitoses might have been in these cells. He thought theca luteal cells probably remained smaller than those derived from the membrana granulosa and found no evidence for their conversion into fibroblasts.

Formation of the corpus luteum of *M. eugenii* (Sharman, 1955*c*) and *M. rufus* (Sharman & Calaby, 1964) were said to be the same as that of *S. brachyurus*. However, examination of a series of more than 30 *M. eugenii* at daily intervals from mature Graafian follicles at oestrus to day 11 post-oestrus has disclosed considerable differences from *S. brachyurus* (C. H. Tyndale-Biscoe, unpublished observations). At ovulation in *M. eugenii* (Fig. 6.2) the follicle wall collapses and thecal elements are drawn in with the folds, the basement membrane remaining intact as in *D. virginiana* and *T. vulpecula*, until day 4 or 5 post-oestrus. Thecal cells do not contribute to the luteal tissue. A few mitoses were seen in luteal cells during the first 5 days after oestrus. By day 8 the corpus luteum had enlarged from 1.5 mm to 2.0 mm in diameter, but maximum size had not been reached by day 11 *p.c.* This is considerably slower than for *S. brachyurus* in which growth of the corpus luteum is complete by day 8 *p.c.* (Sharman, 1955*a*).

The oestrous cycle and pregnancy of M. eugenii, *uninterrupted by lactation*

The oestrous cycle of *M. eugenii* is 30.6 ± 1.7 days (Merchant, 1979) and oestrus lasts for less than 12 h during which time the female may copulate several times (Tyndale-Biscoe & Rodger, 1978). Progesterone has been measured in the peripheral circulation of a group of females through an oestrous cycle and part of the subsequent pregnancy (Hinds & Tyndale-Biscoe, 1982*a*). After the initial decline that preceeded oestrus the level remained low for the next 7 days. Then in each animal progesterone was briefly elevated, usually for only 1 day, before reverting to the basal level (Fig. 6.11). Between day 10 and 15 it increased and remained so until the decline that coincided with the next oestrus 30–31 days later, at which fertilisation occurred. The subsequent profile during the first 10 days of pregnancy was indistinguishable from the first 10 days of the oestrous cycle, with the transient pulse occurring on day 6 or 7. The gestation period of *M. eugenii*, uninterrupted by lactational quiescence, is 29 days (Merchant, 1979), so the early pulse of progesterone occurs 22 days before parturition and precedes the enlargement of the corpus luteum and the expansion of the blastocysts (see Fig. 7.8). No other hormone has been measured through the uninterrupted cycle or pregnancy.

The hormonal patterns at the end of pregnancy and during post-partum oestrus and ovulation have already been described (Fig. 6.1). Post-partum oestrus occurs within 8 h of birth, so that ovulation and the formation of the corpus luteum take place during lactation. For the first 7 days the development is the same as in the non-lactating female, but the progesterone pulse does not occur on day 7 and the luteal cells do not hypertrophy, so that the corpus luteum remains less than 2.0 mm diameter and weighs about 10 mg. No further change occurs in the corpus luteum or in the level of progesterone unless lactation is terminated.

Fig. 6.11. Concentrations (mean ± s.e.m.) of plasma progesterone in *Macropus eugenii* during the oestrous cycle and the first 10 days of pregnancy. The values for the pregnancy were resynchronised to the day of oestrus. Columns – day of oestrus. Redrawn from Hinds & Tyndale-Biscoe (1982*a*).

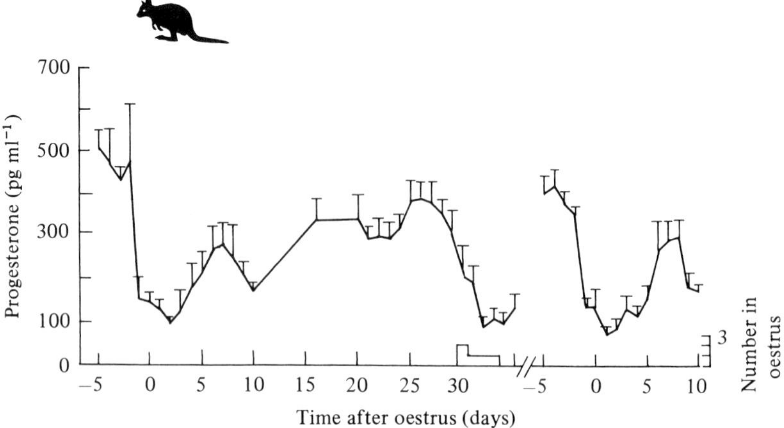

The delayed oestrous cycle and pregnancy

From the foregoing discussion, it is evident that lactation inhibits the corpus luteum at the stage reached 6 days after oestrus, just prior to the transient pulse in plasma progesterone and, in the oestrous cycle uninterrupted by lactation the next oestrus would occur at day 30 or 31. Thus, if resumption were to occur immediately after removing the pouch young, the next oestrus should occur 24–25 days later. The actual interval in the unmated female is 30.4 days (Table 6.4). The additional 5–6 days is the time required for recovery from inhibition and resumption of growth of the corpus luteum. Likewise in pregnancy the interval from oestrus to birth, uninterrupted by lactation, is 29.3 days and, since pregnancy in the lactating female is inhibited at day 6 also, birth could be expected to occur a minimum of 23 days after RPY. The interval is however, 26–27 days, a lag of 3–4 days. It is important to emphasise the point that the first 5 days after removing the pouch young are not equivalent to the first 5 days after oestrus, but represent the sum of a period of declining inhibition followed by a period of recovery of the corpus luteum, endometrium and embryo (see Tyndale-Biscoe, 1979). Therefore, equivalence of the delayed cycle or pregnancy with the undelayed cycle or pregnancy is only reached at day 5 or 6 after RPY. Unless the time course of events in the non-delayed cycle is known, this can only be estimated. As these particular data are only available for *M. eugenii* it is not possible to compare it with other species precisely. In *Setonix brachyurus* the period from removal of pouch young to birth (Shield & Woolley, 1960; Tyndale-Biscoe, 1963*a*) is 2 days shorter than the uninterrupted gestation period and, in both subspecies of *M. rufogriseus*, the differences are 3 days (Table 6.4). It is most likely however, that all three are similar to *M. eugenii* in having a refractory period and a recovery period of several days duration after RPY.

Growth and development of the corpus luteum after removal of pouch young

In *M. eugenii* the corpus luteum increases steadily in diameter from less than 2.0 mm during lactation to 4.0 mm (Renfree & Tyndale-Biscoe, 1973*a*) and in weight from 10 mg to 60 mg (Renfree, Green & Young, 1979; Hinds, Evans & Tyndale-Biscoe, 1983) by day 17 after RPY (Fig. 6.12). The progesterone content also increases during the first week after RPY, mitoses are commonly seen in the luteal cells, but not thereafter (Sharman & Berger, 1969; Renfree & Tyndale-Biscoe, 1973*a*). The same transient hyperplasia was also observed in *Setonix brachyurus* (Tyndale-

Table 6.4. *Duration (days) of the oestrous cycle and pregnancy and the interval from removal of the pouch young (RPY) to oestrus and birth and post-partum oestrus in* Macropus eugenii *and* M. rufogriseus, *to show (a) the influence of lactation on resumption of the reproductive cycle and (b) the influence of pregnancy on the timing of oestrus*

	Non-lactational cycles			Post-lactational cycles		
Oestrus to:	N	Days ± s.d.	RPY to:	N	Days ± s.d.	(*a*) Difference (days)
M. eugenii						
Birth	10	29.3 ± 1.06		10	26.2 ± 0.67	−3.1
Post-partum oestrus	10	29.4 ± 1.26		10	26.4 ± 0.57	−3.0
Oestrus	10	30.6 ± 1.17		10	30.4 ± 0.99	−0.1
(*b*) Difference between oestrus and post-partum oestrus		1.2, $P < 0.025$			4.0, $P < 0.0005$	
M. rufogriseus						
Birth	19	30.5 ± 1.7		5	27.6 ± 1.3	−2.9
Post-partum oestrus	19	30.8 ± 1.7		5	27.6 ± 1.3	−3.2
Oestrus	38	32.9 ± 2.3		5	29.6 ± 3.0	−3.3
(*b*) Difference between oestrus and post-partum oestrus		2.1, $P < 0.02$			2.0, n.s.	
M.r. banksianus						
Birth	12	30.0 ± 1.4		8	27.5 ± 1.3	−2.5
Post-partum oestrus	12	30.5 ± 1.9		8	27.5 ± 1.3	−3.0
Oestrus	14	33.4 ± 2.3		6	30.3 ± 1.9	−3.1
(*b*) Difference between oestrus and post-partum oestrus		2.9, $P < 0.05$			2.8, $P < 0.05$	

From Merchant (1979) and Merchant & Calab (1981).

Biscoe, 1963*a*), in which it coincided with a significant increase in the number of luteal cells from 1.5×10^6 on day 3 to 3×10^6 on day 7, and during the first week after RPY in *M. rufus* (Sharman, 1965*a*), *M. rufogriseus* (Walker *et al.*, 1983) and *Potorous tridactylus* (Shaw & Rose, 1979) and must therefore be considered to be a normal aspect of the development of the corpus luteum after quiescence. There is no evidence at present to say whether a transient phase of mitotic activity also occurs in the normal corpus luteum not delayed by lactation, or is restricted to the recovery phase after RPY. Nevertheless, subsequent growth of the corpus luteum is due in great measure to hypertrophy of luteal cells as it is in the non-delayed corpus luteum. In *M. rufogriseus* the mean cross-sectional area of luteal cells increased from less than 500 μm at day 4 to over 900 μm on days 9–25 (Walker *et al.*, 1983).

Fig. 6.12. Change in wet weight, total progesterone content (solid line) and concentration of progesterone in luteal tissue (broken line) of the corpus luteum (CL) of *Macropus eugenii* during the reproductive cycle initiated by removing the pouch young. Data from Renfree *et al.* (1979).

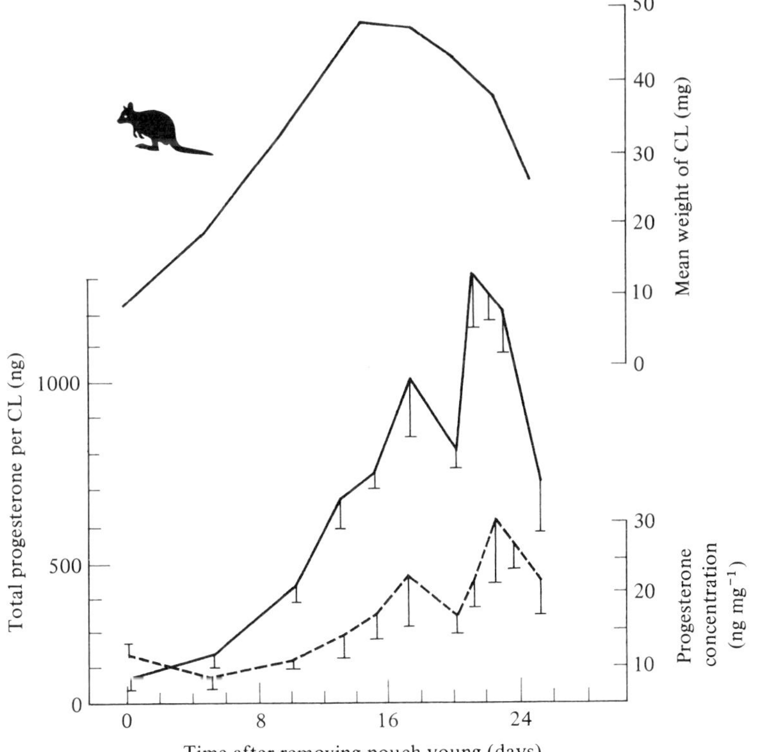

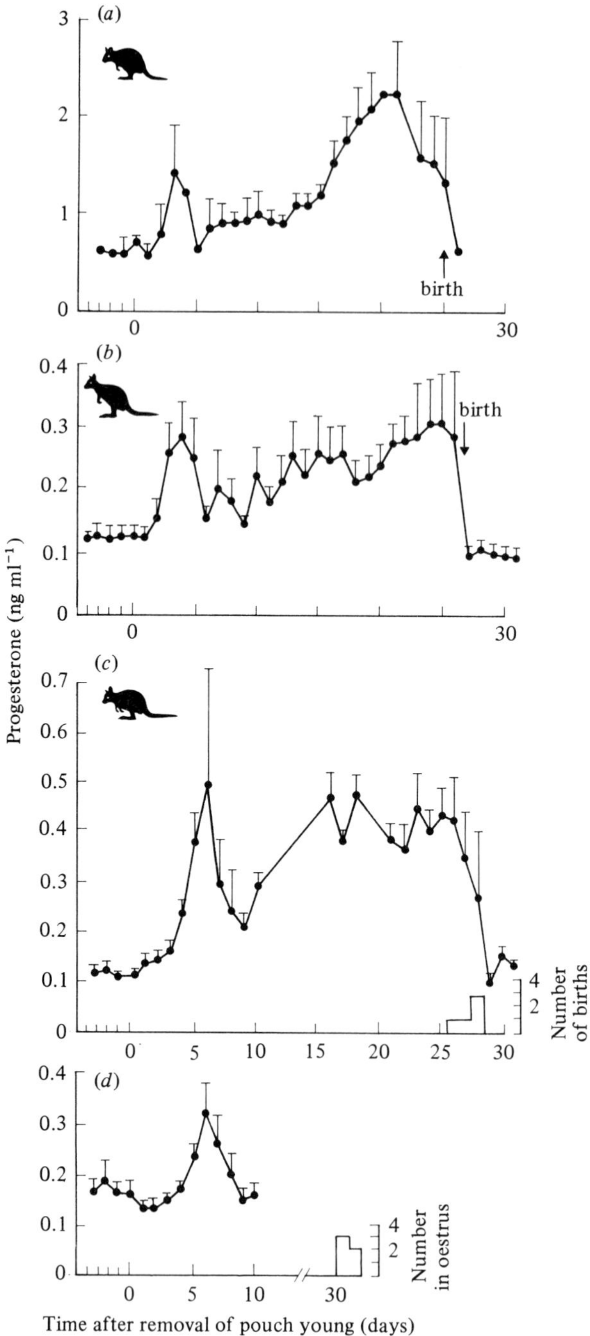
(a)
3
2
1
0
0
30
birth
(b)
0.4
0.3
0.2
0.1
0
0
30
birth
Progesterone (ng ml^{-1})
(c)
0.7
0.6
0.5
0.4
0.3
0.2
0.1
0
0
5
10
15
20
25
30
4
2
Number of births
0.4
(d)
0.3
0.2
0.1
0
0
5
10
30
4
2
Number in oestrus
Time after removal of pouch young (days)

The levels of progesterone have been measured in circulation from RPY to birth and post-partum oestrus and from RPY to oestrus in non-pregnant females (Fig. 6.13*c*, *d*) (Lemon, 1972; Hinds & Tyndale-Biscoe, 1982*a*). In all pregnant animals the concentration was under 200 pg ml^{-1} during lactation and for the first 4 days after RPY. Then on day 5 or 6 there was a transient peak of variable height (341–1270 pg ml^{-1}) followed by a return to the previous level until day 10. From day 16 until day 25 the level was consistently elevated until parturition when it fell precipitously to less than 200 pg ml^{-1} (Hinds & Tyndale-Biscoe, 1982*a*; Tyndale-Biscoe *et al.*, 1983). In the same animals undergoing a non-pregnant cycle the pattern for the first 10 days was the same, with a transient peak on day 5 or 6 (Fig. 6.13*d*), but the decline at the end of the cycle was slower and had not reached basal levels until day 29 (see Fig. 7.27).

As in pregnancy undelayed by lactation, the early peak occurs 22 days before parturition in delayed pregnancy and it coincides with the transient period of hyperplasia in the luteal cells, which precedes their hypertrophy. Cake, Owen & Bradshaw (1980) described a similar transient pulse of progesterone on day 3 or 4 of delayed pregnancy in *S. brachyurus* (Fig. 6.13*a*), which also coincides with the period of hyperplasia in this species (Tyndale-Biscoe, 1963*a*), and Walker & Gemmell (1983*a*) found a similar pulse of progesterone on day 4 after RPY in pregnant *M. rufogriseus* (Fig. 6.13*b*), which also coincides with the time mitoses were seen in the luteal cells (Walker *et al.*, 1983).

Lemon (1972) demonstrated that the elevated progesterone in the second half of pregnancy, in *M. eugenii* is secreted by the corpus luteum, since it was reduced after lutectomy performed on day 10. Findlay *et al.* (1983) and Harder *et al.* (1985) have confirmed that the decline in progesterone after lutectomy on day 17 or 18 is very rapid, reaching basal levels in one day.

Evidence for the corpus luteum providing the transient pulse of progesterone on day 5 is less direct. The progesterone content of the corpus

---

Fig. 6.13. Concentrations (mean ± s.e.m.) of plasma progesterone in three macropodids after removal of pouch young. (*a*) *Setonix brachyurus* during pregnancy, (*b*) *Macropus rufogriseus* during pregnancy, (*c*) *M. eugenii* during pregnancy and (*d*) *Macropus eugenii* during a non-pregnant oestrous cycle. Note all three species show a marked pulse of progesterone 3–6 days after removal of their pouch young, and a sharp fall in progesterone at the time of parturition. Redrawn from Cake *et al.* (1980) (*a*); Walker & Gemmell (1983*a*) (*b*); and Hinds & Tyndale-Biscoe (1982*a*) (*c*, *d*).

luteum of *M. eugenii* during lactation and until day 13 after RPY was 10–11 ng mg^{-1} rising progressively to 32 ng mg^{-1} on day 22 (Renfree *et al.*, 1979; Fig. 6.12). Since the gland increases in weight 6 fold, the total progesterone content per gland rose from 0.1 μg in lactation to 1.2 μg at day 22, and down to half this at parturition. This compares with values obtained by Lindner & Sharman (1966) using gas chromatography on corpora lutea of *M. rufus;* during lactational quiescence they obtained 0–0.4 μg per pair of ovaries and, in active pregnancy, 2.1 μg per pair of ovaries. Although Renfree *et al.* (1979) sampled corpora lutea on day 5, they did not find the progesterone content to be greater than before or after this day to account for the transient pulse. However, Hinds *et al.* (1983) found that the rate of progesterone secretion *in vitro* was significantly higher on day 5 than on day 0 or day 9 (Fig. 6.14), and concluded that

Fig. 6. 14. Secretion of progesterone into the medium by pieces of corpus luteal tissue of *Macropus eugenii* on days 0, 5, 9 and 16 after removal of pouch young. Net production (mean ± s.e.m.) calculated as ng progesterone per milligram tissue per 4 h incubation. Data from Hinds *et al.* (1983).

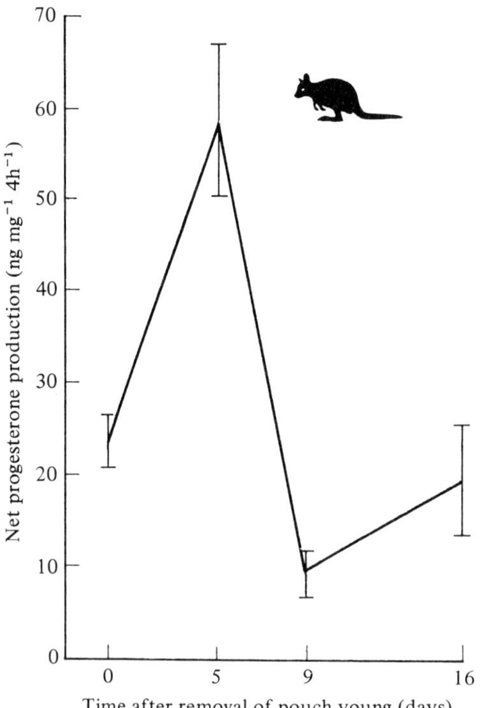

this could account for the transient pulse of progesterone seen in the peripheral circulation. On the other hand, the steady rise in plasma progesterone after day 10 can be accounted for by the increasing size and progesterone content of the older corpora lutea. An alternative hypothesis for the changing plasma concentration is that the metabolic clearance rate of progesterone might alter during the cycle. While Sernia, Hinds & Tyndale-Biscoe (1980) found no evidence for differences in either metabolic clearance rate or secretion rate between intact and ovariectomised females, they did not study animals undergoing an oestrous cycle or pregnancy when marked changes could be expected.

The secretory function of the corpus luteum of *M. eugenii* has also been investigated by incubation with labelled steroid precursors and identification of products by recrystallisation to constant specific activity (Renfree & Heap, 1977; Renfree *et al.*, 1984*b*). Corpora lutea from day 11 to 25 converted pregnenolone to progesterone, the maximum conversion (71.9%) being on day 20, while by day 25 just before parturition the conversion was 37% (3 cases) and as low as 10.7 in another. These authors and Shaw & Renfree (1984) found significant but low concentrations of oestradiol in corpora lutea of lactating *M. eugenii* (139 and 26.5 pg per gland respectively) and after reactivation 90 pg per gland at days 5–7, 69 ± 38 pg per gland at day 10–18, and 29 and 24 pg per gland respectively at the end of pregnancy. Nevertheless, Renfree *et al.* (1984*b*) found no

Fig. 6.15. Concentrations (mean ± s.e.m.) of plasma oestradiol in *Macropus eugenii* between days 0 and 30 after removing pouch young. Note that, after day 24, the values are synchronised to the time of birth rather than to day after RPY. There are two significant rises in oestradiol; at day 5, coincident with the progesterone peak, and one day after birth, coincident with oestrus and mating. Redrawn from Shaw & Renfree (1984).

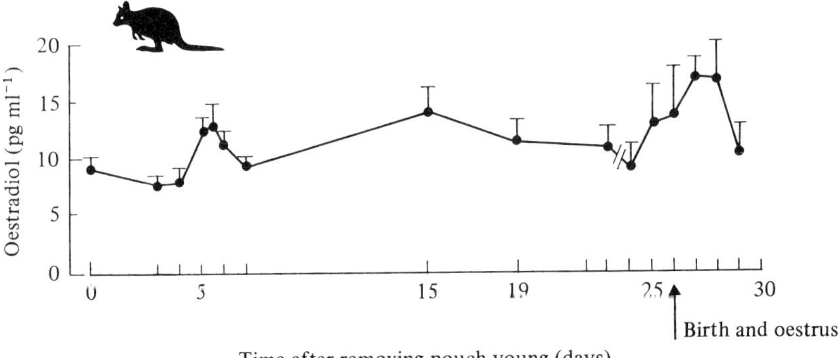

evidence that the corpus luteum can synthesise oestradiol or oestrone itself. Oestradiol has been measured in peripheral plasma of *M. eugenii* by Flint & Renfree (1982), Shaw & Renfree (1984) and Harder *et al.* (1984, 1985). During quiescence and for most of the period after resumption of pregnancy and before oestrus the concentration remains at less than 10 pg ml^{-1}. However, Shaw & Renfree (1984) found a transient elevation to 15 pg ml^{-1} on days 5, 6 or 7, which coincided with the transient elevation of progesterone (Fig. 6.15). There is no direct evidence, like that for the progesterone pulse, that it comes from the reawakened corpus luteum, but this would appear to be the most likely source. On the other hand the subsequent elevation after day 25 at oestrus, as mentioned previously, comes from the Graafian follicle and not the corpus luteum.

Other species of macropodid

Similar patterns of plasma progesterone to that seen in *M. eugenii* have been reported for two other species of macropodids. In *S. brachyurus*, Cake *et al.* (1980) reported an early pulse on day 3–4 after RPY followed by a subsequent rise to 2 ng ml^{-1} on day 20 that was sustained until the day of birth when it fell to 0.5 ng ml^{-1} (Fig. 6.13*a*). However in females undergoing a non-pregnant cycle, no early pulse was detected, but the late rise was observed. They concluded that the occurrence of the early pulse only in the pregnant animals was evidence of a signal from the reawakened blastocyst to the mother. This, if true, would have been the first evidence for such maternal recognition of pregnancy in any marsupial but the data presented are insufficient to make such a firm conclusion; three different animals were used for each condition and the non-pregnant animals were not mated to vasectomised males to control for an influence of copulation. In the light of opposite results in *M. eugenii*, further evidence is required to support this conclusion of Cake *et al.* (1980).

The same pattern of plasma progesterone has been reported for *M. rufogriseus* from Tasmania sampled daily through delayed pregnancy after RPY (Fig. 6.13*b*) (Walker & Gemmell, 1983*a*). A non-pregnant series was not studied but a transient pulse on day 4–5 was found, a late plateau and a rapid decline at birth. The basal and peak levels were about half those reported for *M. eugenii* (Hinds & Tyndale-Biscoe, 1982*a*), which themselves were about half those reported for *S. brachyurus*. Whether these are real differences between the species or are due to differences in antisera and assay procedures between laboratories is not clear at present. Walker *et al.* (1983) correlated changes in the cytology of luteal cells with the levels of progesterone in plasma through pregnancy. The most conspicuous feature in the active luteal cells were the small membrane-bound granules, like those described in *Trichosurus vulpecula* (Shorey & Hughes, 1973*b*) and *Isoodon macrourus* (Gemmell, 1979), which doubled in abundance from day 9, when plasma progesterone was low, to days 14 and 21 when progesterone was maximal. Decline in the number of granules by day 25 preceded the fall in progesterone at parturition. Walker & Gemmell (1983*a*) also reported levels of oestradiol-17β from the same animals in which they measured progesterone. The pattern they reported bears no resemblance to those reported for oestradiol in *M. eugenii* (Renfree & Heap, 1977; Flint & Renfree, 1982; Shaw & Renfree, 1984; Harder *et al.*, 1984, 1985) and the levels were 10 times higher than in these studies and in *Didelphis* (Fig. 6.3) and *Trichosurus* (Curlewis *et al.*, 1985). In view of this, it may be prudent to await confirmation before concluding that *M. rufogriseus* is anomalous in this respect.

A small sample of female *M. giganteus* shot at the beginning of the breeding season has provided some data on levels of progesterone, oestradiol, prolactin and LH in plasma, which can be correlated with the stage of reproduction assessed by the state of the ovaries and embryo of pouch young (C. H. Tyndale-Biscoe, L. A. Hinds & T. P. Fletcher, unpublished results). Although no early pulse of progesterone is evident, this does not preclude it because of the nature of the data. In late stages of pregnancy, progesterone was consistently high until full term, while a female with a new-born young had a low level of progesterone, as did all other lactating females. The oestradiol concentrations were similar to those obtained in *M. eugenii* except that the peripartum rise observed by Flint & Renfree (1982) in a shot sample of *M. eugenii* was not seen in *M. giganteus*. This probably reflects the fact that post-partum oestrus does not occur in *M. giganteus*.

Indirect and rather intriguing evidence for a similar progesterone profile in *Bettongia gaimardi* has been reported by Rose (1985) who measured basal body temperature (BBT) in females from RPY to birth, as in *Vombatus ursinus* (Fig. 2.16). At day 4 there was a transient rise in BBT followed, after day 9, by a sustained rise until parturition when BBT again declined. This bears such a close resemblance to the progesterone profile of other macropodids that they may be causally related in the same way as they are known to be in humans and in the Rhesus monkey.

Role of the corpus luteum in follicular growth, ovulation and oestrus

Sandes (1903) noticed that follicular growth was inhibited in the vicinity of corpora lutea in *Dasyurus viverrinus* and it is a common observation in monovular marsupials that the Graafian follicle grows in the ovary without the corpus luteum (Pilton & Sharman, 1962; Renfree & Tyndale-Biscoe, 1973*a*). These observations suggest that the corpus luteum may have a local effect on neighbouring follicles. As well as this, and more importantly, the corpus luteum has been shown to exercise a systemic inhibition on ovulation and oestrus in several marsupials (Table 6.5). In *T. vulpecula* lutectomy or removal of the ovary bearing the corpus luteum on days 2–7 post-oestrus led to premature oestrus and ovulation 8–9 days later (Sharman, 1965*a*; Shorey & Hughes, 1975). This is the same interval as that from the end of the luteal phase (day 17) to the next oestrus (day 25), or from removing the pouch young to next oestrus in this species (Table 2.2) and suggests that the corpus luteum exercises its inhibition by the same end pathway that operates in lactation. The corpus luteum inhibition begins to wane after day 7, since lutectomy on days 10–12 was usually followed by oestrus at the normal time (day 25) (Shorey & Hughes, 1975). Thus the inhibitory influence of the corpus luteum is strongest before the rise in progesterone, which occurs after day 8, and is absent when plasma progesterone is maximal (see Fig. 6.7).

A very similar pattern can be seen in several species of the Macropodidae (Table 6.5). Ablation of the corpus luteum, or the whole ovary bearing it, during the first 12 days of the cycle led to premature oestrus and ovulation 8–18 days later. On the other hand lutectomy of *M. eugenii* on day 18 (Tyndale-Biscoe & Hawkins, 1977) and unilateral ovariectomy of *S. brachyurus* on day 19 (Tyndale-Biscoe, 1963*b*) was followed by oestrus respectively 9 and 6 days later, the same time as in intact animals. In both the latter species, as in *T. vulpecula*, the period of corpus luteum inhibition coincides with the period of low plasma progesterone, whereas the period when the corpus luteum is without effect on ovulation is the period of high plasma progesterone (Fig. 6.13*c*).

Not only does the young corpus luteum exert a more effective inhibition on follicle growth than the mature corpus luteum but, in macropodids, it exerts this effect during lactational quiescence when it is itself inhibited and not fully grown (Sharman & Clark, 1967; Tyndale-Biscoe & Hawkins, 1977). There is some disagreement between workers on macropodids as to whether ovulation is suppressed solely by the quiescent corpus luteum,

Table 6.5. *Time (days) to next ovulation after removal of the corpus luteum (CLX) or the ovary bearing the corpus luteum (1 – OVX) during the oestrous cycle, lactational (LQ) or seasonal quiescence (SQ) in* Trichosurus vulpecula *and six species of the Macropodidae*

				Time to ovulation		
Species	Operation	*N*	Day of cycle	Mean	Range	Reference
Trichosurus vulpecula	CLX	6	2–7	—	8–9	Sharman (1965*a*)
		4	11	—	14	Shorey & Hughes (1975)
Aepyprymnus rufescens	CLX	2	7, 9	8	7–9	Moors (1975)
Thylogale billardierii	CLX	2	LQ	—	11	Rose & McCartney (1982*a*)
Setonix brachyurus	1-OVX	4	0, 2	12.3	12–13	Tyndale-Biscoe (1963*b*)
	1-OVX	5	4–12	15.8	13–18	
	1-OVX	1	19	—	6	
Macropus giganteus	CLX	1	8	—	< 13	Poole & Pilton (1964)
Macropus rufus	CLX	5	0–11	11.6	8–18	Sharman & Clark (1967)
	CLX	4	LQ	18.0	14–21	
Macropus eugenii	CLX	4	0	17.7	15–18	Tyndale-Biscoe & Hawkins (1977)
	CLX	5	6	12.4	11–15	
	CLX	5	12	11.8	9–15	
	CLX	4	18	9.0	8–10	
	CLX	4	LQ	11.3	11–12	
	CLX	16	SQ	12.5	10–16	

or more directly, as it is in other marsupials such as *D. virginiana* and *T. vulpecula*. The evidence at present available from five species of macropodids could alternatively be interpreted as different degrees of corpus luteum inhibition superimposed on the more primitive direct inhibition of sucking, with *M. fuliginosus* representing one extreme (in which sucking is the sole inhibition) and *M. eugenii* the other (in which the corpus luteum is the sole inhibition).

In *M. fuliginosus* and *M. giganteus* gestation is several days shorter than the oestrous cycle (Table 2.2) (Poole & Catling, 1974), so that post-partum ovulation is inhibited during lactation, as in *T. vulpecula*, and no corpus luteum influence occurs. Removal of the pouch young results in oestrus 8–10 days later, the same interval as occurred after lutectomy on day 8 of the oestrous cycle in this species (Poole & Pilton, 1964). In *M. giganteus*, ovulation may occur after about 180 days of lactation and the corpus luteum then becomes quiescent until about 300 days. As no further ovulation occurs during the period of quiescence, a corpus luteum inhibition may then have been superimposed on the sucking inhibition. In *Setonix brachyurus* a minority of females (11 of 298) failed to ovulate post-partum (Tyndale-Biscoe, 1963*a*) and, in the subsequent lactation, did not have a quiescent corpus luteum; in them lactation alone suppressed ovulation, whereas in the majority the corpus luteum may be inferred to have done so (Sharman, 1955*b*).

In *M. rufus*, Sharman & Clark (1967) concluded that both the corpus luteum and sucking were involved in the inhibition of oestrus and ovulation during lactation. The return to oestrus after lutectomy performed up to 11 days after removing the pouch young occurred 8–18 days (mean 11.6 days) later (Table 6.5), while oestrus occurred 14–21 days (mean 20 days) after lutectomy performed during continuing lactation. Finally, in *M. eugenii* no difference was found in the time to return to oestrus after lutectomy done during lactation or during the reproductive cycle (Tyndale-Biscoe & Hawkins, 1977) and the conclusion was that, in this species, the quiescent corpus luteum is the sole agent of follicular inhibition during lactation. Furthermore, in seasonal quiescence in this species, the corpus luteum is also the main agent of follicular inhibition. This points to there being a common central pathway by which follicular inhibition is effected (see Fig. 9.2).

By contrast to the macropodids, in *Isoodon macrourus* it is the corpus luteum of pregnancy, persisting into lactation, that inhibits ovulation, but – as Gemmell's (1981) experiments show – a sucking inhibition is also effective from the day of birth (Table 6.3). Also, by contrast, this species is at present the only marsupial in which follicular inhibition by the corpus luteum coincides with elevated plasma progesterone (Fig. 6.10). Whether this coincidence is biologically significant remains to be examined.

For the macropodids the evidence at present does not support the idea that progesterone is the agent by which the corpus luteum inhibits ovulation. In *S. brachyurus*, progesterone administered for 5 days after lutectomy did not delay the return to oestrus (Tyndale-Biscoe, 1963*b*), although it did induce a luteal condition in the uteri. Similarly, in *M. eugenii* (Evans *et al.*, 1980) animals treated with progesterone for 7 days after lutectomy responded in the same way as controls, either ovulating or having enlarged follicles, whereas those treated with oestradiol alone or in combination with progesterone did not ovulate and most had only small follicles in their ovaries. Renfree *et al.* (1982*b*) also found that oestradiol inhibited follicular growth in lutectomised *M. eugenii* but neither progesterone nor androstenedione did (Fig. 6.16). Indeed the

Fig. 6.16. Diameters (mean ±s.e.m.) of the three largest follicles in the ovaries of *Macropus eugenii* 18 days after lutectomy and treatment with steroids during seasonal quiescence in August. Sizes shown for the three largest follicles indicate that these were small antral follicles in the pre-gonadotrophin-dependent phase. Treatment: low P, 2 mg progesterone per day for 15 days; high P, 4 mg progesterone per day for 15 days; Oe, 1 μg oestradiol-17β per day for 15 days. Redrawn from Renfree *et al.* (1982).

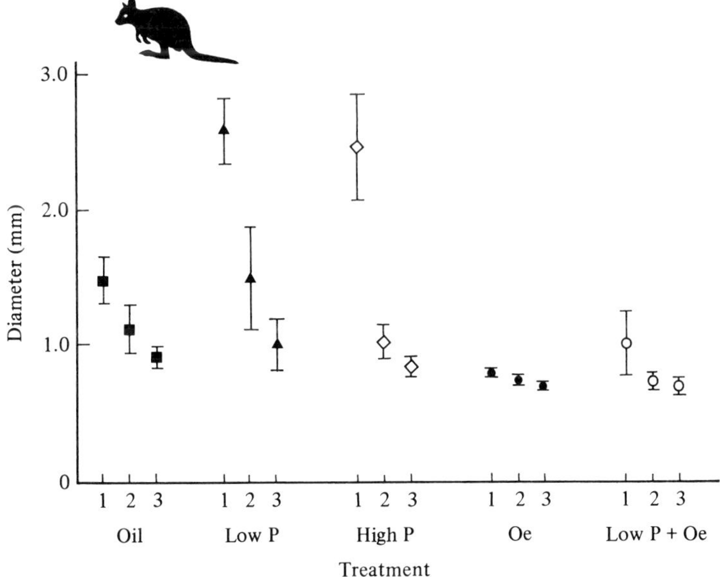

progesterone-treated animals had larger follicles than the controls. If oestrogen rather than progesterone is the agent of corpus luteum inhibition, what evidence is there that the corpus luteum secretes oestrogen? Renfree *et al.* (1984*b*) and Shaw & Renfree (1984) measured oestradiol in quiescent corpora lutea and in those reactivated for 5–7 days (138 and 26.5 pg per corpus luteum respectively) but they found much lower concentrations in later stage corpora lutea. Despite this Renfree *et al.* (1984*b*) could find no evidence of aromatase activity in luteal tissue or interstitial tissue *in vitro* as determined by conversion of androstenedione to oestrone or oestradiol. However, as suggested in the last section, the transient rise in plasma oestradiol early in pregnancy is most probably coming from the corpus luteum (Shaw & Renfree, 1984). Clearly more information is needed before this matter can be resolved.

It will be recalled (Chapter 5) that the pituitary is required in *M. eugenii* for completion of follicular growth and ovulation but, despite this, throughout the period of follicular growth the levels of LH and FSH in intact females are consistently low or undetectable. Nevertheless when females were immunised against GnRH (Short *et al.*, 1985) follicular growth and ovulation were inhibited, although the corpus luteum was not. Thus the low level of the gonadotrophins is necessary for follicular growth and is apparently maintained at this level by negative feedback from the ovary. As in other mammals LH and FSH become elevated after bilateral ovariectomy in *M. eugenii* (Evans *et al.*, 1980; Tyndale-Biscoe & Hearn, 1981; Horn *et al.*, 1985) and this was reversed in females that were implanted with grafts of ovarian cortex in which luteal tissue developed. Likewise graded doses of oestradiol given to ovariectomised females reduced LH and FSH levels in a dose-dependent manner despite the fact that the lowest doses of oestradiol could not be differentiated in peripheral circulation from that of the controls (Horn *et al.*, 1985). Conversely, at the highest dose, LH was first depressed to basal levels and after 24 h there was a transient pulse, similar in concentration and duration to the pre-ovulatory pulse in intact females. These results indicate that the hypothalamus of *M. eugenii* is sensitive to oestradiol at levels that normally occur in intact females, and that at levels that occur at oestrus, oestradiol exercises a positive-feedback effect on the hypothalamus, as it does in other mammals.

Although comparative date are at present insufficient to make firm conclusions, we can propose a working hypothesis for the control of follicle growth and ovulation in marsupials. From results with *M. eugenii* we can assume that both processes depend on pituitary stimulation by means of

FSH and LH, and we can therefore postulate that gonadotrophic secretion is effectively reduced or abolished during lactation by the sucking stimulus, and during the first half of the oestrous cycle by the corpus luteum. Is one mechanism involved in both inhibitions? The corpus luteum probably exerts its inhibition by means of oestradiol secreted into the circulation. In *M. eugenii* and *T. vulpecula* the level of oestradiol in circulation during most of the cycle and lactation is very low, except for the pre-ovulatory surge derived from the Graafian follicles and the peak at day 5, presumably from the corpus luteum.

Consider first a non-macropodid such as *T. vulpecula* in which the ovaries are apparently inactive during lactation. If oestradiol exerts a negative feedback during lactation, the hypothalamus or pituitary must be more sensitive to oestrogen then than it is during the oestrous cycle, when a corpus luteum is present. If sucking increases hypothalamic sensitivity to negative feedback, removing the pouch young would reduce this, so that the low levels of oestrogen would no longer be sufficient to maintain hypothalamo-pituitary inhibition and follicle growth would be stimulated. Then, for the first 10 days of the cycle, the increased level of oestrogen secreted by the young corpus luteum would reimpose the inhibition. As progesterone secretion by the corpus luteum increased, a real or relative decline in oestrogen output by the corpus luteum would prevail and no longer be sufficient to maintain inhibition on an hypothalamus with low sensitivity to oestrogen. If the animal was pregnant and gave birth, sucking would again increase hypothalamic sensitivity and the low level of oestrogen output would become sufficient to maintain inhibition. If the animal was not pregnant and sucking did not occur, hypothalamic sensitivity would remain low and the low level of oestrogen from the declining corpus luteum would be insufficient to maintain the inhibition, thus allowing follicle growth and ovulation to take place.

In those species, like *M. eugenii* with a longer pregnancy, sucking is delayed until the end of the cycle, so there would be no change in hypothalamic sensitivity and the levels of oestrogen in the second half of the cycle, when progesterone is elevated, would be insufficient to inhibit gonadotrophic secretion. After parturition and post-partum ovulation, the same increase in hypothalamic sensitivity would come into play and, now, the small output of oestradiol from the quiescent corpus luteum would provide most of the negative feedback. Differences observed between macropodid species referred to earlier may be accounted for by differences in the degree of hypothalamic sensitivity to oestrogen, and differences in the amount of oestrogen secreted by the quiescent corpus luteum or other ovarian tissue. Clearly this matter can only be resolved by more studies on more species, and particularly important, the development of homologous radioimmunoassays that can detect changes in circulating levels of FSH and LH not possible with existing heterologous assays.

Influence of the corpus luteum on the male

Apart from its possible involvement in oestrous behaviour discussed in the first section of this chapter, the corpus luteum may have another indirect role in successful mating through an influence on the male. Catling & Sutherland (1980) showed with *M. eugenii* that LH and

Fig. 6.17. Effect of breeding females on testosterone concentration (mean ± s.e.m.) in male *Macropus eugenii*. (*a*) Five males isolated from females during the non-breeding and breeding seasons. (*b*) Eight males which associated with 20 females in the non-breeding and breeding seasons. Note that the axes are not the same scale in (*a*) and (*b*). Redrawn from Catling & Sutherland (1980).

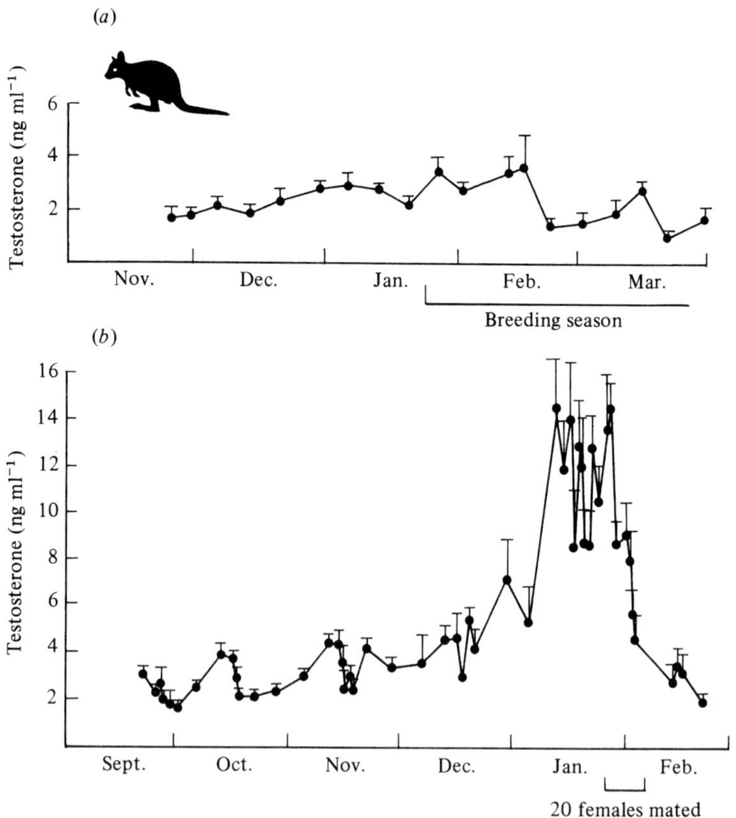

testosterone were increased significantly in males run with females, compared to levels in males isolated from females during the breeding season (Fig. 6.17). The first significant rise in testosterone and LH concentrations in the males occurred 3 weeks before the females came into oestrus and coincided with the first rise in progesterone in the females, and remained high throughout the time that progesterone remained high. Although not measured by Catling & Sutherland (1980), it is likely that a transient rise in oestradiol concentration would have occurred 3 weeks before oestrus also in these females, since Flint & Renfree (1982) observed it to occur in wild *M. eugenii* at this time of year (Fig. 9.3*b*). Males of several species of macropodid, including *M. eugenii*, investigate the pouch and vulva of females and taste their urine and have been observed to show the flehman reaction afterwards (Coulson & Croft, 1981) (see Chapter 2).

The secretions of neither region have been investigated, but if they are controlled by progesterone or oestradiol, alone or in combination, the secretions would reflect corpus luteum activity and be an early indication of forthcoming oestrus. Likewise metabolites of progesterone such as pregnanediol have been reported in greatest concentration in the urine of *T. vulpecula* during the period of elevated plasma progesterone (Pilton & Sharman, 1962) and, if the same holds for *M. eugenii*, this could also be a means of detecting corpus luteum activity and responding, by increasing the output of LH and testosterone. As mentioned in Chapter 4, Inns (1982) has shown a close relationship between prostate weight and plasma testosterone in male *M. eugenii* shot on Kangaroo Island, both being high during the month preceding the main period of mating activity in January–February and the minor period of mating activity in October (Fig. 4.7*a*).

Endocrine control of the corpus luteum

Until 1972 the assumption, based on analogy, was that the corpora lutea of marsupials, like those of Eutheria, depended on luteotrophic stimulation for luteal cell growth and progesterone secretion. Luteinising hormone, alone or in combination with prolactin, secreted by the pituitary, was the most likely agent; the placentae of marsupials were thought unlikely to secrete chorionic gonadotrophin or placental lactogen, because there was no evidence that pregnancy affected the lifespan of the corpus luteum. Indeed, Amoroso (1955) considered that the inability of the marsupial placenta to secrete a luteotrophic hormone and prolong the life of the corpus luteum was the central difference between Eutheria and Metatheria, which led to their dichotomy in the Cretaceous (see Chapter

10). After the phenomenon of lactational quiescence was discovered in macropodids (Sharman, 1954) it was presumed that the quiescent corpus luteum was awaiting a luteotrophic signal from the pituitary to reactivate and complete the delayed cycle (Sharman, 1955*b*, 1970; Tyndale-Biscoe, 1963*a*).

Hearn's (1973, 1974) findings on the effects of hypophysectomy on ovarian function and the corpus luteum in *M. eugenii*, confounded this assumption. His results indicated that, while the pituitary is necessary for folliculogenesis and ovulation, as in Eutheria, the corpus luteum once formed does not require pituitary support for steroidogenesis. Any corpus luteum already formed at the time of hypophysectomy completed development apparently normally and, when lactating animals or those in seasonal quiescence were hypophysectomised, the quiescent corpora lutea resumed growth, induced a luteal phase in the uteri and embryo development proceeded to full term, although parturition did not occur.

Fig. 6.18. Concentrations (mean ± s.e.m.) of plasma progesterone in *Macropus eugenii* females, during 14 days after they underwent hypophysectomy (solid line) or sham hypophysectomy (broken line). In the former group the corpus luteum reactivated as indicated by the pulse of progesterone and by the developing embryos recovered on day 14. From unpublished data in Hinds (1983), by permission.

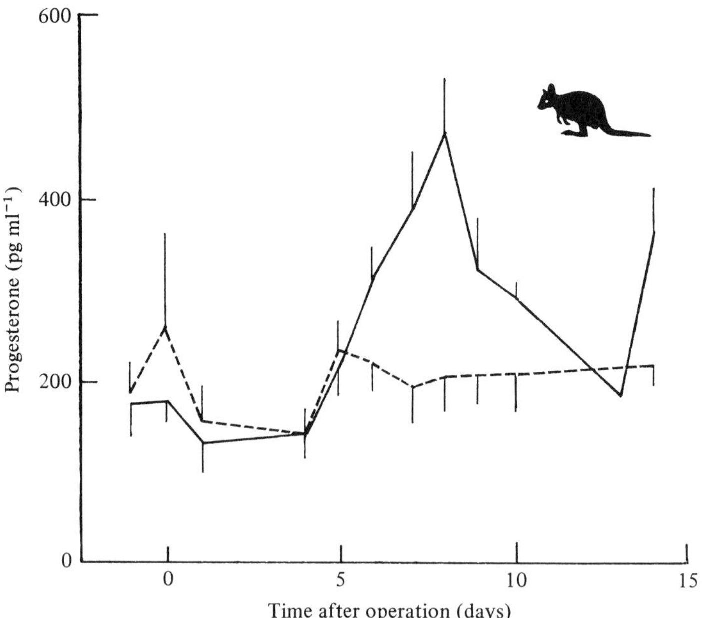

In more recent studies (Hinds, 1983) this conclusion has been confirmed by measuring plasma progesterone daily after hypophysectomy and sham hypophysectomy of lactating animals (Fig. 6.18). In the sham-operated controls the corpora lutea did not reactivate and plasma progesterone remained basal but, after hypophysectomy, the characteristic transient peak of progesterone occurred on day 7 or 8 and this was followed by high levels after day 14. Not only can steroidogenesis thus continue after hypophysectomy but the rate of secretion can also change without LH stimulation.

The conclusion that LH is not necessary for steroidogenesis by the corpus luteum is supported by the work of Short, *et al.* (1985) referred to above. Passive immunisation of *M. eugenii* females with anti-GnRH severely affected gonadotrophin secretion, as judged by failure of folliculogenesis, but had no effect on development of the corpora lutea or the successful completion of pregnancy and parturition.

Studies of specific binding sites for LH and prolactin led to the same conclusion (Fig. 6.19). The granulosa cells of growing and pre-ovulatory follicles contained high concentrations of specific binding sites for LH and lesser concentrations of binding sites specific for prolactin. However, after ovulation, when these same cells become luteal cells, no binding sites for LH could be detected on them (Stewart & Tyndale-Biscoe, 1982) but the concentration of prolactin binding sites was very high indeed, exceeding lactating mammary gland (Sernia & Tyndale-Biscoe, 1979; Stewart & Tyndale-Biscoe, 1982). These results are also consistent with the observation that luteal tissue cultured *in vitro* in the presence of LH or prolactin produced no more progesterone than luteal tissue in culture medium alone (Sernia *et al.*, 1980).

If the corpora lutea of *M. eugenii* lack LH receptors and are unresponsive to LH *in vivo* or *in vitro*, what prevents their development during lactation? Hearn (1973) suggested that the pituitary must provide a tonic inhibition, which once removed (by hypophysectomy), allows the corpus luteum to resume development autonomously. He suggested that either oxytocin or prolactin might be the agent and Tyndale-Biscoe & Hawkins (1977) tested both hormones for this in intact and hypophysectomised females. Intact females from which the pouch young were removed were injected for 7 days with either saline, oxytocin, prolactin or reserpine, a drug known to induce milk secretion in rats (Meites, Nicol & Talwalker, 1959). In all the experimental groups birth and/or oestrus took place 7 days later than in the control group (Fig. 6.20). However, when females were treated with the same doses of prolactin or oxytocin or both together for 7 days after hypophysectomy, only the animals receiving prolactin were delayed for 7 days. This seemed to be strong evidence for prolactin being the main inhibitor of the corpus luteum. Furthermore, because in these animals the corpus luteum did not resume growth until 7 days after hypophysectomy, it was unlikely that any residual gonadotrophin attached to the luteal cells at the time of operation could have been involved.

The presence of high concentrations of prolactin-binding sites on luteal cells could indicate that prolactin exerts its inhibition directly on the luteal cell rather than by more indirect changes to the endocrine milieu of the

Fig. 6.19. Specific hormone binding (expressed as moles $\times 10^{-16}$ per μg DNA) for prolactin (hatched columns) and luteinising hormone (solid columns) in ovarian tissues, corpora lutea and mammary glands of *Macropus eugenii*. Note the absence of LH receptors and abundance of prolactin receptors in corpora lutea. Also the difference in receptor concentration between sucked and nonsucked mammary glands in the same animal. Redrawn from Stewart & Tyndale-Biscoe (1983).

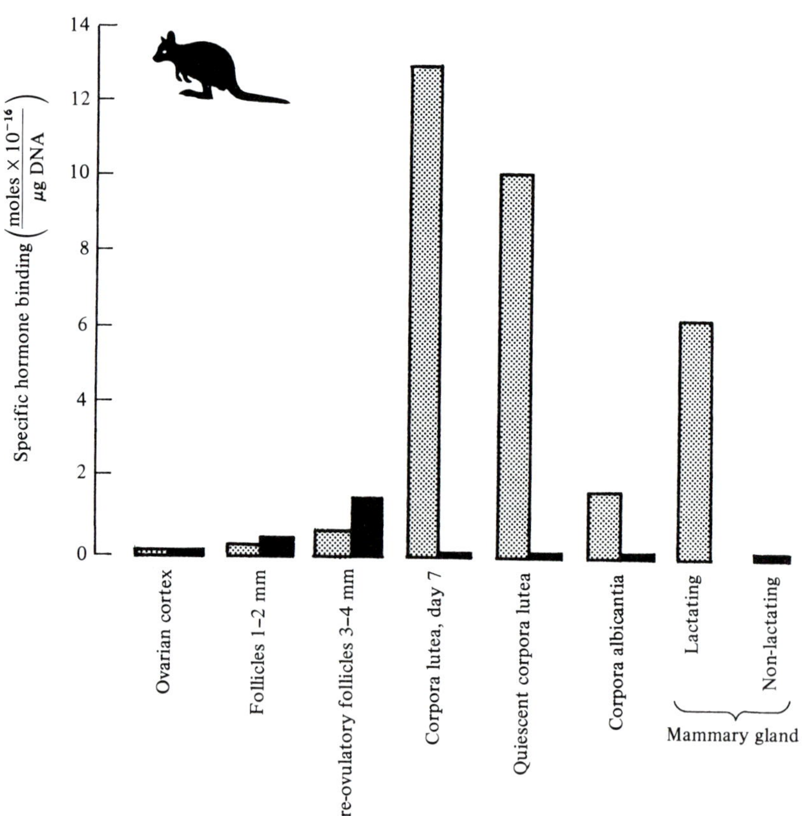

ovary. However, to date, the way in which this is achieved remains unclear. Luteal tissue from quiescent and reactivated corpora lutea secreted the same amount of progesterone into the medium when cultured with prolactin as when cultured without it (Sernia *et al.*, 1979; Hinds *et al.*, 1983), so that prolactin appears not to be acting directly on the steroidogenic pathway. Indeed, what evidence there is seems to favour it effecting a switch in the metabolism of the luteal cell which, once thrown, cannot be reversed. Further discussion of this aspect of corpus luteum control in *M. eugenii* is held over to Chapter 9 (p. 377).

Evidence for or against pituitary dependency of the corpora lutea of other marsupials is very meagre indeed. In *Trichosurus vulpecula*, which does not display lactational quiescence, corpus luteum growth and function continued after hypophysectomy done as early as day 1 post-coitum (Hinds, 1983); plasma progesterone rose after day 8 and reached a peak

Fig. 6.20. Time to birth or oestrus in *Macropus eugenii* from which the pouch young were removed (RPY) and the females variously injected thrice daily for 7 days. The time was extended by 7 days in the animals treated with oxytocin, prolactin and reserpine. Data from Tyndale-Biscoe & Hawkins (1977).

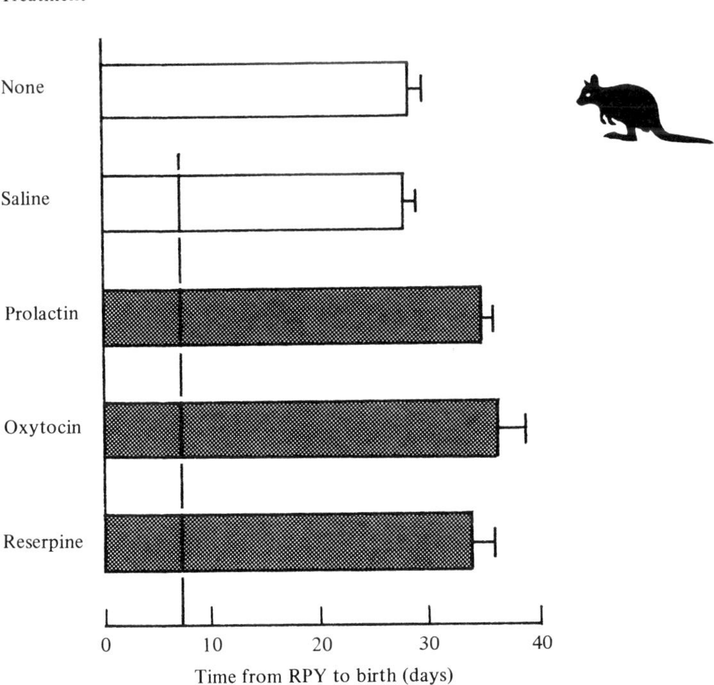

at day 12, the uteri became luteal and embryos developed to full term but were not born. There was some slight indication that progesterone did not rise to the level reached in intact or sham controls but insufficient replicates were available to be certain. However, unlike in *M. eugenii*, luteal cells of *T. vulpecula* have binding sites specific for LH as well as prolactin and so are potentially capable of being stimulated by LH. Further, the concentration of prolactin-binding sites was not as great as in corpora lutea of *M. eugenii* and *M. rufus* (Stewart & Tyndale-Biscoe, 1982).

When Cook & Nalbandov (1968) incubated luteal tissue from ovaries of *Didelphis virginiana*, they found that human chorionic gonadotrophin (hCG) increased progesterone production significantly but that neither FSH nor prolactin had any effect. Nevertheless they concluded that a single release of pituitary gonadotrophin is probably sufficient to induce ovulation and initiate steroidogenesis in this species without any further luteotrophic influence, and that this is the most elementary form of luteal control known in mammals. Their conclusion appears to be fully supported by these later results in *T. vulpecula* and *M. eugenii*, and it may be a feature common to all marsupials. Rothchild (1981) has developed this idea and incorporated it into a general theory of the evolution of corpus luteum regulation to which we will return in Chapter 10. He concludes that the corpora lutea of *Isoodon macrourus* are an exception to the idea of corpus luteum autonomy because they persist into lactation. From Gemmell's (1981) work to which Rothchild (1981) refers, the pattern of secretion of progesterone by the corpora lutea in lactation appeared to be no different from that of non-lactating females and so no special luteotrophic stimulus for steroidogenesis need be invoked. However, Gemmell's (1984) more recent findings do suggest that larger litters may exert an influence on the level of progesterone in circulation. The maintenance of the corpora lutea during lactation, in what appears histologically to be a viable state, is another matter which can only be resolved by applying the same techniques as those used to investigate the corpus luteum of *M. eugenii*.

Demise of the corpus luteum

Within 2 years of Loeb's (1923) demonstration that total hysterectomy after oestrus in the guinea pig prevented luteal regression, Hartman (1925*a*) performed the same operation on *D. virginiana* and found it had no such effect, the corpora lutea degenerating at the same time as in intact females. Clark & Sharman (1965) confirmed Hartman's results in *Trichosurus vulpecula*. No one has re-examined this matter in a marsupial since the discovery in sheep (McCracken *et al.* 1971) that prostaglandins secreted by the endometrium exert a luteolytic effect on the corpora lutea via an arteriovenous anastomosis between the uterine vein and ipsilateral ovarian artery. In sheep, luteal regression appears to be controlled by an interaction of prostaglandin released by the uterus, and oxytocin released by the corpus luteum, and luteal regression is delayed in animals immunised against oxytocin (see review by Flint & Sheldrick, 1983). Luteal oxytocin appears to be involved in stimulating uterine production of $PGF_{2\alpha}$, and bringing about the cessation of each individual episode of $PGF_{2\alpha}$ secretion. In addition, luteal oxytocin may act locally to reduce steroid synthesis by luteal cells. Similar anatomical structures to those in the ewe have been described in marsupials (see p. 190) so that means for direct communication between uterus and corpus luteum are available, but marsupials on present evidence appear to resemble the primates and the monoestrous Carnivora in which the uterus has no influence on the function of the corpus luteum, and the mechanism of luteal regression is unknown, but local production of oestradiol or prostaglandin within the corpus luteum may be responsible for luteolysis in these species (Baird, 1984).

Cook *et al.* (1977*a*) examined the idea that oestradiol from the growing follicles might be responsible for luteal regression in *Didelphis virginiana* by irradiating the ovaries after they had induced corpus luteum formation with gonadotrophins. This led to loss of all growing follicles but did not prolong the life of the corpora lutea.

The only evidence for a shortening of the active life of the corpus luteum comes, again, from *M. eugenii*, Merchant (1979) showed that the intervals from oestrus, or removal of the pouch young, to post-partum oestrus were significantly shorter than the intervals from oestrus, or RPY, to oestrus (Table 6.4), and he suggested that the hormonal events associated with post-partum oestrus might also be affected. Tyndale-Biscoe *et al.* (1983) showed for *M. eugenii* that this is so and that the profiles of four hormones, namely progesterone, prolactin, LH and prostaglandin, differed between the pregnant and non-pregnant cycles of the same group of animals (Fig. 7.27). Progesterone declined on day 25 of the pregnant cycle but not until day 28 of the non-pregnant cycle, which suggests that a luteolytic agent is involved and that it is initiated by the fetus or placenta. A marked pulse of prolactin was observed at the time of the decline in progesterone in all the pregnant but in none of the non-pregnant cycles, and it may be the luteolytic agent, as it is in the rat (Ensor, 1978).

No study has been made to date of prostaglandin synthesis by luteal cells or endometrium of any marsupial and in one study on *M. eugenii* no mesotocin (the oxytocic principle in this species, see Table 8.4) was detected in the corpora lutea (A. P. Flint, A. E. Jetton & M. B. Renfree, unpublished results). However, several attempts have been made to measure the metabolite of prostaglandin $F_{2\alpha}$ (PGFM) in circulation, especially at the time of parturition. In the only species (*Isoodon macrourus*) in which prostaglandin was found to be elevated for 2 days before and one day after parturition (Gemmell *et al.*, 1980) it is not luteolytic; in this species progesterone does not decline at parturition but remains elevated for several days after prostaglandin has fallen to basal levels. Paradoxically, Gemmell (1985) found that administration of an analogue of prostaglandin to *Isoodon macrourus*, during early and late lactation caused a sharp and permanent decline in progesterone concentration.

In *M. eugenii* a very transient pulse of PGFM has been detected at parturition (Tyndale-Biscoe *et al.*, 1983; Shaw, 1983*a*; Lewis, Fletcher & Renfree, 1986; Fig. 7.28) but it occurs after the decline in progesterone (Fig. 7.27) and thus is unlikely to be involved in luteolysis. Walker & Gemmell (1983*a*) did not detect a peripartum pulse in *M. rufogriseus* but this may have been because the daily sampling was too infrequent. Walker *et al.* (1983) disagree with the frequently expressed view that decline of the corpus luteum is due to shrinkage of luteal cells; while they agree that the gland shrinks, their measurements showed no change in size of luteal cells and they conclude that withdrawal of lymph and blood from the gland may be the main factor. This idea certainly is in accord with the rapid decline in circulating progesterone seen in *M. eugenii* and *M. rufogriseus* at this time and with the very rapid decline in progesterone in the venous drainage from the ovary bearing the corpus luteum at parturition (Hinds & Tyndale-Biscoe, unpublished observations). Even a brief pulse of prostaglandin or prolactin could effect such a vascular response.

Conclusions

This review of the corpora lutea of marsupials has fully substantiated Sandes' (1903) perceptive view of the importance and role of this evanescent organ in marsupial reproduction. It has also shown that one must be cautious in speaking of the marsupial corpus luteum; it is now apparent that there are at least three types of corpus luteum and quite probably four. While the general features of formation, growth, progesterone secretion and decline may be the same, there is growing evidence for significant differences between the four types in the duration of each phase and the endocrine control of these phases. The Type 1 corpus luteum appears to be the simplest with little or no variation in its development and decline but the evidence from *D. virginiana* that progesterone secretion is enhanced with human chorionic gonadotrophin and the presence of specific receptors for LH on the corpus luteum of *T. vulpecula* hint at a possible dependence in these species on luteotrophic support. This stands in contrast to the Type 3 pattern of *M. eugenii*, in which the corpus luteum appears to be wholly autonomous and is controlled by tonic inhibition of cell division and growth, probably exercised by prolactin. There is no information at present on how the Type 4 corpus luteum of *Tarsipes rostratus* and *Cercartetus sp.* are regulated during the prolonged periods of diapause but recent work by Gemmell (1985) has provided the first critical evidence for a luteotrophic control of the Type 2 corpus luteum of *I. macrourus*.

The decline of the corpora lutea of marsupials also appears to be largely autonomous. At present the only evidence that prostaglandins cause luteolysis of the corpora lutea of any marsupial is Gemmell's (1985) finding in *I. macrourus* and the only indication that the secretory life of the corpus luteum may be prematurely terminated in any species is that at the end of pregnancy in *M. eugenii*, when the rapid fall in progesterone coincides with a transient pulse of prolactin.

In all these respects the corpora lutea of marsupials appear to be more autonomous than those of all eutherians except the Carnivora, and less regulated by extrinsic influences from the pituitary, the uterus or the placenta. It will be of interest to learn in the future if this greater degree of autonomy is reflected in the endogenous synthesis by marsupial corpora lutea of prostaglandins or of oxytocin, relaxin or other peptides.

7

Pregnancy and parturition

Marsupials give birth to small altricial young after a gestation period which lasts only 12–38 days (Table 2.2). The young then spends a much longer period within a pouch or nest nurtured by milk while it completes its growth and development to the stage of independence at weaning. In recent years there has been much written about the relative merits of marsupial reproduction compared to eutherian reproduction (Lillegraven 1969; 1975; 1979; Sharman, 1970; Cox, 1977; Kirsch, 1977*b*; Renfree, 1981*b*; Padykula & Taylor, 1982). We will enter this debate in Chapter 10 but here we attempt to present a straightforward account of pregnancy and parturition in marsupials uncluttered, so far as possible, by preconceived notions of evolutionary relationships. The process from ovulation to parturition in marsupials can be divided into 10 fairly discrete stages, which are described below.

Sperm transport and fertilisation

Development of the spermatozoa and their maturation in the epididymis have been discussed in Chapter 4 and the varieties of courtship and copulatory behaviour in the several orders of marsupials was considered in Chapter 2. The fate of the spermatozoa after their deposition in the female genital tract and the process of fertilisation itself are less well known. There is a number of isolated observations on different aspects of the phenomena in many species but there are detailed, timed observations on only three species, *Didelphis virginiana* (Rodger & Bedford, 1982*a*,*b*), *Macropus eugenii* (Tyndale-Biscoe & Rodger, 1978) and *Antechinus stuartii* (Selwood, 1982*b*) (Fig. 7.1).

Fig. 7.1. A diagram (not drawn to scale) to show the stages of maturation, fertilisation, cleavage and blastocyst formation in *Antechinus stuartii*. Maturation of the oocyte (oc) surrounded by the zona pellucida (zp) begins in the ovary and is completed in the ampulla of the oviduct, where extrusion of the polar bodies (pb) and fertilization occurs. The eggs receive the mucoid layer (mu) and shell membrane (sm) after fertilisation and enter the uterus in the pronuclear stage. Cleavage and extrusion of the yolk (yb) occurs in the uterus. bm, blastomeres; bl, blastocyst. Redrawn from Selwood (1982*b*) and Selwood & Young (1983).

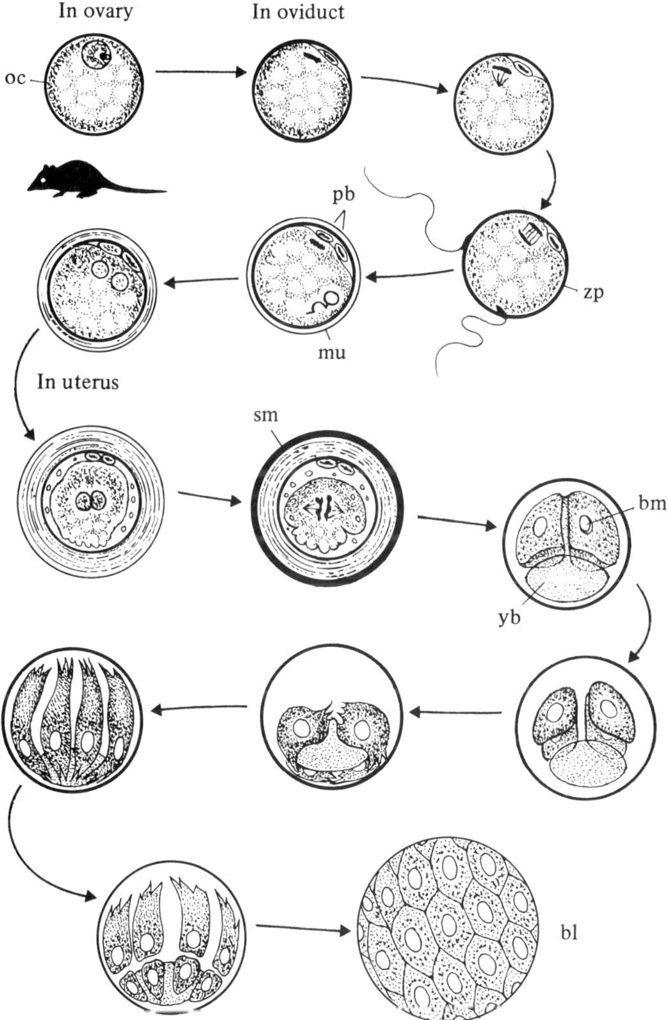

Sperm transport

In *D. virginiana* McCrady (1938) states that the seminal fluid is conveyed via the lateral vaginae (see Chapter 5 for anatomy of female tract) to the vicinity of the cervices 'almost immediately' as a result of the peristaltic action that is pronounced in the lateral vaginae at oestrus but not thereafter (Hartman, 1924). This is very likely induced by the high level of oestradiol in circulation from 4 days to 1 day before oestrus (Fig. 6.3), emanating from the Graafian follicles and which declines to less than 6 pg/ml^{-1} 1 day after oestrus and ovulation (Harder & Fleming, 1981). Hartman (1924) and McCrady (1938) concluded that the spermatozoa must pass into the uterus quickly or become entrapped in the coagulum that results from the mixing of the semen and the secretions of the median vagina.

In *M. eugenii*, 40 min after copulation, the coagulum is devoid of spermatozoa, but thousands are concentrated at both os uteri, and within the two cervical canals (Tyndale-Biscoe & Rodger, 1978). As in *D. virginiana*, oestradiol levels are at their highest at oestrus (Figs 6.1, 6.15) (Flint & Renfree, 1982; Shaw & Renfree, 1984; Harder *et al.*, 1984; 1985) and the rapid transport of spermatozoa may be facilitated by similar contractions of the lateral vaginae. While these have been observed at autopsy, no systematic study has been performed. Unlike *D. virginiana* this species is monovular and oestrus occurs about 8 h after parturition (Fig. 6.1). After post-partum oestrus the distribution of spermatozoa is unequal in the two uteri and oviducts: there are more spermatozoa in the non-parturient uterus and they reach the oviduct on this side 2–5 h after copulation in far greater numbers than in the oviduct associated with the corpus luteum of the foregoing pregnancy (Tyndale-Biscoe & Rodger, 1978). Because ovulation is generally alternate in *M. eugenii*, the greatest concentration of spermatozoa is therefore in the oviduct that is destined to receive the ovum at ovulation. To test the importance of this for fertilisation, females were unilaterally ovariectomised so that ovulation occurred on the same side for 5 successive cycles at each of which the females mated with intact males. The majority had alternately fertile and infertile cycles, as expected if insufficient spermatozoa can traverse the parturient uterus to effect fertilisation (C. H. Tyndale-Biscoe & L. A. Hinds, unpublished results).

In *D. virginiana* the main period of copulation occurs between midnight and 07.00 (Hartman, 1928; Reynolds, 1952; Rodger & Bedford, 1982*a*) and Hartman (1928) and McCrady (1938) observed that copulations that occurred later than this time were less fertile than those that occurred at night. They concluded that this was because ovulation followed oestrus closely and such spermatozoa would not be available to fertilise the newly ovulated eggs. Rodger & Bedford (1982*a*) provided evidence that this is so: in their series, ovulation occurred at 11.00–16.00, or a mean interval of 10 h after copulation, and several thousand spermatozoa were recovered from the oviducts sampled at this time, many being still paired (see Chapter 4, Fig. 4.20). Eggs in process of fertilisation were recovered from the opposite oviducts of the same females a mean interval of 5 h later, by which time all the spermatozoa were separate. Two-celled eggs were recovered from the uterus of one female 22 h after observed copulation. While spermatozoa were found in isthmic crypts for a day or so longer, it is very unlikely that any but those in the infundibulum at ovulation can effect fertilisation.

Fertilisation

Rodger & Bedford's (1982*b*) paper on *D. virginiana* provides the first and only detailed description of fertilisation in any marsupial. Prior

Fig. 7.2. Fertilisation in *Didelphis virginiana*. The fertilising spermatozoon first binds to and penetrates the zona pellucida by its acrosomal face (ac), then attaches by its acrosomal face to the oolemma (oo), some of which has been extruded through the zona (arrow). Note absence of mucoid layer outside zona at this time. From Rodger & Bedford (1982b) with permission.

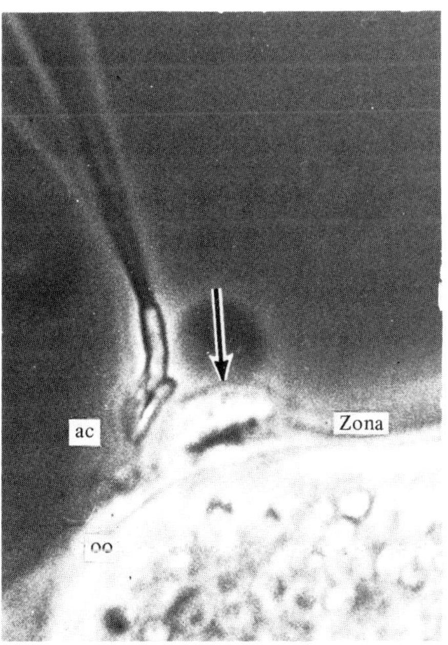

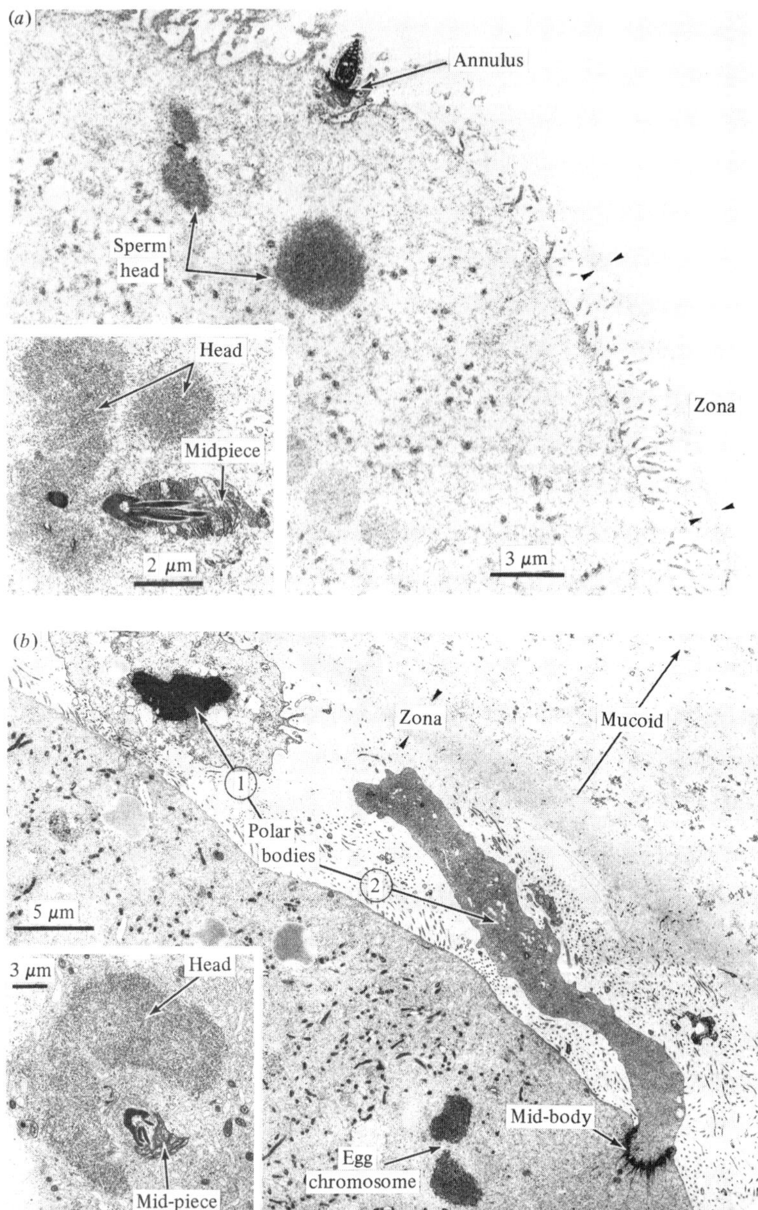

Fig. 7.3. Fertilisation in *Didelphis virginiana*. (*a*) A spermatozoon at a very early stage of incorporation during fertilisation *in vitro*, only the head and mid-piece being within the ooplasm. The zona pellucida (defined by arrow heads) is absent at the point of sperm entry and, for some distance from this point, it appears disrupted. The insert is a detail

to fertilisation the paired spermatozoa separate, probably when housed in the isthmic crypts (Fig. 4.22). Unpairing begins at the anterior end and, as this occurs, the formerly compact acrosome swells and its contents take on the form of membrane vesicles. This reaction does not resemble the acrosome reaction of eutherian mammals and, as mentioned in Chapter 4, the whole function of sperm pairing still remains unclear, except that only separated spermatozoa participate at fertilisation.

The ovum enters the fimbria devoid of cumulus cells but invested by a thin zona pellucida (4 μm) and an extracellular matrix in the perivitelline space (Talbot & Dicarlantonio, 1984). The latter is probably secreted by the granulosa cells as it is present around follicular oocytes. The spermatozoa attach to the zona by the acrosomal face (Fig. 7.2), the site of the previous sperm pairing (see Fig. 4.20), and local dissolution of the zona occurs, presumably effected by enzyme action. Penetration of the zona by the fertilising spermatozoon leaves a characteristic hole, unlike the discrete penetration slit described for many eutherian mammals. Four hydrolytic enzymes have been identified in the acrosome of the spermatozoon of *D. virginiana* – acrosin, arylsulphatase, hyaluronidase and *N*-acetylhexosaminidase (Rodger & Young, 1981) – a crude sonicated preparation of which dissolved the zona in 20 min. Talbot & Dicarlantonio (1984) examined the separate roles of these enzymes *in vitro* and found that while hyaluronidase affected the hyaluronic acid filaments of the extracellular matrix, none of the other enzymes used separately had any effect on this or the zona. However, trypsin removed the zona and granules of the extracellular matrix and they conclude that the acrosomal enzyme acrosin may synergise with proteinases in normal sperm penetration of the zona.

Once inside the zona and the extracellular matrix, the inner acrosomal membrane of the spermatozoon fuses with the oolemma and decondensation of the sperm nucleus begins immediately, even before the rest of the internal structures of the sperm are incorporated into the ovum (Fig. 7.3*a*).

Fig. 7.3 cont.

of the fertilising spermatozoon from another section of the same ovum. In both views, there is no evidence that membrane has been incorporated with the decondensing sperm head. (*b*) Formation of the second polar body (2) during the sperm head decondensation (insert). There is already a substantial coat of mucoid surrounding the fertilised ovum. Cortical granules are absent, and there is no supplementary perivitelline spermatozoon. In this particular egg, fertilised *in vivo*, there were very few spermatozoa trapped in the inner layers of the mucoid, and none in the region illustrated here. From Rodger & Bedford (1982*b*), with permission.

Plasma membranes of the sperm head were not incorporated, a point of contrast with fertilisation in eutherian mammals in which these membranes enshroud the sperm head, due to the phagocytic manner of incorporation. At the same time, the second maturation division of the ovum is completed and the second polar body extruded (Figs 7.1, 7.3*b*), in preparation for syngamy. Eggs containing two pronuclei were recovered from the oviduct of one female 26 h after copulation and likewise Hartman (1916) recovered eggs with pronuclei from the uteri of two females. While Rodger & Bedford (1982*b*) observed two holes in the zonae of some ova incubated *in vitro*, they did not find more than one spermatozoon inside the zona and they suggest that cortical-granule release which accompanied fertilisation may provide a first block to polyspermy, as it does in the rabbit (Aitken, 1981). However, when viewed with the electron microscope, all supernumerary spermatozoa were found to be separated from the zona by a thin layer of the mucoid coat, secreted by the oviducal cells. They conclude that the timing of fertilisation and the secretion of mucoid are finely regulated and that the mucoid coat may have an important role as a secondary block to polyspermy, as it does in the rabbit (see Austin, 1961).

In the light of these new findings, Selenka's (1887) and Hartman's (1923*a*, 1928) conclusion, that copulations late on the day of oestrus are infertile is explicable; by the time the spermatozoa reach the upper oviduct the eggs would already be covered by the mucoid coat and be impenetrable to spermatozoa. This may also explain the failure of many eggs to be fertilised in those species that ovulate large numbers at a time. For instance, in *Didelphis virginiana* the average number of eggs shed is 16 but actual values from 30 to 60 have been recorded by Rafferty-Machlis & Hartman (1953) who also reported that 28% of 1378 eggs examined were unfertilised and a further 10% were abnormal. Likewise in *Sarcophilus harrisii* the mean number of eggs shed by 6 females was 39, of which 20% were undeveloped and one-third of cleaving eggs failed to develop past the 9-cell stage (Hughes, 1982). Similarly in *Antechinus stuartii* 24% of eggs were unfertilised or had failed to complete cleavage (Selwood, 1983). Asynchrony between ovulation and arrival of spermatozoa at the site of fertilisation could also be the cause of infertility in monovular species. Among the monovular Macropodidae, a proportion of the adult females in wild populations fail to conceive at post-partum oestrus as judged by recovery of unfertilised eggs (Tyndale-Biscoe, 1963*a*; Frith & Sharman, 1964; Sadleir, 1965; Renfree & Tyndale-Biscoe, 1973*a*; Catt, 1977).

On the other hand, if oestrus occurs 1 or more days before ovulation and the spermatozoa can be retained in a viable state, fertilisation is more likely to occur. In *Trichosurus vulpecula* ovulation is reported to occur up to 2 days after oestrus (Pilton & Sharman, 1962; Hughes & Rodger, 1971) while in *Dasyurus viverrinus* and *Antechinus stuartii* ovulation occurs about 5 days and 10 days respectively after the onset of oestrus and spermatozoa survive in the oviductal crypts (Bedford *et al.*, 1984) throughout this time (Hill & O'Donoghue, 1913; Woolley, 1966*b*). Nevertheless, unfertilised eggs occur among normal fertilised eggs in both these species (Hill, 1910; Selwood, 1983), so that high failure rate may be a feature of fertilisation in polyovular marsupials.

Egg membranes

The egg membranes of marsupials and monotremes were reviewed by Hughes (1977; 1984) and compared to those of other vertebrates. As well as the primary plasma membrane of the oocyte itself, the egg is invested by a secondary membrane, the zona pellucida laid down in the developing follicle, and by tertiary membranes secreted by the epithelia of the oviduct and uterus (Figs 7.4, 7.5, 7.6). Tertiary egg membranes are not found in the majority of eutherian species but all monotremes and marsupial eggs are surrounded by a mucoid layer and a keratin-like shell membrane.

The primary vitelline membrane

The primary vitelline membrane of the oocyte of marsupials projects outwards as a nap of microvilli (Fig. 7.3*a*), which interdigitate with microvilli arising on the granulosa cells of the surrounding follicle (Lyne & Hollis, 1983; Hughes, 1984). In *D. virginiana* this is surrounded by an extracellular matrix in the perivitelline space which was secreted by the granulosa cells (Talbot & Dicarlantonio, 1984).

The secondary egg membrane – the zona pellucida

As it is deposited between the granulosa cells of the developing follicle and oocyte, the zona pellucida isolates the egg from the maternal tissues; the oocyte microvilli remain however within the inner, less dense, layers of the zona and remain so even after ovulation in *Trichosurus vulpecula* (Hughes, 1977), *Didelphis virginiana* (Rodger & Bedford, 1982*b*) and *Macropus eugenii* (Fig. 7.4*a*). The fully formed zona varies in thickness from about 1 μm in *D. virginiana* to 6 μm in *T. vulpecula* and *M. eugenii* (Table 7.1) and is thus considerably thinner than in eutherian mammals in which it is 10–30 μm thick (Austin, 1961). The width of the zona in marsupials is generally less than the length of the respective sperm head

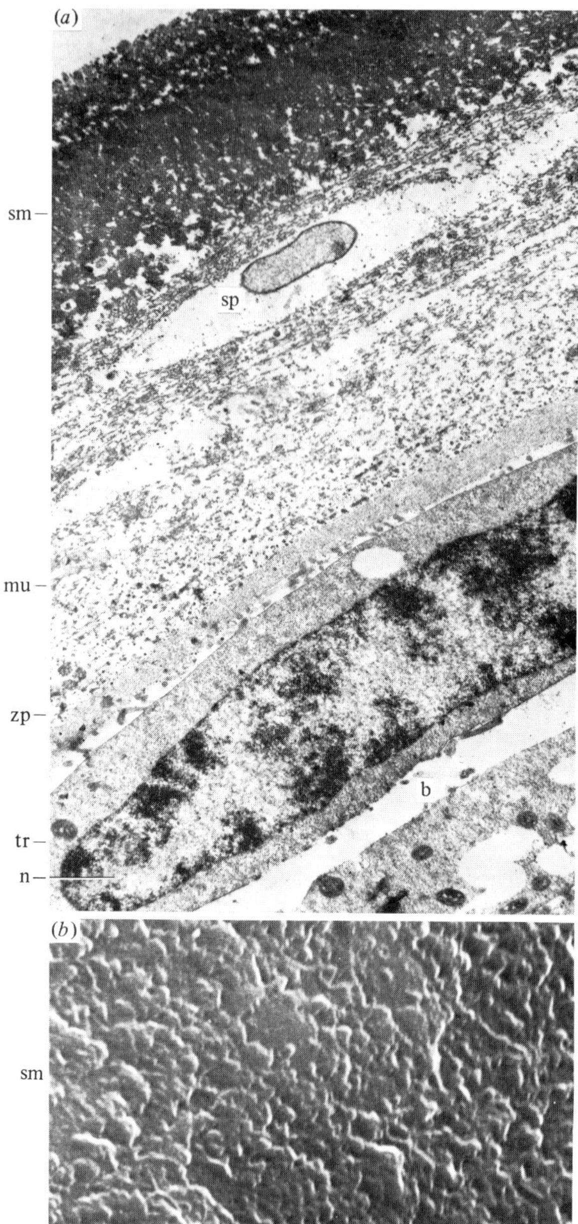

Fig. 7.4. Structure of the egg coats of *Macropus eugenii*. (*a*) Transmission electron micrograph to show shell membrane (sm), mucoid layer (mu) with spermatozoon (sp), zona pellucida (zp) and trophoblast cell (tr) and nucleus (n), and blastocoele (b). (*b*) Scanning electron micrograph of the surface of the shell membrane of a unilaminar blastocyst.

and presumably offers less obstacle to sperm penetration than do the thicker zonae of eutherian species in which sperm are frequently found embedded. By transmission electron microscopy Rodger & Bedford (1982*b*) could find no evidence that spermatozoa remained lodged in the thin zona of *D. virginiana* but Sharman (1961*b*) and Selwood (1982*b*) reported them to be within the somewhat thicker zonae of *T. vulpecula* and *Antechinus stuartii*. As the latter authors relied on light microscopy, their findings need to be confirmed with the electron microscope.

The histochemical composition and properties of the zona of *T. vulpecula* has been investigated by Hughes & Shorey (1973) and Hughes (1974). It is composed of a matrix of weekly acidic glycoproteins and is readily permeated by large molecules such as ferritin (460000 daltons).

The zonae of marsupials probably do have an important role during cleavage (Hill, 1910) in retaining the separate blastomeres in loose contiguity (see Figs 7.6*b*, 7.7*b*) for several days until tight junctions develop between them at blastocyst formation, as in Eutheria. However, this has not been tested experimentally in any marsupial species. As the unilaminar blastocyst expands, the zona disappears (Hughes, 1974; 1984).

The inner tertiary egg membrane

The mucoid coat or mucolemma occurs in all marsupials (Figs 7.3*b*, 7.4*a*, 7.5) and in monotremes but in only a few species of eutherian mammals, such as the rabbit. The mucoid coat of *T. vulpecula* has the histochemical properties of a strongly sulphated glycoprotein and is secreted in the ampulla and isthmus regions of the oviduct (Hughes, 1974). Hughes & Shorey (1973) showed that it was permeable to horseradish peroxidase (40000 daltons) but to only a limited extent to the much larger molecule of ferritin. This membrane was formerly referred to as an albumin coat but, Hughes (1974) considers it preferable to refer to it as the mucoid coat or mucolemma. It is laid down in concentric layers as the egg passes through the oviduct (see Fig. 7.6*b*, 7.7*c*) and supernumerary spermatozoa become entrapped in it and remain there until its final disappearance later in development. When Hartman (1923*a*) placed *Ascaris* eggs at the fimbria of oestrous *D. virginiana*, they became covered by a mucoid coat and he also observed clusters of spermatozoa to be similarly engulfed. The fully formed mucoid coat varies in thickness in marsupials from less than 10 μm in *Perameles* to 230 μm in *Didelphis* (Table 7.1). The coat is markedly thinner around newly formed blastocysts than at earlier cleavage stages (e.g. *A. stuartii*, Selwood & Young, 1983) and in all species disappears, with the zona, when the blastocyst expands.

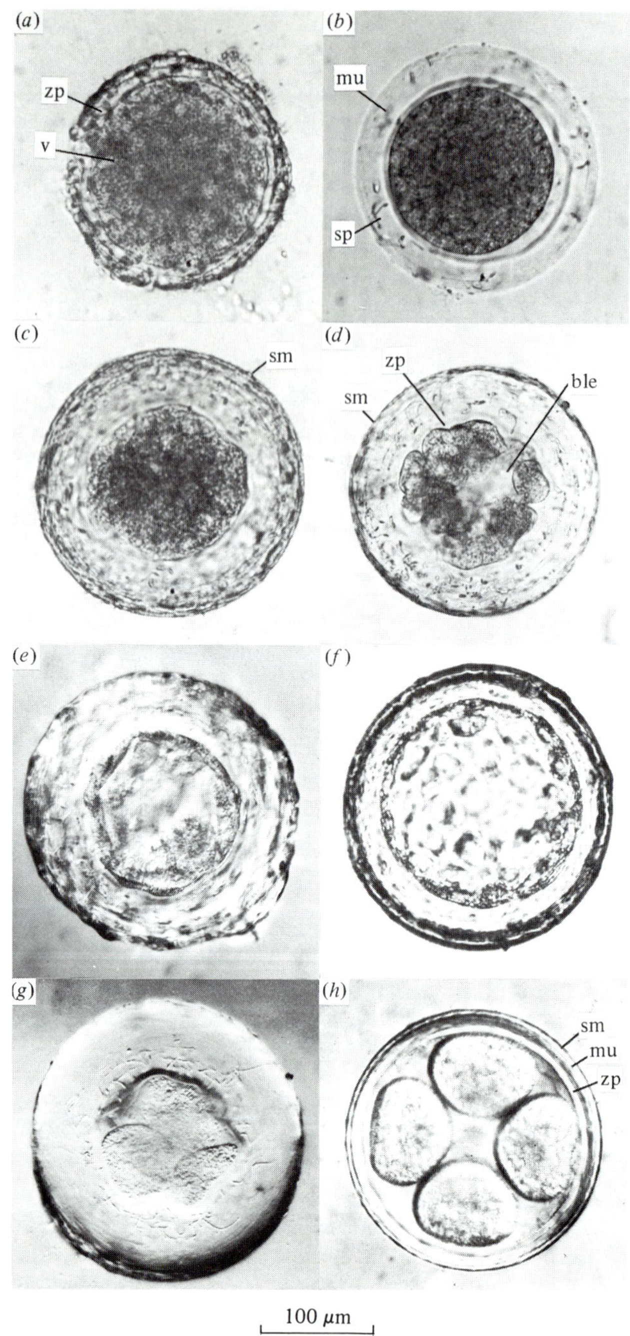

Fig. 7.5. For legend see opposite.

This was most clearly shown by Hartman (1928) for *D. virginiana* where the blastocyst is eccentrically placed inside the shell membrane, and he concluded that the embryo was drawing upon both the zona and the mucoid coat for its nutrition prior to its absorption of uterine secretions.

The shell membrane

The shell membrane was first recognised to be an homologue of the basal layer of the monotreme shell by Caldwell (1887) and confirmed by Hill (1910) and Hartman (1916) and reviewed by Hill (1933). Hill (1910) described the shell membrane of *D. viverrinus* as:

> a transparent, perfectly homogeneous layer, highly refractive in character and of a faint yellowish tint. It is distinguishable at once from the albumin [= mucoid coat] by its optical characters and staining reactions, so that there is not the slightest justification for the supposition that it may represent simply the specially differentiated outermost portion of that layer.

The first trace of it is seen on eggs recovered from the uterotubal junction, where it is secreted by shell glands of the lower oviduct (Hartman, 1916; Andersen, 1928; McCrady, 1938).

Hughes (1974, 1984) and Hughes & Hall (1984) state that in *T. vulpecula* the shell glands only occur in the endometrium and they conclude that the material for the shell membrane surrounding eggs recovered from the oviduct must have passed back from the uterus. In our experience the uterotubal junction does not allow such retrograde flow (see p. 173). On the basis of his conclusion that the shell material is exclusively secreted by uterine glands, Hughes (1984) has proposed that the shell membrane of marsupials is not homologous with the basal layer of the monotreme shell, because that is secreted by oviducal glands, but with the outer layer which is secreted by uterine glands. This homology disregards the evidence from *D. virginiana* and is less parsimonious than Caldwell's homology.

Fig. 7.5. Living stages in the early development of *Macropus eugenii*, *M. giganteus* and *D. viverrinus*. (*a*) Preovulatory follicular oocyte at oestrus, to show vitellus (v) and zona pellucida (zp). (*b*) Fertilised egg recovered from the oviduct less than 1 day *p.c.*, surrounded by the mucolemma (mu) and entrapped spermatozoa (sp). (*c*) Cleaving 4-cell egg recovered from the uterus 30 h *p.c.* with shell membrane (sm) covering the mucolemma. (*d*) Late cleavage stage on day 5 *p.c.*, comprising 12 cells and beginning formation of blastocoele (ble). (*e*) Newly formed blastocyst on day 7*p.c.* (*f*) Quiescent unilaminar blastocyst. (*g*) 4-cell egg of *M. giganteus*. (*h*) 4-cell egg of *Dasyurus viverrinus*.

Deposition continues after the egg enters the uterus as the maximum thickness is attained around unilaminar blastocysts, which have not yet expanded (Fig. 7.5*f*). This thickness ranges from 1 to 1.6 μm for *D. virginiana* to 6 μm for *M. eugenii* (Table 7.1). Secretion from the glands must cease at this stage, in species of Macropodidae at least, because the shell

Fig. 7.6. Cleavage, 'yolk' extrusion and blastocyst formation in (*a*, *b*) *Didelphis marsupialis* and (*c*) *D. virginiana*. (*a*) 2-cell stage with first extrusion of 'yolk' (y. sph.), surrounded by zona pellucida (zp. (*b*) 14-cell stage at first appearance of cleavage cavity (cc) which will later become the blastocoele. Note the thick mucoid layer (mu) and folded shell membrane (sm). (*c*) Early blastocyst with inward migration of first endoderm cells (en) and establishment of polarity. From Hill (1918) and Hartman (1919).

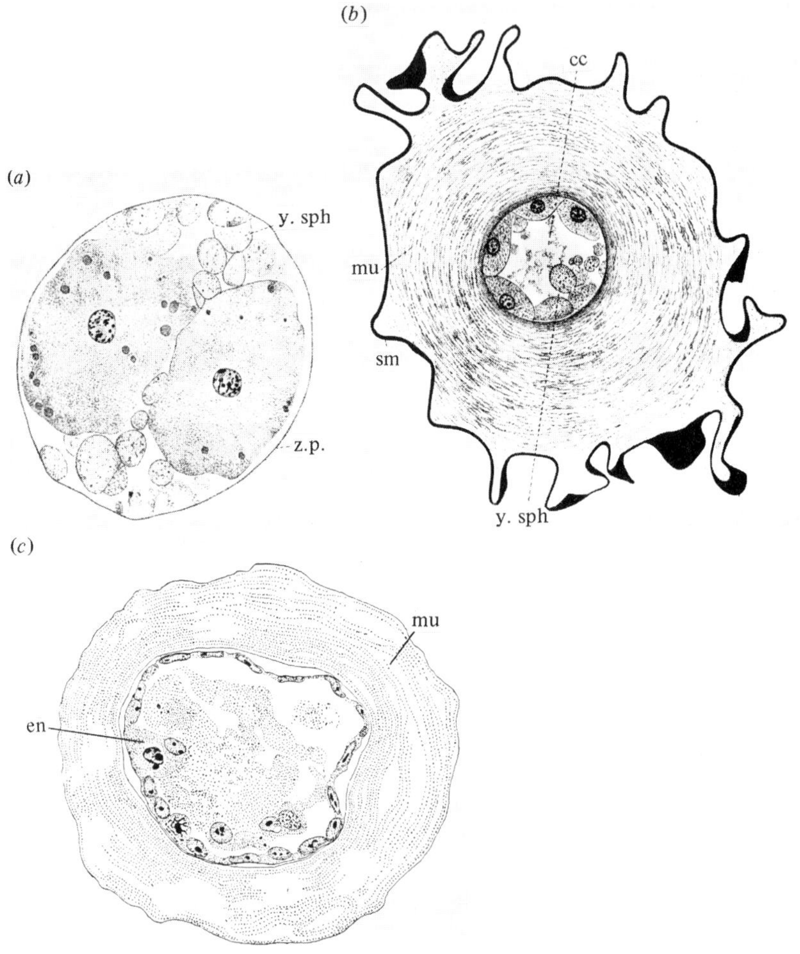

Table 7.1. *Dimensions of tubal eggs, egg coats and spermatozoa of marsupials*

Species	Vitellus (1 cell) (μm)	Zona pellucida (μm)	Mucoid coat (μm)	Shell membrane (μm)	Sperm head		References	
					Length (μm)	Width (μm)	For eggs	For spermatozoa
Didelphis virginiana	135–165	1	140	1.2	7	3.5	Hartman (1916)	J. C. Rodger (personal communication)
	113	1	230	1.0	—	—	McCrady (1938)	
D. marsupialis (aurita)	144	1	140	1.2	—	—	Hill (1918)	
Dasyurus viverrinus	240	1.6	15–22	2–8	11.0±0.09	1.9±0.08	Hill (1910)	Harding *et al.* (1982*b*)
Antechinus stuartii	158±2	5.5±0.4	16.6±0.6	2.2±0.2	—	—	Selwood & Young (1983)	
Sarcophilus harrisii	210	4	28	5	11.1±0.4	2.2±0.2	Hughes (1982)	Hughes (1965)
Perameles nasuta	220[a]	1.3	6.7	1.0	5.7±0.2	3.0±0.1	Lyne & Hollis (1976)	Hughes (1965)
Isoodon macrourus	250[b]	1.3	—	3.1	6.0±0.1	3.3±0.2	Lyne & Hollis (1976)	Hughes (1965)
Trichosurus vulpecula	229±2	5.9±3.1	50±2.3	4.8	3.7±0.2	1.6±0.1	Hughes & Hall (1984)	Harding *et al.* (1982*b*)
Macropus eugenii	126±13	6.3±1.4	24.3±7.8	5.9±1.8	—	—	Renfree & Tyndale-Biscoe (1978)	
Macropus parma	—	—	—	—	5.3+0.3	1.4		Harding *et al.* (1982*b*)

[a] 4-cell uterine egg.
[b] 8-cell uterine egg.

Fig. 7.7. 'Yolk' extrusion and blastocyst formation in (*a*,*b*) *Antechinus stuartii*, (*c*) *Macropus eugenii* and (*d*) *M. rufus*. (*a*) 4-cell egg with extruded yolk mass (ym), membraneous bodies (arrows) and microvilli (mv) on surface. (*b*) Meridional section of a 16-cell embryo showing that the blastomeres of the yolky-hemisphere (arrows) are more rounded and have less fibrous material than those of the non-yolky-hemisphere. The cells are attached to the zona pellucida by a number of cell processes (p). From Selwood & Young (1983). (*c*) Section of early blastocyst at comparable stage to that in Fig. 7.5 (*e*) to show extrusion of cytoplasm and arrangement of blastomeres (bm) around the inner surface of the zona pellucida (zp), which is surrounded by the mucoid layer (mu) and shell membrane (sm). (*d*) Quiescent blastocyst lying in a crypt of the uterine mucosa of a female at day 50 of lactation. The unilaminar blastocyst is seen partly in section and partly in surface view. From Sharman (1963) with permission.

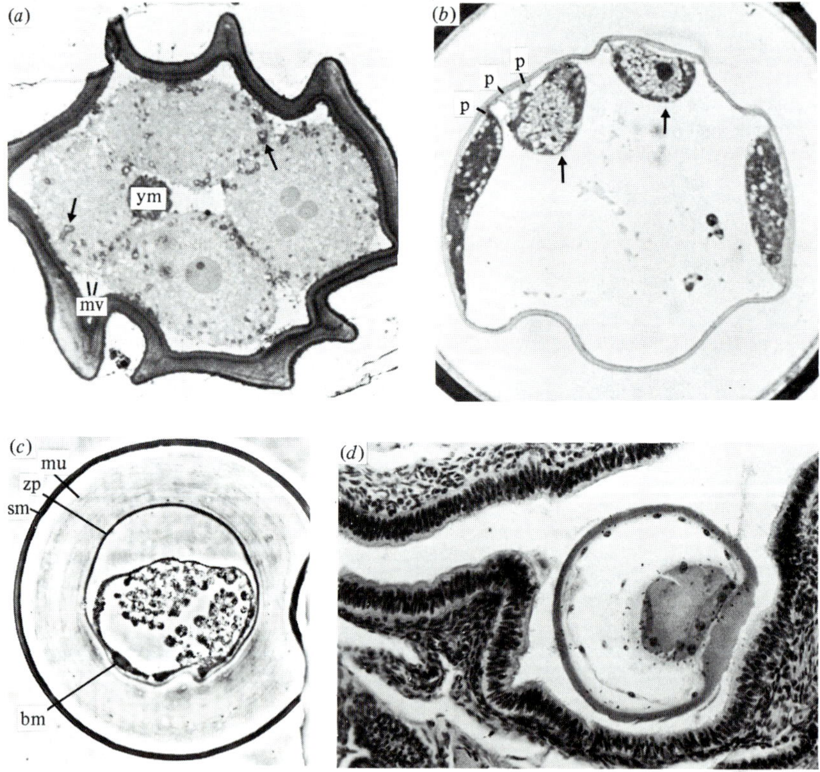

membranes of unilaminar blastocysts retained in diapause for many months have the same thickness as newly formed ones (Smith, 1981). In all species examined the shell membrane becomes progressively thinner as the blastocyst expands and the two inner membranes (zona pellucida and mucoid coat) are absorbed. This thinning is probably due to the greatly increased area that the shell must cover without rupture, for it does not break down until the last third of pregnancy when embryogenesis is complete (Fig. 7.15) (Hughes, 1974).

The shell membrane of several species was shown by Hughes (1974, 1977) to be a resistent proteinaceous substance, rich in disulphide bonds and with the histochemical properties of an ovokeratin. He identified flakes of a similar material in uterine glands, and the surface appearance of the shell of *D. virginiana* (Krause & Cutts, 1983) and *Macropus eugenii* as seen with the scanning electron microscope (Fig. 7.4*b*) supports the idea that it is built up by accretion of such flakes. It also appears to be porous, which is in agreement with the conclusion of Hughes & Shorey (1973) that the shell membrane of *T. vulpecula* is freely permeable to peroxidase but less so to ferritin. In section, the mature shell membrane of *D. virginiana* (day 8–9) and *M. eugenii* consists of a homogeneous mat of interwoven branching fibres with no apparent substructure (Krause & Cutts, 1983) (Fig. 7.4*a*). The fibres are somewhat loose, in contrast to the tightly packed structure seen in *Isoodon macrourus*, *Perameles nasuta* (Lyne & Hollis, 1977*b*) and *T. vulpecula* (Hughes, 1977) but Krause & Cutts (1983) attribute this to the earlier stages of development examined in the latter species.

Sharman (1961*b*) observed that the shell membrane of *Setonix* first broke down on day 19 in the region of the non-vascular yolk sac (Fig. 7.17*a*) and persisted for longer in the opposite vascular region. Denker & Tyndale-Biscoe (1986) have demonstrated the presence of two proteinase enzymes in the non-vascular yolk sac membrane and adjacent uterine epithelium of *Macropus eugenii*. One enzyme is an SH dependent endopeptidase of the cathepsin type with an acid pH optimum; the other, present only on day 19 where the shell membrane was in process of dissolution, is an SH-proteinase with an alkaline pH optimum, analogous to a similar enzyme observed in rabbit blastocysts at the time of dissolution of the egg coverings of that species.

The function of the shell membrane is not clearly established but its persistence for two-thirds of gestation in all marsupials examined (Hughes, 1974; Renfree, 1977) and its disappearance at the beginning of organogenesis may be significant. Impressed by Lillie's (1917) interpretation of the cattle freemartin being due to the shared circulation of the twins, Hartman (1920*b*) suggested that the main role of the shell membrane might be to prevent the fusion of the fetal membranes of adjacent embryos. In *Philander opossum* Enders & Enders (1969) showed that such a fusion does occur after breakdown of the shell membrane (Fig. 7.18) but there is no evidence to suggest that this species has a high proportion of freemartins. Sharman (1963) was the first to suggest that the shell membrane may have an immunoprotective role in separating fetal tissues from maternal tissues and this idea was further developed by Moors (1974), Tyndale-Biscoe (1973) and Lillegraven (1975), and will be discussed later in this chapter.

'Yolk' extrusion, cleavage and blastocyst formation

The eggs of marsupials vary quite considerably in size (Table 7.1) but even the smallest at 135 μm diameter (*D. virginiana*) exceed the largest eutherian eggs (Austin, 1961). The ooplasm contains clear material and lipid droplets that has long been termed 'yolk', although recent examination with the electron microscope (Lyne & Hollis, 1976) has shown it to be composed of other subcellular organelles as well.

In *Dasyurus viverrinus*, which has an exceptionally large egg, the yolk is clearly demarcated in the oocyte and in the fertilised egg (Hill, 1910) and is segregated as a distinct body during the first and second cleavages. A similarly distinct yolk body is segregated in *Antechinus stuartii* (Figs 7.1, 7.7*a*) by the second cleavage (Selwood & Young, 1983) and in both these species the cleavage planes are at right angles, so that a ring of four blastomeres is formed, with the yolk body eccentrically placed between them. In both species the third cleavage is also meridional (Fig. 7.1) so that a ring of eight equal but detached blastomeres surround the yolk body, each being adpressed to the inner surface of the zona (Fig. 7.7*b*). The fourth cleavage is parallel to the equator and divides these eight cells into two tiers of unequal size. The upper tier of eight small cells surround the yolk body while the lower tier are pear shaped and extend into the lower hemisphere. At this stage therefore the embryo has a distinct polarity but, in both species, the divisions then become asynchronous and the small cells indistinguishable, so that in embryos of 30 cells or more, polarity is lost (Fig. 7.1). By this stage the blastomeres are joined at their margins and form a unilaminar blastocyst which is complete by the 60-cell stage. Up to this stage there has been no increase in the overall diameter of the embryo, but hereafter the blastocyst expands.

The eggs of *Didelphis virginiana* and *D. marsupialis* are smaller than those of *D. viverrinus* and *A. stuartii* (Table 7.1) and the yolk less abundant.

In the unfertilised egg it is distributed in a submarginal zone (Hartman, 1916) and after fertilisation it is not concentrated at one pole as in the previously described species. The yolk is extruded in a more diffuse manner during the first few cleavage divisions (Fig. 7.6*a*). The first cleavage results in unequal-sized blastomeres so that polarity can be said to be established then. The second cleavage plane is oblique, so that the four blastomeres form a tetrad surrounded by the extruded yolk material. After the third and fourth cleavage divisions the blastomeres migrate to the periphery where they come to lie against the inner surface of the zona pellucida (Fig. 7.6*b*). The yolk material is displaced by these movements to the centre and thus becomes enclosed within the early blastocyst. Because of the small size and the predominance of the blastomeres within the zona, it is not surprising that the process has been termed yolk extrusion. However, the process, especially as described for *D. virginiana* by Hartman (1916, 1919) and McCrady (1938), and for *D. marsupialis* by Hill (1918), could be interpreted in the opposite way as a segregation of a germinal disc from the yolk, in the manner described for monotremes (see Griffiths, 1978).

The blastocyst is completed at the 16-cell stage in *D. virginiana* at a diameter of 110 μm, which is much smaller than that of the dasyurids. At this stage there is a distinct polarity surviving from the unequal first cleavage, but both Hartman (1919) and McCrady (1938) say that this is lost and the cells of the unilaminar blastocyst are indistinguishable one from another. However, the first endoderm mother cells differentiate at a very early stage and are found attached to the inner surface of one pole of the 50–60-cell blastocyst (Fig. 7.6*c*). While Hartman (1919) and Hill (1918) believed that the primary and the secondary polarity were related, McCrady (1938) was not so convinced.

There is very little information on cleavage and yolk extrusion in other marsupials. Hughes (1982) has illustrated cleavage stages in *Sarcophilus harrisii*, which closely resembles that of *D. viverrinus* and *A. stuartii*, except that the first two blastomeres are unequal in size and a yolk body is not distinct. More material would be needed to determine whether these are significant differences.

Lyne & Hollis (1976, 1977*b*) examined 4-cell, 8-cell and 75-cell embryos recovered from the uteri of *Perameles nasuta* and a 75-cell embryo recovered from *Isoodon macrourus*. In the 4-cell stage the blastomeres were equal in size and were each still connected to a central mass of yolk, which thus resembles the dasyurid rather than the didelphid pattern of yolk extrusion. At the 75-cell stage the embryo was a unilaminar blastocyst and with the electron microscope the cells were seen to be joined by tight junctions. In the blastocyst from *I. macrourus* at the same stage the cells are thicker at one pole than the other, reminiscent of the early blastocyst of *A. stuartii* and *D. virginiana*. In a somewhat later stage, this difference is more apparent and may possibly represent the initial stage of endoderm formation.

Cleaving eggs have been illustrated for five species of diprotodont marsupial. Sharman (1961*b*) collected a 4-cell embryo from the uterus of *T. vulpecula* and Hughes (1974) figures a 16-cell embryo of this species, which was found in the oviduct. It is the only cleavage stage of any marsupial to be reported from the oviduct. It is not clear from either of these reports how yolk is extruded in this species but in *Petauroides* (*Schoinobates*) *volans* Bancroft (1973) figured a 2-cell egg in the uterus that shows some separated material within the zona that may be extruded yolk. In cleaving eggs of *Macropus eugenii* (Fig. 7.5*c*, *d*) the blastomeres do not separate distinctly as in the dasyurids and separated yolk is not evident. Similarly, yolk is not clearly separated in the 4-cell egg of *Macropus giganteus* (Fig. 7.5*g*).

There is some information on the rate of early development in the species mentioned above. All authors agree that the eggs of marsupials traverse the oviduct in less that 48 h and probably in less that 24 h. With the exception of Hughes' (1974) case referred to above, all authors agree that the eggs reach the uterus before the first cleavage. Hill (1910), Hartman (1916) and Sharman (1955*c*) recovered pronuclear eggs from the uterus and Godfrey (1969*a*; 1975), Tyndale-Biscoe & Rodger (1978), Evans *et al.* (1980) and Selwood (1982*b*) recovered 1-cell eggs from the uterus. Rodger & Bedford (1982*a*) recovered 2-cell eggs in the uterus 24–30 h after copulation.

In contrast to the rapid passage through the oviduct, the first six cleavage divisions leading to formation of the unilaminar blastocyst are remarkably slow (Table 7.2). Thus embryos of 16–32 cells, i.e. those that have undergone four or five cleavage divisions, are found on day 3 (*Didelphis virginiana*, *Perameles nasuta*), day 4 or 5 (*Trichosurus vulpecula*), day 6 (*Antechinus stuartii*) and day 7 (*Macropus eugenii*, *Dasyurus viverrinus*) and the fully formed unilaminar blastocyst on days 4, 5, 8, 8 and 9 respectively. While these times are quite comparable to the intervals from copulation to blastocyst formation of 4–10 days among eutherian mammals (Austin, 1961; Wimsatt, 1975), the main differences are that this development occurs in the uterus in marsupials and the embryo is invariably enclosed by three egg membranes. In *D. virginiana* cleavage is rapid, averaging 12 h for each division but in *M. eugenii* the first cleavage

Table 7.2. *Estimated developmental times during cleavage in culture and* in vivo *in* Antechinus stuartii

Condition		Duration in culture (h:min)	Cumulative time (h:min)	
From	To		In culture	*In vivo*
1-cell (yolk polarization and elimination)	1-cell (onset of 1st division)	4:15	4:15	—
1-cell (onset of 1st division)	2-cells (onset of 2nd division)	19:15	23:30	24:00
2-cells (onset of 2nd division)	4-cells (rounded)	24:00	47:30	36:00
4-cells (rounded)	4-cells (flattened – onset of 3rd division)	54:00	101:30	108:00
4-cells (flattened – onset of 3rd division)	6–8 cells	9:00	110:30	120:00
6–8 cells	9–16 cells	12:00	122:30	132:00
9–16 cells	*c.* 32-cell unilaminar blastocyst (slight expansion)	54:30	177:30	—

From Selwood (1980) and Selwood & Young (1983).

takes 2 days, the subsequent divisions less than 1 day each (Tyndale-Biscoe, 1979, and published observations), whereas in *A. stuartii* the third cleavage division takes 3 days to complete, while those before and after take less than 1 day each (Selwood, 1980, 1981).

The unilaminar blastocyst and the phenomenon of diapause

Development to the unilaminar blastocyst stage is achieved without increase in diameter (Fig. 7.5) and limited evidence from culture *in vitro* and ovariectomy suggest that the marsupial embryo, like the eutherian, can reach this stage on endogenous resources but that subsequent expansion and differentiation depends absolutely on the provision of sufficient uterine secretions. Many writers from Selenka (1887) and Hill & O'Donoghue (1913) onwards have commented upon the close association of expanding blastocysts and a highly secretory endometrium (Hartman, 1928; McCrady 1938; Sharman, 1955*b*; Pilton & Sharman, 1962; Tyndale-Biscoe, 1963*a*; Godfrey, 1969*a*; Renfree & Tyndale-Biscoe, 1973*a*). In some species such as *Didelphis virginiana* (Fig. 6.4) and *Trichosurus vulpecula* the increase in weight is marked and the secretion is so abundant that it can be pipetted out of the lumen of the opened uterus (Renfree, 1975). In other species, such as *M. eugenii* the secretion is less abundant but endometrial growth is nonetheless pronounced (Fig. 7.24).

The evidence for autonomy to the unilaminar blastocyst stage derives from one study on culture *in vitro*. Selwood & Young (1983) cultured eggs of *Antechinus stuartii* in Dulbecco's modified Eagle medium containing 10% fetal calf serum. Eggs from seven litters were successively cultured from 1-cell fertilised eggs to the unilaminar blastocyst of 32 cells (Fig. 7.1 and Table 7.2). The rate of development at each stage was comparable to the rate *in vivo*, including a long period at the 4-cell stage. So far no eggs have developed from 1-cell to blastocyst, and it may prove to be difficult to culture eggs past the 4-cell stage because of the natural stasis of 2 or 3 days, which Selwood & Young (1983) have suggested may require a signal from the corpora lutea to overcome. No attempts have been made to culture the eggs of any other species so far, but attempts by Renfree & Tyndale-Biscoe (1978) to culture unilaminar blastocysts of *Macropus rufus* or *M. eugenii* in diapause to later stages of development were unsuccessful. However, expanded bilaminar blastocysts of *M. eugenii* survived *in vitro* in either Eagle's medium with 10% maternal serum for 6 days and developed from the pre-somite stage to 20-somite embryos with limb buds and heart beat (C. H. Tyndale-Biscoe & L. A. Hinds, unpublished observations) or in Whittinghams T6 medium with 10% wallaby serum for

2–6 days and similarly developed to somite embryos (M. B. Renfree & A. E. Trounson, unpublished observations).

Similar results were obtained by New & Mizell (1972) and New, Mizell & Cockroft (1977), who incubated embryos of *D. virginiana* under several different culture conditions for up to 30 h. Although most embryos showed some development in culture, the youngest embryos (8½–9 days), which had intact shell membranes and yolk sacs, grew best.

These few observations suggest that the unilaminar blastocyst has peculiar properties or resistance, which may be associated with the phenomenon of embryonic diapause, but that subsequent development is more tolerant of conditions *in vitro*.

All experimental work on the initiation of blastocyst expansion has been done with several species of the Macropodidae, which display embryonic diapause associated with a quiescent corpus luteum described in Chapters 2 and 6. Here we will consider the factors that initiate and maintain diapause and the process of reactivation; environmental factors in the control of diapause well be considered in Chapter 9.

Embryonic diapause

As discussed in Chapter 2, all species of the two sub-families of the Macropodidae, except *M. fuliginosus*, exhibit the phenomenon of embryonic diapause (Sharman & Berger, 1969; Tyndale-Biscoe *et al.*, 1974; Renfree, 1981*a*), and the appearance and stage of the embryo is remarkably uniform in all (see Smith, 1981). In 12 species the outer diameter of the shell membrane surrounding the embryo varied from 0.25 mm to 0.33 mm (Smith, 1981) and the diameter of the trophoblast of 4 species varied from 0.20 mm to 0.25 mm (Tyndale-Biscoe, 1963*a*, Renfree & Tyndale-Biscoe 1973*a*). The blastocyst is invariably unilaminar and consists of 70–100 uniform cells (Smith, 1981) in which mitoses are never seen (Figs 7.5*f*, 7.7*d*). This conclusion is reinforced by the observation that blastocysts collected from females throughout lactation have the same dimensions and cell number (Clark, 1966; Tyndale-Biscoe, 1963*a*). Conversely, in one species *Potorous tridactylus*, there is some evidence (Smith, 1981) that the number of cells decreases during lactation, but how this occurs is not clear. There is no part that can be recognised as the presumptive embryo or formative region, nor can any endoderm mother cells be distinguished.

Blastocysts in diapause are always associated with a quiescent corpus luteum and some experiments have been done to examine whether the corpus luteum is essential to its survival. In *Setonix brachyurus* the quiescent corpus luteum disappears from the ovary during anoestrus and the blastocyst degenerates. Progesterone has not been measured during anoestrus but, from the appearance of the anoestrous uterus (Sharman & Berger, 1969), the level must be very low. However, in *M. eugenii* blastocysts have survived in the uteri for 4 months after bilateral ovariectomy or lutectomy and, in the latter cases, resumed development when a new corpus luteum or luteinised follicle formed and grew in the remaining ovary (Tyndale-Biscoe & Hearn, 1981). In this species, plasma progesterone does not decline (Sernia *et al.*, 1980) after ovariectomy so that the corpus luteum is not required to maintain diapause.

Likewise, the corpus luteum is not required for the embryo to enter diapause. Using *M. eugenii* Sharman & Berger (1969) removed both ovaries on day 2 *p.c.*, when the fertilised egg could be expected to be in the uterus (Fig. 7.8*a*), and autopsied the animals on day 12. At this time the control animals had expanded vesicles of 1 mm (Berger, 1970), whereas the ovariectomised females contained unexpanded unilaminar blastocysts (0.25 mm), indistinguishable from those in diapause. This experiment showed that the embryo is not dependent on the corpus luteum for any part of cleavage or blastocyst formation but that it cannot proceed beyond this stage, which is reached by day 8, in the absence of the corpus luteum.

The precise time of entry into diapause has been determined by examining matched series of embryos on days 1–10 *p.c.*, taken from lactating females with quiescent corpora lutea and from non-lactating females with active corpora lutea (Fig. 7.8*a*) (Tyndale-Biscoe, 1979 and unpublished results). The rate of development up to formation of the unilaminar blastocyst of 80 cells (7th cleavage division) on day 8 was the same in both series but on day 10 the blastocysts from non-lactating females were markedly expanded, while those from lactating females were still small. The endometrial weights of the non-lactating females were significantly heavier than those of the lactating females on day 8 and thereafter, that is to say shortly before the expansion began. It will be recalled (Fig. 6.11) that a transient pulse of progesterone occurs on day 7 *p.c.* in non-lactating females but is blocked in lactating females. It is tempting to consider these three phenomena to be causally related and that the blastocyst expansion is dependent on endometrial growth, which is in turn dependent on progesterone and oestrogen from the corpus luteum. However, it is possible that the pulse of progesterone (and/or oestrogen) acts directly on the blastocyst or that specific uterine proteins secreted in response to progesterone act upon it and so synchronise its development with that of the endometrium. This has been investigated in the process

Fig. 7.8. Growth of the embryo of *M. eugenii* before and after lactation. (*a*) Growth of the conceptus after oestrus uninhibited by the presence of (▲) a sucking pouch young compared with (●) lack of growth in the presence of a neonatal young. ▲ represents the mean day of transient pulse of progesterone in non-lactating animals; ⊙, mean values for pregnant animals ovariectomised (lower point) or sham operated (higher point) on day 2 by Sharman & Berger (1969). (*b*) Growth of the diapausing blastocyst stimulated by 10 days of progesterone injection (○) or (*c*) after RPY (●). △, start of progesterone treatment: ⧍, mean day of transient pulse of progesterone after RPY. Time of birth of treated animals shown in top panel (*a*) ▼ (day 29 post-oestrus), (*b*) ▽ (day 23 after start of progesterone treatment) and (*c*) ⧩, (day 27 after RPY). Note the 3–4 day advance in time of birth after progesterone treatment compared to RPY. From Renfree & Tyndale-Biscoe (1973*a*) and additional unpublished data from both authors.

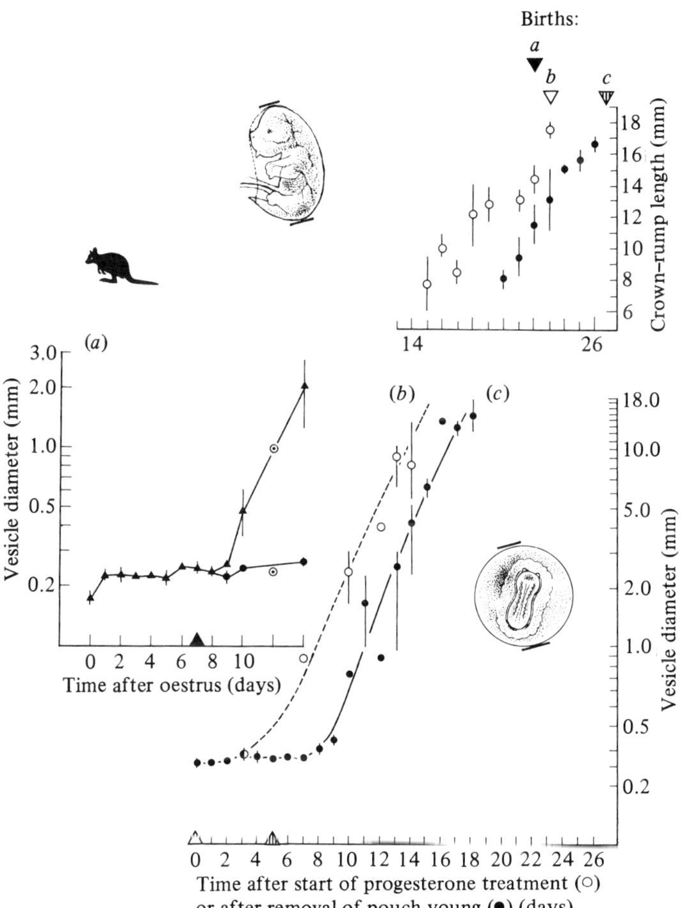

of reactivation that follows removal of the pouch young from lactating females.

Reactivation after diapause

Role of the corpus luteum

In the last section we showed that the embryo of *M. eugenii* reaches the unilaminar blastocyst stage by day 8 *p.c.* and that no further development occurs in either a lactating female, which has a quiescent corpus luteum, or in one ovariectomised on day 2. In the intact non-lactating female birth occurs on day 29 *p.c.* (Merchant, 1979; Table 6.4), so the minimum time required to complete gestation after the blastocyst is reactivated should be 21 days. Such a short interval has never been observed, the normal interval from removal of pouch young to birth being 26.4 days (Table 6.4). While several papers give values for delayed gestation that exceed this it is evident from Table 7.3 that much of the discrepancy can be accounted for by the frequency of observations in the peripartum period. The other factor is the time from removal of the pouch young to the early pulse of progesterone (Fig. 6.13) which is variable (see Table 9.1). On the other hand, the interval from the early pulse of progesterone to birth is the same as in a non-delayed pregnancy; in the non-lactating female the pulse occurs on day 7 *p.c.* or 22 days before birth

Table 7.3. *Delayed gestation in* Macropus eugenii: *a comparison of the recorded intervals from removal of pouch young to birth in relation to the frequency of observation at the time of birth*

	N	$\bar{x}$	S.D.
a. Daily observations:			
Berger (1970)	14	27.5	0.7
Renfree & Tyndale-Biscoe (1973*a*)	11	26–28	—
Tyndale-Biscoe & Rodger (1978)	16	27.6	1.2
Merchant (1979)	10	26.2	0.7
Young & Renfree (1979)	10	26.9	0.6
Hinds & Tyndale-Biscoe (1982*a*)	6	27.4	1.0
Ward & Renfree (1984)	30	27.3	1.0
b. 12-hourly observations:			
Shaw & Renfree (1984)	6	26.7	1.0
Harder *et al.* (1985)	5	25.9	1.0
	5	26.3	1.0
c. 8-hourly observations:			
Tyndale-Biscoe *et al.* (1983)	7	26.1	0.5
d. Continuous observations:			
M. B. Renfree (unpublished)	24	26.4	1.0

and in the delayed cycle it occurs on day 5–6 after removing the pouch young, also 21–22 days before birth. As the pulse is generated by the corpus luteum, this suggests that the discrepancy of 5 days is due to the time the corpus luteum requires to recover from the inhibition imposed by lactation. This aspect will be discussed in Chapter 9 (Table 9.1).

The conclusion that the blastocyst awaits a signal from the corpus luteum is supported by a number of experiments done with *M. eugenii* and *Setonix brachyurus* before the progesterone profile was known for either species. More recently similar results have been obtained in *Potorous tridactylus* (Bryant & Rose, 1986) and these are summarised for each species in Table 7.4. There is a critical period after day 2 and before day 6 in *S. brachyurus* (Tyndale-Biscoe, 1963*b*) and in *M. eugenii* (Sharman & Berger, 1969; Berger, 1970; Tyndale-Biscoe, 1970) when ovariectomy is followed by resumption of blastocyst growth and subsequent collapse and by failure of the luteal phase to develop in the endometrium; ovariectomy before this stage blocks reactivation and the blastocyst remains in diapause, while ovariectomy after this stage does not prevent the appearance of a luteal uterus and fetal development to full term. In both species the critical period includes the time when the early progesterone pulse occurs but it excludes the second rising phase of progesterone.

The results of treatment with exogenous progesterone and oestradiol-17β also support the idea that there are two phases in reactivation. Progesterone given for 5 days (5 mg day^{-1}) to ovariectomised *S. brachyurus* (Tyndale-Biscoe, 1963*b*) or for 3 days (1, 2 or 4 mg day^{-1}) to lactating or ovariectomised *M. eugenii* (Berger & Sharman, 1969) reactivated the blastocysts but, when examined 2–4 weeks later, gestation had not been maintained and only enlarged collapsed vesicles were recovered. However, when higher daily doses (10 mg day^{-1}) were given for 7 days to ovariectomised *S. brachyurus*, development continued to full term (Tyndale-Biscoe, 1963*b*); likewise in a series of 114 tammars given the same dose for 10 days during seasonal quiescence, 39% developed normal embryos and one gave birth on day 23 (Renfree & Tyndale-Biscoe, 1973*a*). When compared to the growth of embryos reactivated by removing the pouch young (Fig. 7.8*c*) the growth induced by progesterone treatment was advance by 3–4 days (Fig. 7.8*b*). However, the proportion that gave birth was considerably lower than could have been expected to be carrying blastocysts (73–86%) and a large part of the failure was ascribed to blastocysts that resumed development and then collapsed, perhaps because of inadequate progesterone after day 10.

Experiments by Clark (1968*a*) with *Macropus rufus* bear this out (Table

Table 7.4. *Effects on the embryo of removing the corpus luteum or both ovaries during the period of reactivation from diapause after removal of the pouch young (RPY) in* Setonix brachyurus, Macropus eugenii *and* Potorous tridactylus

Lutectomy/ovariectomy (time after RPY in days)	*N*	Number reactivated	Number died before term	Number developed to full term	Number born	Number survived > 1 day	Reference
Setonix brachyurus							
0–2	9	1	1	0	—	—	(1)
4–6	11	11	7	4	0	—	(1)
7–10	11	11	2	9	0	—	(1)
13–22	10	10	5	5	1	1	(1)
Macropus eugenii							
0	5	0	0	0	—	—	(2)
2	4	1	1	0	—	—	(2)
4	6	5	5	0	—	—	(2)
6	8	8	0	8	0	—	(2)
6	4	—	—	4	0	—	(4)
8	4	4	0	4	—	—	(3)
10–15	12	—	—	12	0	—	(4)
17, 21	10	—	—	10	4	0	(4)
23, 25	10	—	—	10	3	3	(4)
Potorous tridactylus							
0–3	3	—	—	0	—	—	(5)
6–21	7	—	—	5	0	—	(5)
25	2	—	—	2	2	0	(5)
27	1	—	—	1	1	1	(5)

Data from (1) Tyndale-Biscoe (1963*b*), (2) Berger (1970), (3) Tyndale-Biscoe (1970), (4) Young & Renfree (1979), (5) Bryant & Rose (1986).

Table 7.5 *Interval (days) from removal of pouch young (RPY) to birth and oestrus in* Macropus rufus *pretreated with progesterone (P) or oestradiol (E_2) on days 1–3*

Hormone treatment days 1–3	Day of RPY after start of hormone treatment	Number	RPY to birth (mean ± s.e.m.)	RPY to oestrus (mean ± s.e.m.)
Not injected	4	8	31.3 ± 0.4	34.0 ± 0.89
Oil injected	4	8	32.4 ± 0.3	33.1 ± 0.40
5 mg P	4	5	29.0 ± 1.5	—
10 mg P	4	10	28.3 ± 0.8	33.6 ± 0.80
20 mg P	4	6	30.8 ± 2.0	—
10 mg P	11	1	19	35
10 mg P	18	2	12, 33	34, 33
20, 60, or 100 μg E_2	4	5	27.6 ± 1.7	33.2

Data from Clark (1968*a*).

7.5). In one experiment lactating kangaroos were injected with progesterone (10 mg day^{-1}) for 3 days and the blastocysts examined on days 4 and 8. On day 4 they comprised 145 and 148 cells compared to the average of 85 cells for blastocysts in diapause and on day 8 they comprised several hundred cells each. In a second experiment the same treatment was given but the pouch young were removed on day 4 and the pregnancy allowed to run its course. The 10 treated animals gave birth 4 days earlier than 8 control animals but all showed post-partum oestrus on the same day, from which it can be concluded that the blastocysts were reactivated by the exogenous progesterone and were subsequently maintained by the corpora lutea reactivated 4 days later by removal of the pouch young.

In the same study, Clark (1968*a*) induced premature development with relatively high doses of oestradiol benzoate (50–100 μg) injected on 3 days before removing the pouch young and three of these kangaroos likewise gave birth 4 days before the controls (Table 7.5). Similar studies by Berger & Sharman (1969) and Smith & Sharman (1969) in *M. eugenii* were done to determine if oestrogen had stimulated blastocyst reactivation more indirectly by stimulating the corpus luteum to secrete progesterone. As in *M. rufus*, oestradiol benzoate (50 or 100 μg day^{-1} for 3 days) resulted in reactivation of blastocysts, but normal development was not maintained in the absence of a developing corpus luteum or in animals treated with oestradiol 7 days after ovariectomy (Smith & Sharman, 1969). It is now known that such high doses of oestradiol would result in increased vascularity and metabolic rate of the endometrium (Shaw & Renfree, 1986). It is also now known that progesterone remains in plasma at basal levels after ovariectomy or lutectomy (Lemon, 1972; Sernia *et al.*, 1980), so it is not excluded as an agent in these experiments. Since a transient pulse of oestradiol occurs at the same time as the transient pulse of progesterone in *M. eugenii* (Fig. 6.15) (Shaw & Renfree, 1984), both steroids may act in concert to provide the early signal to the blastocyst. However, in preliminary experiments by M. B. Renfree and A. E. Jetton, the incidence of reactivation was not increased when both steroids were used, compared to using progesterone alone, and when oestradiol was given at physiological level (1 μg day^{-1}) for 10 days no blastocyst reactivated.

First changes in the blastocyst

What was not resolved by these studies was whether the corpus luteum influences the blastocyst directly or whether its effect is mediated through increased or specific secretion from the uterus. Undoubtedly, the later expansion of the embryo is dependent on adequate endometrial secretion, but this does not begin until several days after the period under consideration. This question has still not been resolved but there are now some data on the early changes in the blastocyst and on the uterine secretions during the first week. In *M. eugenii* there is no change in the outer diameter of the blastocyst up to day 6 after RPY (Fig. 7.8*c*) but cell division has begun by day 4 (Berger, 1970). This is reflected in a slight but significant increase in diameter by day 8 and a three-fold increase by day 10 (Renfree & Tyndale-Biscoe, 1973*a*). Between day 5 and day 10 the volume increased 45-fold (Table 7.6) and these rates of increase continue through day 15 by which time the volume has increased about 10 000-fold (Renfree & Tyndale-Biscoe, 1973*a*; Pike 1981).

Moore (1978), Thornber, Renfree & Wallace (1981) and Shaw & Renfree (1986) examined two aspects of RNA activity in blastocyst cells before and after reactivation and all found the first significant changes to have occurred on day 5 (Table 7.6). Moore (1978) incubated fixed blastocysts with [^{3}H]UTP for 30 min and detected the incorporation of [^{3}H]UMP by autoradiography as the density of grains overlying the nucleoli and nuclei of blastocyst cells. Nucleolar label was considered to represent RNA polymerase I activity and nucleoplasmic label to represent predominantly RNA polymerase II activity. While some polymerase activity was detected in blastocysts from lactating females and on days 1–4, a marked increase in both nucleolar and nuclear grain counts did not occur until day 5. When females were treated with progesterone however, the increase occurred 48 h after the first injection. This is in agreement with other evidence (Renfree & Tyndale-Biscoe, 1973*a*) that progesterone treatment causes reactivation 3–4 days earlier than after removal of the pouch young (Fig. 7.8). Nevertheless, the delay in response of 48 h is much longer than the response to oestrogen injection of delayed rat and mouse blastocysts in which [^{3}H]uridine incorporation occurred within 1 h (Mohla & Prasad, 1971; Holmes & Dickson, 1975). This would suggest that in *M. eugenii* the progesterone is not acting directly on the blastocysts but indirectly through stimulating endometrial secretions.

Thornber *et al.* (1981) attempted to test these conclusions in quiescent blastocysts taken from lactating *M. eugenii* females and those reactivated for 5 and 10 days. They incubated the live blastocysts for 5 h in medium containing [^{3}H]uridine and measured the trichloroacetic acid (TCA)-soluble and insoluble fractions in the washed blastocysts. The insoluble fraction was considered to have been incorporated into the cell nuclei, whereas the TCA-soluble fraction was considered to have been transported into the

Table 7.6. *Summary of changes observed in embryos of* Macropus eugenii *during reactivation from diapause after removal of the pouch young (RPY)*

Days after RPY	Outer diameter of embryo (μm)	Volume of embryo (nl)	Surface area of trophoblast (mm^2)	Nucleo-plasmic grain number	Nucleolar grain number	[^{3}H]uridine incorporation (dpm per embryo per 5 h) $\times 10^{-2}$	[^{3}H]uridine uptake (dpm per embryo per 5 h) $\times 10^{-2}$	Glucose incorporation (pg atoms C per embryo h^{-1})	Glucose incorporation (μg atoms per embryo per 4 h) $\times mm^{-2}$
Reference	(1, 3)	(2)	(4)	(1)	(1)	(3)	(3)	(2)	(4)
0	264 ± 4	8.03	0.18 ± 0.02			2.0 ± 0.7	94 ± 14	32	0.70 ± 0.39
1	260 ± 8			14.0 ± 2.7	3.8 ± 0.8				
2	264 ± 8			19.1 ± 1.9	5.8 ± 0.6				
3	283 ± 18			13.9 ± 2.3	3.7 ± 0.7				
4	244 ± 4			13.3 ± 2.1	2.2 ± 0.5				
5	271 ± 5			24.6 ± 2.9^a	6.4 ± 0.9^a				
5	255 ± 24	7.45	0.18 ± 0.02			9.2 ± 3.4	88 ± 21	50	1.10 ± 0.17
10	1067 ± 379^a	361.4^a	2.44 ± 0.25^a			72.3 ± 15.6	1574 ± 260	454	0.75 ± 0.15
15			86.95 ± 7.10					27983	1.29 ± 0.27

[a] Difference from previous value significant $P < 0.01$.

Data from (1) Moore (1978), (2) Pike (1981), (3) Thornber *et al.* (1981), (4) I. L. Pike & M. B. Renfree (unpublished).

blastocoele fluid but not yet incorporated. While no change in uptake was found between day 0 and day 5, a 17-fold increase had occurred by day 10. This probably reflects the large increase in volume by day 10 but not necessarily an increase on a per cell basis (see below). The pattern of [^{3}H]uridine incorporation, however was most significant. The incorporation was low in the day 0 blastocysts and equally low in four of the six day 5 blastocysts. In the other two there was a 4–5 fold increase in incorporation despite no change in the diameter of the blastocysts. By day 10, the incorporation had increased 35 times when compared to day 0. The results for day 5 suggest that blastocysts are beginning to reawaken on this day, and thus support Moore's (1978) conclusion.

Shaw & Renfree (1986) incubated another series of blastocysts at days 0, 5, 7 and 9 with [^{3}H]uridine to assess RNA synthesis. The day 0 blastocysts had very low uptake and incorporation of uridine (Fig. 7.9*a*, *b*). The day 5 and day 7 blastocysts had not expanded (Fig. 7.9*c*) but they incorporated uridine at a much greater rate, and all but one day 5 blastocyst had a much increased rate of uridine uptake compared to day 0. Two of the three day 9 blastocysts had expanded significantly and all three showed high rates of uptake and incorporation, although the rate of incorporation was similar to those on day 5 and 7. The quantities of uridine taken up and incorporated by the blastocyst were small compared to the concentration in the medium and imply that the blastocyst is not freely permeable to uridine. The rate of transfer of uridine across the blastocyst membrane [expressed as cpm (uptake+incorporation) per square millimeter of trophoblast] increased from about 250 cpm mm^{-2} at day 0 to about 2000 cpm mm^{-2} at day 5 and day 7. Interestingly, at day 9 the rate was lower in the larger blastocysts than in the blastocyst that had not yet expanded.

A similar analysis of glucose incorporation per square millimetre of surface area (μg atoms per embryo per 4 hr period) has been made (I. L. Pike & M. B. Renfree, unpublished observations) (Table 7.6), which showed that even though the total incorporation was increasing, the rate per unit (or per cell) was not markedly different at the different times. Pike (1981) measured the uptake of glucose carbon for storage as glycogen and for the synthesis of nucleic acids and proteins. While the overall incorporation in blastocysts of *M. eugenii* was higher than in mouse blastocysts in diapause, this could be accounted for by the larger size of the marsupial blastocyst. Very little increase was seen in day 5 blastocysts compared with day 0, but a 9–10-fold increase was found at day 10, when the number of cells would also have increased to the same extent. However by day 15 the incorporation far exceeded the preceding stages and it was only in blastocysts of this stage that glucose was being incorporated into glycogen pools. Prior to this stage, none was detected and it was concluded that all the glucose was being incorporated into the carbon skeletons of proteins, nucleic acids and lipids or precursors of lower molecular weight. While Pike (1981) finds no evidence from his study to support the idea that nucleic acids are being synthesised by day 5 it is not excluded by his methods, so we return to the original question, now more sharply focussed: if the first measurable response of the blastocyst cells on day 5 is a response to the awakened corpus luteum, is the progesterone-oestrogen pulse on day 5–6 the signal and is it direct or is it mediated by specific uterine proteins?

Fig. 7.9. Uridine (*a*) incorporation and (*b*) uptake by individual blastocysts of *Macropus eugenii* during the early stage of reactivation. Blastocyst diameters are shown in (*c*). Data derived from Shaw & Renfree (1986).

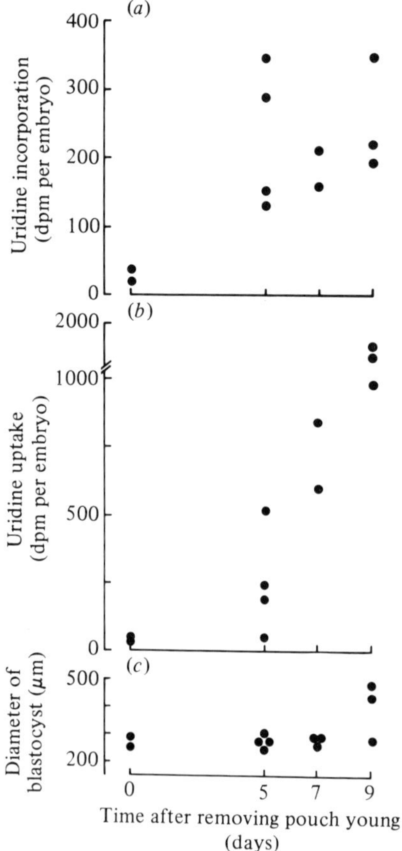

The role of uterine secretions

The evidence for the importance of uterine secretions in reactivation is equivocal. The reactivation is probably not triggered by a change in the ionic content of potassium, calcium or magnesium in uterine fluid. In *M. eugenii* the concentration of sodium and potassium in uterine flushings increases between day 7 and day 9, that is to say after reactivation and coincident with the first observed expansion of the blastocysts (Shaw, 1983*b*). No change was seen at day 5, when metabolic reactivation of blastocysts starts but more samples should be assessed over the period from day 3 to day 7 to substantiate this. The potassium:sodium ratio in uterine fluid is much higher than in serum and thus, as in other (eutherian) species, the uterine epithelium of *M. eugenii* maintains an ionic gradient. No striking increase was seen in either the calcium or magnesium content of flushings over the period studied, despite their potential importance as metabolic regulators (Shaw, 1983*b*). Thornber *et al.* (1981) observed a two-fold increase of calcium concentration in uterine exudates between days 0–2 and day 10. The differences between the two studies probably reflect differences of the stages sampled, although flushings and exudates may also give a different pattern of change.

Renfree (1973*a*) found a marked increase in the number and concentration of proteins in endometrial exudates between day 0 and day 6 (Fig 7.23*a*) and suggested that some of these, particularly the low molecular weight pre-albumins, might be providing the earliest signal to the blastocyst because they could easily pass through the egg coats. Support for this view was provided by the evidence (Tyndale-Biscoe, 1970) that quiescent blastocysts transferred to the uteri of recipient day 8 females, which were ovariectomised at the time of transfer, developed at the same rate as day 8 blastocysts transferred simultaneously to the other uterus. These results implied that the blastocyst responds solely to the appropriate uterine milieu. To test these two aspects Thornber *et al.* (1981) examined the response of mouse blastocysts to incubation in endometrial exudate collected from *M. eugenii* females on days 0, 5, 10 and 15. Despite equal concentrations of protein in all incubation wells, the response to endometrial exudates was never more than 7% of the response to bovine serum albumin. This is similar to the response of mouse blastocysts to mouse uterine secretion (Surani, 1977). However, within these limitations, there was a greater response to exudates from day 10 than from day 0 or 5. These authors also attempted to induce development of *M. eugenii* blastocysts *in utero* by instilling endometrial exudate from reactivated animals but without success, possibly because of the trauma caused to the uterus by the injection. Thus the matter of a specific protein signal still remains unresolved.

Although the protein content of uterine fluids of *M. eugenii* undoubtedly changes in the pre-implantation period (Fig. 7.23*a*), the metabolism of the endometrium might be expected to show the first change. Shaw & Renfree

Fig. 7. 10. Leucine incorporation by gravid (●—●) and non-gravid (○--○) endometrium during pregnancy in *Macropus eugenii*, determined as disintegrations min^{-1} (dpm) μg^{-1} tissue DNA, Incorporation into (*a*) secreted protein and (*b*) tissue protein increased between day 0 and day 12. At day 26 gravid uteri were more secretory than non-gravid uteri. Tissue incorporation decreased between day 12 and day 26 in both uteri. From Shaw & Renfree (1986), with permission.

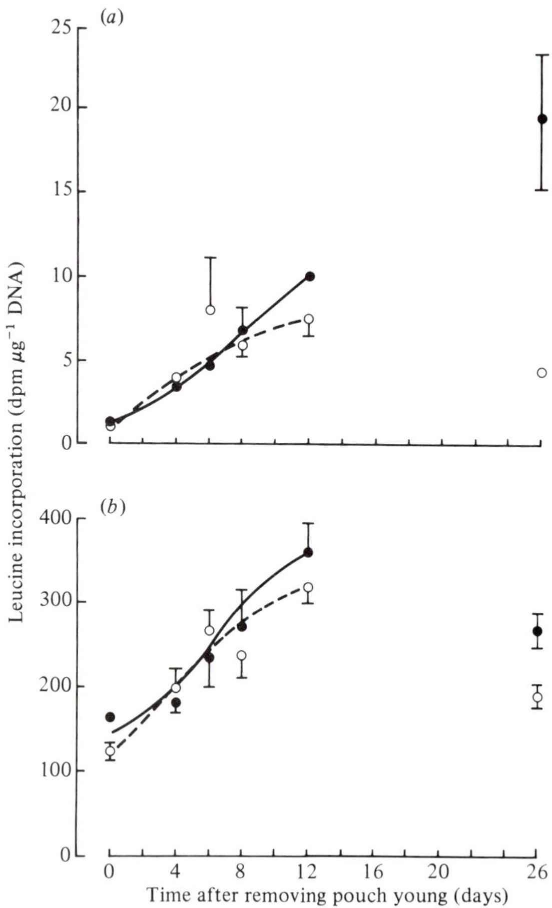

(1986) have assessed uterine tissue protein and secreted protein on days 0, 4, 6, 8, 12 and 26 (Fig 7.10). Endometrial explants were incubated in medium containing [^{3}H]-leucine and analysed for incorporation, as well as for DNA and protein content. Incorporation into secreted protein by the endometrium had significantly increased by day 4 (Fig. 7.10), and incorporation into tissue protein had also increased by day 4. This is one day before metabolic reactivation of the blastocyst and is consistent with the view that a component(s) of uterine secretion triggers blastocyst reactivation. This component may not be a protein, as no significant changes in the electrophoretic pattern of protein secreted *in vitro* occurred over the reactivation period. However, proteins of MW $< 10\,000$ would not have been detected due to the procedure used and there are many smaller proteins and polypeptides such as the somatomedins which have regulatory properties. In addition, Shaw (1983*b*) demonstrated that progesterone and oestradiol given for 3 days to bilaterally ovariectomised *M. eugenii* stimulated protein synthesis and secretion by the endometrium to rates equal to or greater than the levels at day 26. It is interesting that the early peak of progesterone (Hinds & Tyndale-Biscoe, 1982*a*) and oestradiol-17 β (Shaw & Renfree, 1984) in plasma lasts only 2–3 days, and it may be that these changes are critical for induction of uterine secretion, although it should be remembered that the peak concentrations occur after uterine reactivation (Shaw & Renfree, 1986).

Conclusions concerning reactivation

For *M. eugenii* these several pieces of evidence can be draw together. Although Berger (1970) observed a few mitoses in the blastocyst on day 4, the first measurable changes in the metabolism of the embryo occur on day 5, preceding the peak of the transient pulses of progesterone and oestradiol. However, as shown above, it seems unlikely that these are causally related because Moore (1978) showed that 48 h are required for the blastocyst to respond to progesterone and because blastocysts were reactivated after removal of the corpus luteum on day 4, presumably before the pulse had occurred (Berger, 1970). The first signal to awaken the blastocyst must therefore occur after day 2 and before day 5, which coincides with Renfree's (1973*a*) and Shaw & Renfree's (1986) evidence of enhanced endometrial secretory activity on day 4.

The involvement of the corpus luteum in this initial signal is not clear; if it does stimulate the first changes in the endometrium it must be via arterio-venous exchange, since no changes in progesterone are detectable in the peripheral circulation at this time. If it is not involved it is difficult to see why the response of the blastocyst to ovariectomy should be different on day 2 and day 4. The involvement of the corpus luteum in the second step is clear; its presence is essential until day 8 for the continued development of the embryo and the most probable factor in this is the pulse of progesterone and oestradiol on day 5–6. It is possible that the pulse stimulates the synthesis of progesterone receptors on endometrial gland cells (see p. 188), which can then respond to subsequent lower concentrations of progesterone by maximal growth and secretion, but this has not been investigated.

These somewhat tentative conclusions from *M. eugenii* nevertheless point to the corpus luteum being crucially involved during the first 8 days in synchronising endometrial secretory capacity with embryo development, especially the rapid expansion that follows the unilaminar blastocyst stage. While the Macropodidae show this role to a marked degree, because of the evolution of a mechanism for inhibiting the corpus luteum, there is evidence for a similar role in other species of marsupial. In the Burramyidae and in *Tarsipes rostratus* embryo development is held in abeyance until the corpora lutea enlarge; in *Antechinus stuartii* the most variable stage of gestation is the unilaminar blastocyst, with expansion coinciding with growth of the corpora lutea and an elevated concentration of progesterone. Whether it is common to all marsupials, as suggested by Tyndale-Biscoe (1968), must await further work on species other than macropodids.

Primary endoderm formation and the bilaminar blastocyst

Formation of the bilaminar blastocyst begins with the differentiation of endoderm mother cells in the trophoblast and by rapid expansion of the blastocyst through absorption of fluid into the blastocoel (Fig. 7.11). Both Hill (1910, 1918) and Hartman (1916, 1919) believed that there was continuity between the early polarity and the subsequent differentiation of the formative region from which the first endoderm cells are differentiated and the medullary plate is formed. The only point of reference for this in *Dasyurus viverrinus* was the position of the yolk body, which remained attached to one point inside the blastocyst. In *Didelphis virginiana* the first endoderm cells differentiate in the 60-cell blastocyst from the hemisphere that contains thicker trophoblast cells (Fig. 7.6*c*). McCrady (1938) reviewed all of Hartman's material and Hill's papers and came to the conclusion that the identity of the early polarity could not be maintained, but he conceded that better methods of identifying particular cells might yet disclose a formative region. Selwood (1982*b*) agreed with McCrady.

The formation of the primary endoderm is similar in all seven species

Fig. 7.11. Timetable of development in *Antechinus stuartii* from fertilisation in the oviduct to birth at 27 days, to illustrate the slow development to day 18 and then rapid period of embryogenesis. This is compared to the rate of development in four other species by indicating the day post-oestrus that comparable stages are reached. From Selwood (1980), with permission, additional data for *Didelphis virginiana* from Hartman (1928), *Isoodon macrourus* from Lyne & Hollis (1977*b*), *Trichosurus vulpecula* from Hughes & Hall (1984) and *Macropus eugenii* from Renfree & Tyndale-Biscoe (1973*a*) and Renfree (1973*b*).

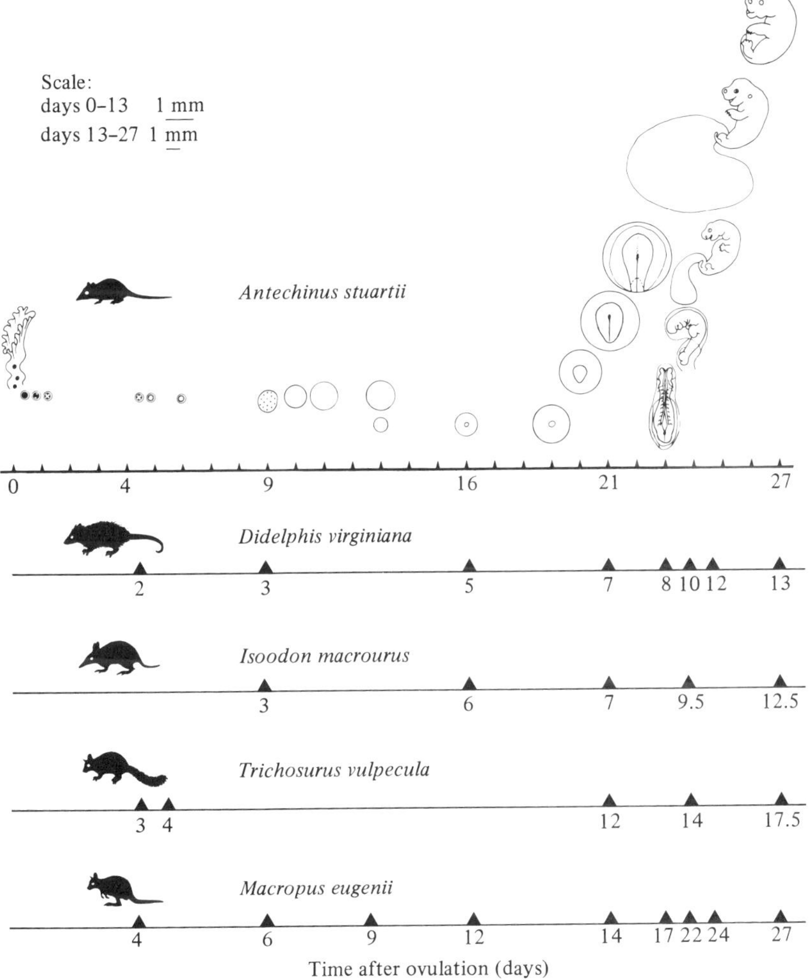

Table 7.7 *Summary description of McCrady's (1938) 35 stages of intrauterine development in* Didelphis virginiana *to show the days post-oestrus on which they occur in that species and approximate days on which the stages occur in five other species*

Stage	Description	*Didelphis virginiana*	*Antechinus stuartii*	*Dasyurus viverrinus*	*Trichosurus vulpecula*	*Macropus eugenii*	*Macropus rufogriseus*
1	Fertilisation; extrusion of polar bodies; tubal transport	1		4		1	
2	1st cleavage; yolk extrusion; entry to uterus	2		5–6			
3	2nd cleavage commences	3	3				
4	4-cell			5–11		4	
5	3rd cleavage commences		5				
6	8-cell		6				
7	4th cleavage, 16-cell				3–4	6	
8	5th cleavage, 32-cell						
9	6th cleavage; unilaminar blastocyst	4	8	(1.2) 11	4	8	
10	Endoderm mother cells form		9				
11	Ectodermal thickening over endoderm						
12	Blastocyst expansion begins	(0.11)	(1.0) 11	(5.0) 13	(0.75) 7		9
13	Bilaminar blastocyst; medullary plate appears	(0.5) 5	15	13–17		9	
14	Zona and mucoid coat absorbed	(0.75) 6					
15	Late bilaminar blastocyst	(1.00) 7	(2.2) 20		(3.2) 9		
16	Primitive streak appears	(1.80)	21	17	(4.5) 11		
17	Mesoderm crescent appears	(2.00)					(4.8) 15
18	Primitive groove and Hensen's node appear	(2.2) 8	(3.5)			(6.0) 16	
19	Elongation of medullary plate posteriorly		(3.8)				
20	Extension of mesoderm into yolk sac membrane						

21	Notochord forms	(3.7)	22		(7.5) 12		
22	First somites appear; parietal mesoderm						
23	Coelomic cavities form; subcephalic fold forms						
24	6–7 somites; heart tubes and first blood vessels; 1st branchial pouches; sensory anlagen; pronephros; proamnion	9					
25	12–13 somites; neural folds close; lung anlagen					17	
26	Headfold; fusion of heart tubes; formation of amnion		23		12		
27	Cervical flexure; heart beats; lung buds; tail amnion	10				19	
28	Primary lumbar flexure; mesonephros and duct; liver anlagen; allantois appears; shell lost		(9.0)				18
29	Naso-oral groove; anterior limb bud; posterior limb ridge; thyroid and pancreas anlagen; amnion closed		24			21	20
30	Secondary lumbar flexure; liver cords differentiate	11				22	
31	Hind limb buds; digits on forelimb; allantois enlarged		25		13		
32	Hind limb club shaped; adrenal cortex differentiated	12					
33	Eyelid folds distinct; tongue protruding; digits distinct				16	24	25
34	Oral shield; claws on digits; mammary anlagen; cloaca opens	13	26				
35	Resorption of oral shield; parturition	13	27	19	17.5	26.4	26

Figures in brackets are vesicle diameter (mm) at that stage. For the other species the sequence of development is not precisely the same so it is not always possible to establish a strict equivalence to *D. virginiana*. For the two species of *Macropus* the post-blastocyst stages have been dated from removal of pouch young rather than oestrus. This Table should be read in conjunction with Fig. 7.11.
Data from Selwood (1980) for *A. stuartii*, Hill (1910) and Hill & O'Donoghue (1913) for *D. viverrinus*, Hughes & Hall (1984) for *T. vulpecula*, Renfree (1972*a*) and Tyndale-Biscoe (1979) for *M. eugenii* and Walker & Rose (1981) for *M. rufogriseus*.

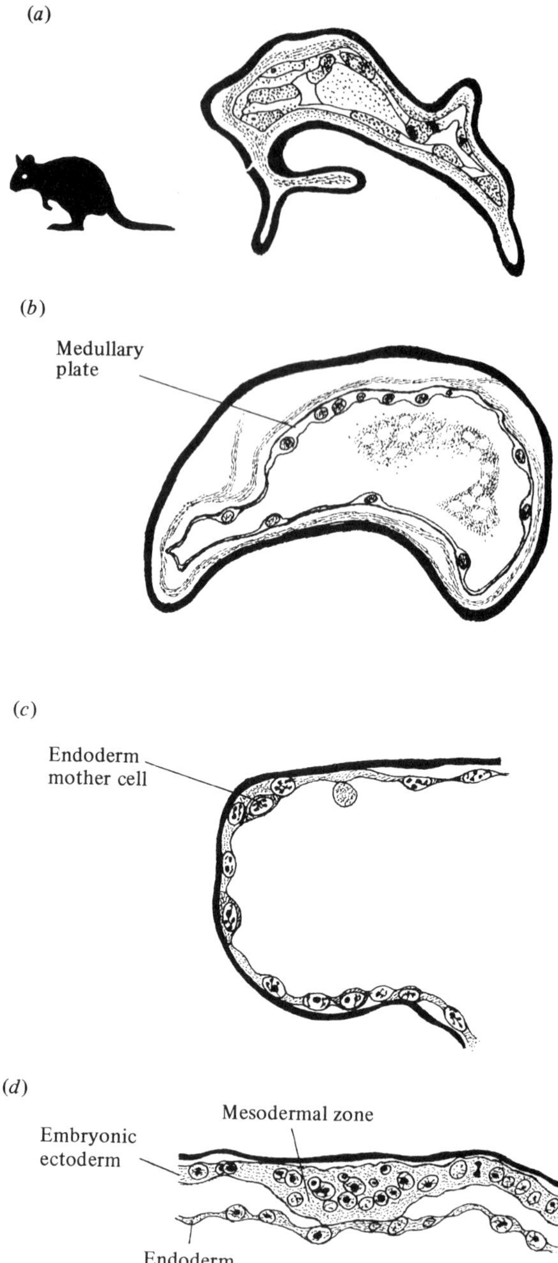

Fig. 7.12. Endoderm formation in *Bettongia gaimardi*. (*a*) Quiescent unilaminar blastocyst; (*b*) first stage of differentiation of thickened medullary plate or formative area, vesicle diameter 0.27 mm; (*c*)

in which it has been described, the differences between species being in relation to the size of the blastocyst at its appearance and at completion of the layer (Table 7.7). Thus in *D. virginiana* the diameter is only 0.11 mm while, at the other extreme, the blastocyst of *Da. viverrinus* is between 4.5 and 5 mm in diameter before the formative area differentiates and endoderm cells appear (Hill, 1910). Other species lie within those extremes; for *Bettongia gaimardi* (Kerr, 1935) and *Macropus rufogriseus* (Hill, 1910; Walker & Rose, 1981) the first endoderm cells appear in the formative region of blastocysts of 0.3–0.4 mm in diameter (Fig. 7.12) and for *Perameles nasuta* and *Isoodon macrourus* (Hill, 1910; Hollis & Lyne, 1977) in blastocysts of 1.0 to 1.5 mm and for *Petauroides volans* (Bancroft, 1973) at 1.5 to 2.0 mm. The endoderm mother cells are recognisable by their darker-staining, more granular, cytoplasm and their compact shape (Fig. 7.12). While they are invariably associated with the formative area, they are not formed by inwardly directed cell divisions of trophoblast cells, but by migration and transformation of trophoblast cells themselves. After migrating inwards separately, they put out pseudopodial filaments which connect adjacent endoderm cells and so form a loose meshwork beneath the trophoblast which, by further cell division, becomes complete with the formation of the bilaminar blastocyst. This occurs at the end of day 6 in *D. virginiana* (Table 7.7), when the blastocyst is 0.75 mm in diameter (McCrady, 1938) and in blastocysts of the same size in *Bettongia gaimardi* (Kerr, 1935), but it occurs in *Antechinus stuartii* when the blastocyst is about 3.0 mm in diameter and the endoderm layer is not complete until day 20 (Selwood, 1980). The much longer period in this species is due to a period of quiescence, or possibly diapause at the unilaminar blastocyst. The bilaminar blastocyst of *Da. viverrinus* is 4.5–6.0 mm (Hill, 1910) and this occurs on or about day 15 *p.c.* or day 10 after ovulation. In *P. nasuta* and *I. macrourus* the blastocysts were fully bilaminar when 1.5–1.9 mm (Hollis & Lyne, 1977) and this was estimated to have occurred 6 days *p.c.* (Lyne & Hollis, 1977*b*).

The medullary plate, primitive streak and embryogenesis

After completion of the endoderm, expansion accelerates and the zona pellucida and the mucoid layer (with its included spermatozoa)

Fig. 7.12. *cont.*
differentiation of endoderm cell in medullary plate and its inward migration, vesicle diameter 0.39 mm; (*d*) differentiation of mesoderm at margin of primitive streak, vesicle 1.32 × 0.99 mm. Redrawn from Kerr (1935).

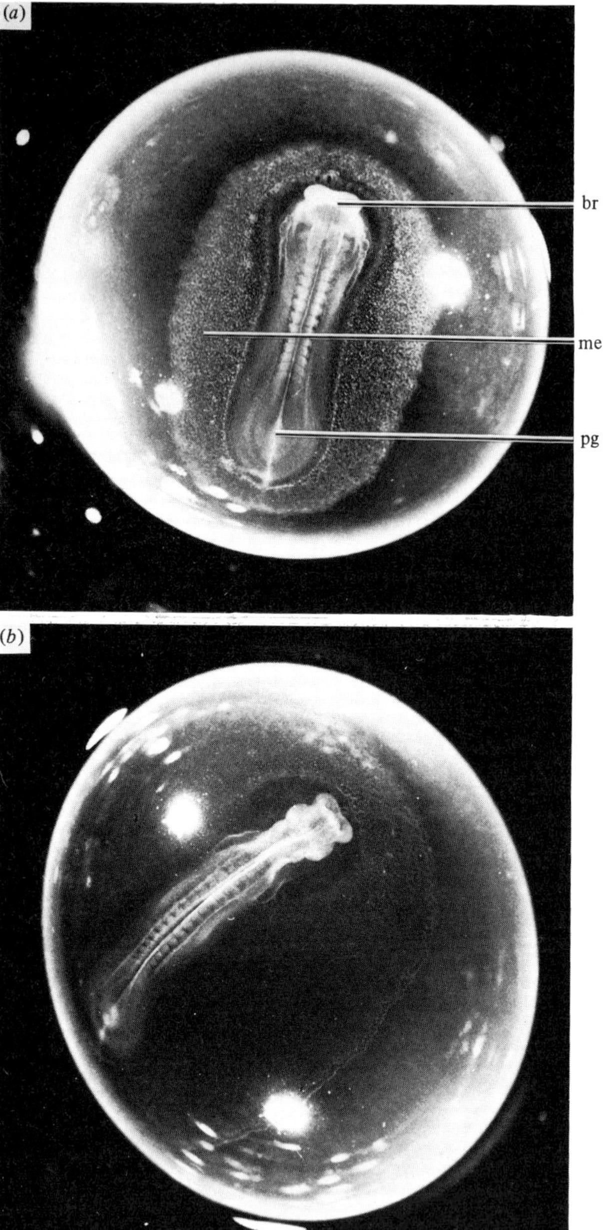

Fig. 7.13. Differentiation of the embryo of *Macropus eugenii* on day 17 after removing pouch young. (*a*) Embryo of 12 somites. Mesoderm tissue (me) has not yet spread far, and no blood can be distinguished. The primitive groove (pg) can be clearly seen, and the early stages of

disappear, presumably through absorption by the trophoblast cells. Such a process is illustrated by Hollis & Lyne (1977). As this is proceeding, the cells in the formative region are becoming more compact and thicker, so that a distinct medullary plate is visible on the surface of living blastocysts (Fig. 7.12). As the endoderm is completed, the first sign of the primitive streak appears (Kerr, 1936; McCrady, 1938). This is followed by elongation of the medullary plate into a pear shape, as the primitive streak lengthens and the primitive groove develops within it (Fig. 7.13*a*). The embryo now has bilateral symmetry and the anterior end has now been established by the appearance of the primitive knot, or Hensen's node, and the prechordal plate (Fig. 7.13). In transverse section, the mesodermal tissue is seen to be proliferating laterally (Fig. 7.14) and the margin of mesoderm can also be seen in surface view (Fig. 7.13*b*).

In *M. eugenii*, the appearance of the medullary (= mesodermal) plate and formation of the primitive streak and early somites occur in vesicles of quite variable sizes, so that vesicle diameter *per se* is not necessarily a good index of stage of development reached. Thus, some vesicles between 4 and 6 mm in diameter have no mesodermal plate, whilst others do on the same day of gestation (Renfree, 1972*a*, 1973*b*). These early stages of development are the most variable in terms of size reached by a certain day; many day 15 vesicles are about 8 mm diameter, by which size all embryos have differentiated the primitive streak, which is about 6 mm long (Renfree, 1972*a*). The smallest vesicles observed to have a primitive streak are around 5 mm diameter. A similar pattern has also been observed in *Macropus rufogriseus* (Walker & Rose, 1981). In *D. virginiana*, a clear spot appears at the anterior tip of the differentiating primitive streak in vesicles of around 3.5 mm diameter (McCrady, 1938). The primitive streak shortens anteriorly, and at the same time the notochord and the medullary groove appear. When the medullary plate is about twice as long as the primitive groove, two slight condensations of the mesoderm beneath the medullary plate on either side of the central mesoderm-free strip represent the first somites (McCrady, 1938). Mesoderm formation begins in 2 mm vesicles of *Isoodon macrourus* (Hughes, 1984) and the first somites of other species are seen in vesicles between 4 and 8 mm in diameter; an 11 mm vesicle of *M. eugenii* had 3 somites (Renfree, 1972*a*) while a 4.4 m vesicle

Fig. 7.13. *cont.*

differentiation of the brain (br). (*b*) Embryo of 20 somites, clearly showing the spread of the vascular area, and an enlarged area of pro-amnion around the head. The paired heart tubes have begun to form just posterior to the brain. From Renfree (1972*a*).

Fig. 7.14. Detail of a 15-somite embryo of *M. eugenii* (day 17 RPY), together with representative transverse sections at positions along it. ec, endocardium; eec, extra embryonic coelom; fb, forebrain; hb, hind brain; lu, lung anlage; m, mesonephros; mc, myocardium; np, neural plate; op, optic placode; ot, otic placode; p, pharynx; pa, pro-amnion; pg, primitive groove; s, somite; vv, vitelline vein; ys, yolk sac. From Renfree (1972*a*).

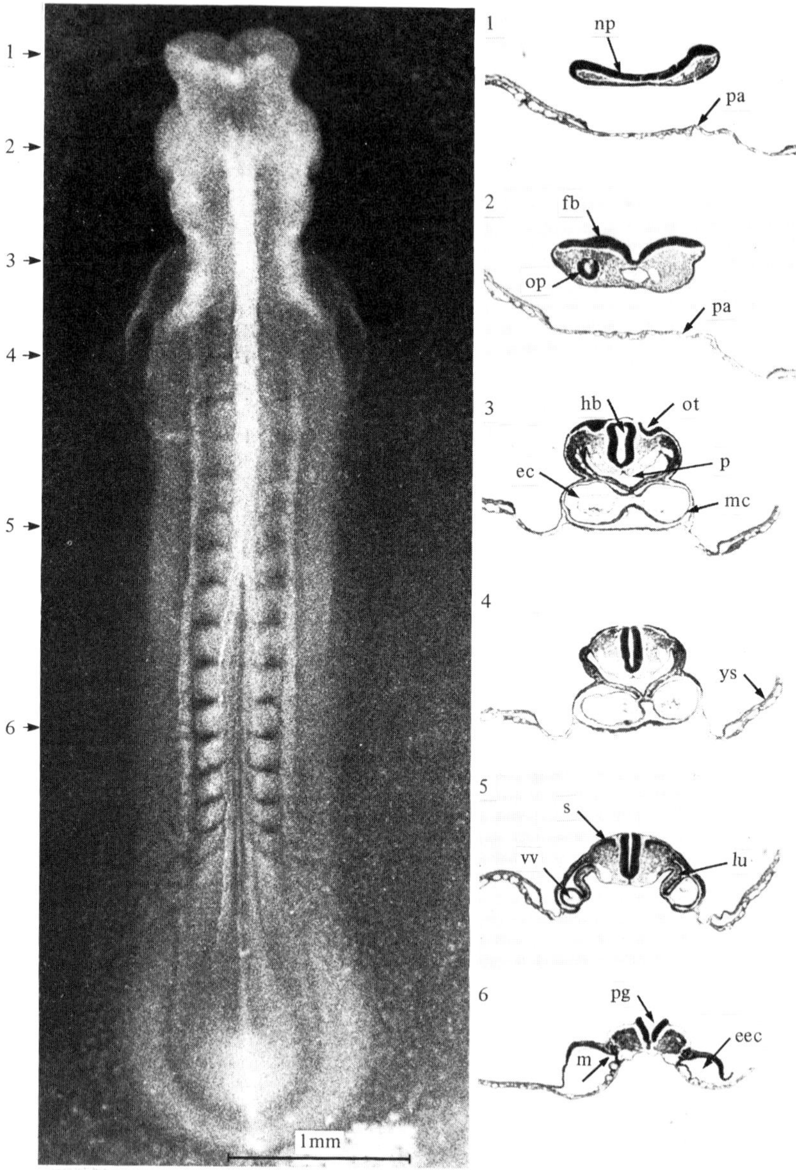

of *Petauroides volans* had 2 somites (Bancroft, 1973). These species differences in timing are summarised in Table 7.7.

Compared to eutherian mammals, formation of the amnion occurs relatively late in marsupials; soon after the formation of somites the anterior pro-amnion (Fig. 7.13*b*) and then the tail amnion appear. In *D. virginiana*, the pro-amnion forms when the embryo has 7–12 somites (McCrady, 1938), and cephalic flexure in embryos of 17–19 somites (day 9) results in the distinct formation of the pro-amniotic headfold, and the (somatopleuric) amniotic tailfold is evident at 20–25 somites (Selenka, 1887; McCrady, 1938). In *P. volans* the pro-amnion and tail amnion were apparent in an embryo with 10 somites (Bancroft, 1973), whereas an embryo of *Perameles nasuta* with 15 somites lacked any distinct evidence of amniogenesis (Luckett, 1977). However, in a 23-somite embryo (4.7 mm greatest length) of *Dasyurus viverrinus*, both amniotic folds were present, as in *D. virginiana* (Luckett, 1977). Amniotic head and tailfolds were similarly observed in 4.8 mm (G.L.) *Setonix brachyurus* (day 19), 4.0–5.0 mm *M. eugenii* (day 18), 4.2 mm *M. rufogriseus* (day 18) and 5.4 mm *Pseudocheirus peregrinus* embryos (Sharman, 1961*b*; Renfree, 1973*b*; Walker & Rose, 1981).

The amniotic folds approach each other and the exocoelom invades the caudolateral margins so that the amnion and chorion are separated by it. The pro-amnion of *D. virginiana* persists throughout intrauterine life (Selenka, 1887; McCrady, 1938) but not in phalangerids and macropodids (Sharman, 1961*b*; Bancroft, 1973; Renfree, 1973*b*).

By the time the marsupial embryo has about 18–20 somites, cervical flexure has occurred (Fig. 7.11) and the neural tube is closed along most of the length of the embryo. The anlagen of the olfactory, optic and otic organs are differentiated (Fig. 7.14). The heart tubes are fused, and branchial pouches differentiated. In the pharyngeal region, Rathke's pouch develops at the anterior end. Lung anlagen are also present, and pronephric as well as mesonephric tubules are formed. This occurs about 4–5 days before birth (day 23–24) in *Antechinus stuartii* (Selwood, 1980), 3–4 days (day 9–10) in *D. virginiana* (McCrady, 1938), 7–8 days in *M. eugenii* (Renfree, 1973*b*), *M. rufogriseus* (Walker & Rose, 1981) and *S. brachyurus* (Sharman, 1961*b*; Tyndale-Biscoe, 1963*a*).

The notochord is complete at day 10 in *D. virginiana* (McCrady, 1938), day 22 in *A. stuartii* (Selwood, 1980), day 18–19 in *M. eugenii* and *M. rufogriseus* (Renfree, 1972*a*; Walker & Rose, 1981). It is also about this time that the shell membrane, which has become more and more attenuated, ruptures, and close and direct apposition of the trophoblast to the uterine epithelium can occur. Whereas all development up to this stage has been relatively slow and occupied two-thirds of the active gestation period (Fig. 7.11), the remaining period of organogenesis is rapid (Hughes, 1974; Renfree, 1977). Embryogenesis can be said to be complete with the appearance of the notochord, and organogenesis occupies the remaining third of gestation.

Organogenesis

The duration of organogenesis is remarkably constant for species of very different adult body sizes (Fig. 7.15). Selwood (1980) suggests that there may be a relationship between the length of organogenesis, the type of placenta, and the weight of the neonate, so that the longer the period

Fig. 7.15. Relative duration (days) of the pre-attachment and post-attachment phases in six different marsupials. Species from top to bottom are *Antechinus stuartii, Perameles nasuta, Didelphis virginiana, Trichosurus vulpecula, Macropus eugenii* and *Potorous tridactylus*. Numbers given are days of gestation. The period of organogenesis is relatively constant (thick line) amongst these divergent species which range in size from less than 50 g to over 6 kg. The pre-attachment phase when embryogenesis occurs is however very variable, and may be extended even further by embryonic diapause. Redrawn from Renfree (1980*b*).

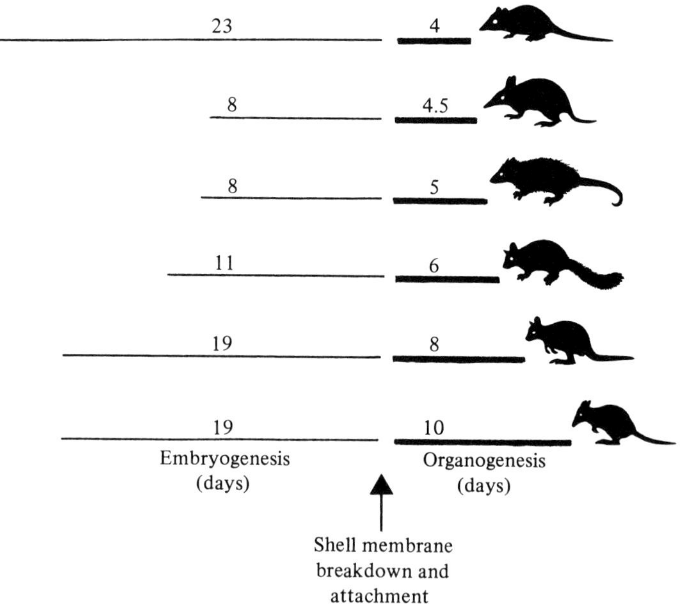

of organogenesis or the more invasive the placenta, the greater the birthweight. Walker & Rose (1981) make a similar suggestion with regard to the degree of functional differentiation of the mesonephros at birth (see below. However, relatively few studies have been made of the development of particular organ systems of marsupial species or of placental structure and ultrastructure to enable these ideas to be taken further yet. *Trichosurus vulpecula* was well studied in the early 1900s (Fraser & Hill, 1915; Parker, 1917; Tribe, 1918; 1923; Buchanan & Fraser, 1918; Fraser, 1919) but the most comprehensive and detailed account is that of McCrady (1938) for *Didelphis virginiana*. He recognised 35 stages of intrauterine development, which have often been used by later workers to refer to embryos of other species of marsupial (Table 7.7). Here we summarise the intrauterine development of the main organ systems of marsupials.

The excretory system

The non-functional pronephros first appears during the free vesicle phase but, shortly afterwards, the mesonephros differentiates into distinct tubules, complete with glomeruli and Bowman's capsules. In *D. virginiana* the pronephros consists of solid cords connected to the pronephric duct (McCrady, 1938). It appears at about stage 24 (7 somites), with the most anterior pronephric anlage opposite the 7th somite and, by the 11 somite stage (end of stage 24), the nephrogenic ridge includes all of the pronephric and the beginning of the mesonephric region. Pronephric tubules are found at about the same stage in *Petauroides volans* (Bancroft, 1973) and in *Setonix brachyurus* and *M. eugenii* at day 16–17 (Sharman, 1959; Renfree, 1972*a*). Mesonephric tubules (Fig. 7.14) appear at day 18–19 in *M. eugenii* and *M. rufogriseus* (Renfree, 1972*a*; Walker & Rose, 1981). At day 20 in *M. eugenii* the degenerating, non-patent pronephric duct connects with the anterior end of the Wolffian duct (Alcorn, 1975), which is patent from the mesonephric tubules to the allantois (Fig. 3.6), but beyond it the proctodaeum is closed (Renfree, 1972*a*). This results in an accumulation of execretory products, notably urea (Renfree, 1973*b*) in the allantois which, by the end of gestation, is considerably enlarged and occupies a space equivalent to that of the embryo itself (Fig. 7.16*b*).

In *Philander opossum*, the allantois is also large at full term and is more vascular than that of *D. virginiana* (Enders & Enders, 1969). However, in *D. virginiana* the cloacal membrane breaks through 12–14 h before birth and releases urine into the amniotic cavity. From this time on, the allantois looks collapsed and expels its contents back through the urachus and cloaca into the amniotic cavity (McCrady, 1938). The composition of allantoic fluid and amniotic fluid taken from *M. eugenii* (Fig. 7.22) and *Setonix brachyurus* just before birth suggest that this process does not occur in these species (Renfree, 1973*b*; G. I. Wallace, unpublished observations). Also, in *M. rufus* and *M. eugenii* the allantois is expelled at birth as an intact sac (Fig. 2.23*c*) which does not appear to be collapsed in any way (Sharman & Calaby, 1964; M. B. Renfree, personal observations).

The mesonephros functions as an active excretory organ for the first week after birth (Buchanan & Fraser, 1918; McCrady 1938; Bancroft, 1973; Krause *et al.*, 1979*a*, Walker & Rose, 1981; Wilkes, 1984). In *D.*

Fig. 7.16. Fetal stages of *Macropus eugenii.* (*a*) Day 20 RPY, around the time of attachment. The yolk sac (ys) has a well-developed vascular area, and the embryo is at the late headfold stage (McCrady stage 30–31). The embryo has been removed from the uterus but is still enclosed within the (collapsed) yolk sac. (*b*) Fetus at day 22 RPY showing the attachment of the yolk sac (ys) to the uterus (ut), which is very vascular and oedematous. The vitelline veins (vv) are prominent, and there is a much enlarged allantois (al). The amnion can just be seen around the embryo itself. (*c*) Fetus at day 25, 1–2 days before term, showing well-developed forelimbs with claws, open mouth and nostrils, but poorly developed hind limbs. (*a*) From Renfree (1983) and (*b* & *c*) from Renfree (1973*b*), with permission.

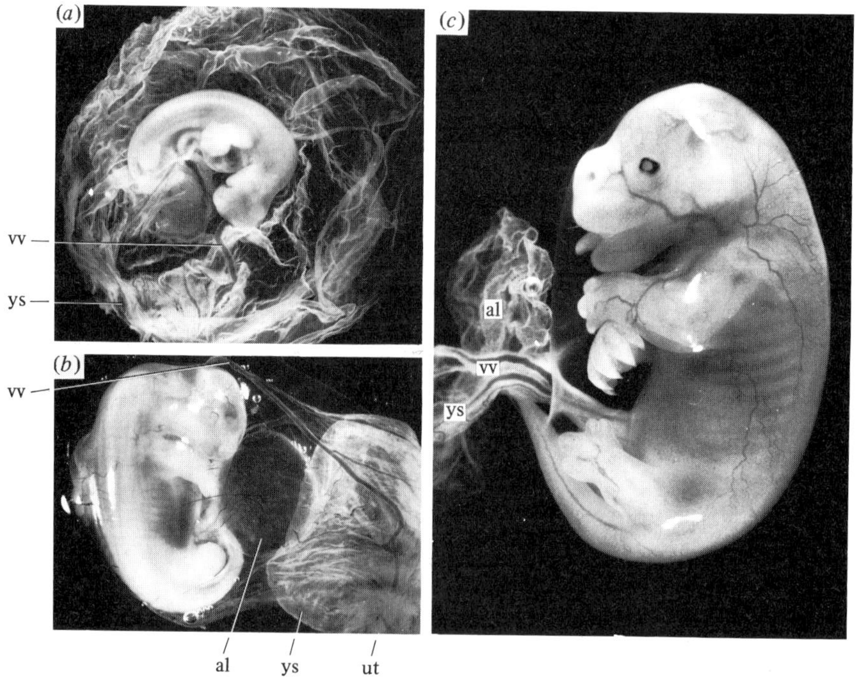

virginiana there are several nephron units with a renal corpuscle, proximal and distal segments but no loop of Henle is present (Krause *et al.*, 1979*a*). A similar structural arrangement is found in *T. vulpecula* (Fig. 3.8) and in *M. eugenii* in which degeneration of the mesonephros is first apparent 5 days after birth (Wilkes, 1984).

The metanephric blastema of *D. virginiana* appears at stage 35, just before birth (McCrady, 1938), and Krause *et al.* (1979*b*) record that the neonatal metanephros consists of only a few collecting tubules and a few immature nephrons. However, in *M. rufogriseus* (Walker & Rose, 1981) and *M. eugenii* (Renfree, 1972*a*; 1973*b*) it appears several days before birth and, by full term, the embryo is more advanced than that of *D. virginiana* (see Fig. 3.7), since it has Mullerian ducts and a well-developed metanephros (Fig. 3.6) (Alcorn, 1975). In *M. eugenii* it is smaller than the adjacent mesonephros, and it has little structural organisation, consisting of collecting ducts, developing glomeruli and differentiating tubules separated by connective tissue (Wilkes, 1984). The Wolffian ducts are patent in both sexes at birth, and enter the urogenital sinus separately from the ureters. The early development of the gonads and genital ducts have been discussed in Chapter 3.

Derivatives of the endoderm

The lung undergoes early development (Fig. 7.14), and the diaphragm separates the heart and lungs from the peritoneal cavity well before birth. The lung of the neonatal *D. virginiana* shows pecularities which make it appear that respiration is carried on in specially modified bronchi and bronchioles before the alveoli and the infundibular portion of the lungs is developed (Selenka, 1887; Bremer, 1904; McCrady, 1938). More recent studies in *Dasyurus viverrinus* (Hill & Hill, 1955), *Didelphis virginiana* (Krause & Leeson, 1973), *Isoodon macrourus* (Gemmell & Little, 1982) and *M. rufogriseus* (Walker & Gemmell, 1983*b*) confirm that the peripheral regions of the lung lobes consist of terminal sacs lacking the septae which characterise true alveolar sacs, but these are lined with an extensive capillary network which presumably facilitates respiration. As in other mammals, it appears that marsupials may also utilise surfactant as an essential agent in allowing the lungs to be functional at birth although the mucous-secreting cells (Goblet and submucosal cells) are absent (Krause & Leeson, 1937*b*; Krause, Cutts & Leeson, 1976; Gemmell & Little, 1982; Walker & Gemmell, 1983*b*).

The buccal cavity shows structural adaptations that permit continuous attachment to the teat, and breathing and feeding to proceed simultaneously (Owen, 1834). The tongue is very large and there is a complex epiglottis. In *Da. viverrinus* and *Isoodon obesulus* the epiglottis is intra-pharyngeal, but in *Trichosurus vulpecula* and in *Macropus rufus* the epiglottis just pierces the soft palate so that the glottis opens into the nasopharynx, giving these two species an intra-nasopharyngeal epiglottis (Sharman, 1937*b*). In *M. eugenii* there is a similar anatomical arrangement, so that milk can apparently pass from the tongue on either side of the trachea at the same time as the young is breathing (Renfree, 1972*a*).

The pancreatic anlage is present before a recognisable stomach rudiment, but does not enlarge until the liver and stomach are fully differentiated (Tribe, 1918; McCrady, 1938; Renfree, 1972*a*; Walker & Rose, 1981). At birth stomach and intestine lack the degree of biochemical differentiation of the adult. In *M. eugenii* the liver is probably functional at or soon after the yolk sac placenta attaches to the uterus, because the cells are clearly organised and breakdown products, presumably from red blood cell catabolism, are found in the yolk sac fluid at this time (Renfree, 1973*b*). It is certainly capable of regulating gluconeogenesis from an early post-natal age (Janssens *et al.*, 1977), and the presence of urea at higher concentrations in the allantoic fluid than in the maternal serum (Fig. 7.22*b*) is suggestive of urea cycle enzyme activity before birth.

Endocrine organs

There is growing evidence that the adrenal glands of marsupials are differentiated by birth and capable of steroid synthesis. McCrady (1938) observed their appearance in *D. virginiana* 1 day before birth and Owen (1834) in a full term fetus of *M. giganteus*. In *M. eugenii* the adrenal cortex is recognisable by day 22 as a discrete group of cells adjacent to the mesonephros by which stage 3 β-hydroxysteroid dehydrogenase activity was detected (Call, Catling & Janssens, 1980). By day 25, 1–2 days before birth, the cells are organised into cords and appear to be fully functional (Renfree, 1972*a*). The zonation of the adrenal cortex occurs post-natally, but a distinct medulla does not appear until day 60 (Call *et al.*, 1980). The presence of cortisol in the plasma of two full term fetuses (9 and 15 ng ml^{-1}) supports the indication from histology of cortical function prior to birth (Catling & Vinson, 1976). Similarly a degree of structural differentiation at birth was evident in *M. rufogriseus*, and the adrenals contained 0.58 ng of cortisol each (Walker & Gemmell, 1983*b*). In the smaller *Isoodon macrourus*, 0.094 ng of cortisol per adrenal was measured in the newborn and the ultrastructure was characteristic of steroid-secreting cells of functional cortex (Gemmell *et al.*, 1982). In addition, cells similar in structure to the catecholamine-secreting cells of the adrenal medulla were observed.

Parker (1917) described the early development of the pituitary and related structures in seven species of Australian marsupial and there have been more recent studies with the electron microscope. Rathke's pouch appears in the 10½ day embryo of *Isoodon macrourus*, 2 days before birth (Hall & Hughes, 1985). In *M. eugenii*, Rathke's pouch closes off from the pharynx at about day 21 and the pituitary is a well-developed, glandular structure by day 24, 2–3 days before birth (Renfree, 1972*a*). In *M. rufogriseus* Rathke's pouch remains connected to the pharyngeal roof by a thin line of cells which serve to separate the pars distalis (the anterior lobe of the pituitary) from the pars intermedia (Walker & Rose, 1981). This remnant persists in the 5 day old pouch young of *M. eugenii* (Leatherland & Renfree, 1983), and also in *Isoodon macrourus* (Hall & Hughes, 1985). At birth, the pars distalis makes up the bulk of the pituitary and in *M. rufogriseus* many cells contain electron-dense granules (Walker & Gemmell, 1983*b*). Non-granulated cells compose 70% of the gland at birth in *M. eugenii* (Leatherland & Renfree, 1982). Presumptive somatotrophs were found in most regions of the pars distalis, while presumptive mammotrophs, corticotrophs, thyrotrophs and gonadotrophs together represented less than 3% of the pars distalis (Leatherland & Renfree, 1983). Similar cell types were seen in neonatal *I. macrourus* (Hall & Hughes, 1985). The various structural observations suggest that the fetal pituitary may be capable of secreting hormones before birth but, for the adenohypophysis, this has not been determined. However, in the neurohypophysis of *M. eugenii*, vasopressin has been assayed at about 0.05 $\mu g\ mg^{-1}$ pituitary in 7 day old pouch young (Wilkes, 1984).

The neonatal marsupial has a mixture of 'altricial' and 'precocial' features: although very small its neuromuscular coordination is sufficiently developed to enable it to climb to the pouch unassisted by the mother, using its well-developed forelimbs (Figs 2.23*e*, *f*, *g* and 7.16*c*). These are supported by a well-developed coracoid articulation to the sternum, which later breaks down (Broom, 1900; Watson, 1917; Cheng, 1955). Its lungs are functional, the nostrils open and the olfactory centre of its brain is well developed. The mouth, tongue, and digestive system (including liver and pancreas) are sufficiently developed to cope with the change to a milk diet. By contrast, features such as the eyes, the hind limbs and the gonads remain undifferentiated; pouch and scrotum can be distinguished only after several days post-partum. The mesonephric kidney remains functional for the first few days after birth; the metanephros is differentiated but not functional immediately. Nevertheless, the entire period of differentiation before birth is accomplished in a short period which prepares the neonate for its more lengthy stay in the pouch, where the rest of development is completed.

The marsupial placenta

In no mammal does the ovum contain sufficient nutriment to sustain the developing embryo through pregnancy and all must rely on uterine secretions and later on the uterine vasculature to provide the main supply. The extraembryonic or fetal membranes play an important role in transporting these nutrients to the embryo, as well as for respiration and as a route for subsequent excretion. Although these structures are only transitory, a more or less intimate contact is developed between the outer cell layers of the fetal membranes and maternal uterine tissues. This apposition of fetal and maternal tissues is termed implantation and the structure so formed is termed a placenta.

There has been much debate about whether marsupials possess a true placenta. In his paper on the placenta of *Perameles* and *Isoodon* Flynn (1923) wrote that 'the term placenta should be applied to all organs consisting of an intimate apposition or fusion of the foetal membranes with the uterine wall for the purpose of carrying out physiological processes destined for the well-being of the embryo'. Mossman (1937) observed that this definition could not apply to the placentae of ectopic embryos or to those placenta-like connections seen in certain non-mammalian chordates and he modified Flynn's definition to the one that is currently accepted: 'An animal placenta is any intimate apposition or fusion of the foetal organs to the maternal (or paternal) tissues for physiological exchange'. For normal mammalian conditions this was rephrased to 'The normal mammalian placenta is an apposition or fusion of the foetal membranes to the uterine mucosa for physiological exchange.'

Despite Mossman's clear inclusion of all marsupials as well as eutherians as 'placental' mammals, Hartman's (1925*b*) apellation of *D. virginiana* as an 'aplacental mammal' has led to an uncritical acceptance of the misnomer when referring to marsupials. Padykula & Taylor (1982) explain that this probably arose because the functional activity and importance of the yolk sac placenta in eutherian mammals was unappreciated for a long time, in contrast to the great importance of the chorioallantoic placenta. Nevertheless, a yolk sac placenta is widespread among eutherian mammals (Luckett, 1977) (see Fig. 1.3), and even in the higher primates, where the yolk sac does not establish contact with the chorion, it still retains its primary role as the first producer of embryonic blood cells (Padykula & Taylor, 1982) and primordial germ cells (Byskov, 1983). Furthermore, in the rodents and lagomorphs it retains an important function in the transfer of passive immunity from mother to fetus (see Brambell, 1970). Conversely, in most marsupials the choriovitelline placenta forms the major organ of physiological exchange and only in *Phascolarctos cinereus* and the Peramelidae does the allantois form a true placenta as well (Figs. 1.3, 7.17). For the majority the allantois has only an excretory storage function, as the respiratory function is also performed by the choriovitelline placenta (Renfree, 1977).

Placentation

The placental structures of marsupials (Fig. 7.17) have been grouped according to the condition of the allantois relative to the chorion (Sharman, 1959; Hughes, 1984), and according to the intimacy of the contact between fetal and maternal tissues and circulations after the break down of the shell membrane (Hughes, 1974).

Type 1 The most common placental type (Fig. 7.17*a*) was first described by Owen (1834) for *Macropus giganteus*. He noted that the uterus produced abundant secretion 'for the increase of the ovum', and that the allantois was relatively small and did not reach the chorion. Owen made no comment as to whether he considered these features 'primitive' or 'advanced': it was left to later workers to suggest evolutionary status on the basis of placental structure. The well-vascularised yolk sac was in intimate association with the endometrium, and the membranes apposed to the uterine wall consisted of vascular and non-vascular portions of the yolk sac and only a small area of true chorion, while the allantois, which becomes a large vessel at the end of gestation (Fig. 7.16*b*), remained enclosed in the folds of the yolk sac (Owen, 1834) and poorly vascularised. Subsequently similar arrangements were described in the Didelphidae, Phalangeridae and Macropodidae. In the Didelphidae these were *D. virginiana* (Selenka, 1887; New *et al.*, 1977; Krause & Cutts, 1984) and *Philander opossum* (Enders & Enders, 1969); among the Phalangeridae they were *T. vulpecula* (Sharman, 1961*b*; Hughes, 1974), *Pseudocheirus peregrinus* (Sharman, 1961*b*; Hughes *et al.* 1965), *Petauroides volans* (Bancroft, 1973) and *Petaurus norfolcensis* (Semon, 1894); and among the Macropodidae they were *Aepyprymnus rufescens* (Semon, 1894), *Bettongia gaimardi* (Selenka, 1892; Flynn, 1930), *Potorous tridactylus* (Sharman, 1961*b*; Shaw & Rose, 1979), *Macropus giganteus* (Chapman, 1882), *M. rufogriseus* (Caldwell, 1884; Sharman, 1961*b*; Walker & Rose 1981), *M.*

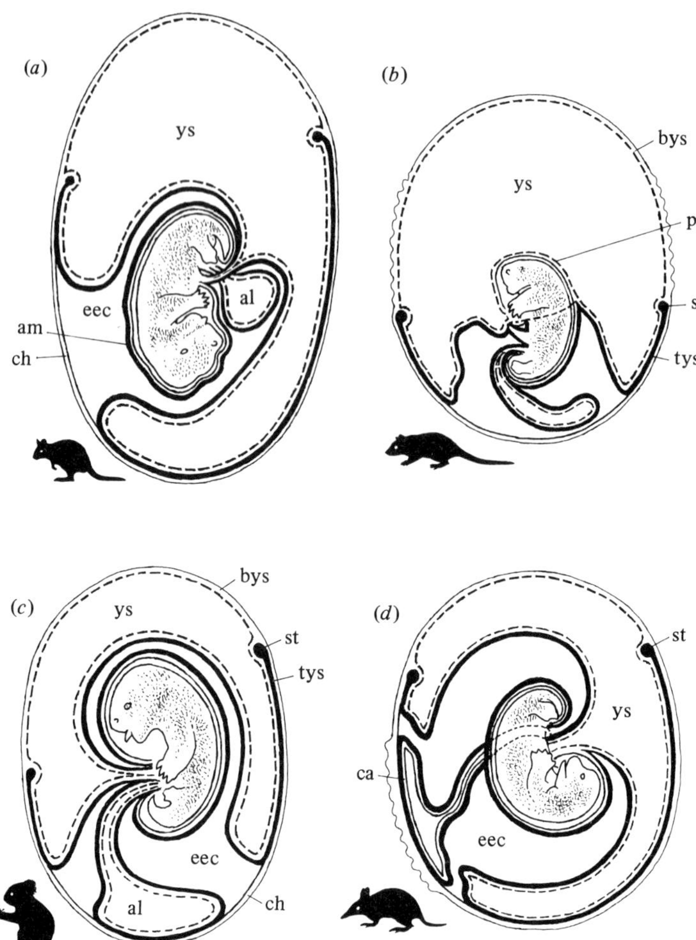

Fig. 7.17. The arrangement of the fetal membranes in species representing the four types. (*a*) Type 1: *Setonix brachyurus* at 21 days after RPY. At later stages the allantois is larger and both it and the fetus are more completely enclosed in the folds of the yolk sac. Redrawn from Sharman (1961*b*). (*b*) Type 2: *Dasyurus viverrinus*. The allantois approaches the chorion but does not fuse with it. As it retreats from the chorion at later stages its blood vessels degenerate. Over part of the bilaminar yolk sac close attachment to the endometrium occurs. Redrawn from Hill (1900*b*). (*c*) Type 3: *Phascolarctos cinereus*. The allantois makes contact with the chorion to form a chorioallantoic placenta that attaches to the uterine epithelium. Redrawn from Semon (1894). (*d*) Type 4: *Perameles nasuta*. Extensive choriovitelline and chorioallantoic placental attachments. Redrawn from Hill (1898). al, allantois; am, amnion; bys, bilaminar yolk sac; ca, chorioallantois; ch, chorion; eec, extra

robustus (Selenka, 1892), *M. parma* (Hill, 1900*a*), *M. eugenii* (Renfree, 1972*a*, 1973*b*) and *Setonix brachyurus* (Sharman, 1961*b*).

Type 2 A few species differ from this general pattern, with regard to the allantois (Fig. 7.17*b*). For example, in *Dasyurus viverrinus*, the allantois becomes apposed to the chorion and then retreats as development proceeds and does not take part in the formation of the placenta (Hill, 1900*b*); its vascularity degenerates and it lies in the extraembryonic coelom as a vestigial structure. In a single specimen of *Sminthopsis crassicaudata*, the only other species of dasyurid so far examined, the allantois in an advanced-stage fetus was small and not fused with the chorion (G. K. Godfrey in Hughes, 1974), so it is not possible to say whether other dasyurids share this feature with *Da. viverrinus*.

Type 3 This is represented by *Phascolarctos cinereus* (Fig. 7.17*c*) in which the allantois becomes apposed to the chorion but never develops villi, although it is highly vascular in the later stages (Caldwell, 1894; Semon, 1894; Hughes, 1974; 1984). The only attachment of the embryo to the uterus is entirely non-vascular and is effected by an amoebic growth of giant cells of the sub-zonal membrane of the bilaminar yolk sac immediately outside the sinus terminalis. It is easily disrupted (Hughes, 1984). The attached area is next to the opening of the uterus into the cervix (Caldwell, 1884). Amoroso (1952) considered the allantois of this species and also that of *Vombatus ursinus* to vascularise a placenta, and more recent study of specimens of both species in the Hill Collection by Hughes (1984) has confirmed this as a potential organ of respiratory exchange, although the yolk sac placenta is probably the major organ for nutritive absorption and respiration.

Type 4 Placentation in which a true chorioallantoic placenta is present in addition to a choriovitelline placenta (Fig. 7.17*d*). During the later stages of gestation the chorioallantois forms a discoidal placenta with a long, thick umbilical cord (Fig. 2.12*b*). This has been described from a total of 14 late-stage placentae from four species of the Peramelidae, namely *Isoodon obesulus* (3), *I. macrourus* (4), *Perameles nasuta* (6) and *P. gunnii* (1) by Hill (1895, 1898, 1899, 1900*a*), Flynn (1923) and Padykula & Taylor (1976*b*, 1977, 1982).

Fig. 7.17. *cont.*

embryonic coelom; pro, proamnion; st, sinus terminalis; tys, vascular yolk sac; ys, yolk sac. The ectoderm is represented by a thin line, the endoderm by a dotted line, somatic mesodern medium line and the splanchnic mesoderm by a thick line.

The relative importance to the fetus of the yolk sac and allantois were assessed by Tribe (1923) by comparing the size and disposition of the vitelline and allantoic veins after their entry into the body of the fetus. In *Macropus*, *Trichosurus* and *Didelphis*, which have Type 1 placentae, the vitelline veins are large and enter the posterior vena cava direct, while the allantoic veins are very small and indistinguishable from veins of the body wall. In *Perameles* the vitelline veins are likewise large in fetuses up to a length of 8.5 mm but thereafter decline, while the allantoic veins are prominent vessels that penetrate the liver and the left one continues to grow larger in fetuses greater than 8.5 mm. Tribe (1923) considers that the change reflects the increasing importance of the chorioallantoic placenta for respiration but because the vein does not anastomose in the liver, the nutritional role of the liver may be slight.

In *Phascolarctos cinereus*, with a Type 3 placenta, the allantoic veins have a similar disposition as in *Perameles* but in *Dasyurus viverrinus* and surprisingly *Vombatus* they are inconspicuous, as in the Type 1 species. This casts doubt on the importance of the supposed allantoic connection for respiration of the fetus in these species.

Feto-maternal contact

After the shell membrane breaks down there is considerable variation in the degree of invasiveness of the yolk sac placenta, from simple apposition to erosion and penetration of the uterine epithelium in different regions to actual fusion of fetal and maternal syncytia (Hughes, 1974). In *D. virginiana*, the yolk sacs of embryos older than 10 days are fused to each other in the avascular region and cannot be pulled apart, and by $11\frac{1}{2}$ days are so firmly attached to the endometrium they cannot be removed intact (McCrady, 1938; New *et al.*, 1977). Krause & Cutts (1984, 1985) state that the close adhesion of the trophoblast to the uterine mucosa occurs in the vascular region and that the cells of this area have abundant microvilli which may also aid in absorption of the uterine secretion (Fig. 7.19*a*). Similar features distinguish *Philander opossum* (Enders & Enders, 1969), *Da. viverrinus* (Hill, 1900*b*) and *Tarsipes rostratus* (M. B. Renfree, unpublished observations). The walls of the bilaminar yolk sacs of adjacent embryos become fused together and the chorion becomes closely applied to the uterine epithelium (Fig. 7.18). However, in both species, a more intimate contact than that described for *D. virginiana* is established between the yolk sac adjacent to the sinus terminalis and the uterus.

In *Da. viverrinus* (Fig. 7.17*b*) the trophoblast cells of the bilaminar yolk sac become enlarged and penetrate between the epithelial cells to reach the subepithelial tissue where maternal capillaries are abundant. Later the trophoblast cells fuse to form a syncytium which encloses the maternal capillaries and blood from these passes into spaces within the syncytium. Hill (1900*b*) concluded that this may have a nutritive function particularly in providing iron for the near-term fetus. In *P. opossum*, trophoblast cells

Fig. 7.18. Fetal membranes in *Philander opossum* – diagrammatic representation of a cross-section of a pregnant uterus. The trophoblast is heavily stippled; endoderm is a solid black line; the vessels of choriovitelline placenta are solid black ovals; the vessels of the allantois are small circles. Note that a large vessel, the sinus terminalis, marks the edge of the two choriovitelline placentae shown, and that the trophoblast penetrates the uterine epithelium at the tips of endometrial folds near the margin. In the lower left, a portion of the residual shell membrane is folded between the trophoblast and the endometrium. On the right side is a small residual portion of the fused bilaminar yolk sacs which once separated the embryos. From Enders & Enders (1969) by permission.

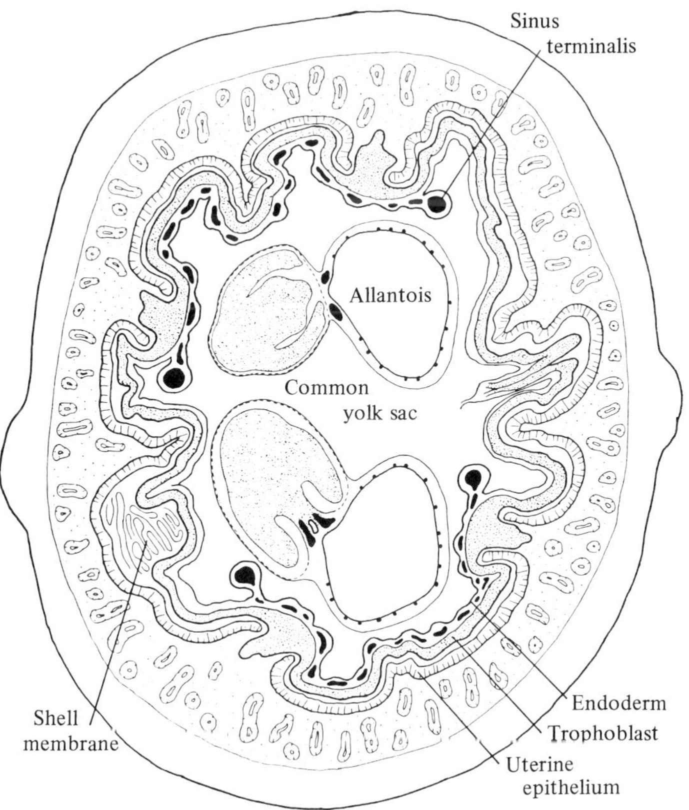

in the trilaminar yolk sac penetrate between uterine epithelial cells as far as the basement membrane surrounding uterine blood vessels but do not rupture them. The fetal cells form desmosomal junctions with the uterine cells and Enders & Enders (1969) consider that the main role is to aid in adhesion of the fetal membranes to the uterine epithelium (Fig. 7.18).

No such penetration of the uterine epithelium has been reported for any of the diprotodont marsupials, although very close apposition of the apical surfaces of the two tissues does occur and attenuation of the cells themselves brings the fetal and maternal circulations into close proximity. Thus in *Bettongia gaimardi* Flynn (1930) describes the fusion of fetal and maternal tissues as being most intimate in the region of the trilaminar yolk sac, where fetal and maternal bloodstreams are separated by uterine and fetal epithelia, by the two endothelia and by maternal connective tissue. He states that 'the impression is given that glandular secretion does not play a very important role in embryotrophic processes', but that 'whenever a gland is found pouring its secretions into the uterine lumen, the trophoblast cells can be seen actively absorbing the secreted material by means of pseudo-podial processes'. He thought that the non-vascular or bilaminar yolk sac was important for nutrition, and noted that 'cells of the bilaminar omphalopleure are tremendously vacuolated, which points to the active absorption of carbohydrates in this portion of the yolk sac placenta', while 'definite union is particularly evident in the region of the vascular omphalopleure, and between maternal and foetal tissues, where separated, there occurs abundant "uterine milk" consisting of cell debris, leucocytes, haematids, lymph and other material all being actively absorbed and *Setonix brachyurus* (Sharman, 1961*b*).

In *Potorous tridactylus* and *T. vulpecula*, the yolk sac is not attached to the uterine epithelium at any point (Hughes, 1974) and no uterine invasion occurs in these species, or in *Pseudocheirus peregrinus*, *Macropus rufogriseus* and *setonix brachyurus* (Sharman, 1961*b*).

These conclusions contrast with those of Hill (1900*b*), who thought that the poorly vascularised trilaminar yolk sac of *Da. viverrinus* served mainly for gaseous exchange, and that the bilaminar region is concerned with nutrition. He suggested that this region must be of considerable functional importance because of the comparatively poor development of the capillary system of the vascular region. Semon (1894) also considered the persistence of the bilaminar part of the yolk sac to have a physiological importance, and be the route for uterine secretions to reach the vitelline vessels.

In *Macropus parma*, the larger surface area of the vascular region than in *Perameles nasuta* led Hill (1900*a*) to conclude that this correlates with its importance to the embryo. Thus, although all agree that the yolk sac is an organ of exchange between mother and fetus, the relative importance of the two parts for respiration and nutrition remains uncertain. The advent of the electron microscope has helped to clarify this by identifying sites of active absorption and transport.

In *M. eugenii* the lumenal surfaces of the trophoblast and uterine epithelium are covered with thick naps of microvilli, which interdigitate (Fig. 7.19*b*). We have not observed any erosion of the epithelium but maternal capillaries lie close beneath the junctional complexes of adjacent epithelial cells so bringing them into closer relationship with the fetal circulation. The fetal trophoblast cells contain numerous coated vesicles and there is evidence of pinocytosis on the lumenal surface, suggestive of active absorption. Similarly, in *M. rufogriseus*, the fetal extraembryonic ectoderm becomes hypertrophied at the time the shell breaks down between days 17 and 21; it is extensively microvillous and active in endocytosis, intracellular transport and synthesis (Walker, 1983). From about day 17 an extensive subepithelial capillary layer is established in the endometrium. The yolk sac endoderm cells have few organelles that might be associated with steroidogenesis but they have extensive microfilaments and intercellular desmosomal contact, usually involved in structural support. These cells have few vesicles but are associated with vesicular material of presumed ectodermal origin, especially in the bilaminar yolk sac (Walker, 1983).

In *Philander opossum*, the ectoderm cells at the margin of the vascular region of the yolk sac penetrate folds of the endometrium (Fig. 7.18), as already mentioned, but elsewhere the microvilli of the yolk sac and uterine epithelium do not interdigitate (Enders & Enders, 1969). There is less contained vesicular material in the cells of the outer surface of the vascular yolk sac and, as in *M. eugenii*, the ectoderm has numerous coated vesicles and inclusions, suggestive of absorptive activity (Enders & Enders, 1969). The endoderm of *P. opossum*, on the other hand, has an appearance of synthetic activity.

In *Perameles nasuta* and *Isoodon macrourus*, the trophoblastic cells of both the vascular and non-vascular yolk sac are covered with microvilli which interdigitate with those of the uterine epithelium (Fig. 7.20*a*) (Padykula & Taylor, 1976*b*; 1982). At the same time multinucleate homokaryons appear in the luminal epithelium, presumably by fusion of adjacent epithelial cells (Fig. 7.21*a*), and subepithelial capillaries of large diameter protrude between them. With the attenuation of the trophoblast cells, particularly of the vascular yolk sac, the fetal capillaries are thus

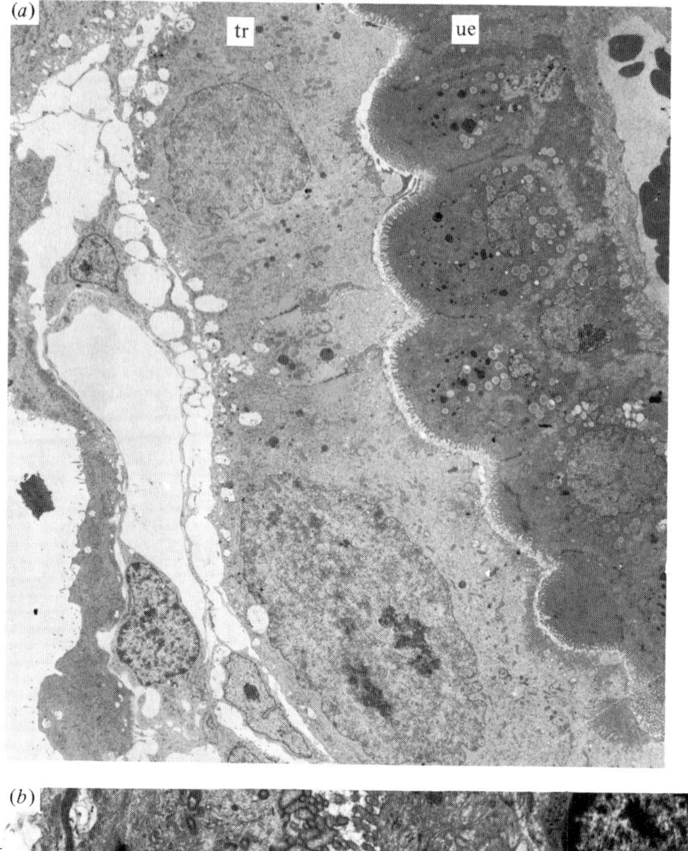

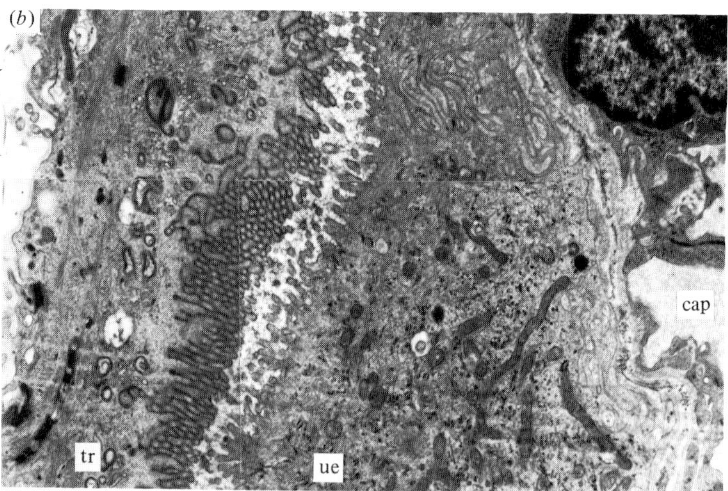

Fig. 7.19. The feto-maternal contact in late pregnancy. (*a*) *Didelphis virginiana* at day 11 after breakdown of the shell membrane when the trophoblast (tr) of the vascular yolk sac comes into close contact, but

brought close to the maternal ones. As well as this the trophoblast cells, particularly of the bilaminar yolk sac, have conspicuous endocytotic systems which are indicative of macromolecular uptake (Padykula & Taylor, 1976*b*; 1982). This is in contrast to rodents where it is the visceral endodermal cells which have the regulatory endocytotic mechanisms (Deren, Padykula & Wilson, 1966; Padykula, Deren & Wilson, 1966). Thus, the marsupial yolk sac placenta, particularly the bilaminar yolk sac, has structural characteristics consistent with absorption, intracellular transport, and synthesis and the close apposition of uterine capillaries and fetal capillaries of the vascular yolk sac must facilitate respiratory exchange.

In the chorioallantoic placenta of the Peramelidae, fusion of fetal and maternal tissues reaches its most profound degree in marsupials (Fig. 7.20*b*). When the allantois makes contact with the chorion, the initial apposition to the uterine epithelium is similar to that of the yolk sac placenta. However the uterine syncytial masses in this region contain nuclei of two distinct types, one resembling the nuclei of the homokaryons of the uterine epithelium in the yolk sac region and the other the nuclei of adjacent trophoblast cells (Fig. 7.21*b*,*c*). Padykula & Taylor (1967*b*) have called these heterokaryons. When first observed, they are intimately associated with trophoblast cells below which are the fetal capillaries. However, at full term the trophoblast has disappeared and the endothelia of the fetal and maternal capillaries are separated only by attenuated portions of the syncytial heterokaryons and endometrial stroma (Fig. 7.20*c*). So thin is the barrier that Amoroso (1952) considered this to represent an endothelio-endothelial placenta.

The disappearance of the trophoblast was first recognised by Hill (1898) and confirmed by Flynn (1923), and they both supposed that it degenerates in the late stage of pregnancy. However, Padykula & Taylor (1982) consider this unlikely because, at its last appearance, the cells contain abundant mitochondria and large nuclei. They suggest, to the contrary, that the trophoblast cells incorporate, or become incorporated into, the uterine homokaryons to form the heterokaryons and their nuclei may even redirect the activities of the heterokaryons. However, final resolution of

Fig. 7.19. *cont.*

does not fuse with, the uterine epithelial cells (ue) covered with microvilli. From Krause & Cutts (1985) with permission. (*b*) *Macropus eugenii* at day 24, vascular yolk sac trophoblast with dense microvillous surface interdigitating with microvilli of uterine epithelium. Note complex junction between epithelial cells overlying a surface capillary (cap). Electronmicrograph by C. H. Tyndale-Biscoe.

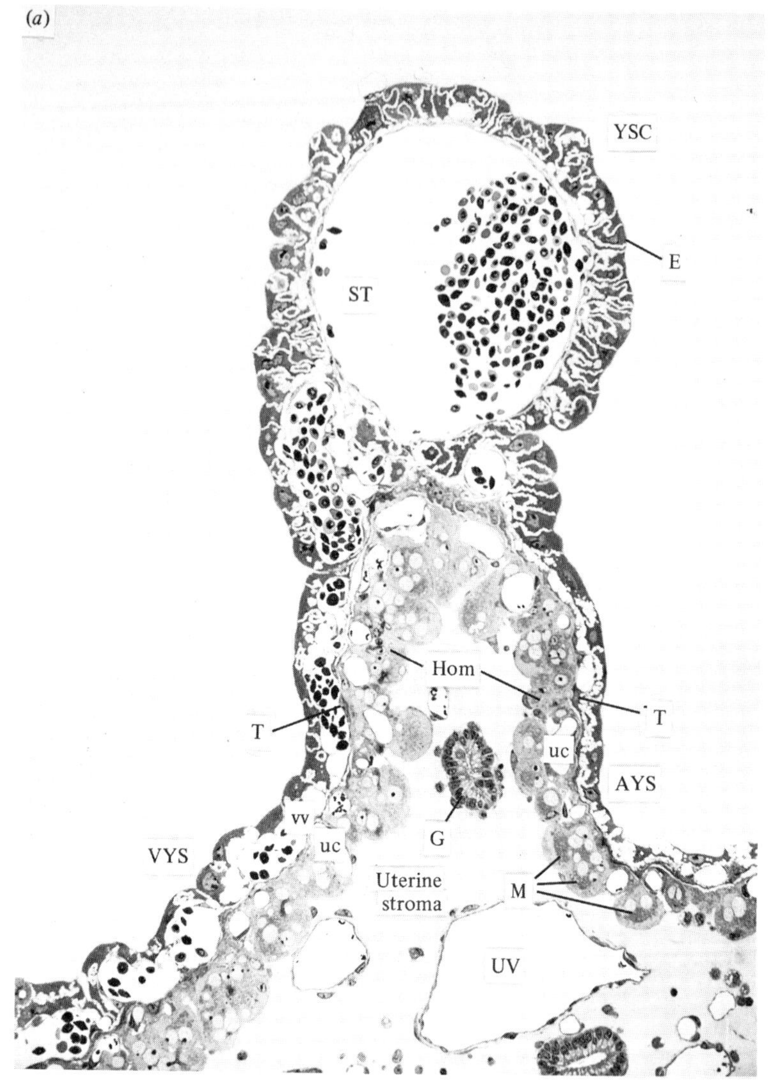

Fig. 7.20. (*a*) Feto-maternal contact at sinus terminalis of the yolk sac placenta of *Isoodon macrourus*. The thin-walled sinus terminalis (ST) is surrounded by a tall simple columnar endodermal epithelium (E), which lines the yolk sac cavity (YSC). In the vascular region of yolk sac (VYS) at the left, the endoderm is underlain by large thin-walled vitelline vessels (vv); an attenuated trophoblastic layer (T) is in contact with the uterine homokaryons (Hom), which are partly separated from each other by large thin-walled uterine capillaries (UC). The avascular yolk-sac (AYS) at the right resembles closely that of the

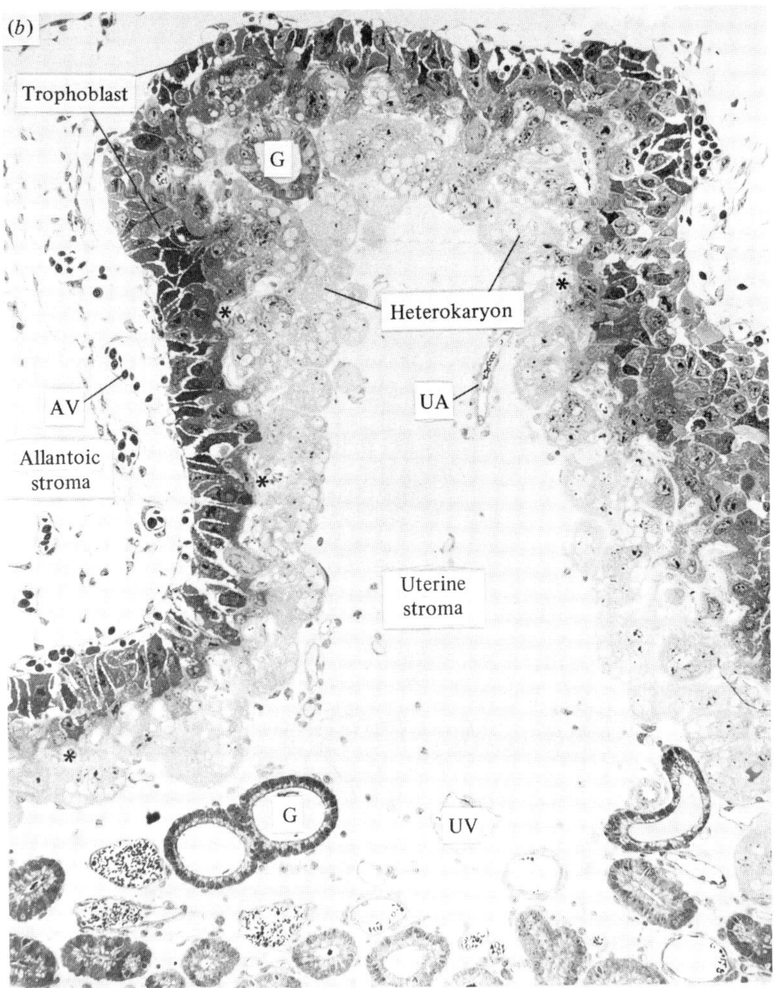

vascular region, except for the absence of the vitelline vessels. G, uterine gland; UV, uterine venules; M, mitochondrial aggregates in the homokaryon. (*b*) Feto-maternal contact at the chorio-allantoic placenta of the same specimen. A darkly stained trophoblast layer of irregularly columnar cells is apposed to the uterine luminal epithelium, which consists of heterokaryons with euchromatic and heterochromatic nuclei. Note the nucleated fetal red blood cells within the allantoic vessels (AV). The superficial uterine microvasculature located in slender stromal prongs, which extend among the large heterokaryons is difficult to recognise because this specimen was fixed by vascular perfusion; asterisk, large empty maternal capillaries, G, uterine glands; UA, uterine arteriole; UV, uterine venule. From Padykula & Taylor (1982), with permission.

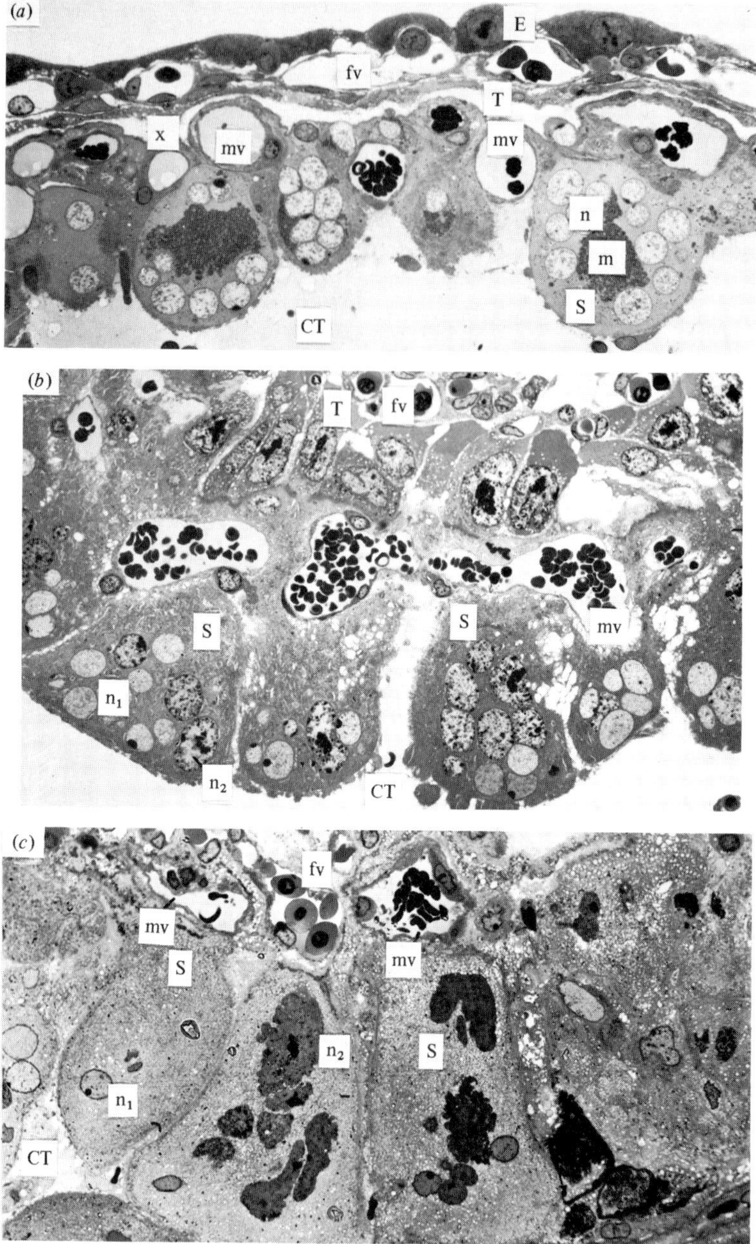

Fig. 7.21. Ultrastructure of the placenta in *Perameles nasuta* and *Isoodon macrourus*. (*a*) Visceral layer of the yolk sac placenta of *P. nasuta*. The large syncytial masses (S) or homokaryons that comprise

this question must await the use of nuclear-labelling techniques as suggested by Padykula & Taylor (1982). They point out that this pattern of fusion of fetal and maternal syncytia resembles the first transient phase of implantation in the rabbit. It is equally transient in the peramelids, being present for fewer than 3 days during which the fetus undergoes very rapid organogenesis.

Although peramelids have the shortest gestation of any marsupial, their neonates are relatively well developed at birth (Sharman, 1965*c*) and their post-natal growth is the most rapid, which suggests that their chorioallantoic placenta is a more efficient organ of exchange than the yolk sac placenta of other marsupials (Taylor & Padykula, 1978; Padykula & Taylor, 1982).

Placental functions

The morphological studies provide an indication of how the placental structures may function but the definitive evidence for nutritive, metabolic and excretory functions of the marsupial placenta, and the evidence for its endocrine and immunosuppressive functions can only come from biochemical analyses.

Fig. 7.21. *cont.*
the luminal epithelium of the uterus contain many pale nuclei (n) and central aggregations of mitochondria (m). Maternal blood vessels (mv) have penetrated between syncytial masses to acquire an extremely superficial position. The trophoblastic surface of the visceral yolk sac has separated from the uterine surface leaving an artefactual space (x). Fetal blood vessels (fv) occur in the mesenchyme between the attenuated trophoblast (T) and the visceral endoderm (E). CT, endometrial stroma. (*b*) The chorioallantoic placenta, of the same specimen. The syncytial masses (S) are distinctly separated by stromal septa that place the maternal blood vessels (mv) into a most superficial location. Note the two types of nuclei in the syncytial masses, pale ones (n_1) with peripheral nucleoli, as in the yolk sac placenta, and larger, distinctly heterochromatic ones (n_2) with large nucleoli. Thus these syncytial masses are heterokaryons in comparison with those of the yolk sac placenta (Fig. 7.21*a*). At this placental site, the trophoblast cells (T) are conspicuous and tall. Compare the nuclear structure of the trophoblastic cells with that of the larger, heterochromatic nuclei of the heterokaryons. Fetal blood vessels (fv) occur in the mesenchyme beneath the trophoblast. (*c*) The chorioallantoic placenta of *I. macrourus* at a later stage when the trophoblast has disappeared altogether and the maternal (mv) and fetal (fv) blood vessels are in close apposition, as a consequence. Note the two distinct kinds of nuclei in the heterokaryons as in (*b*). From Padykula & Taylor (1976*b*).

Biochemical functions of the placenta

The most detailed study of the biochemical function of the fetal membranes of a marsupial has been made on *Macropus eugenii* (Renfree, 1970; 1973*b*; 1974*b*; Renfree & Tyndale-Biscoe, 1973*b*), with less complete observations on *Setonix brachyurus* (G. I. Wallace, personal communication), *Didelphis virginiana* (Renfree, 1975; Renfree & Fox, 1975), *Perameles nasuta* (Renfree, 1977) and *Trichosurus vulpecula* (Tyndale-Biscoe *et al.*, 1974). In *M. eugenii* the constituents of the fluids from the yolk sac, allantois, amnion and fetal circulation were compared to those of maternal serum and endometrial secretion at successive stages of pregnancy.

During the pre-attachment phase, nutrient and respiratory requirements must be met by exchanges with uterine secretion across the yolk sac membrane and at this stage the concentrations of glucose and urea in yolk sac fluid resemble those in maternal serum (Fig. 7.22). However, the concentrations of free amino acids are 10 times higher than in serum and 3 times higher than in uterine fluid, although the protein concentration is much lower than in either (Renfree, 1973*a*, *b*). The proteins at this stage comprise albumin and pre-albumin bands with identical electrophoretic mobility to those of the uterine secretion (Fig. 7.23) and it is therefore likely that the great increase in volume of yolk sac fluid at this time is achieved by direct uptake of these components across the non-vascular yolk sac membrane.

After attachment, which occurs on day 19, 7–8 days before birth, marked changes occur in the yolk sac fluid; its protein components (Renfree, 1970; 1973*b*) now resemble those of serum, although there is no increase in the limited number of proteins in endometrial fluid (Fig. 7.23); the concentration of glucose increases steadily (Fig. 7.22*a*) but that of some amino acids, notably glutamine, decreases; the colour of yolk sac fluid changes from clear, through straw, to yellow by full term. The yellow colour results from breakdown products of haemoglobin, notably bilirubin. By contrast, allantoic and amniotic fluids remain clear and contain the same limited number of proteins as early yolk sac fluid (Fig. 7.23). Their glucose concentration is low but the urea concentration of the allantois increases progressively to a maximum 4–5 times that of the other compartments and fetal serum. This coincides with the development of the mesonephros to which it is connected by the Wolffian ducts.

The changes in yolk sac fluid after attachment could be due to a greater ease of transport of maternally derived components, especially proteins, across the vascular yolk sac from the closely applied maternal capillaries or they may reflect the developing autonomy of the fetus and its ability to synthesis its own proteins. In three experiments to examine this, the results supported the fetal origin of at least some of the proteins: first, when maternal serum proteins labelled with ^{125}I were injected into the mother's circulation during late pregnancy, less than 0.1% of protein in the yolk sac fluid was subsequently labelled (Renfree, 1972*a*); second, by

Fig. 7.22. The concentrations of (*a*) glucose and (*b*) urea in yolk sac (●), allantoic (○) and amniotic fluids (■) of late-stage fetuses of *Macropus eugenii*. Values plotted in relation to the size of the fetus, the age of which can be determined on Fig. 7.8. Redrawn from Renfree (1973*b*).

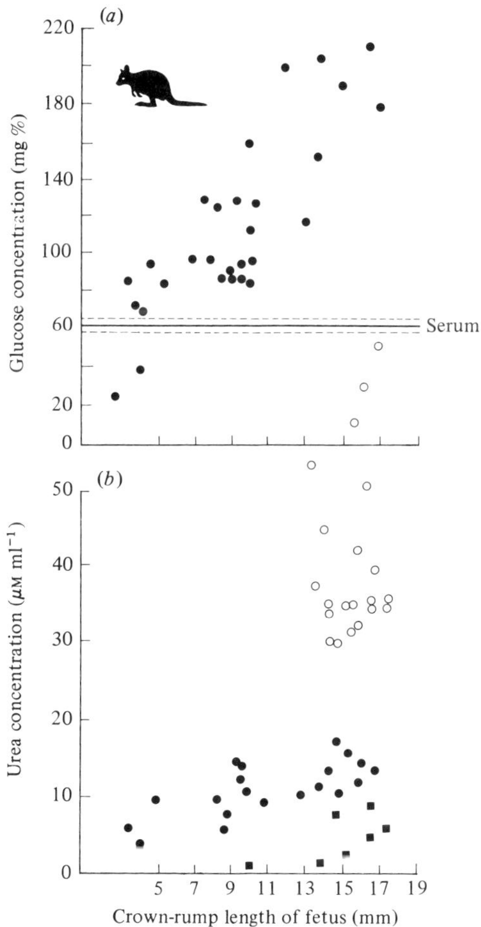

gel electrophoresis the isozymes for lactic acid and malic acid dehydrogenase (LDH and MDH, respectively) in yolk sac fluid were found to differ in proportion (LDH) and number (MDH) from those in maternal serum, whereas glucose-6-phosphate dehydrogenase was the same in both (Renfree, 1974*b*); third, the mobility of fetal transferrin differed from both the adult

Fig. 7.23. Proteins in the endometrium and fetal fluids of *Macropus eugenii*. (*a*) Disc electrophoresis of endometrial exudates from non-gravid (NG) and gravid (G) uteri collected at days 0, 6, 13, 20 and 24. Identical volumes of exudates were applied to all gels, but samples from gravid uteri were more concentrated than from the non-gravid sides and there was a marked increase between days 0 and 6. Note that the NG 0 sample has had to be over exposed to enhance the very faint bands. (*b*) Representative disc acrylamide gels of yolk sac fluid collected between days 13 and 25, and serum (S), amniotic (Am) and allantoic fluid (Al) collected on day 25. Attachment occurs between days 18 and 20, after which time the protein composition of yolk sac fluid changes qualitatively and quantitatively, to become similar to serum; transferrin appears at about this time and albumin and pre-albumin bands become denser. By contrast, amniotic and allantoic fluids collected at day 25 were similar to early stage yolk sac fluid. From Renfree (1973*a*,*b*).

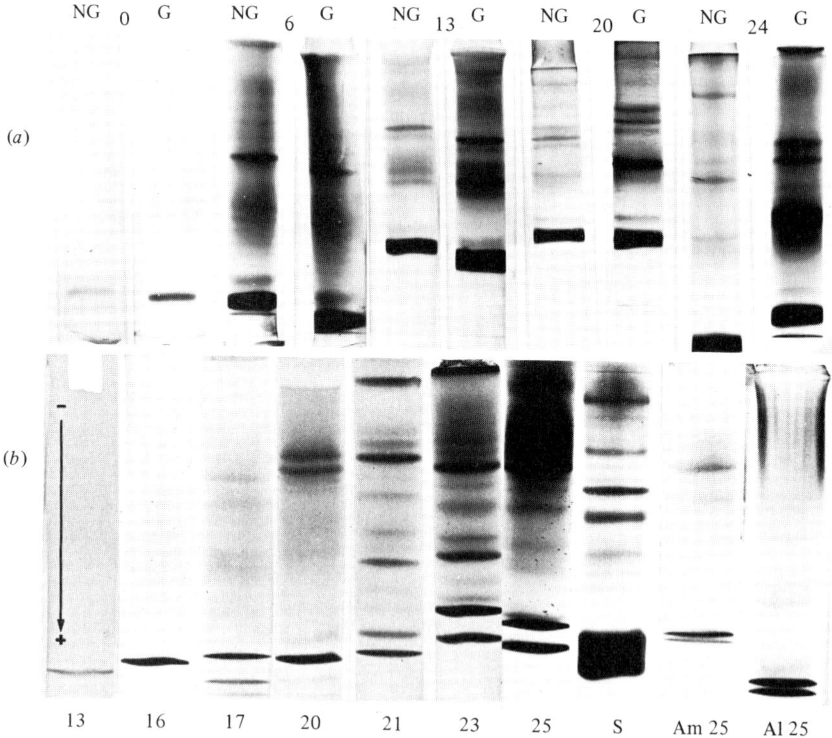

forms (Renfree & Tyndale-Biscoe, 1973*b*). On the other hand, gamma globulins in fetal serum (day 24) reacted with antibodies raised to adult gamma globulin (Renfree, 1973*b*; E. M. Deane, D. W. Cooper & M. B. Renfree, unpublished results), which may imply their transfer across the yolk sac to provide passive immunity after birth.

If the proteins in the yolk sac fluid are not maternal in origin, the late-stage fetus must have developed the necessary mechanism for synthesis. As most serum proteins are synthesised in the liver, it may be no coincidence that the fetal liver differentiates at the time of the greatest increase in proteins in yolk sac fluid. This is also the stage when the yolk sac fluid begins to turn yellow. Similar colour changes of yolk sac fluid occur in *D. virginiana* on day 11, about 1 day after attachment (Renfree, 1975) and in *T. vulpecula* on day 14, 2 days after attachment (Tyndale-Biscoe *et al.*, 1974) and differentiation of the liver (Hughes & Hall, 1984). In addition, the yolk sac membrane and fluid of *M. eugenii* contain several enzymes of glycolysis, of transamination and of the tricarboxylic acid cycle (Renfree, 1974*b*) with which the amino acid precursors and energy substrate could be used for protein synthesis. The rapid rise in glucose concentration and expansion of the vascular yolk sac suggests that the metabolic rate of the fetus increases in the late stage of gestation. Nevertheless the non-vascular yolk sac membrane continues to extend over two-thirds of the surface to the end of gestation. If the vascular yolk sac membrane is solely respiratory in function, the non-vascular yolk sac membrane may persist because it is the route of absorption of precursors and substrates for fetal metabolism and synthesis. This conclusion, like that from electron microscopy, supports Hill's (1900*b*) idea that there is a 'division of labour' between the two parts of the yolk sac of marsupials.

Maternal recognition of pregnancy

Apart from its nutritive functions, the placenta may elicit responses in the uterus locally or by the systemic alteration of the maternal endocrine system and it may induce a host versus graft reaction through the expression of transplantation antigens at the surface of the placenta. Because of the relatively brief periods of gestation, more especially the post-attachment portion, in marsupials and the absence (until recently) of evidence that the maternal endocrine system is redirected in pregnancy, it has generally been assumed that the marsupial feto-placental unit does not have an endocrine or immunosuppressive role. There is some evidence now that both roles may exist in species of the Macropodidae.

Morphogenetic effects of the placenta.

Owen (1834) noted that the endometrium of the gravid uterus of *Macropus giganteus* was twice as thick as the endometrium of the contralateral, non-gravid uterus (see Fig. 1.2) and Flynn (1930) showed that there were marked histological differences between the gravid and non-gravid uteri of *Bettongia gaimardi* but these observations were largely overlooked until the finding in *M. eugenii*, that the wet weight of the endometrium of the gravid uterus was consistently and significantly greater than that of the adjacent non-gravid one (Fig. 7.24*a*), the composition of the secretions was different from the two uteri (Fig. 7.23*a*) (Renfree, 1972*b*; Renfree & Tyndale-Biscoe, 1973*a*), and the rate of protein synthesis (Fig. 7.10) was greater in the gravid uterus (Shaw & Renfree, 1986).

The importance of the feto-placental unit was demonstrated by three observations. First, it was shown that the greater weight was not due to a local effect of the adjacent corpus luteum of pregnancy, such as von der Borsch (1963) and Curlewis & Stone (1986) have shown in *T. vulpecula*, because the transfer of a blastocyst to a cyclic, non-pregnant animal, or to the side contralateral to the corpus luteum in a pregnant one, initiated similar endometrial proliferation in each uterus in which a fetus developed and did not relate to the position of the corpus luteum. Second, as mentioned earlier in this chapter (p. 283) blastocyst development can be initiated during seasonal quiescence with exogenous progesterone and this also induced endometrial growth of both uteri for the duration of treatment (Fig. 7.24*b*). However, after the injections ceased, only the uterus containing the developing embryo continued to enlarge while the non-gravid uterus declined to a size usually observed in the non-pregnant cycle (Renfree & Tyndale-Biscoe, 1973*a*). Since, in this experiment, the corpus luteum remained inactive the continuation of endometrial stimulation in the gravid uterus after progesterone withdrawal again must have been due to the presence of the fetus or its placenta. The third observation differentiated between the fetus and the placenta as the main site of the influence. In four females, which were treated with progesterone during seasonal quiescence to induce blastocyst reactivation, a vesicle grew but no embryo or vascular mesoderm developed (Renfree & Tyndale-Biscoe, 1973*a*), but these vesicles, consisting only of the non-vascular yolk sac membrane, stimulated the endometrium and the uterine secretions as in a normal pregnancy.

The fact that the endometrium of the non-gravid uterus declines in weight and secretory activity during the second half of pregnancy or the cycle, even though peripheral progesterone concentrations are at their highest (Fig. 6.13), suggests that there is a declining sensitivity to progesterone. It is possible that the local stimulation of the gravid uterus may reverse this by stimulating increased progesterone-receptor synthesis. As mentioned in Chapter 5, Owen *et al.* (1982) have observed a difference in

Fig. 7.24. Changes in wet weight of the endometria of gravid (●) and non-gravid (○) uteri of *Macropus eugenii* sampled (*a*) during pregnancy induced by removing pouch young and (*b*) during pregnancy induced by 10 days of progesterone treatment. Individual values are presented as a weighted quadratic regression with 90% confidence limits for the regression lines. From Renfree & Tyndale-Biscoe (1973*a*).

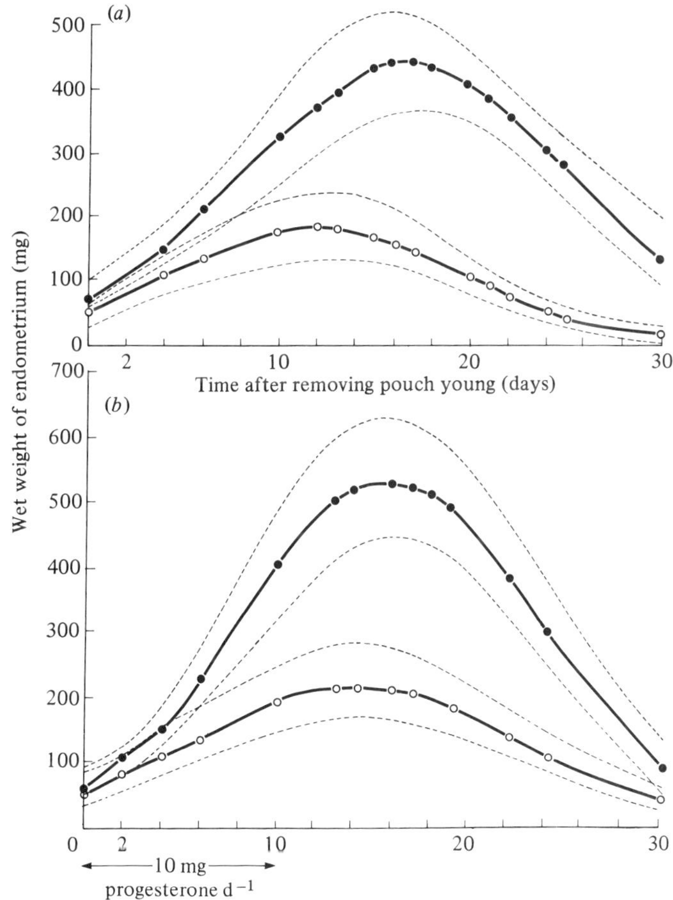

progesterone-receptor concentration between the two uteri of *S. brachyurus*, and a similar unilateral response of the uterus to pregnancy has been observed in this species (Wallace, 1981) and in *Potorous tridactylus* by Shaw & Rose (1979). So far these effects are only known in macropodids; in monovular non-macropodids, such as *Trichosurus vulpecula*, both gravid and non-gravid uteri respond to a similar extent, and in polyovular species, like *Didelphis virginiana*, no assessment can be made since both uteri are gravid, but a few values from non-pregnant animals (Renfree, 1975; Fleming & Harder, 1981*a*, *b*) suggest that the response is the same during the oestrous cycle and pregnancy (see Fig. 6.4).

Endocrine functions of the placenta

While the morphogenetic changes just described appear to be associated with the presence of the placenta, they do not necessarily imply an endocrine role for it. Lemon (1972) offered some evidence that the concentration of progestins in peripheral plasma of *M. eugenii* was higher in late pregnancy than at the same post-oestrous stage of the non-pregnant cycle but later measurements have not confirmed this (Fig. 6.11 and Fig. 6.13*c*). Nevertheless, the placenta could provide a local endocrine stimulation to the adjacent uterus without it being detected in the peripheral circulation, and this could be important in enabling pregnancy to continue to full term after ablation of the corpus luteum or ovariectomy in several marsupials (see Table 7.4).

Incubations of the meagre amount of placental tissue available (< 100 mg) from *S. brachyurus* and *M. eugenii* have shown that they have incipient endocrine activity by converting pregnenolone to progesterone (Bradshaw *et al.*, 1975; Renfree & Heap, 1977). A more detailed study of the yolk sac membrane of *M. eugenii* showed that it can convert a range of steroid precursors into a variety of products (Fig. 7.25), although levels of conversion were generally low (Heap, Renfree & Burton, 1980). Small amounts of pregnenolone were converted to progesterone and pregnanediol; androstenedione was metabolised to 5 α-androstane-3, 17-dione and androsterone, but was not incorporated into oestrogens. There was therefore no evidence for the enzymes arylsuphatase or sulphotransferase in the placenta. It thus appears that the yolk sac membrane, while capable of limited progesterone synthesis, has enzymes associated predominantly with steroid catabolism (Heap *et al.*, 1980). By contrast, endometrial and ovarian tissues of *M. eugenii* have a much greater capacity for steroid synthesis and metabolism (Heap *et al.*, 1980; Callard, Petro & Tyndale-Biscoe, 1982; Renfree *et al.*, 1984*b*). Although some parallels can be seen between steroid metabolism in the yolk sac placenta of *M. eugenii* and those of some eutherian mammals with superficial implantation, an important difference is that aromatase activity, which is pronounced in the placentae of pig and horse (see Gadsby, Heap & Burton, 1980), is undetectable in *M. eugenii* (Heap *et al.*, 1980; Renfree *et al.*, 1984*b*).

It has generally been assumed that the marsupial placenta has no gonadotrophic activity because the corpus luteum is autonomous. However, preliminary results, using the mouse uterine-weight assay of Hobson (1983), show that placental material from *M. eugenii* has biological activity

Fig. 7.25. Suggested interactions of steroid hormones in *Macropus eugenii* during pregnancy, based on incubation *in vitro*. Tissue was obtained at various days after removal of pouch young and incubated for 3 h at 37 °C with labelled steroids. After incubation, tissue and medium were extracted and the steroid fractions applied to thin-layer chromatograms which were developed in the solvent system, methylene dichloride: methanol (75:30, v/v). AND, androstenedione; AS, androsterone; ASD, androstanedione; CL, corpus luteum; DHA, dehydro-epiandrosterone; E_1, oestrone; $E_2\beta$, oestradiol-17β; Preg, pregnenolone; P, progesterone. Based on data in Heap *et al.* (1980) and Renfree *et al.* (1984*b*). Redrawn from Renfree (1980*c*).

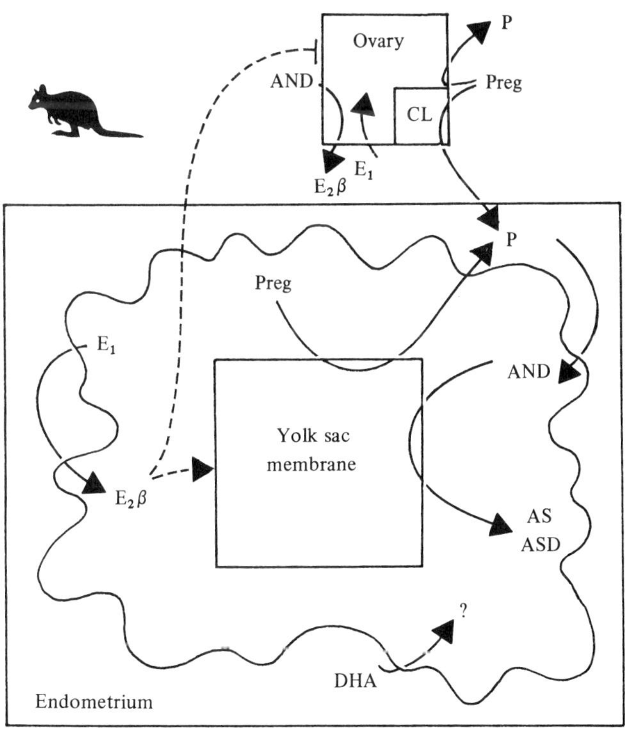

(M. B. Renfree & B. M. Hobson, unpublished results). Whether this is due to a non-specific effect or a genuine gonadotrophin must await confirmation by more specific radioimmunoassay.

No study has yet been made of the endocrinology of the chorioallantoic placenta of the Peramelidae. It is highly invasive, as in many eutherian mammals, and it is retained in the uterus after parturition. Close (1977) suggested that it may have a luteotrophic role on the corpora lutea during lactation but this has not been investigated yet.

Two other observations indicate that the feto-placental unit of macropodid marsupials may influence the maternal hormones systemically. First, the lengths of gestation in two species of kangaroo, *Macropus giganteus* and *M. fuliginosus* are respectively 36 and 31 days (Table 2.2) but the gestation period of 22 *M. giganteus* females mated to *M. fuliginosus* males was intermediate at 34 days (Kirsch & Poole, 1972; Poole, 1975), reflecting the hybrid genotype of the fetuses. The female progeny themselves had oestrous cycles intermediate in length between the parent species and, when back-crossed to *M. fuliginosus* males, had even shorter gestations (32 days, $N = 12$) (Poole, 1975). Second, Merchant's (1979) demonstration that gestation advances the time of oestrus by up to 3 days in *M. eugenii* and other species (Table 6.4) and that associated changes in four hormones (Fig. 7.27) are also advanced (Tyndale-Biscoe *et al.*, 1983) is strong evidence for a feto-placental influence operating in the first instance on the corpus luteum and indirectly on the pituitary thereafter.

All these effects may be said to be a maternal recognition of pregnancy and some or all may be due to endocrine signals from the placenta, although direct evidence for this has not yet been obtained. As the vascular anatomy of the reproductive tract (Fig. 5.11) would enable some of these to be local effects (see Chapter 5) high concentrations of hormones may not be necessary. It is of interest that laparotomy alone at the time of attachment of the yolk sac placenta (days 18 and 19) interferes with subsequent parturition in *Setonix brachyurus* and *M. eugenii*, which may further implicate the placenta as an important component in normal gestation (Young & Renfree, 1979). This form of maternal recognition of pregnancy is probably restricted to the Macropodidae, as recent studies on *D. virginiana* (Harder & Fleming, 1981) and *Dasyuroides byrnei* (Fletcher, 1983) disclosed no evidence for an endocrine recognition of pregnancy in either species. However, behavioural aspects of late pregnancy indicate some maternal recognition of pregnancy in many marsupials; the behaviour of females at parturition is specialised (Fig. 2.23*a*) and additional to that seen in the non-pregnant cycle, as discussed in Chapter 2.

Immunosuppressive function of the placenta

As we have seen the shell membrane, a maternally derived structure, is not ruptured until late in gestation, whereas the subsequent period, when fetal and maternal tissues are in close apposition, is very brief (Fig. 7.15). Yet it is in the latter period that organogenesis sufficient for life outside the uterus is accomplished. Moors (1974) suggested that marsupials may not have evolved a trophoblast in which transplantation antigens are masked, and the brief period of organogenesis was necessary for gestation to be completed before any immune rejection of the now-exposed fetal tissues could occur. A corollary of this is that females mated successively to the same male, or females made hypersensitive to a male mate by prior skin grafting, should develop an accelerated second set rejection of their late-term fetuses and thus become progressively infertile. Both these consequences were examined in *M. eugenii* by Walker & Tyndale-Biscoe (1978) and Rodger, Fletcher & Tyndale-Biscoe (1986). Even though the females in both studies became hypersensitive to the males after two successive skin grafts, there was no impairment of their fertility, nor was there any impairment of fertility in a group of females mated to the same male for 5 years. These results are comparable to the results of similar experiments with eutherians, designed to compromise pregnancy by antipaternal immunisation (see Billingham & Head, 1983) and provide no support for the hypothesis that the marsupial trophoblast lacks immunosuppressive properties. Rodger *et al.* (1986) conclude that this unique property of the trophoblast must have appeared in mammalian evolution before the dichotomy of Eutheria and Metatheria, a point to which we will return in Chapter 10.

Parturition

The behavioural responses which precede parturition presumably reflect the numerous, internal endocrine signals which are essential for preparation for birth. Although the full sequence of endocrine changes are still to be defined, the present evidence suggests that the control of parturition and the initiation of lactation are similar in marsupials and eutherians.

In marsupials, the corpus luteum and both anterior and posterior pituitary are necessary for successful parturition and there is growing evidence, in some species at least, that the fetus and placenta may also be involved. Marsupial neonates are very small (Table 2.1), ranging between 4 mg in *Tarsipes rostratus* (Renfree, 1980*a*) to 750 mg in the large kangaroos (Poole, 1975), but the full-term fetus has to travel a long way after it has left the uterus and before it emerges at the urogenital opening (see Chapter 5). Removal of the inhibition to myometrial contraction, as well as softening of the cervix and pseudo-vaginal canal are important prerequisites for this process. Thus, understanding parturition requires an examination of the events preceding it. These may be divided into myometrial activity during pregnancy, preparation of the birth canal, and the combined endocrine effects of the maternal and feto-placental systems.

Myometrial activity during pregnancy in M. eugenii

Myometrial activity was assessed by electro-uteromyography (Renfree & Young, 1979) or by measuring the intrauterine pressure (Shaw, 1983*b*) of the gravid and non-gravid uteri of pregnant animals at different stages of pregnancy and immediate post-partum. During quiescence there were frequent (3–10 min) small contractions of the myometrium, but up to mid-pregnancy (days 13–18) there was little spontaneous activity from either the gravid or non-gravid uterus of pregnant animals, although non-gravid uteri were more active. Whereas the quiescent uterus showed no response to injected oxytocin, the gravid uterus at day 20 gave a small response (Fig. 7.26*b*). In late pregnancy no response was seen in the non-gravid uterus, but the gravid uterus showed increased contractility and sensitivity to oxytocin (Fig. 7.26*b*) but, by day 1 post-partum, both uteri were inactive and refractory to oxytocin (Shaw, 1983*b*).

The prostaglandin analogue Cloprostenol also caused contractions in the gravid uterus on day 20 and 25 with an immediate rise in intrauterine pressure (Fig. 7.26*a*; Shaw, 1983*a*). Again, the myometrium was refractory to Cloprostenol after parturition. Prostaglandin $F_{2\alpha}$ induced a rise in tonus of both gravid and non-gravid uteri, sufficient to cause abortion (Renfree & Young, 1979; Shaw, 1983*b*). These effects are similar to those seen in Eutheria, where prostaglandin $F_{2\alpha}$ is important in stimulating uterine contractions at parturition (Thorburn, Challis & Robinson, 1977). However, prostaglandin release in *M. eugenii* is very brief as we shall see in a later section.

Preparation of the birth canal

At parturition the young passes through the cervix to the median vaginal cul de sac and thence via the pseudo-vaginal canal or median vagina to the urogenital sinus (see p. 179).

In eutherians, softening of the cervix and vaginal canal may be induced by relaxin (see Bryant-Greenwood, Niall & Greenwood, 1981), but the role of relaxin in marsupials is less clear. Corpora lutea of *M. eugenii* and *T. vulpecula* contained high concentrations of relaxin activity, by the mouse pubic-symphysis bioassay, in the second half of gestation but not in early gestation or 1–5 days post-partum (Tyndale-Biscoe, 1969; 1981). *S. brachyurus* females ovariectomised in late pregnancy and treated with high doses of oestradiol or of Pituitrin® (an oxytocin preparation) did not give birth, but 2 of 9 treated with lutrexin (a porcine relaxin preparation) gave birth, but the young did not survive (Tyndale-Biscoe, 1963*b*). However, this result may have been compromised by the time of ovariectomy, which was done on days 18–20 in the relaxin-treated group, compared with day

Fig. 7.26. Myometrial activity in *Macropus eugenii*. (*a*) Effect of Cloprostenol on intrauterine pressure during late pregnancy (day 20 and day 25) and 1 day post-partum. Recordings were made from the gravid uterus or the recently evacuated uterus. Cloprostenol (0.5 μg) was injected into the lateral tail vein (arrow). The vertical scale is the same for all traces but the base line differs. Redrawn from Shaw (1983*a*). (b) Response of the myometrium of gravid (G) and non-gravid (NG) uteri to 50 mU or 100 mU oxytocin injection. At day 20, neither uterus showed much spontaneous activity, but the gravid uterus responded strongly to oxytocin both *in vivo* (shown here) and *in vitro*. At day 25, spontaneous activity in both uteri is slight. Oxytocin injections into the tail vein induced strong contractions in the gravid uterus but the non-gravid uterus responded weakly. In a female with a 1 day old pouch young, spontaneous myometrial activity was slight. Oxytocin (100 mU) produced an increase in tonus of the post-partum uterus, but the stimulation was less pronounced than seen in pre-partum uteri. Redrawn from Shaw (1983*b*).

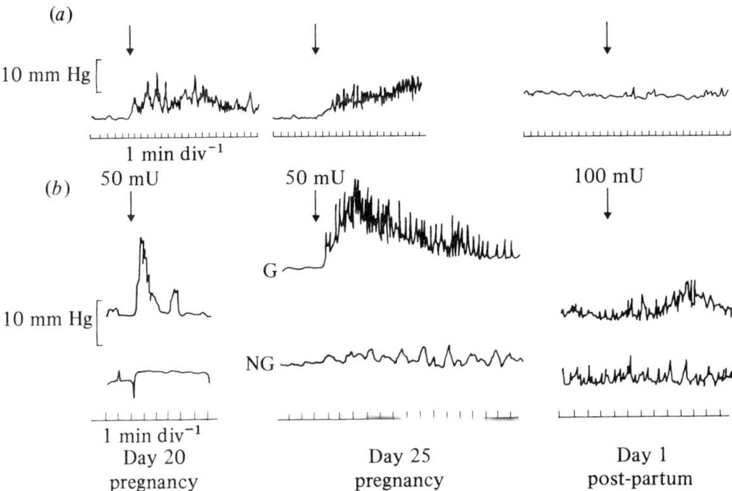

16–18 for the oestradiol-treated group and day 17–18 in the oxytocin-treated group. In *M. eugenii*, lutectomy (Table 7.4) or even laparotomy, at mid-pregnancy will cause loss of the fetus or failure to survive (Young & Renfree, 1979). Also, both oestradiol and oxytocin promote uterine contractions which may have caused the observed death of the embryos *in utero*.

In *M. eugenii*, no convincing effect of porcine relaxin on parturition could be demonstrated (Renfree & Young, 1979) and in *T. vulpecula* there was no apparent effect of relaxin, either following oestrogen or in combination with progesterone, on the urogenital strand, where the pseudo-vaginal canal should form (Tyndale-Biscoe, 1966). Although these results do not support a role for relaxin in parturition, the experiments have had to assume that porcine relaxin, the hormone preparation administered, will bind to relaxin receptors in marsupial tissues.

There is better evidence, however, that progesterone, perhaps acting with oestrogen (and relaxin?), may soften the connective tissue surrounding the birth canal before parturition. Treatment with oestradiol and progesterone caused substantial loosening of the connective tissue of the urogenital strand in ovariectomised *D. virginiana* (Risman, 1947) and *T. vulpecula* (Tyndale-Biscoe, 1966). Conversely, ovariectomy performed in late pregnancy prevented parturition in *D. virginiana* (Hartman, 1925*b*; Renfree, 1974*a*) although fetal development continued. Lutectomy or ovariectomy likewise prevented parturition in the three Australian marsupials so far studied.

Sharman (1965*a*) reported that *T. vulpecula* gave birth after lutectomy on day 11 of the $17\frac{1}{2}$ day gestation but that lutectomy at earlier stages led to death of the intrauterine embryos. However, when females were injected with progesterone for 2–3 days after lutectomy on day 7 they gave birth. It is unclear from these experiments whether progesterone facilitated parturition or merely allowed pregnancy to continue to term. In *S. brachyurus*, ovariectomy or lutectomy from day 6 to 18 did not prevent development to full term but invariably prevented birth (Table 7.4), and fetuses were recovered dead in the vaginal culs de sac (Tyndale-Biscoe, 1963*b*). This implied a role for the corpus luteum in facilitating passage through the birth canal, a conclusion supported by other evidence from *M. eugenii*. When embryos in diapause were reactivated by exogenous progesterone, the associated corpus luteum remained inactive and, although development in many cases went to term, parturition usually failed (Renfree & Tyndale-Biscoe, 1973*a*). Again fetuses were impounded in the vaginal cul de sac, reinforcing the idea that the corpus luteum is essential for preparation of the genital tract, and the median vaginal canal in particular, for birth.

The timing of the luteal influence for normal parturition was narrowed down by Young & Renfree (1979) who showed that the corpus luteum is essential at least up to day 21 of pregnancy, that is to say until 5 days before birth (Table 7.4). After lutectomy on days 17 or 21, some pregnant animals delivered, but none of the young survived beyond 24 h post-partum, whereas animals lutectomised on days 23 or 25 had produced neonates by day 27, which survived. As well as emphasising the importance of the corpus luteum for parturition, these results also suggest that the corpus luteum is necessary at least until day 23 for normal development of the mammary gland (see Chapter 8).

Endocrine changes at parturition

The essential role of the corpus luteum is presumably due to the influence of progesterone (perhaps acting with relaxin). Prior to birth in all species so far studied, progesterone is elevated and is the major luteal steroid hormone of *D. virginiana* (Cook & Nalbandov, 1968), *T. vulpecula* (Shorey & Hughes, 1973*b*; Curlewis *et al.*, 1985), and *M. eugenii* (Renfree *et al.*, 1979). Progesterone concentration in peripheral plasma falls prior to parturition in all species studied (Chapter 6), except in *Isoodon macrourus* (Fig. 6.10). In *M. eugenii*, the fall in progesterone is very rapid (Fig. 7.27) and takes less than 8 h (Hinds & Tyndale-Biscoe, 1982*a*; Tyndale-Biscoe *et al.*, 1983), and it was suggested that progesterone withdrawal might be an essential signal for the onset of parturition.

This idea was tested by artificially elevating the progesterone levels during the peripartum period, by either implants (up to 2 ng ml^{-1}) or injections (up to 50 ng ml^{-1}) (Ward & Renfree, 1984). However, neither treatment prevented parturition occurring at about the expected time in about two-thirds of the animals and the neonates were of normal size. The remaining third of treated animals either retained their fetuses or lost them, presumably by abortion, so elevated progesterone can interfere with the normal course of development and birth in a proportion of animals.

Several workers have suggested that the oestrogen:progesterone ratio may be more important at parturition than the actual level of either hormone (Liggins *et al.*, 1977), and that an oestrogen-dominated uterus would be more sensitive to the action of prostaglandin and oxytocin. In *M. eugenii* the progesterone fall coincides with a rise in oestradiol, so there is a marked change in the ratio of the two hormones (Shaw & Renfree, 1984; Harder *et al.*, 1985). However, parturition still occurred when this ratio was not permitted to change, either by elevating progesterone (Ward & Renfree, 1984) or by abolishing the oestradiol rise (Harder *et al.*, 1985).

Although oestradiol is elevated in *M. eugenii* at post-partum oestrus, it is clearly not essential for parturition to occur since oestrus does not invariably follow parturition (Tyndale-Biscoe *et al.*, 1974) and parturition occurred in animals that had been immunised against GnRH and lacked large follicles (Short *et al.*, 1985). Likewise in other species, such as *Macropus giganteus* and *M. fuliginosus*, follicular oestradiol presumably is not necessary for parturition since oestrus does not occur post-partum (Poole, 1975), and parturition occurred in *M. rufus* that had entered anoestrus during pregnancy (Newsome, 1964*a*).

From the progesterone profiles and the effects of lutectomy, it might be expected that the four marsupial species examined are dependent on the corpus luteum for parturition, as the goat is (Currie & Thorburn, 1977),

Fig. 7.27. Hormones at parturition and oestrus in *Macropus eugenii*. Concentrations in plasma (mean ± s.e.m.) of progesterone (●), prolactin (▲) and prostaglandin (PGFM, ■) in six females sampled 48–72 h before and after (*a*) post-partum oestrus and (*b*) oestrus. The occurrence and size of the transient LH peaks for each animal are shown as separate vertical bars and the occurrence of birth in relation to oestrus as a histogram. Redrawn from Tyndale-Biscoe *et al.* (1983).

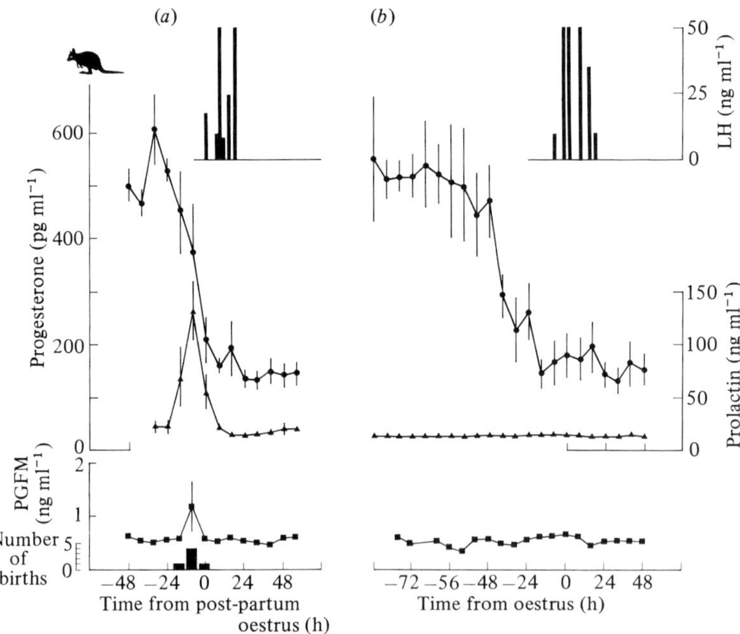

and that prostaglandins are important for uterine evacuation. If so, the feto-placental unit might also be involved. There is some evidence on both of these aspects.

A metabolite of prostaglandin $F_{2\alpha}$ (PGFM) has been found in peripheral plasma of three marsupials – *Isoodon macrourus* (Gemmell *et al.*, 1980), *M. eugenii* (Shaw, 1983*a*; Tyndale-Biscoe *et al.*, 1983) and *M. rufogriseus* (Walker & Gemmell, 1983*a*). In *I. macrourus*, which does not have a peripartum decline in progesterone, there was a well-defined peak of PGFM, extending at least 1 day either side of parturition (Gemmell *et al.*, 1980). In *M. rufogriseus*, none of the four animals measured once a day had a significant increase of PGFM at parturition (Walker & Gemmell, 1983*a*). However, in one-third of *M. eugenii* sampled every 6–8 h around the time of parturition an elevation of PGFM was detected, suggesting that in this species prostaglandin is only very transiently elevated (Tyndale-Biscoe *et al.*, 1983; Shaw, 1983*a*) (Fig. 7.27).

In another study plasma was collected at short intervals (10 min) before and after birth (Fig. 7.28) and peak concentrations of PGFM (up to 3 ng ml^{-1}) were only observed in the 20 min before birth and had declined to 200 pg ml^{-1} within 45 min, and were undetectable by 2 h post-partum (Lewis, Fletcher & Renfree, 1986). Clearly therefore in this species prostaglandin is elevated during parturition, but the peak is very short-lived.

Both the endometrium and myometrium of *M. eugenii* contain the prostaglandins $PGF_{2\alpha}$ and PGE_2, and the concentrations increase between day 25 and day 27, immediately pre-partum (Shaw, 1983*b*). At day 26 the extremely high concentrations, especially of PGE_2, suggest that these tissues can rapidly synthesise large quantities of this hormone (Shaw 1983*a*). There is therefore a potential source of prostaglandin available to act as a myometrial stimulator at parturition.

At present it is not known how the secretion of prostaglandin is regulated and whether the fetus or placenta is involved. There is some evidence, referred to earlier in this chapter (p. 308), that the adrenal cortex of the fetus has differentiated sufficiently to synthesise cortisol by the end of gestation, but whether it has a similar role to that of the fetal adrenal of the sheep in initiating parturition, is not known. The one attempt to delay parturition in *M. eugenii* with Metapyrone, an inhibitor of cortisol synthesis, gave equivocal results (Renfree & Young, 1979).

The fetus or placenta of *M. eugenii* may provide a stimulus to the anterior pituitary immediately before parturition, which is important in initiating parturition. About 4 h before parturition in this species there is a pronounced but transient peak of prolactin (Fig. 7.27), which does not occur at the equivalent stage of the non-pregnant cycle, and may therefore be evoked by the conceptus (Tyndale-Biscoe *et al.*, 1983). As it coincides with the precipitate fall in progesterone, the prolactin pulse may be luteolytic, as suggested in Chapter 6 (p. 255), but how this might be involved in parturition is not clear. Since lutectomy also causes a precipitate fall in progesterone (Findlay *et al.*, 1983) without initiating parturition, some additional factor must be involved.

Both parts of the pituitary are essential at parturition in *M. eugenii*. After hypophysectomy in mid-pregnancy the corpora lutea continued to grow and the fetuses are carried to full term but die in the uterus (Hearn, 1973, 1974). Since the levels of progesterone are not affected by hypophysectomy (Fig. 6.18), the failure of parturition is not due to lack of luteotrophic

Fig. 7.28. Prostaglandin metabolite in the plasma of peripartum *Macropus eugenii*. Concentrations (mean ± s.e.m.) were undetectable or low until 1 h before birth, but 10 min before, during and after birth they reached 4 ng ml^{-1}. By 1 h post-partum the concentrations were approaching the assay sensitivity of 50 pg ml^{-1}. From Lewis, Fletcher & Renfree (1986).

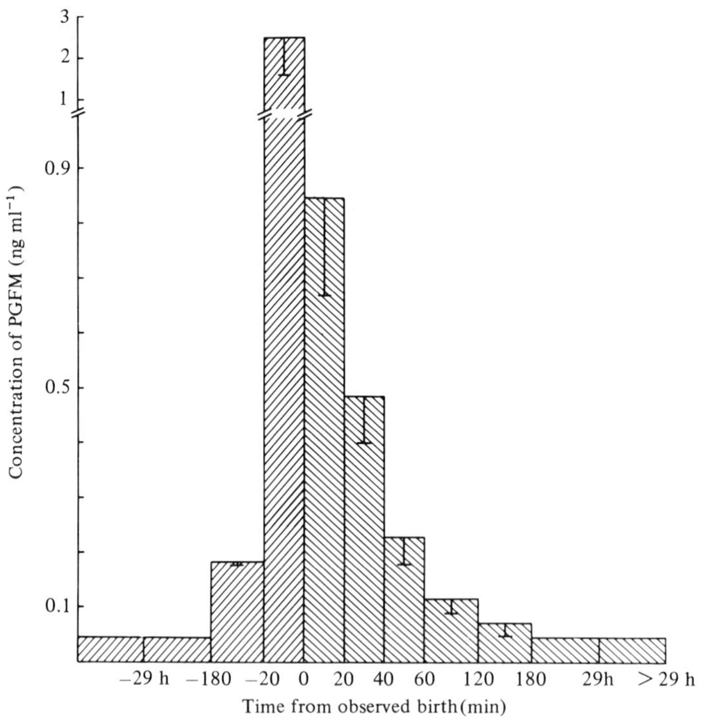

support; nor it is due to lack of gonadotrophic stimulation, since females immunised against GnRH delivered their young (Short *et al.*, 1985). This lends more support to the conclusion that the pre-partum pulse of prolactin is important. In animals in which the neurohypophysis was left intact the uteri were evacuated but parturition was not successfully completed, which suggests that oxytocin, or a similar peptide, is necessary for this first part of parturition and, as discussed earlier, the pre-partum uterus is very responsive to circulating oxytocin (Fig. 7.26*b*).

Only one other species of marsupial (*T. vulpecula*) has been hypophysectomised during pregnancy and the results were similar to those in *M. eugenii*, in that pregnancy went to full term and the fetuses died in the uterus or median vaginal culs de sac. There was some suggestion in this species, however, that the corpus luteum may have been affected (Stewart & Tyndale-Biscoe, 1983).

Conclusions

Despite its relatively short duration and the small size of the product, gestation in marsupials is not a passive incubation but a vigorous interaction between embryo and mother, as in other mammals. The rates of cleavage and embryogenesis are comparable to those of other mammals and the main distinctions are the absence of early differentiation of presumptive embryo and extraembryonic tissue and the retention of the shell membrane until embryogenesis is complete. The phenomenon of diapause involves complex interactions, especially during reactivation, and the evolutionary or adaptive significance of diapause appears to be the need to synchronise embryo development with endometrial sufficiency to sustain it. In this respect, the phenomenon may be widespread even in species that appear to have continuous development. Apart from the intimate attachment in Peramelidae, implantation in most marsupials is rather tentative but the roles of the fetal membranes are complex and may encompass endocrine function and the masking of antigenic determinants, as in eutherian mammals.

Organogenesis in the last third of gestation is remarkably rapid, nowhere more so than in *Perameles* with its chorioallantoic placenta. So fast are the processes and so short is gestation that it is only in the larger species of a macropodid that a fetal role in the initiation of parturition can be demonstrated, but the way in which this is achieved, remains unclear.

Throughout the whole process the profound importance of the corpus luteum is clear; from its initial signal to the blastocyst, through its effects on the endometrium and myometrium, to its involvement in parturition, it is the primary endocrine organ of gestation. As we will show in Chapter 8, its role extends to the stimulation of mammogenesis prior to parturition as well. Despite its profound importance, the corpus luteum is largely autonomous. The pituitary is necessary for its formation but neither it nor the placenta are required for luteal function, nor is the uterus luteolytic. This contrasts with eutherian mammals in which the corpus luteum is usually subordinate to the pituitary, the placenta or the uterus, and it may reflect the relative importance of pregnancy and lactation in these two groups of mammals.

8

Lactation

Lactation is a fundamental characteristic of all mammals and has been a major factor in the evolution of mammalian patterns of reproduction (Pond, 1977, 1983). Despite their separate evolution the cellular processes of mammary development appear to be very similar in monotremes, marsupials and eutherians (Bresslau, 1912, 1920; Griffiths *et al.*, 1972; Griffiths *et al.*, 1973; Griffiths, 1978; Cowie, 1984), although the particular lactational strategies of each taxon differ. The young of monotremes and of marsupials are much less developed than the most altricial young of eutherian species and lactation is proportionally much longer. Monotremes lack teats, the galactophores opening on two areolae, and both the mammary glands secrete milk during lactation; among marsupials, however, each young initially has exclusive use of one teat and only the mammary glands associated with these teats lactate. After the first phase of continuous attachment the young begin to release and reattach to the teat and littermates may move from one to another of these teats. During this phase the young may be carried in the pouch or left in a nest, depending on the species (Russell, 1982*a*).

The composition of the milk of marsupials, unlike that of eutherians, changes substantially through lactation (Green, 1984) and the size of the lactating mammary gland enlarges manyfold, these changes being, presumably, necessary for the considerable growth and development that the young undergoes during this phase of its life. Three broad stages of lactation in marsupials can thus be recognised: development of the mammary glands during pregnancy; initiation and maintenance of lactation during the first, dependent phase; and increased growth of lactating glands and their production of mature milk during the later rapid growth phase of the young before weaning.

Mammary gland development and growth

Differentiation of the mammary gland

Bresslau (1912, 1920) reviewed his own and other studies on the differentiation of the mammary glands in marsupials and established the relationships between the pouch region, the mammary glands and their teats and the underlying musculature. In all marsupials the teat anlagen are formed in genetically female young a few days after birth. Rarely they may form in males of some species, such as *Didelphis*, *Monodelphis* and *Trichosurus* illustrated by Bresslau (1912). The teats do not differentiate from a mammary line, as in Eutheria, but each begins as a separate anlage, the number varying from 2 to 25 in concordance with the adult

Fig. 8.1. Diagram of teat formation in marspials. (*a*) Teat formation begins as a thickening of the epidermis (black) to form the mammary anlage (hatched); asterisks show the boundary between anlage and epidermis. (*b–d*) Teat formation and eversion. The main part of the surface of the teat is produced from the anlage which first forms a pocket and then everts. gl, galactophore; h, mammary hair; k, keratin capping; mg, mammary gland anlage; sg, sebaceous gland anlage. Redrawn from Bresslau (1912).

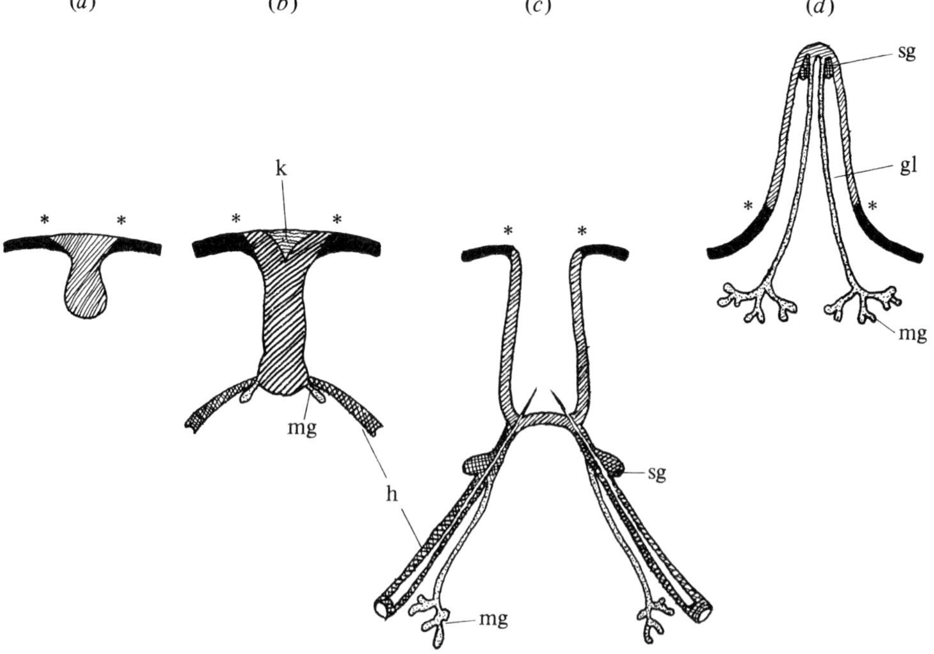

teat number for the species. The anlage begins as a local proliferation and consequent thickening of the epidermis in the presumptive pouch region (Fig. 8.1*a*) and then the special mammary hair follicles and their associated glands are formed by downgrowths of epidermal cells into the subdermal tissue. At the periphery of each anlage these glands become sebaceous glands, while the central ones become the primordia of the mammary glands (Fig. 8.1*c*). At first they are solid cords but later become tubular and form the ducts and acini. Such specialised mammary hairs do not occur in the developing mammary glands of eutherian mammals and are peculiar to marsupials and monotremes. They are retained until puberty, when they are shed at teat eversion (Fig. 8.1*d*).

In polytocous species, such as *Dasyurus viverrinus*, O'Donoghue (1911) observed that there were six mammary hairs in each teat anlage and, since each hair follicle gave rise to a mammary duct, the number of ducts or galactophores in the teats of adults was also six. Similarly, in *Marmosa robinsoni* there are three galactophores in each teat (Barnes, 1977) and in *Didelphis* eight (Bresslau, 1920). Conversely in monotocous species, which have fewer teats, the number of galactophores is much greater; in *Phascolarctos cinereus* there are 24 (Bresslau, 1920). In the macropodids (Fig. 8.2) the number varies, being 15–20 in *M. eugenii* (Findlay, 1982*a*), 24–26 in *M. rufus* (Griffiths *et al.*, 1972) and 20–33 in *M. agilis* (Lincoln & Renfree, 1981*b*). From each duct tubules branch off, which are embedded in connective tissue stroma and a fat pad, and they are lined by a double layer of cuboidal cells. Thus in the adult each galactophore drains one lobule of the mammary gland, a fact which can be used for experiments on hormonal response *in vivo* (see Fig. 8.12 and 8.13).

Before puberty the mammary glands consist of a few branching ducts without alveoli and the teats are inverted (Fig. 8.1 d). Eversion of the teats and loss of the mammary hairs heralds the onset of sexual maturity (Bresslau, 1920). In anoestrous females the glands never revert wholly to the pre-pubertal state; the teats remain everted and the glands comprise primary and secondary ducts with some small alveoli (Sharman, 1962; Findlay, 1982*a*, Stewart, 1984).

The major development of the mammary gland, however, begins during the first half of pregnancy or the equivalent stage of the non-pregnant cycle. These changes were first described in *Dasyurus viverrinus* by O'Donoghue (1911) and were subsequently described in *Didelphis virginiana* (Hartman, 1923*a*), *Trichosurus vulpecula* (Sharman, 1962), *Macropus rufus* (Griffiths *et al.*, 1972) and *M. eugenii* (Findlay, 1982*a*; Stewart, 1984). Some proliferation occurs before ovulation (Hartman, 1923*a*) but according to

Fig. 8.2. (*a*) Longitudinal and (*b*) transverse section of a teat from *Macropus eugenii* during lactation, showing 19 galactophores and lack of an obvious teat sphincter. Photomicrographs kindly provided by Dr A. T. Cowie and A. Turvey.

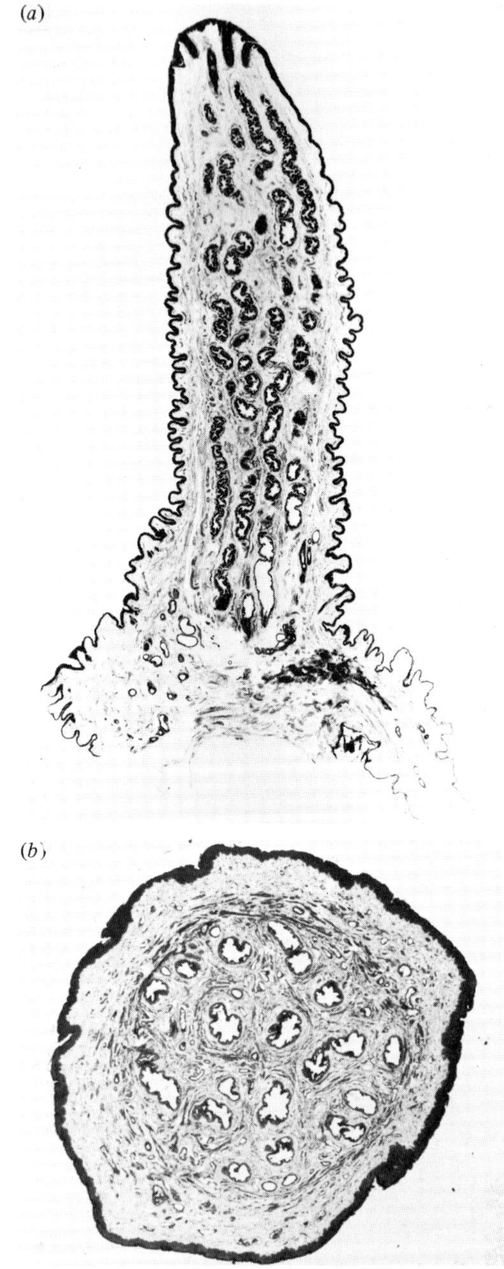

O'Donoghue (1911) proliferation of the epithelial cells is most intense in the post-ovulatory stage, when the embryos are unattached vesicles, while in late pregnancy most of the growth is due to hypertrophy and distension of the alveoli with secretory products. By day 10 of pregnancy in *Macropus eugenii*, a lobular pattern of development is apparent, with clusters of alveoli surrounding central bilaminar ducts (Findlay, 1982*a*). By day 15 this structure is well defined, and the alveoli have small lumena lined by a single layer of cuboidal epithelial cells. Towards the end of pregnancy (day 27) secretion is accumulating in the lumena of some alveoli. A similar condition was described in the mammary glands of *M. rufus* at oestrus and at parturition by Griffiths *et al.* (1972). The tubules are grouped into lobules surrounded by connective tissue and myoepithelial cells and

Fig. 8.3. Growth of the mammary gland during lactation. (*a*) The weight of the sucked mammary gland during lactation and post-lactation in *Trichosurus vulpecula*. Most had weaned by day 130–140. (*b*) Mammary gland weight of lactating *Macropus eugenii* showing sucked (●) and non-sucked (○) glands. Redrawn from (*a*) Smith *et al.* (1969) and Sharman (1962), (*b*) Stewart (1984).

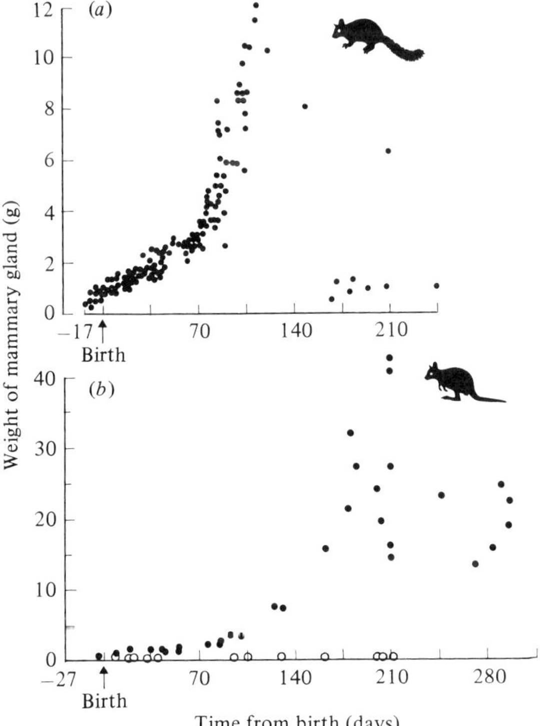

swathes of striated fibres of the muscle, *m. ilio-marsupialis*, penetrate to half the thickness of the glands (Fig. 8.4*a*).

These histological changes are reflected in the increase in size of the mammary glands at the end of pregnancy (Fig. 8.3) or the equivalent post-oestrous stage (Hartman, 1923*a*; Sharman, 1962; Tyndale-Biscoe, Stewart & Hinds, 1984), and the secretion of a clear or pellucid fluid from the teats (Morgan, 1829; O'Donoghue, 1911; Sharman, 1962; Griffiths *et al.*, 1972).

In all these species the histological changes, as well as the enlargement of the glands was found to be the same in non-pregnant animals as in pregnant animals at the same post-oestrous stage. O'Donoghue (1911) was the first to remark on it and Sharman (1962) the first to demonstrate experimentally that the non-pregnant gland is functionally developed and can sustain a young transferred to the teat of a non-pregnant female on the day of its birth. This has since been done in other species and between species (Sharman & Calaby, 1964; Merchant & Sharman, 1966; Clark, 1968*b*; Findlay, 1982*b*; Tyndale-Biscoe *et al.*, 1984).

After parturition only those glands to which the young become attached, enlarge and lactate. The other glands quickly regress to the anoestrous size (Sharman, 1962; Stewart, 1984). For the first 2 or 3 days after parturition the non-sucked glands have the same appearance as the sucked gland but, by day 7, regressive changes have begun (Fig. 8.4*a*); by day 9, the alveoli have become very small with shrunken epithelial cells (Griffiths *et al.*, 1972; Findlay, 1982*a*; Stewart, 1984). In the adjacent sucked gland the alveolar lumena expand, the epithelial cells enlarge and there is an increased rate of cell division at parturition and for the first week of lactation (Fig. 8.4*a*). However, in *M. eugenii* between days 21 and 65 of lactation hyperplasia was again low, as judged by infrequent mitoses and the reduced uptake of [^{3}H]thymidine *in vivo* (Findlay & Renfree, 1984) (Table 8.1). This may be reflected in the relatively small change in weight of the lactating gland in this species until about 140 days (Fig. 8.3*b*) (Stewart, 1984) and in *T. vulpecula* until 85 days (Smith *et al.*, 1969). Thereafter in both species the mammary gland grew rapidly from 3 g to 12 g at 140 days in *T. vulpecula* (Fig. 8.3*a*) and from 2.5 g to 35 g at 240 days in *M. eugenii* (Fig. 8.3*b*).

In longitudinal studies the growth of the mammary gland has been assessed in these and other species by use of a 'mammary index', usually the product of the two diameters but sometimes including a component for thickness as well. In *D. virginiana* (Reynolds, 1952), *T. vulpecula* (Sharman, 1962), *M. eugenii* (Findlay, 1982*a*; Green, 1984) and *M. agilis* (Lincoln & Renfree, 1981*a*) these indices showed corresponding increases

Fig. 8.4. Histology of the mammary gland of *Macropus rufus* during lactation. (*a*) Section of adjacent sucked (lower right) and non-sucked glands (upper left) at day 6 of lactation. (*b*) Section of a sucked mammary gland at day 312 of lactation, at same magnification. From Griffiths *et al.* (1972), by permission.

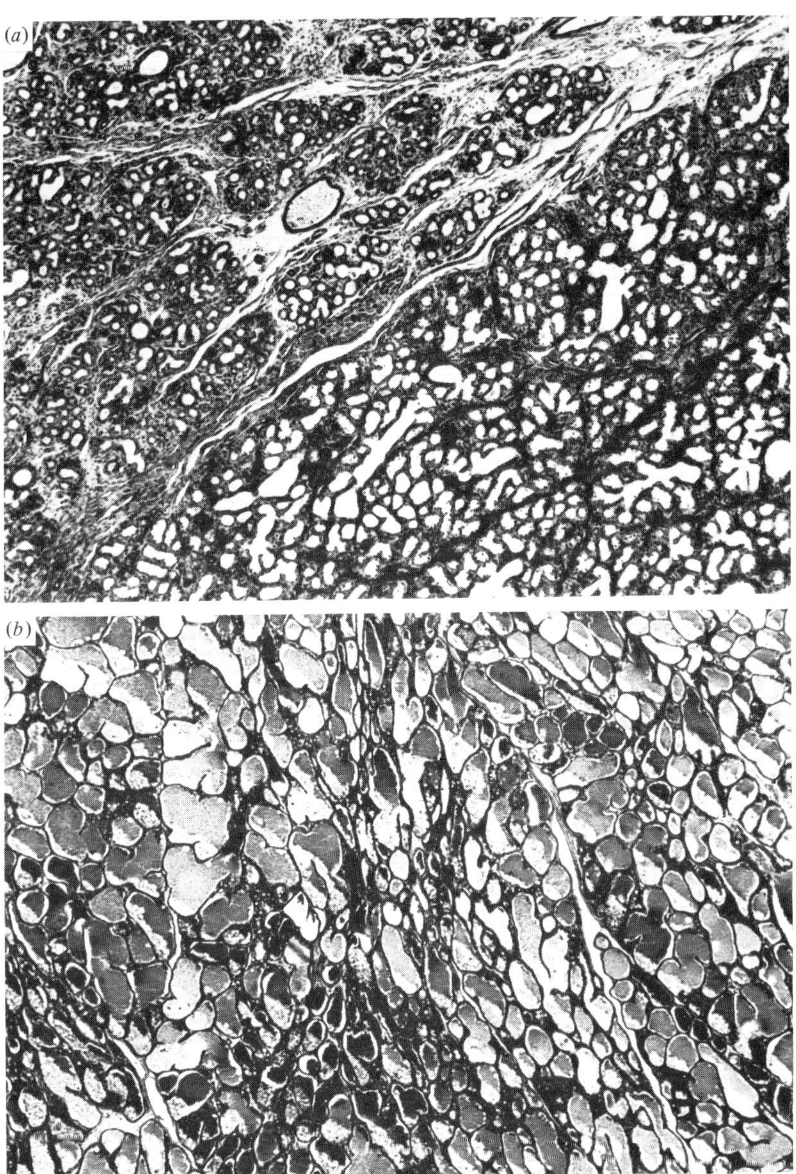

to a maximum 4–6 times greater than in early lactation. The large increase at mid-lactation in *M. eugenii* is associated with increased volume of milk secreted and with accelerated growth of the young and its development of homeothermy (Green, 1984). The question it raises is how this growth is achieved in a gland that is actively synthesising and secreting milk. Hyperplasia has not been observed at this stage and O'Donoghue (1911) expressly states that it is not involved in *D. viverrinus*. He believed that enlargement of the alveoli was the sole cause, and later studies in other species have confirmed that the alveolar diameter does increase markedly in late lactation (Fig. 8.4*b*; Fig. 8.5) when the young ceases to be permanently attached (Griffiths *et al.*, 1972, 1973; Griffiths, 1978; Findlay, 1982*a*). However, this may reflect the change from continuous to intermittent emptying of the gland with consequent accumulation of milk and alveolar distension. Lincoln & Renfree (1981*b*) and Tyndale-Biscoe *et al.*

Table 8.1. *Proportion of labelled to unlabelled cells in mammary glands exposed to tritiated thymidine* in vivo[a]

	Score of labelled cells			
	Sucked gland		Non-sucked gland	
Day of lactation	Alveolar cells	Stromal cells	Alveolar cells	Stromal cells
Day 1	1.0	< 0.5	2	0.5
Day 7	4.0	3.0	2	0.5
Day 7	2.0	1.0	3	2.0
Day 7	4.0	4.0	3	2.0
Day 21	2.0	1.0	< 0.5	< 0.5
Day 21	2.0	1.0	< 0.5	< 0.5
Day 65	< 0.5	< 0.5	< 0.5	< 0.5
Day 65	< 0.5	< 0.5	< 0.5	< 0.5
Non-lactating (lactational quiescence)			< 0.5	0.5
Non-lactating (seasonal quiescence)			< 0.5	< 0.5

[a] The mammary gland sections were examined using autoradiography. A score of 4 represents a labelling index of about 20 labelled nuclei per 100 cells. A score of 0.5 represents a labelling index of less than 2.5 labelled nuclei per 1000 cells. The labelled stromal cells included fibroblasts, endothelial cells and (probably) myoepithelial cells.
From Findlay & Renfree (1984).

(1984) considered that alveolar distension could not account for the whole growth of the gland and that hyperplasia as well as hypertrophy must occur. Stewart (1984) supported this conclusion from measurements of DNA concentration in mammary gland tissue. Although the glands at 240 days were many times larger than at 140 days the concentration of DNA per gram of tissue remained constant, which would not be the case if the growth was due to hypertrophy alone.

Not only does the mammary gland grow during lactation but so does the associated teat. In *D. viverrinus* it increases from 2–3 mm to 10 mm (O'Donoghue, 1911) and in *D. virginiana* from 1.5 mm to 35 mm at day 80 when the young first emerge from the pouch (Reynolds, 1952). Similarly, in *M. eugenii* (Fig. 8.7 *b*) it increases from 2–3 mm to 23 mm at day 180 (Findlay, 1982*a*) and in *M. agilis* from 6 mm to 40–50 mm at day 100 (Lincoln & Renfree, 1981*b*).

Fig. 8.5. The volume densities of alveolar cells (□) and connective tissue stroma (■) in mammary glands of *Macropus eugenii* during pregnancy and lactation. The average diameters of 35 alveoli measured at the same stages are also shown (●). Redrawn from Findlay (1982*a*).

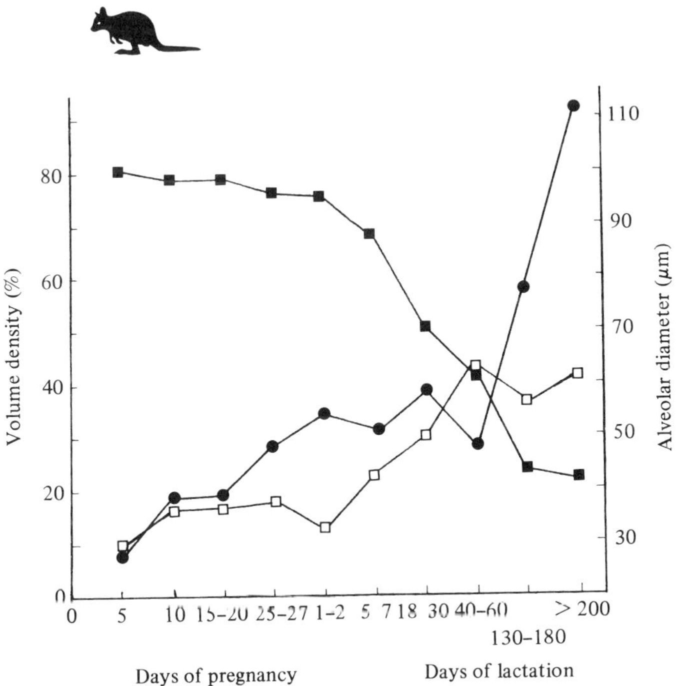

Milk composition

The marked change in the growth of the mammary gland in the second half of lactation, as seen for instance in *T. vulpecula* and *M. eugenii* (Fig. 8.3), coincides with changes in all the constituents of the milk secreted (Fig. 8.6) and this seems to be common to all marsupials (Green, 1984). It is in marked contrast to eutherians in which the composition of milk in any one species remains much the same throughout lactation, although differing widely between species (see Chapter 10 for further discussion of this point, p. 411). In the first half of lactation the milk of all marsupials studied is low in total solids, lipids and proteins; the carbohydrate moiety, which is relatively high, comprises oligosaccharides. In the second half, the amount of lipid and protein increases considerably, while the oligosaccharides disappear and are replaced by a low concentration of monosaccharides. Other constituents such as sodium, iron and several amino acids also change at this stage of lactation and several new proteins appear in late-stage milk (Lemon & Barker, 1967; Lemon & Poole, 1969; Green & Renfree, 1982). This implies that the differentiated secretory cells of the mammary epithelium express other genes at this stage or that newly differentiated cells are recruited into the epithelium. This has not yet been investigated critically, although some studies have examined the related question of whether the pouch young exercises a controlling influence (Findlay, 1982*b*).

Fig. 8.6. Changes in milk of *Macropus eugenii* through lactation. Relative proportion (mean ± S.D.) of lipids (●), proteins (■) and carbohydrates (△) in the solids fraction of milk. Redrawn from Green (1984).

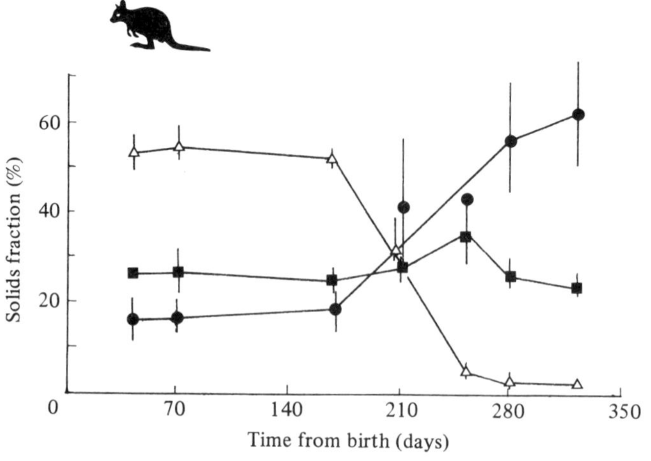

It seems likely that the changes in the composition and production rate of the milk during lactation are closely correlated with the nutritive requirements of the pouch young. For example, Renfree *et al.*, (1981) noted a correlation between sharp increases in the amount of sulphur-containing amino acids in the milk and the time when hair follicles develop on the young of *M. eugenii*. Growth of the brain occurs at different rates in the pouch young of this species before and after the development of endothermy, and different kinds and amounts of nutrients may be required for these growth phases (Renfree *et al.*, 1982*a*).

The nursing pattern also alters as lactation progresses: after an initial phase of continuous attachment to the teat the time spent on the teat gradually decreases, until sucking is reduced to brief intermittent bouts after the young leave the pouch. Ben Shaul (1962) was the first to suggest that differences in milk composition between species is related to the frequency of nursing, those species that suckle frequently having dilute milk and those suckling infrequently having concentrated milk. The pattern within marsupial species appears to conform to this hypothesis and leads to the idea that the changes in marsupial milks and glands are regulated by the sucking young.

An experiment by Reynolds (1952) with *D. virginiana*, referred to in the next section, appears to support it so far as growth in size of the gland is concerned but a more recent study in *M. eugenii* indicated that changes in composition of the milk are an intrinsic characteristic of the mammary epithelial cells and occur as the cells age (Findlay, 1982*b*). In these experiments twin, single or serial transfers resulted in the ages of the pouch young differing from the stage of lactation of the female, or by serial transfer of similar-aged young, maintaining an unchanging sucking stimulus to the gland. Transitory effects on milk composition were produced during the first few weeks after transfer (i.e. depressed fat and protein concentrations), which may have resulted from the altered rate of milk withdrawal and/or a change in the intensity of the sucking stimulus, but these effects were not maintained against the apparently over-riding inherent control of the cellular biochemical processes. Many of the pouch young receiving milk from glands at different stages of lactation (earlier or later) showed disturbed growth rates, and died; there were indications that some of the lactating glands were increasing their production at the maximum possible rate as they developed, and were unable to respond to increased demands by larger pouch young. Other pouch young sucking 'older' glands were apparently ingesting more milk than normal and showed accelerated weight gain for a time, but many of these young also died. The same effect was observed by Merchant & Sharman (1966) in the

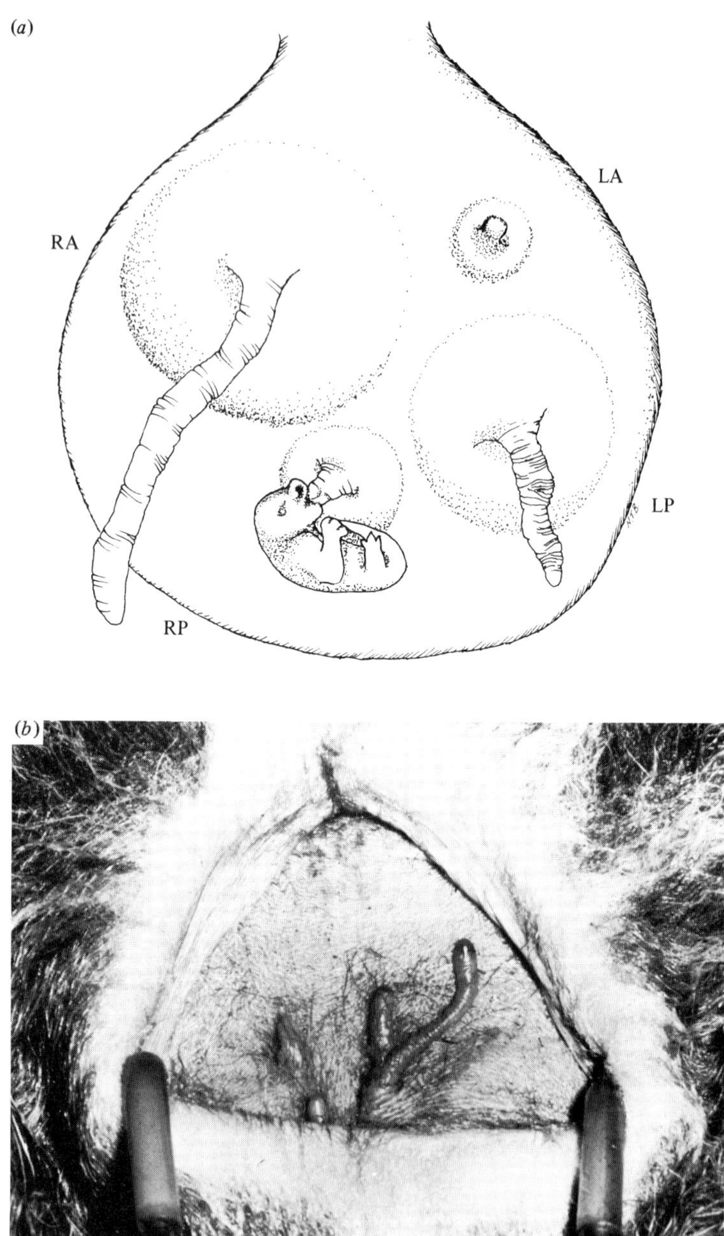

Fig. 8.7. Ventral view of pouch and mammary glands of macropodids. (*a*) Diagram of the pouch of *Macropus rufus* to illustrate a possible situation in which each gland is at a different stage of development:

young of smaller species (*Wallabia bicolor*) transferred to the pouch of larger species (*M. rufus*) but in these the young became obese and reached puberty unusually early.

Mammary regression and successive lactation

Regression of the gland and teat after lactation varies between species. In *Didelphis virginiana* females return to oestrus at about day 90 when the litter are starting to be weaned and the glands regress and the teats shorten from 25 to 3 mm in the few days before the next litter are born. Reynolds (1952) was able to delay regression in this species by transferring 60 day old pouch young to a female at day 94 of lactation, whereupon the glands decreased to suit the smaller young and subsequently enlarged again to meet their demands, so that the glands lactated for 154 days. *Didelphis* may be exceptional since in other species different glands are used by successive young. In the polytocous Peramelidae, regression of the post-lactational gland is slow and young of the subsequent litter generally attach to unused teats (Fig. 2.12*c*); in the rare instances when a young attaches to a regressing teat it fails to grow normally and dies (J. C. Merchant, personal communication). Among monovular species such as *T. vulpecula*, regression is also slow and different teats are used by successive offspring. Sharman (1962) observed that the rate of regression of the previously sucked gland was slower in females lactating from the other gland than in non-lactating females, but no milk could be expressed from such regressing glands.

By contrast in continuously breeding macropodids, such as *M. rufus* (Fig. 2.25), *M. agilis* or *M. eugenii* under experimental conditions, where the female gives birth immediately after the pouch has been vacated by the older young, two of the four mammary glands simultaneously secrete milk of entirely different composition (Figs 8.7, 8.12*a*); one gland, fully grown with large alveoli, secretes mature milk for the young-at-heel, while

Fig. 8.7. *cont.*

RA, right anterior gland is secreting mature milk for a young at foot recently out of the pouch (Fig. 8.4*b*); RP, gland with neonate feeding on early milk (Fig. 8.4*a*); LP, gland in post-lactational regression from weaned young > 400 days old; LA, non-sucked gland regressing from progesterone-stimulated condition during recent pregnancy (Fig. 8.4*a*). (*b*) Pouch of *Macropus eugenii*, showing the four teats: the two small teats had never been sucked, the third was sucked during a previous lactation and the fourth, elongated teat was supporting a 40 day old young from its lactating gland, (*b*) From Findlay & Renfree (1984), with permission.

the other, in which lactogenesis has just been initiated, secretes early-stage milk for the neonate. Furthermore, of the two remaining glands, one may still be regressing from supporting the previous, weaned young and the fourth regressing from the prepartum condition (Fig. 8.7*a*). Well before the next birth, however, these two unsucked glands redifferentiate independently of the two lactating glands in preparation for the next lactation. This is achieved by day 60 of lactation in *M. eugenii* (Findlay, 1982*a*).

The mechanisms which allow the four mammary glands to differentiate and regress independently during concurrent asynchronous lactation are unclear. Local factors such as sucking and milk removal are obviously important for the maintenance of lactation in the lactating gland, and the differing responses of the four individual glands could be the result of differing hormone-receptor concentrations in the glands. In eutherians, high concentrations of progesterone fail to inhibit established lactation, but can inhibit lactogenesis (Schmidt, 1971; Kuhn, 1977), and this is because progesterone-receptor concentrations are low in lactating glands (see Cowie, Forsyth & Hart, 1980; Topper & Freeman, 1980). It would be interesting to measure progesterone-receptor concentrations in the glands of macropodids during pregnancy and also in the two lactating and two non-lactating glands of animals lactating asynchronously.

This is the most complex form of lactation known in any mammal and poses fundamental questions about cellular differentiation in the mammary epithelium (already alluded to), about the endocrine control of lactogenesis in each gland, and about the control of concurrent milk secretion from two glands. The latter two questions will be discussed in the next sections of this chapter.

Hormonal control of lactation

Control of mammary development before lactation

The preparation of the mammary gland for lactation is termed lactogenesis Stage I in eutherians (Cowie *et al.*, 1980) and many hormones, including progesterone, placental lactogen and prolactin, are essential to its completion (Topper & Freeman, 1980). Much less is known of this stage in marsupials except that the corpus luteum and pituitary are necessary and the placenta is not.

O'Donoghue (1911) was the first to recognise the close correlation between the development of the mammary gland and the corpora lutea during pregnancy in *Da. viverrinus*, and Hartman (1925*b*) demonstrated in *D. virginiana* that bilateral ovariectomy during pregnancy or the non-pregnant cycle prevented mammary development. Indirect evidence in two other species supports the idea that an active corpus luteum must be present at the end of pregnancy for effective lactogenesis to occur. In Clark's (1968*a*) experiments with *M. rufus*, described in Chapter 7 (Table 7.5), neonates failed to survive when born to females in which the development of the corpus luteum had been delayed by 9 or more days, and in those animals no milk could be expressed from the teat. Similarly in *M. eugenii*, females lutectomised 10 days or 6 days before parturition delivered young that failed to survive whereas the neonates of those lutectomised 3–4 days before did survive (Young & Renfree, 1979). Tyndale-Biscoe *et al.* (1984) confirmed these findings in *M. eugenii* and reported that the weight of mammary glands on day 29 of those animals lutectomised on day 17 were significantly lighter than the glands of those lutectomised on day 23.

Fig. 8.8. Prolactin during lactation in *Macropus eugenii*. (*a*) Concentrations (mean ± s.e.m.) of prolactin in plasma throughout lactation in six females. (*b*) Concentrations of prolactin receptors on epithelial cells of sucked (●) and non-sucked (△) mammary glands during pregnancy and lactation. Compare with Fig. 8.3*b*. Data for (*a*) from Hinds & Tyndale-Biscoe (1985) and (*b*) Stewart (1984).

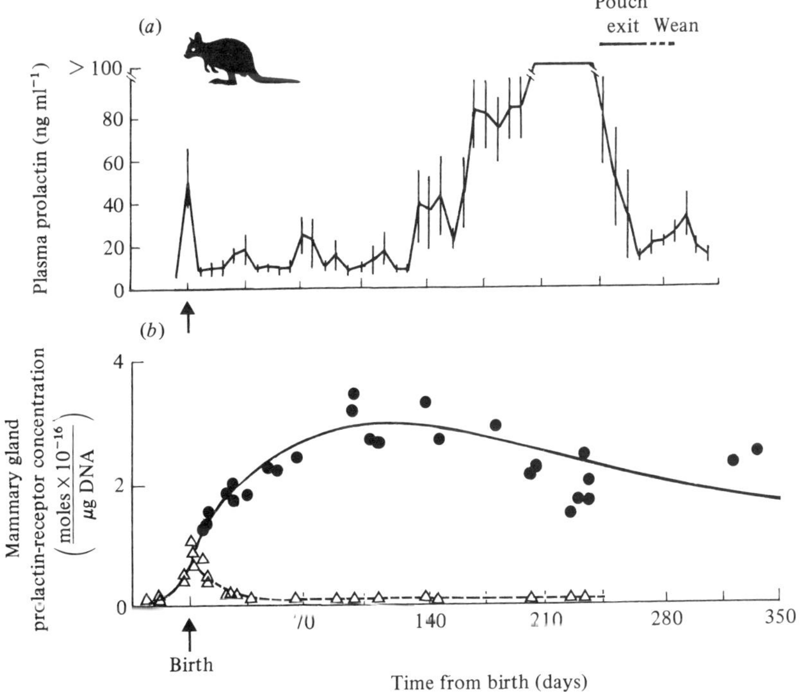

Because in macropodids parturition is so closely followed by oestrus (Fig. 6.1), elevated oestradiol at this time could be important. To distinguish between these two factors and to exclude any possible placental effects, Stewart (1984) removed the corpus luteum 2 weeks before expected oestrus in non-pregnant *M. eugenii*, and after oestrus occurred she measured the glands and determined the concentration of prolactin receptors on the cell membranes of mammary secretory tissue. In the lutectomised animals the weight (0.38 g) and receptor concentration (0.62×10^{-6} moles prolactin per microgram of DNA) were equivalent to totally inactive glands (see Fig. 6.19) and less than half those found in the non-lutectomised group (0.82 g and 1.85×10^{-6} moles respectively) (Fig. 8.8*b*). From this we may conclude that the corpus luteum, but not the Graafian follicle, induces mammary growth and the synthesis of specific membrane receptors for prolactin.

The concentration of prolactin in circulation is low and does not increase during pregnancy in *M. eugenii* (Hinds & Tyndale-Biscoe, 1982*b*), except for a very brief prepartum pulse (Fig. 7.27; 8.8*a*). However, the increasing concentration of prolactin receptors in the mammary gland, induced by the corpus luteum, may enable prolactin to provide a necessary stimulus to the mammary gland or they may be required only after parturition for the initiation of lactogenesis.

Removal of the pituitary during pregnancy prevented mammary gland development (Hearn, 1972*a*) but immunisation against GnRH did not (Short *et al.*, 1985). Since the corpus luteum does not require pituitary support (see p. 250) these results suggest that a deficiency of prolactin was the cause or that adrenocortical and thyroid hormones, reduced by the lack of pituitary ACTH and TSH, play a role in mammary gland development before lactation begins.

Initiation of lactation

In all six species that have been investigated progesterone is elevated in the circulation during the last third to half of pregnancy (Figs 6.3, 6.5, 6.6, 6.7, 6.13) and in all, except *Isoodon macrourus* (Fig. 6.10), falls shortly before birth and the onset of lactation. This pattern is similar to that of several eutherian species in which progesterone has been shown to play an essential role in lobulo-alveolar development of the mammary gland, and to inhibit the synthesis of specific milk proteins, such as casein and α-lactalbumin, and lactose until after parturition. For this reason it was thought that the initiation of lactation in marsupials might also be induced by progesterone withdrawal at parturition. Two recent studies have investigated this idea.

Findlay *et al.*, (1983) measured lactose in mammary tissue before and after parturition and found a 10-fold increase in animals immediately after parturition, even in those from which the neonate was prevented from attaching to a teat (Fig. 8.9*a*). In non-pregnant females the increase was only 2–3 fold and occurred later (Fig. 8.9*b*). These results suggest that the prepartum pulse of prolactin, which does not occur in non-pregnant females (Fig. 7.27) might be involved. This conclusion is supported by another finding of Findlay *et al.* (1983), that maintaining a high concentration of progesterone through the peripartum period did not prevent the

Fig. 8.9. Concentrations of lactose in mammary glands of *Macropus eugenii* (*a*) during pregnancy (●) and day 1 lactation (■), and (b) in non-pregnant animals (▲). Note broken scale on both axes. *Value from an animal which had given birth but the pouch young had not yet attached to any of the teats; †value from non-sucked gland of an animal which had given birth and the young was being suckled from another gland (‡). Redrawn from Findlay *et al.* (1983).

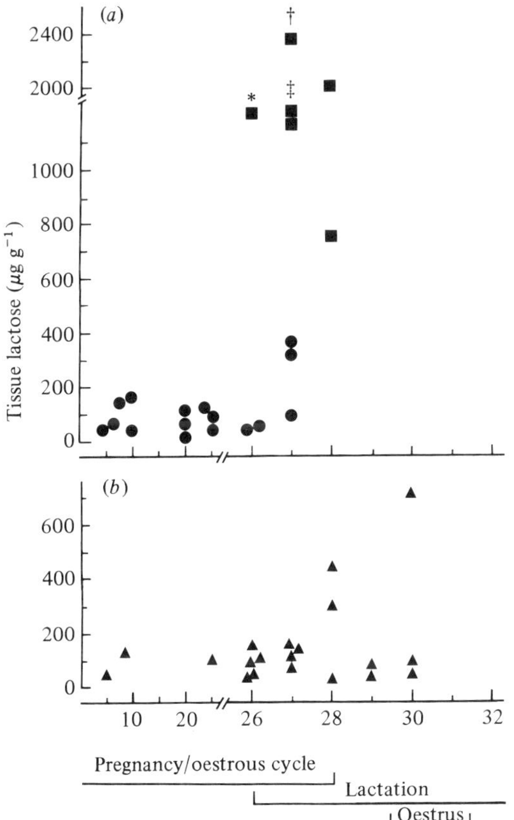

rise in lactose concentration seen in the untreated animals. The converse experiment, where progesterone was reduced prematurely by luctectomy did induce an early rise in mammary lactose as predicted but, since the sham-operated animals also showed a rise, this was probably a non-specific effect. Therefore we conclude that a factor other than progesterone withdrawal must trigger synthetic activity in the mammary gland of this species.

Nicholas & Tyndale-Biscoe (1985) investigated the requirements for synthesis of α-lactalbumin by mammary gland taken from female *M. eugenii* 2 days before parturition and cultured *in vitro*. Their results supported the studies *in vivo* in showing that progesterone could not inhibit the synthesis of α-lactalbumin, whereas it does in rat tissue similarly prepared. More interestingly, they found that the sole hormone required for induction of synthesis of α-lactalbumin is prolactin at concentrations equivalent to the basal levels in the pregnant female. No other hormones, such as cortisol, thyroxine or insulin, which all enhance synthesis in rat and rabbit tissue had any effect, either alone or when combined with prolactin. Another experiment which supports this was reported by Findlay (1982*b*) and Findlay & Renfree (1984). Prolactin or saline was injected via the teat ducts into individual galactophores of female *M. eugenii* on day 15 of pregnancy (11–12 days before parturition), and a large proportion of the alveoli of the prolactin treated tissue subsequently contained secretion (Table 8.2). These several results suggest that the

Table 8.2. *Percentage of alveoli containing secretion after intraductal injection of about* 200 *μg of ovine prolactin or saline*

Prolactin treated		Saline treated	
Injected area	Non-injected area	Injected area	Non-injected area
62	6	12	12
64	8	0	0
62	10	0	0
64	14	30	26
64	8		
58	18		
8	8		

From Findlay & Renfree (1984).

prepartum prolactin peak may play an important role in initiating lactation. However, when neonatal young were transferred to non-pregnant females at the same stage, which had not had such a prolactin peak (Tyndale-Biscoe *et al.*, 1984), they were able to suckle the young adequately, so it cannot be essential to successful lactogenesis.

Galactopoiesis and the maintenance of lactation

It is unlikely that the ovaries play any part in the maintenance of lactation in marsupials; in many species, such as *T. vulpecula*, they are much regressed at this time and in the macropodids that have a quiescent corpus luteum during lactation, ovariectomy in at least two species, *Setonix brachyurus* (Tyndale-Biscoe, 1963*a*) and *M. eugenii* (personal observations) did not affect lactation. On the other hand the pituitary is essential for the maintenance of both phases of lactation. In early lactation, pouch young ceased to grow and died in less than 2 weeks (Hearn, 1972*a*; L. A. Hinds, C. A. Horn & C. H. Tyndale-Biscoe, unpublished results). As no milk could be expressed after 3 days it is likely that the young died from starvation. Similar results were obtained in late lactation, when the alveolar diameter was reduced 25-fold by 20 days after hypophysectomy (Hearn, 1972*a*). Since in all these studies both parts of the pituitary were removed, it is not clear whether the failure was due to lack of prolactin or oxytocin or a combination of both. It is now known that oxytocin is important for milk ejection (see below), while other evidence suggests that prolactin is important for lactogenesis.

The concentration of prolactin receptors in mammary cell membranes of *M. eugenii*, increases in the sucked mammary gland for the first 12 weeks after parturition (Fig. 8.8*b*), whereas it has fully declined in the unsucked glands after 2 weeks (Stewart, 1984). After 12 weeks the concentration does not increase, but since the gland enlarges manyfold (Fig. 8.3*b*) the total number of receptors increases accordingly, and this is maintained until the end of lactation. In the same species the concentration of prolactin in peripheral circulation follows a pattern very different from that seen in eutherian species (Fig. 8.8*a*). After the transient prepartum pulse the concentration reverts to a consistently low level of about 40 ng/ml^{-1} which is the same as in non-lactating females, until 140 days of lactation (Hinds & Tyndale-Biscoe, 1982*b*, 1985).

During the period 140–168 days, the concentration fluctuates upwards and after 168 days remains at more than 100 ng ml^{-1} until about 280 days. When measured at hourly intervals at this time marked fluctuations are seen. When the young were removed, the level fell to basal within 7 h and recovered equally quickly after they were returned (Hinds & Tyndale-Biscoe, 1985). From this it appears that the sucking of the young is a necessary stimulus to maintain the secretion in late lactation, but is not involved in early lactation. However, the sucking stimulus in early lactation does stimulate receptor synthesis locally which, in turn, presumably enables the lactating gland to bind more of the available prolactin in circulation. This may be essential for galactopoiesis and is in agreement with the finding that early-stage lactation was maintained and the young grew normally after denervation of the lactating gland (Renfree, 1979).

Fig. 8.10. Concentrations of plasma prolactin (mean ± s.e.m.) in (*a*) *Dasyurus viverrinus* and (*b*) *Trichosurus vulpecula* throughout lactation. Note the increase during the second half of lactation (when the young are in the nest or riding on the mother's back) followed by a tailing off as the young begins to eat solid food and is weaned. Compare the profile in (*b*) to Fig. 8.3*a*. Note that the concentration of prolactin is higher in *D. viverrinus* with 5–6 young than in either of the monotocous species. (*a*) From Hinds & Merchant (1986). (*b*) From Hinds & Janssens (1986).

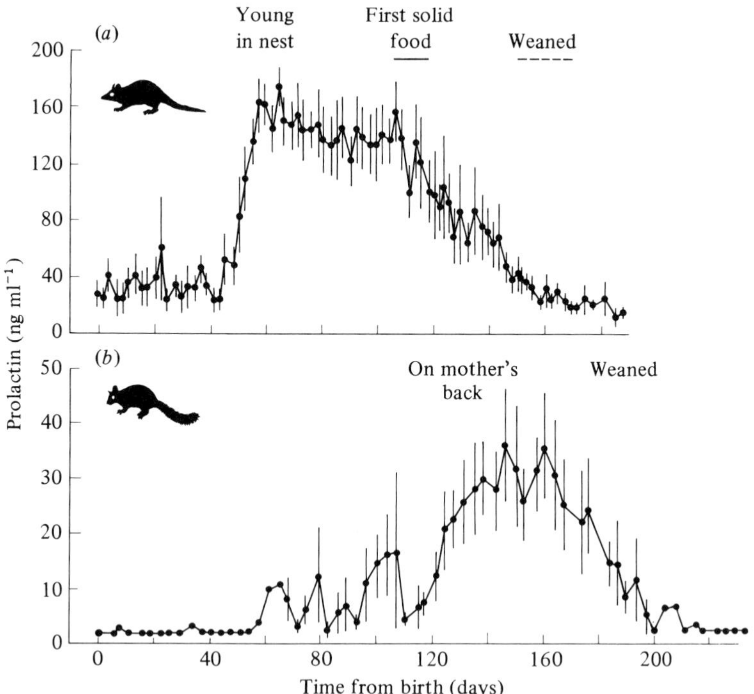

When the young vacates the pouch but is still feeding on milk, the concentration of prolactin declines to the basal level (Fig. 8.8*a*). So far prolactin has not been measured in any animals that delivered a new young after the older one vacated the pouch, as described in the earlier part of this chapter, but these results suggest that a low prolactin level might already prevail when the young is born, which would be appropriate to a gland in early lactation.

Similar profiles have been described in two other species, *Trichosurus vulpecula* (Hinds & Janssens, 1986) and *Dasyurus viverrinus* (Hinds & Merchant, 1986), although in both the time at which prolactin becomes elevated is different from *M. eugenii*. In both species it coincides with the increase in size of the mammary gland and the accelerated growth of the young (Figs. 8.10 and 8.11). Likewise the later decline coincides with weaning. Thus this pattern would appear to be common to marsupials and reflect their different and longer lactation compared to eutherians.

Fig. 8.11. Young of *Dasyurus viverrinus*, 75 days old, at the end of pouch occupation, just before their eyes open and they are left in the nest. This is when the maternal prolactin concentration rises (see Fig. 8.10*a*). Photograph by S. L. Bryant, with permission.

Control of milk secretion

Role of striated muscle

A widespread belief, first suggested by Seiler (1828) and Morgan (1829) and reinforced by Owen (1868), is that the marsupial neonate is too small to suck milk from the teat, as other mammals do. Instead the milk was thought to be forcibly ejected from the mammary glands into the young by contraction of the striated muscle fibres of *m. ilio-marsupialis*, the homologue of *m. cremaster* in the male and termed by Owen (1868) *m. compressor mammae* (see p. 122). Later authors, such as McCrady (1938), Hartman (1952) and Enders (1966), observed that neonatal *D. virginiana* and *Marmosa* can suck actively, and Enders (1966) provided the first experimental evidence that milk cannot be expressed by contraction of striated muscle. He exposed the motor innervation of *m. ilio-marsupialis* and several other associated muscles in lactating females of *D. virginiana*, *D. marsupialis* and *Marmosa* and subjected the nerves and the muscle fibres in the gland to sufficient electrical stimulation to induce maximum contraction of the muscle. No milk was expressed by this means, although it had been expressed manually before electrical stimulation, and the young reattached and grew afterwards. This experiment has been repeated by M. E. Griffiths (unpublished results) in lactating *M. eugenii* with the same result. Despite Enders' (1966) results this function for *m. ilio-marsupialis* is still advanced (Barnes, 1977) but M. E. Griffiths (unpublished results) considers that the more likely function of the muscle is for mammary support and retraction of the teats and attached young, especially in species that lack a pouch, in the same way that *m. cremaster* retracts the testis in the scrotum (see p. 129). *M. ilio-marsupialis* is a flat triangular sheet of muscle, which takes origin on the pelvis and inserts on the dorsal or deep surface of the mammary gland. The degree to which it penetrates the gland varies between pouched and pouchless species; in *T. vulpecula* (Bolliger & Gross, 1960; Barbour, 1963) and in *M. rufus* (Griffiths *et al.*, 1972) branches of the muscle diffusely invade the gland, and in *Tarsipes rostratus* reach each of the 4 teats (M. E. Griffiths, unpublished results). In all the pouchless species examined by M. E. Griffiths (unpublished results) discrete branches of the muscle attach at the base of the teats. These species are *Antechinus stuartii* and *A. swainsoni*, *Dasyuroides byrnei* and *Dasyurus viverrinus* as well as the South American species *Caenolestes obscurus*. Barnes (1977) described the same arrangement in *Marmosa robinsoni*. When *A. stuartii* and *D. viverrinus* were anaesthetised the attached young hung from flaccid teats but as the mothers regained consciousness the teats retracted and the young were again held close to the body as seen in Figs 2.9, 2.11 and 8.11.

The conclusion from these studies is that *m. ilio-marsupialis* does not have the special function in marsupials of expressing milk and that the secretion of milk in these mammals is under the same control as in eutherians.

Oxytocin and milk ejection

Milk removal is facilitated by the milk-ejection reflex in eutherian mammals (Cross, 1977). The contraction of the myoepithelium investing the alveoli is caused by oxytocin released from the neurohypophysis during suckling. The sucking stimulus is transmitted by an afferent arc to the hypothalamus and this contraction raises intra-alveolar pressure, leading to expulsion of the milk into the ducts, which can then be removed from the glands by the sucking of the young.

Exogenous oxytocin will induce milk ejection in all eutherian species (Cross, 1977), and also in monotremes and marsupials (see Griffiths, 1978) in which the injection of heterologous oxytocin is routinely used to obtain milk samples.

It has been known for some time that the neurohypophyses of marsupials contain oxytocic and vasopressor activity but only recently has the amino sequence of these polypeptides been determined in three species of Didelphidae, five species of Macropodidae and *T. vulpecula* (Chauvet *et al.*, 1981, 1983*a*, *b*, *c*, 1984*a*, *b*, 1985; Hurpet *et al.*, 1982) (Table 8.3). In all the macropodids two pressor hormones occur, namely lysine vasopressin and phenypressin, which each differ from arginine vasopressin by one substitution at position 8 and position 2 respectively (Table 8.4). By contrast only arginine vasopressin occurs in *T. vulpecula*, as in most eutherian species, and arginine and lysine vasopressin in the didelphids. In all six Australian species investigated the oxytocic hormone is mesotocin, which occurs in lower tetrapods but not in the Eutheria or the Monotremata, which have oxytocin. The three American species have oxytocin and *D. virginiana* has mesotocin as well. Mesotocin differs from oxytocin by a single amino acid, having isoleucine instead of leucine at position 8 (Table 8.4), and it had been supposed that this change was associated with the evolution of lactation and the milk-ejection role of this neuropeptide. However, these amino acid alterations seem to have no functional significance, because heterologous (eutherian) oxytocin injected, or homo-

Table 8.3. *Occurrence of neurohypophyseal peptides in mammals*

	Arginine vasotocin	Arginine vasopressin	Lysine vasopressin	Phenyl vasopressin	Oxytocin	Mesotocin
Sauropsida	1					1
Monotremata		1			1	
Eutheria		1	1		1	
Metatheria:						
Philander opossum		1	1		1	
Didelphis marsupialis		1	1		1	
D. virginiana		1	1		1	1
Trichosurus vulpecula		1				1
Setonix brachyurus			1	1		1
Macropus eugenii			1	1		1
Macropus rufus			1	1		1
Macropus giganteus			1	1		1
Macropus fuliginosus			1	1		1

Data from Chauvet *et al.* 1981, 1983*a*, *b*, *c*, 1984*a*, *b*, 1985; and Hurpet *et al.* 1982.

logous (marsupial) mesotocin released after stimulation of the presumptive oxytocinergic neurones both stimulate a milk-ejection response (Lincoln & Renfree, 1981*a,b*).

Despite this evidence for oxytocic activity, experiments by Renfree (1979) cast doubt on the milk-ejection reflex being crucially involved in early lactation in *M. eugenii*. When the entire innervation of the sucked mammary gland was cut, lactation continued apparently unimpaired since the young grew normally for the duration of the experiment. Prolactin concentrations were not affected by denervation (M. B. Renfree, unpublished observations) but, as already mentioned, the important role of sucking at this stage is the induction of prolactin-receptor synthesis not the release of prolactin, so that was not surprising. How then is milk let-down accomplished in early and late lactation and, in species that have concurrent asynchronous lactation, how is let-down controlled to release very different volumes of milk from the two glands? This was investigated by Lincoln & Renfree (1981*a*, *b*) in *M. agilis* by measuring the intraductal pressure within the mammary gland that occurred in response to injections of oxytocin (Fig. 8.12) or electrical stimulation of the presumptive oxytocin neurons in the brain (Fig. 8.13). They found that the basic dose-response curve is very similar to that of eutherian mammals but, early in lactation, the threshold doses of oxytocin are lower and the peak intraductal pressures obtained much higher than in late lactation (Lincoln & Renfree, 1981*b*). This was clearly seen when recordings were made simultaneously in females suckling 2 young of different ages (Fig. 8.12*b*).

Table 8.4. *Structures of vertebrate neurohypophysial hormones*[a]

I. Oxytocin-like hormone family	
	1 2 3 4 5 6 7 8 9
Oxytocin	Cys-Tyr-Ile-Gln-Asn-Cys-Pro-**Leu**-Gly (NH_2)
Mesotocin ([Ile[8]]-oxytocin)	Cys-Tyr-Ile-Gln-Asn-Cys-Pro-**Ile**-Gly (NH_2)
II. Vasopressin-like hormone family	
	1 2 3 4 5 6 7 8 9
Arginine vasotocin	Cys-Tyr-**Ile**-Gln-Asn-Cys-Pro-Arg-Gly (NH_2)
Arginine vasopressin	Cys-**Tyr**-Phe-Gln-Asn-Cys-Pro-**Arg**-Gly (NH_2)
Phenypressin ([Phe_2]-arginine vasopressin)	Cys-**Phe**-Phe-Gln-Asn-Cys-Pro-**Arg**-Gly (NH_2)
Lysine vasopressin	Cys-**Tyr** Phe-Gln-Asn-Cys-Pro-**Lys**-Gly (NH_2)

[a]A disulfide bridge binds half-cystines in positions 1 and 6.

Fig. 8.12. Response to oxytocin in adjacent mammary glands of *Macropus agilis* supporting young of two different ages. (*a*) The smaller teat and gland (left) is supporting the 38 day pouch young shown attached; the larger teat and gland (whose area occupies half of the photograph) is supporting the 248 day young-at-heel shown out of the pouch (right). (*b*) Simultaneous recordings of intramammary pressure from galactophores of the left mammary gland (upper trace) at 38 days, and the right mammary gland (lower trace) at 248 days of lactation to systemic doses of oxytocin from 1 mU to 20 mU. Note that the 38 day gland responded to lower doses of oxytocin and generated higher peak pressures than the 248 day gland. From Renfree (1983), by permission.

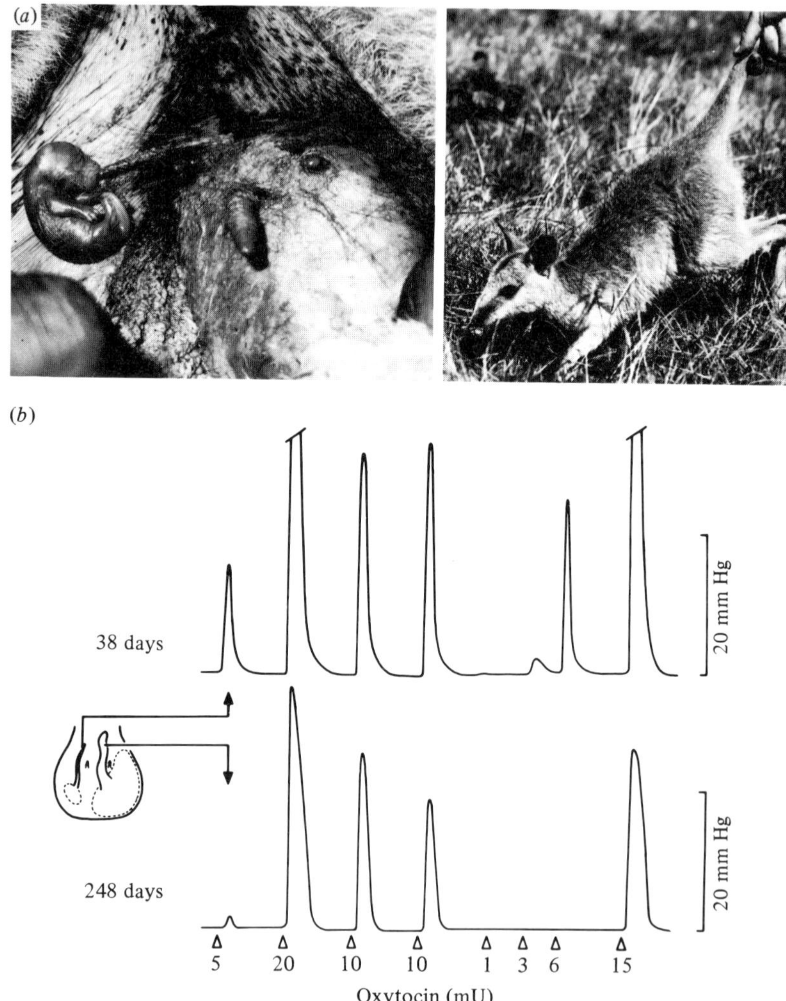

Fig. 8.13. Response to electrical stimulation of the presumptive oxytocinergic neurons of the hypothalamus of *Macropus agilis*. (*a*) A stimulatory electrode was lowered into the basal part of the hypothalamus from a point of entry in the skull 28 mm anterior to the bregma suture. Stimulation was for 1 ms at an amplitude of 700 μA. (*b*) Recordings of intramammary pressure from a galactophore at day 82 of lactation after three injections of oxytocin (5, 7.5, and 10 mU) and stimulation of the neurons by the electrode positioned as in (*a*). A 50 Hz × 20 s train of pulses (solid triangles) was applied at intervals as the electrode was gradually lowered until a position was obtained within the brain at which milk ejection was induced. (*c*) Pressure recordings at this position obtained with different combinations of stimulus frequency (Hz) and duration(s). From Renfree (1983) by permission.

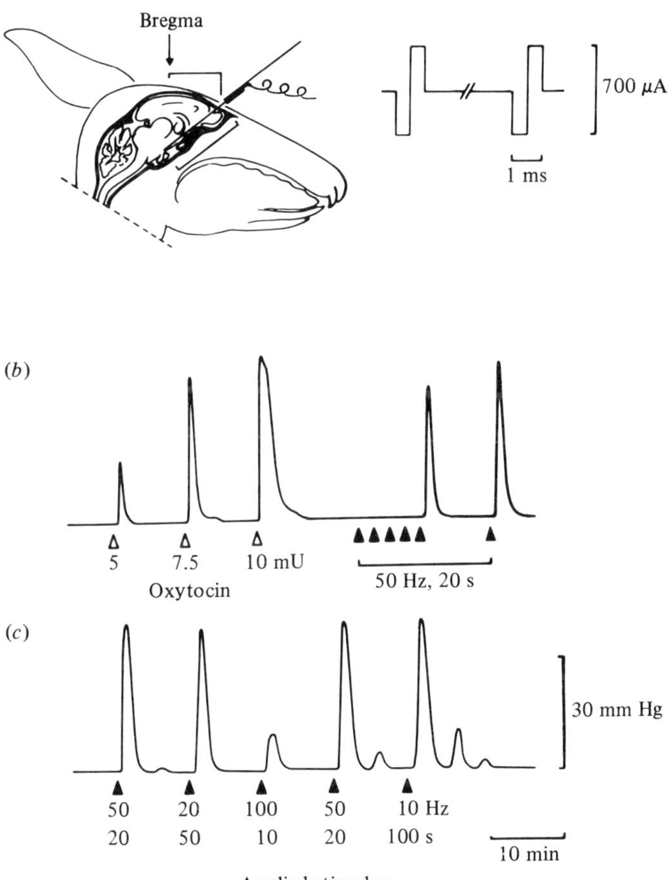

As well as the initially lower threshold required for milk ejection in *M. agilis*, when compared with rats and rabbits (Table 8.5), there were differences in the response to electrical stimulation of the presumptive oxytocinergic neurones in the basal hypothalamus (Lincoln & Renfree, 1981*a*). Electrical stimulation at the frequency which produced optimal oxytocin release in these two eutherian mammals (50 Hz × 20 s) was also optimal in *M. agilis* (Fig. 8.13) but, unlike these mammals, milk ejection also occurred when the neurones were stimulated at much lower frequencies, suggesting that the process of hormone release may differ in some way. The decline in responsiveness of the gland to oxytocin during lactation would thus allow the milk ejection caused by small releases of the hormone to be confined to the gland supporting the small young, while milk ejection would occur in both glands in response to the larger hormonal releases presumably triggered by the more vigorous sucking of the young-at-foot.

The afferent arc of the reflex may be regulated by changes in sensitivity of the teat. The extremely small neonate may not provide sufficient sensory input to cause a release of oxytocin, as has already been shown for

Table 8.5. *A comparison of the characteristics of milk ejection in* Macropus agilis *and two eutherian mammals*

	M. agilis		
	Early lactation	Late lactation	Rat and rabbit
Peak intramammary pressure at milk ejection (mm Hg)	60	20	10
Threshold dose of oxytocin for the production of a recordable contraction [mU (kg body weight)$^{-1}$]	0.2	1	1
Minimum frequency of electrical stimulation of oxytocin-producing neurones for induction of milk ejection (Hz)	4.5		25
Effectiveness of the teat sphincter to prevent the escape of milk		Poor	Good
Milk ejection response to vasopressin (vasopressin to oxytocin ratio)		30:1	4:1

From Lincoln & Renfree (1981*b*).

prolactin. However, a low or basal level of oxytocin, such as might be released after the very low-frequency stimulations which approach 'background noise' levels of firing neurones, could, due to the extreme sensitivity of the myoepithelium evoke recurring contractions. As *M. agilis* (and presumably macropodids in general) have no teat sphincter (Fig. 8.2), milk could effectively pump into the attached neonate (Lincoln & Renfree, 1981*a*; Lincoln & Paisley, 1982). The fact that lactation can continue for several weeks after teat denervation (Renfree, 1979) suggests that milk ejection in early lactation, like maintenance of lactation, is also relatively independent of sensory stimuli. Thus the original idea that milk was forced into the neonate might not be so far from the truth, although the mechanism is due to contraction of the myoepithelium rather than contraction of *m. ilio-marsupialis*.

Conclusions

Marsupials have evolved a different but highly successful reproductive strategy when compared with eutherian mammals, in that their major reproductive investment is placed in lactation rather than gestation and placentation, with consequent refinement of their mammary physiology. This is most pronounced in those macropodids which exhibit concurrent, asynchronous lactation. Nevertheless, there are fundamental similarities between lactation in marsupials and other mammals.

The structure and development of the mammary glands of all mammals are essentially the same, with differences introduced by the temporal course of development. Mammary lactose synthesis in *M. eugenii*, and probably in other marsupials, is initiated around the time of parturition and is independent of the sucking stimulus, as it is in eutherians. The specific hormonal control of lactogenesis in marsupials remains unclear: although lactogenesis usually occurs at parturition to coincide with the demands of the neonate, in other circumstances the glands can commence milk secretion in response to a variety of stimuli such as sucking alone, or abnormal changes in hormonal concentrations. There are many similarities to the endocrine control of eutherian lactation, both prepartum (lactogenesis Stage I) and later in lactation at the stages termed lactogenesis Stage II, when copious milk secretion begins. Mammary gland development before parturition is dependent on luteal progesterone and on the pituitary, but the fetus and/or placenta are not essential. In late lactation the sucking of the young maintains hyperprolactinaemia and galactopoiesis. However, there is no real eutherian equivalent to the early phase of lactation in marsupials.

Marsupials differ from eutherians in the pattern of control of early lactation in that sucking by the young has little or no effect on prolactin levels (though milk removal and basal prolactin levels are important for maintenance of lactation) but the sucking does induce and maintain the synthesis of specific prolactin receptors on mammary cells.

Although the composition of the milk during the suckling period changes more in marsupials than in eutherians, these changes appear to be an intrinsic character of the mammary epithelial cells and, at least in early to mid-lactation, occur regardless of alternations in the local stimuli applied to the glands. The hypertrophy of the second phase of lactation, when the sucked mammary gland increases dramatically in size, must however be related to the different sucking pattern of the larger young, since it is not dependent on ovarian hormones. The differing pattern of oxytocin release and the refinement of the milk-ejection response to oxytocin permits the mother to nourish the minute neonate in the months immediately after birth, and in some macropodids allows her intermittently to suckle an older sibling on an adjacent mammary gland. Much more information is needed on the role of oxytocin.

9

Neuroendocrine control of seasonal breeding

Most species of marsupial are seasonal breeders and, as shown in Chapter 2, the emergence of the young at the end of lactation coincides with the most favourable season of the year. For small species with a short gestation and pouch life (Russell, 1982*a*) the environmental factor that determines the mating season may be some component of favourable conditions such as increasing temperature or protein rich food (e.g. in the Peramelidae, see p. 53) although photoperiod may be an additional factor (e.g. in *Sminthopsis crassicaudata* (Godfrey, 1969*b*), see p. 41). For larger species with a longer pouch life, however, mating may occur at a time when environmental conditions are highly unfavourable (e.g. in *Setonix brachyurus*, see p. 85) and, for these species, photoperiod may be an important factor. By far the best-studied marsupial in this regard is *Macropus eugenii* and in this chapter we will discuss the evidence for control of seasonal breeding by daylength in this species.

The highly seasonal nature of breeding in *M. eugenii* was described in Chapter 2 (Fig. 2.26) and other aspects of the interrelationship between the corpus luteum, ovary and blastocyst were referred to in Chapters 6 and 7 respectively. Superficially the reproductive pattern resembles that of the domestic sheep in that both species mate during declining photoperiod; for that reason they are said to be 'short-day' breeders. During the period of increasing daylength the ewe is anoestrous (Legan, Karsch & Foster, 1977) and the ram has small testes in which spermatogenesis is much reduced or absent (Lincoln & Short, 1980). Declining photoperiod after the summer solstice stimulates a resumption of folliculogenesis and ovulation in the ewe and testis growth and spermatogenesis in the ram. In *M. eugenii* by contrast the female is never truly anoestrus; removal of the corpus luteum at any time of year will result in a resumption of ovulation, and the male is always fertile. Although the prostate enlarges during the breeding season (Fig. 4.7*a*) this is probably in response to elevated testosterone levels stimulated indirectly by the presence of females with active corpora lutea (see Fig. 6.17). This marked difference between the breeding seasonality of *M. eugenii* and other seasonally breeding species raises several questions which will be addressed in this chapter.

First, if true anoestrus does not occur, is there any change in hypothalamo-pituitary sensitivity to steroid feedback from the ovary or corpus luteum during the year? Second, what is the nature of the inhibitory control of the corpus luteum during lactation? Third, is the control of the corpus luteum in seasonal quiescence the same as that in lactation? Finally, if photosensitivity is indeed restricted to females, at what stage of sexual differentiation does it develop?

Seasonal change in pituitary and hypothalamus

Hearn (1974) could not detect any difference in the mean level of gonadotrophin in peripheral plasma of females between the breeding and non-breeding season but in a later paper (Hearn, Short & Baird, 1977) suggested that the frequency of short-term pulses of LH, rather than the basal level may be important, as they have been shown to be in the Soay ram (Lincoln, 1976). However, C. A. Horn & C. H. Tyndale-Biscoe (unpublished observations) could discern no consistent differences in the circadian pattern of peripheral plasma concentrations of LH or FSH, obtained each 2 months through 1 year that would suggest differences in pulse frequency at any time. They did observe that the basal level of LH, but not of FSH, was significantly lower (< 2 ng ml^{-1}) during the period of declining daylength in the first half of the year, which includes the breeding season, than during the second half (> 2 ng ml^{-1}) from the winter solstice to the summer solstice (Fig. 9.1). To determine if this was due to a changing sensitivity to gonadotrophin-releasing hormone (GnRH), as observed by Lincoln, Peet & Cunningham (1977) in the Soay ram, another group of females was challenged every 2 months with an injection of 10 μg of GnRH and the responses in plasma LH measured. The duration and amplitude of LH response remained the same throughout the year, from which it was concluded that there is no change in pituitary sensitivity to GnRH, nor any change in the storage of LH in the pituitary, through the year. This conclusion is consistent with the observed responses of females to removal of the corpus luteum.

A mentioned in Chapter 6, females lutectomised during lactation or during the first half of the oestrous cycle or pregnancy ovulate again 12–18

Fig. 9.1 Summary of hormone profiles and responses to experimental manipulations through the highly seasonal annual cycle of reproduction in the female *Macropus eugenii*. Refer to Fig. 2.26 for additional information on annual cycle. Preg, active pregnancy; LQ, lactational quiescence; SQ, seasonal quiescence; horizontal shading, lactation superimposed on non-lactating condition. Other data as follows: prolactin, Hinds & Tyndale-Biscoe (1982*b*; 1985); LH and FSH, C. A. Horn & C. H. Tyndale-Biscoe (unpublished results); progesterone, Tyndale-Biscoe & Hinds (1984); RPY, removal of pouch young, Sharman & Berger (1969); mammary denervation, Renfree (1979); bromocriptine, Tyndale-Biscoe & Hinds (1984 and unpublished); hypophysectomy, Hearn (1974); melatonin implant, Renfree & Short (1984).

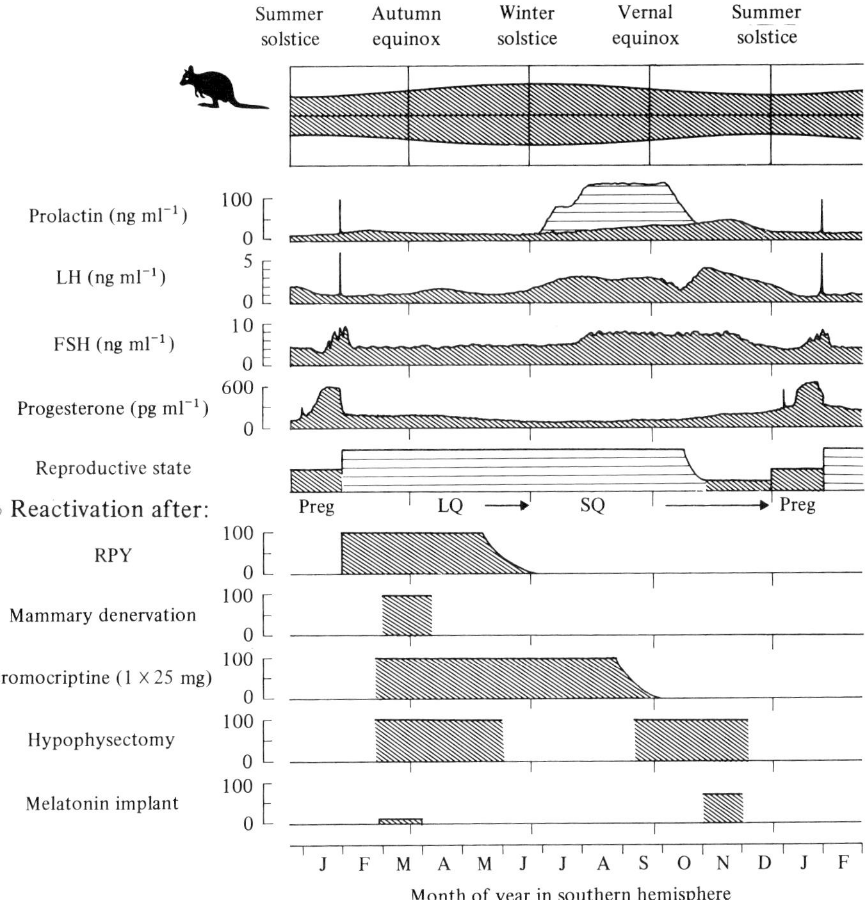

days later. The same response has been observed after lutectomy in seasonal quiescence (Tyndale-Biscoe & Hawkins, 1977) but these animals did not display oestrous behaviour, presumably because of the low level of progesterone preceding ovulation (see p. 207). Furthermore, females that attain sexual maturity in the non-breeding season will ovulate and their corpora lutea become quiescent (Fig. 2.26). Thus ovulation can occur at any time of the year, provided a quiescent corpus luteum is not present in either ovary.

Two aspects of ovarian steroidal feedback on the hypothalamo-pituitary axis have been examined in females during the breeding and non-breeding seasons. The responses of plasma LH and FSH concentrations to bilateral overiectomy have been studied and also the degree of inhibition of LH and FSH in ovariectomised females injected with different doses of oestrogen. In the first studies the levels of LH and FSH in intact females was low at both times of the year, but rose several-fold after ovariectomy. The speed of the response and the final concentrations of gonadotrophins attained was the same when females were ovariectomised during the non-breeding season (Tyndale-Biscoe & Hearn, 1981; Horn, Fletcher & Carpenter, 1985) as during the breeding season (Evans *et al.*, 1980; Horn *et al.* 1985). The depression of LH and FSH concentrations by oestradiol was dose dependent and on the highest dose of oestradiol the depression of LH only was followed by a brief LH surge, similar to the pre-ovulatory surge (Horn *et al.*, 1985). There was no significant difference within treatments between the animals injected in the breeding season, and those injected during the non-breeding season, from which it was concluded that the female *M. eugenii* shows no seasonal change in either negative- or positive-feedback response to oestradiol. This again is in marked contrast to the ovariectomised ewe in which elevated LH was depressed by oestradiol implants only during the non-breeding season and not at all during the breeding season (Webster & Haresign, 1983; Karsch *et al.*, 1984).

From these several experiments and observations we may conclude that there is no important change in the hypothalamus of the female through the year and no essential change in the pituitary gland's sensitivity to oestrogen nor in its secretion of LH or response to GnRH. The elevation in the concentration of LH seen in the non-breeding season may reflect other factors operating on the pituitary, such as those from the pineal, but it appears to be unimportant to the control of seasonal breeding. The central question is therefore to know what controls the corpus luteum which, in turn, controls ovulation.

The control of lactational quiescence

Lactational quiescence is a condition common to the great majority of Macropodidae (Table 2.2), whereas seasonal quiescence is only known to occur in *M. eugenii* and the Tasmanian subspecies of *M. rufogriseus*. It seems probable therefore that lactational quiescence is phylogenetically the older condition and that seasonal quiescence has been acquired in these two species more recently.

The mechanism for maintenance of lactational quiescence stems from the presence of a young in the pouch. Using *M. rufus*, Sharman (1965*a*, *b*) showed that the essential factor in this is the stimulation of sucking, rather than the volume of milk being taken. Renfree (1979) confirmed the importance of the stimulus of sucking more directly by denervating the sucked teat and associated mammary gland in *M. eugenii* during early lactation (Fig. 9.1). In most of the denervated and control females the pouch young continued to feed and grow for the full duration of the experiment, but only in the experimental animals did the corpora lutea resume growth and support advanced pregnancies equivalent to the stage that would have been reached if the pouch young had been removed at surgery. The probable means by which lactation was able to continue after denervation was discussed in Chapter 8, but it is clear from these results that the afferent arc from the mammary gland or its teat plays an essential role in the control of the corpus luteum and the response to severing it is as rapid as the response to removing the pouch young (Fig. 9.2).

Removal of the pituitary during lactation also leads to resumption of corpus luteum growth and embryonic development at the same rate as after removing the pouch young (Hearn, 1974; Hinds, 1983) but, in contrast to mammary denervation, lactation ceases and the pouch young die within 2 weeks. Furthermore, follicular growth and ovulation do not resume.

In Chapter 6 (p. 242) we showed that follicular growth and ovulation are inhibited by the quiescent corpus luteum and that oestradiol-17β is probably the steroid agent involved (Fig. 9.2). Taken together these several findings suggest that ovarian activity is suppressed during lactation in a two-step process: under the influence of the sucking stimulus, the pituitary inhibits the growth of the corpus luteum but not its capacity to secrete steroids at a low level, and these steroids, probably oestradiol, exert a negative feedback on the pituitary or hypothalamus to prevent gonadotrophic stimulation of follicular growth (Fig. 9.2). One difficulty with this idea is that no change has been detected in gonadotrophins during lactation or after removal of the pouch young, as this hypothesis predicts.

Fig. 9.2. Hypothetical scheme showing pathways for the control of quiescence of *Macropus eugenii*, B, blastocyst; CL, corpus luteum; DRG, dorsal root ganglion; $E_2\beta$, oestradiol-17β; F, follicle; FSH. follicle-stimulating hormone; GnRH, gonadotrophin-releasing hormone; LH, luteinizing hormone; P, progesterone; PIF, prolactin-inhibiting factor; PRL, prolactin; SCG, superior cervical ganglion; SCN, suprachiasmatic nucleus. ↓ Stimulation; ⊥ inhibition. Redrawn from Renfree (1981*b*).

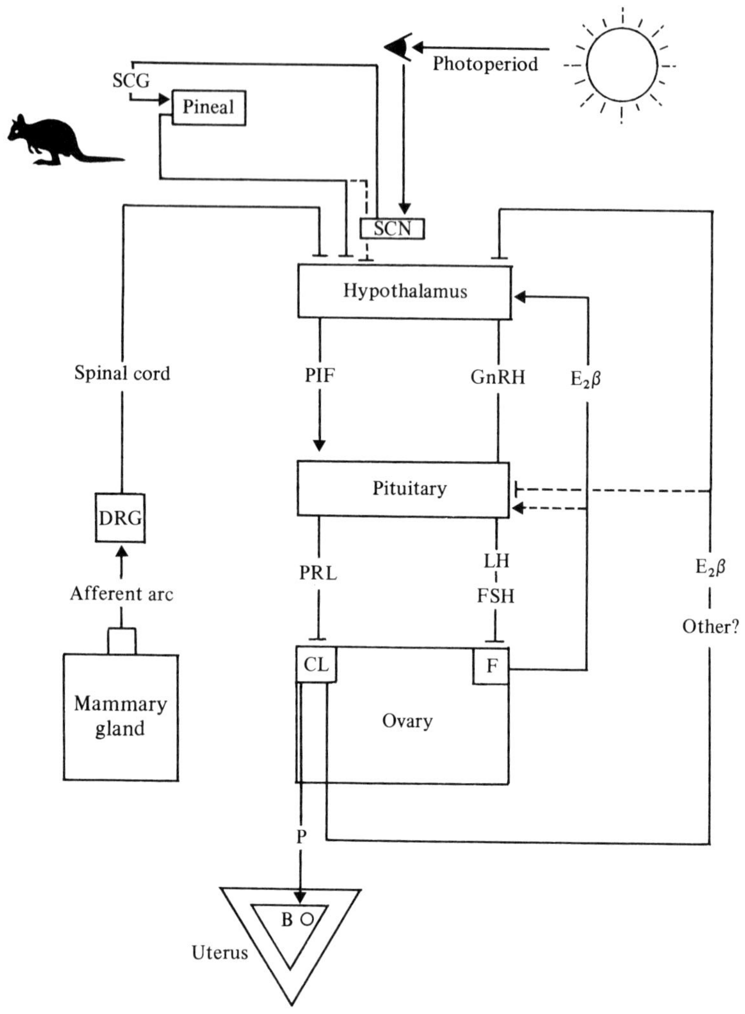

Since exogenous oestradiol (Horn *et al.*, 1985) or ovarian grafts (Evans *et al.*, 1980) can reduce the high levels of FSH and LH in ovariectomised females, these negative findings may be because the heterologous assays for LH and FSH are insufficiently sensitive to detect such changes. A second difficulty is that the quiescent corpus luteum has not been shown to have the capacity to synthesise oestradiol (Renfree *et al.*, 1984). The apparent absence of aromatase in luteal tissue requires further study.

The other aspect of this model that has received attention is the nature of the inhibition exerted by the pituitary on the corpus luteum. Since it is evident from the results of hypophysectomy (Hearn, 1974*b*; Hinds, 1983) and immunization against GnRH (Short *et al.*, 1985) that reactivation of the corpus luteum does not depend on a pituitary luteotrophic stimulus (see p. 251), an inhibitory factor from one or other part of the pituitary must be postulated.

Prolactin and oxytocin were examined for their ability to delay reactivation after the pouch young were removed because both are associated with lactation in other mammals and oxytocin had been shown to inhibit luteal function in heifers (Armstrong & Hansel, 1959). In *M. rufus* (Sharman, 1965*b*) and *M. eugenii* (Tyndale-Biscoe & Hawkins, 1977) oxytocin delayed reactivation for an equivalent time but in *M. eugenii* so did prolactin and reserpine, an alkaloid that stimulates prolactin release in rats (Fig. 6.20). However, after hypophysectomy only prolactin delayed reactivation (Tyndale-Biscoe & Hawkins, 1977). Furthermore, reactivation was provoked by removing only the anterior pituitary, leaving the neurohypophysis and pars intermedia intact, all of which suggests that prolactin is the main inhibitor of the corpus luteum and that oxytocin exerts its effect in *M. eugenii* and *M. rufus* by stimulating the release of prolactin.

If prolactin is providing a tonic inhibition to the corpus luteum and the effect is not due to a reciprocal fall in gonadotrophin, it must be acting on the corpus luteum directly and the luteal cells should have specific membrane-bound receptors for prolactin. As discussed in Chapter 6, the quiescent corpus luteum does bind prolactin and the concentration of prolactin receptors in luteal tissue exceeds all other tissues tested, except fully lactating mammary glands (Fig. 6.19). Moreover, no difference in the receptor concentration per cell was found between quiescent corpora lutea of females in early lactation, late lactation or seasonal quiescence (Stewart & Tyndale-Biscoe, 1982). We will return to this latter point when considering changes in responsiveness of the corpus luteum throughout the year.

The dopamine agonist bromocriptine depresses plasma prolactin in eutherian mammals. Therefore treatment with this drug should allow reactivation to occur in *M. eugenii* if prolactin tonically inhibits the corpus luteum, but the results were paradoxical; it did not have any demonstrable effect on the level of plasma prolactin in *M. eugenii* in early lactation, in late lactation (when prolactin is elevated, see Chapter 8) or in seasonal quiescence (Tyndale-Biscoe & Hinds, 1984). Nevertheless, it is a highly effective inducer of corpus luteum activity. During lactation a single injection will induce reactivation of the corpus luteum and blastocyst, with birth and post-partum oestrus occurring 27 days later (Table 9.1), precisely the same interval of time as after removing the pouch young (Tyndale-Biscoe & Hinds, 1984).

As in the experiments involving hypophysectomy and mammary denervation, some animals lost their pouch young 1 or 2 weeks after injection of bromocriptine but, in others, the corpus luteum was reactivated concurrently with continuing lactation and the neonate entered an occupied pouch and attached to one of the remaining three teats (Tyndale-Biscoe & Hinds, 1984). The interval from injection to birth was the same in females that lost their young as in those that retained them, so the response to bromocriptine cannot be due to failure of lactation.

An important conclusion from the experiments with bromocriptine is that the tonic inhibition need only be removed briefly for the corpus luteum to be reactivated. This is in contrast to other species, such as the sheep, in which treatment with bromocriptine, which reduces plasma prolactin, must be continued for many days to affect reproduction. In *M. eugenii* a single injection causes the delayed reproductive cycle to resume as quickly as it does when the pouch young is removed, the mammary gland denervated or the pituitary gland removed (Table 9.1).

While *M. eugenii* females do not give birth after hypophysectomy and none of the animals that were subjected to mammary gland denervation were given the opportunity to give birth, the initial rate of corpus luteum and embryo development is very similar to that after removing the pouch young or after an injection of bromocriptine (Table 9.1). This suggests that all four treatments are affecting the phenomenon in the same way and that 26–27 days is the minimum time required for the corpus luteum to recover from inhibition and cause the embryo to be reactivated and develop to full term. As mentioned in Chapter 6 (p. 237) the period from the transient pulse of progesterone to birth is 22 days and the same is so after bromocriptine (Table 9.1), so we are here really concerned only with an effect during the first 4–5 days after treatment. As the last 2 or 3 days of

Table 9.1. *The rate of response to various procedures that lead to reactivation in* Macropus eugenii, *as reflected in the intervals (days) to the progesterone pulse and birth*[a]

Treatment	Interval to progesterone pulse	Interval to birth	Progesterone pulse to birth	Reference
Removal of pouch young	5.6 ± 0.3 (5)	27.4 ± 0.4 (5)	21.8	Hinds & Tyndale-Biscoe (1982*a*)
Bromocriptine, 5 mg $kg^{-1} \times 1$, August	5.2 ± 0.2 (6)	26.3 ± 0.3 (4)	21.1	C. H. Tyndale-Biscoe & L. A. Hinds (unpublished results)
Hypophysectomy	7.3 ± 0.2 (6)	—	—	Hinds (1983)
Photoperiod change, 15L:9D to 12L:12D	10.0 ± 1.3 (5)	33.5 ± 1.1 (6)	23.5	Hinds & den Ottolander (1983)
Melatonin given 2.5 h before dark on 15L:9D	8.3 ± 0.3 (9)	29.9 ± 0.3 (9)	21.6	C. H. Tyndale-Biscoe & L. A. Hinds (unpublished results)
After summer solstice, 22 December	20.6 ± 0.7 (19)	42.6 ± 0.8 (11)	22.0	C. H. Tyndale-Biscoe & L. A. Hinds (unpublished results)

[a] Mean $\pm$ s.e.m. (N). Note that there is little variation in the interval from the pulse to birth, so that most variation is in the time taken to the pulse.

this period are known to be involved in the responses of the blastocyst to reactivation (see p. 287), it means that only 2–3 days are required for the corpus luteum to awaken from quiescence after the inhibition has been removed by one or other treatment. This gains further support from the observation that reactivation will not occur if the young is temporarily removed from the teat for 72 h or less (K. Gordon and M. B. Renfree, unpublished results).

So far we have considered the control of lactational quiescence only and in the majority of the Macropodidae this is the only condition in which the corpus luteum is restrained and the embryo held in diapause. However, in *M. eugenii* this process has superimposed on it a seasonal component.

The control of seasonal quiescence

From February to May (late summer to autumn in the southern hemisphere) all females will respond to removal of their pouch young by reactivation of the corpus luteum (Fig. 9.1), and females isolated from males at this time will undergo successive oestrous cycles. The proportion that will respond in these ways declines in May and, from June through December (winter and spring) it is very rare for a female to reactivate, whether her young is small, large or weaned (Fig. 9.1). During this period all adult females are in seasonal quiescence and carry a small, quiescent corpus luteum and most have an embryo in diapause. Young females that are weaned in October may undergo their first oestrus then and mating and fertilisation may take place (Andrewartha & Barker, 1969, Tyndale-Biscoe & Hawkins, 1977; Inns, 1982), but their corpora lutea also cease to develop until after the summer solstice. This is strong evidence that it is the corpus luteum and not follicular growth that is the primary target of inhibition by the pituitary.

The response of females to bromocriptine, like that to removing the pouch young, also varies with the season of the year (Fig. 9.1). In one study (Tyndale-Biscoe & Hinds, 1984) 19/20 females given a single injection of bromocriptine in February, March and June reactivated but those treated in September, October and November did not. In December, at the summer solstice, when animals were treated with bromocriptine none reactivated but about 14 days later all of them and the control animals spontaneously reactivated without the benefit of bromocriptine. In a second study (C. H. Tyndale-Biscoe & L. A. Hinds, unpublished results) groups of females were injected every 2 weeks from 22 June to 22 September. Up to August 19, all those expected to be carrying blastocysts (Renfree & Tyndale-Biscoe, 1973*a*) gave birth to a new offspring 27 days later. All those injected on 25 August reactivated while only one-third of those injected on 9 September and none of those injected on 22 September reactivated. These results indicate a similar pattern of response to bromocriptine as to removing the pouch young, except that females continue to respond to bromocriptine after the winter solstice and then cease to respond abruptly at the vernal equinox (Fig. 9.1). From the vernal equinox until shortly after the summer solstice bromocriptine, like removing pouch young, is almost wholly without effect and it would appear that an additional extrinsic factor over-rides the basic control of corpus luteum function. The most probable extrinsic factor is clearly changing photoperiod, which will be discussed in the next section, but this still leaves unresolved the way in which the response to bromocriptine is effected. Two possibilities have been examined but neither appears at this stage to be satisfactory.

The first possibility, suggested by Sernia & Tyndale-Biscoe (1979), is that the responsiveness of the quiescent corpus luteum to prolactin might change through the year, as it ages or as lactation proceeds, as a result of changes in the concentration of prolactin receptors on the luteal cell membranes. However, Stewart & Tyndale-Biscoe (1982) found no differences in prolactin receptor concentrations between corpora lutea of females in early (3 months) or late (8 months) lactation or in those not lactating and in seasonal quiescence. The second possibility to be considered was that the level of prolactin in peripheral plasma might be much higher in seasonal quiescence than before the winter solstice and that the dose of bromocriptine administered was either too low or too brief to reduce it sufficiently to elicit reactivation. The level of prolactin in peripheral plasma was determined before injection in all the animals used in the bromocriptine trials done at six times of the year by Tyndale-Biscoe & Hinds (1984). The lowest level was found at the winter solstice (Fig. 9.1); when compared to this value, the levels were significantly higher in September, October and November, when the females did not respond to bromocriptine. The levels were not significantly different in February and March when the females did respond to bromocriptine. This appears to support the hypothesis except that prolactin was the same concentration in February and March as at the summer solstice when none of the animals responded to bromocriptine. Another problem with this hypothesis, as mentioned earlier, is that no depression of prolactin could be discerned after bromocriptine treatment in either March or September even though the doses given were some 1200 times greater than those that will reduce plasma prolactin to undetectable levels in sheep (McNeilly & Land, 1979).

While the way in which bromocriptine affects the corpus luteum of *M. eugenii* remains unresolved there can be little doubt that it has an effect on the same pathway as that affected by the sucking stimulus and the pituitary. Extrinsic factors associated with photoperiodic change interfere with this.

Influence of daylength on seasonal quiescence

The period of birth after the summer solstice is remarkably constant from year to year and in different places (Table 9.2). Collie (1830) described late stage pregnancy, parturition and recently born young in six females on Garden Island off Western Australia and the estimated dates of birth are 10–27 January. On the same island 147 years later, 8 females examined by one of us (M.B.R.) on 7–10 December 1976 had quiescent corpora lutea while 7 females examined on 17–18 January 1977 had developing embryos, whose estimated dates of birth would have been 20 January to 4 February. Maynes (1977) recorded birth in *M. eugenii* on Kawau Island, New Zealand between 18 January and 11 February. This population originated in 1870 from South Australia and the females have retained the same pattern unchanged from that seen on Kangaroo Island, South Australia (Berger, 1966; Andrewartha & Barker, 1969; Renfree & Tyndale-Biscoe, 1973*a*; Flint & Renfree, 1982), and in the colonies studied in Canberra and Perth, where the mean date of birth on successive years has been 28 January to 3 February (Table 9.2).

The actual date of reactivation of the corpus luteum can be determined precisely by measuring the concentration of progesterone or oestradiol in plasma. The transient peak of progesterone occurs about 22 days before birth and is followed by the main rise in progesterone subsequently (see Fig. 6.13). In 17 females the transient pulse of progesterone was observed between 9 and 17 January and births occurred 22 days later between 31 January and 8 February (Table 9.1). As the early pulse usually occurs on day 5 after removing the pouch young this places the minimum date of reactivation at about January 4, or about 2 weeks after the summer solstice. Likewise Flint & Renfree (1982) observed high values of oestradiol in the plasma of females shot on Kangaroo Island on 3 January but not before or immediately after (see Fig. 9.3). From Shaw & Renfree's (1984) results, discussed in Chapter 6 (Fig. 6.15), this pulse of oestradiol occurs on day 5 or 6 and thus corresponds to the time of the progesterone pulse too and implies that reactivation began in this population on 29 December.

These several results show that corpora lutea, which may have been quiescent for up to 11 months, resume their activity in a highly synchronised

Table 9.2. *The main season of birth in* Macropus eugenii *in several southern hemisphere localities*

Locality	Year	N	Range	Mean date	Reference
Garden Is., Western Australia	1830	6	10J–27J	22 Jan[a]	Collie, 1830
	1977	7	20J–4F	27 Jan[a]	M. B. Renfree, unpublished observations
Kangaroo Is., South Australia	1964–9	246	7J–14F	end Jan[b]	Andrewartha & Barker, 1969
	1979	69	17J–6F	26 Jan[a]	Flint & Renfree, 1982
Kauau Is., New Zealand	1973	9	18J–11F	30 Jan	Maynes, 1977
Rotorua, New Zealand	1983	200	26J–14F	3 Feb	R. M. F. S. Sadleir, personal communication
Canberra, Captive colony, ex-Kangaroo Is.	1978	20	25J–5F	31 Jan	Tyndale-Biscoe & Hinds, 1984
	1979	25	14J–10F	28 Jan	C. H. Tyndale-Biscoe & L. A. Hinds, unpublished observations (1979–1983)
	1980	8	19J–2F	27 Jan	
	1982	17	31J–8F	3 Feb	
	1983	12	24J–9F	1 Feb	

[a] Estimated from stage of fetal development in pregnant females.
[b] Estimated from age of pouch young captured in later months.

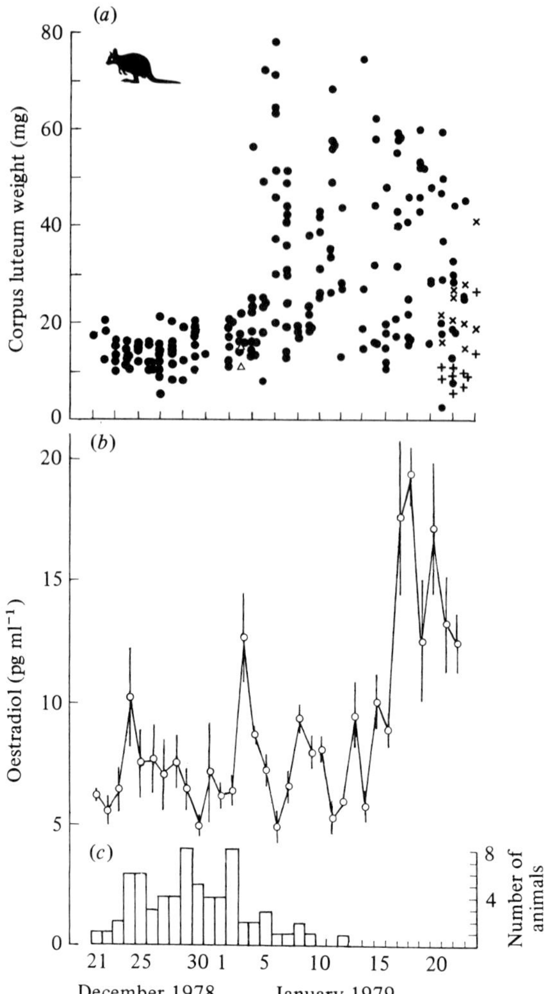

Fig. 9.3. Changes at reactivation in 201 female *Macropus eugenii* shot between 20 December 1978 and 23 January 1979. (*a*) Weight of the corpus luteum of each animal, showing the onset of growth in early January, coincident with the expansion of the reactivated blastocyst. One animal killed on 2 January carried two blastocysts and had two corpora lutea (△). In animals carrying newborn pouch young on or after 20 January, the corpora lutea of pregnancy are indicated by × and those resulting from the post-partum ovulation are shown by +. (*b*) Concentrations (mean ± s.e.m.) of oestradiol-17β in cardiac plasma from the 143 pregnant and post-partum of these animals. Note the peak on 3 January and the longer period of elevated concentrations at the time of parturition and post-partum oestrus. Six of the animals sampled on 21 and 22 January carried newborn pouch young. (*c*) Date

way shortly after the longest day of the year. Between the summer solstice and the estimated dates of reactivation is 7–13 days and, in this time, the daylength on Kangaroo Island decreases by about 6 min. If, indeed, the animals are sensitive to this small change of daylength it is possible that the time of full moon at or near to the solstice might moderate the animal's response and so the time of reactivation. When the date of reactivation of 299 tammars was compared by analysis of variance a significant positive correlation was found ($F = 28.498$, $p < 0.01$), which suggested that moon phase may affect the time of reactivation (Sadleir & Tyndale-Biscoe, 1977; Tyndale-Biscoe, 1980*b*).

To test this, the day of birth or oestrus was recorded for each of 20 females on successive years, when the full moon nearest to the solstice was on 13 December and 29 December, the maximum possible difference in relation to the solstice. The mean dates of birth were 3 February and 5 February respectively and therefore do not support the idea that the moon may be involved in the photoperiod response of *M. eugenii*.

Response to experimental change in photoperiod

Photoperiod influence has been investigated experimentally in four studies. Hearn (1972*b*) subjected six females to an accelerated photoregimen so that they experienced summer solsticial daylength 1 month early. They were compared to a group of animals exposed to normal outside conditions without artificial light. The six experimental animals cleaned their pouches, a preliminary to parturition (see Chapter 2 p. 76), 1 month earlier than the controls but both groups gave birth at the usual time in early February. Sadleir & Tyndale-Biscoe (1977) and Hinds & den Ottolander (1983) held females on 3 photoperiod regimens for 4 months from the vernal equinox (21 September). One group experienced constant daylength of 15 h light, 9 h dark (15L:9D), approximately equivalent to that at the summer solstice. The control group received artificial light equivalent to the changing pattern of natural daylength and the third group were held on equinoctial daylength (12L:12D). The females on constant long daylength gave birth at the same time as the control groups in late December or early January, from which it was concluded that long

Fig. 9.3. *cont.*
of reactivation calculated for 69 of these animals from the diameter of reactivated vesicles flushed from uteri between 1 and 20 January and crown-rump lengths of fetuses obtained between 14 and 22 January, using the growth curve (Fig. 7.8) derived from embryos reactivated after removing pouch young. Redrawn from Flint & Renfree (1982).

daylength itself is not the stimulus for reactivation of the corpus luteum. Since these animals had no other photoperiod cue it is possible that they were responding to an endogenous circannual rhythm. On the other hand, the third group (on 12L:12D) gave birth at various times before December but none at the end of December, when the control group gave birth. This seemed to suggest that this photoregimen is neither inhibitory nor stimulatory but permissive.

A preliminary experiment by Sadleir & Tyndale-Biscoe (1977) was more informative for understanding the actual signal to which tammars may be responding. In August 6 weeks after the winter solstice, non-lactating females were exposed to 15L:9D for 6 weeks, that is for almost twice the length of gestation. Thus there was ample time for the animals to reactivate and give birth if the primary stimulus is long day itself. None did so, and at the vernal equinox they were placed in an outside yard, thereby experiencing a 3 h reduction in daylength. All 6 gave birth 29–36 days after the change. Subsequently, the group in the main experiment on 12L:12D that did not give birth by the time the control group had, were exposed to 9L:15D, a similar drop in daylength, and 4 of them gave birth 30 or 31 days later. These two sets of results suggested that a sudden shortening in photoperiod is a potent signal for reactivation of the corpus luteum in *M. eugenii*. Since the interval (29–36 days) is significantly longer than that observed from removal of pouch young or bromocriptine injection to birth (26–27 days) some additional factor may be involved, such as changing levels in prolactin or progesterone.

To examine this, Hinds & den Ottolander (1983) repeated this design, except that the animals were under artificial light during the first phase on 15L:9D and the second phase on 12L:12D and prolactin and progesterone were measured in the females. The response was the same with births occurring 34 days after the change of photoperiod (Table 9.1). Plasma prolactin in these animals declined from 80 ng ml^{-1} during 15L:9D to about 25 ng ml^{-1} on 12L:12D and the transient peak of progesterone occurred about 10 days after the photoperiod change (Table 9.1). This is longer than the interval (5–6 days) after removing the pouch young, after hypophysectomy or after injecting bromocriptine in lactational quiescence. Furthermore, all the variation observed between females was in the interval before the peak, while the interval from the peak to birth was 23 days, about the same as that seen after removing pouch young, after bromocriptine and after the summer solstice.

In a second group, initially on 12L:12D and then on 9L:16D, all the animals gave birth at erratic times before or after the change to 9L:16D, presumably because this photoregimen was permissive as previously suggested. However, when all these animals had given birth their new young were removed. On this second occasion the progesterone pulse occurred 6 days and, birth 28 days later, which are the same intervals as those after removing pouch young during the first half of the year on declining photoperiod. Together these results indicate that there is an additional photoperiod-controlled component in the inhibition of the corpus luteum during increasing daylength or long days, which is not present on decreasing daylength or short days. Furthermore, the factor is abolished when the animals experience artificially short days in the second half of the year.

Role of the pineal gland

There is now much evidence that the pineal gland is involved in transducing photoperiod information and thereby the control of seasonal breeding (Lincoln & Short, 1980; Reiter, 1981). The pineal can be inactivated by destroying its sympathetic innervation, so that it is unable to complete the synthesis, and hence secretion, of the hormone melatonin. This is done by surgically removing the two superior cervical sympathetic ganglia (Fig. 9.2). However, this also destroys the sympathetic innervation of the whole head and so the effects are more pervasive than direct surgical extirpation of the pineal gland itself. The activity of the pineal can be monitored by measuring melatonin in the plasma by radioimmunoassay and this may reflect responses of the animal to altered photoperiod.

In *M. eugenii*, as in all other mammals in which it has been measured, the concentration of melatonin in peripheral plasma shows marked circadian fluctuations (Renfree *et al.*, 1981a) with the rise coinciding with the onset of the dark phase and the fall with the onset of the light phase each day (Fig. 9.4). In the Soay ram the duration of elevated melatonin is inversely related to the peak concentration (Lincoln & Short, 1980) and in female *M. eugenii* there is a similar pattern, with nighttime values being lowest and of longest duration in winter and spring, and highest and of shortest duration in summer and autumn (McConnell, 1986). As in the ewe (Kennaway *et al.*, 1977) both pinealectomy and sympathetic ganglionectomy abolished the nocturnal rise in *M. eugenii* (Fig. 9.5) (Renfree *et al.*, 1981*a*; McConnell & Hinds, 1985).

Since the response of the corpus luteum to change in daylength during seasonal quiescence takes 6–7 days longer than the response to removal of the pouch young in lactational quiescence (see Table 9.1), that interval might be required for the diurnal pattern of melatonin to change

Fig. 9.4. Concentrations of plasma melatonin (mean ± s.e.m.) in female *Macropus eugenii* on successive photoperiod regimens. (*a*) On winter photoperiod of 10 h light: 14 h dark, then (*b*) after change to the inhibitory photoperiod of 15 h light: 9 h dark, and (*c*) on the day of change (lower profile) in the stimulatory photoperiod of 12 h light: 12 h dark and on day 5 after the change (upper profile). (*d*) Profiles of melatonin on 2 successive days in a female on 15 h light: 9 h dark, the first without (●), and the second after (○) a subcutaneous injection of melatonin 2.25 h before lights off. The stippled bars indicate the dark phase and asterisks denote a significant difference ($P < 0.05$) from the corresponding value at the same time of the previous photoperiod. From McConnell & Tyndale-Biscoe (1985) with permission.

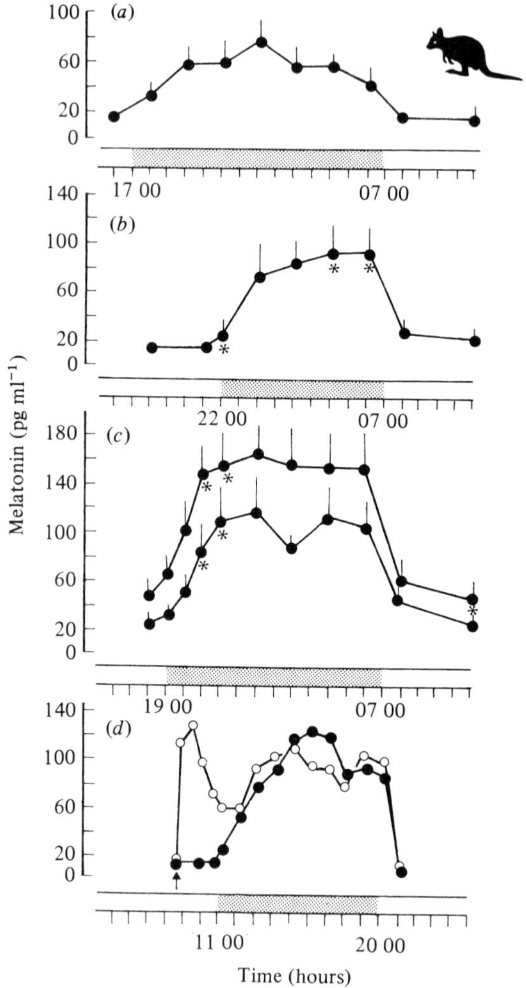

in response to the photoperiod change. McConnell & Tyndale-Biscoe (1985) examined this using a similar experimental design to that of Hinds & den Ottolander (1983). After 6 weeks on summer daylength (15L:9D) the daylength was reduced to equinoctial (12L:12D). From the first occasion when the dark phase increased to 12 h, the nocturnal rise of melatonin was advanced and became fully synchronised with the new photoregimen by day 5 (Fig. 9.4*b*, *c*). These animals gave birth 32 days after the photoperiod change, as in the previous experiments of Sadleir & Tyndale-Biscoe (1977) and Hinds & den Ottolander (1983). To test the idea that the shift in melatonin is the signal that initiates reactivation of the corpus luteum another group on the 15L:9D photoperiod were injected with either melatonin or arachis oil 2.5 h before dark each day for 15 days (Fig. 9.4*d*). Only the animals injected with melatonin reactivated and gave birth 32 days after the start of melatonin injections.

In another experiment (Renfree & Short, 1984) females in seasonal

Fig. 9.5. Concentrations of plasma melatonin (mean ± s.e.m.) in sham-operated (*a*), intact (*b*) and ganglionectomised (*c*) female *Macropus eugenii* bled at 09.00, 18.00 and 24.00 h. Ganglionectomy abolishes the nighttime increase in melatonin concentration. Redrawn from Renfree *et al.* (1981*a*).

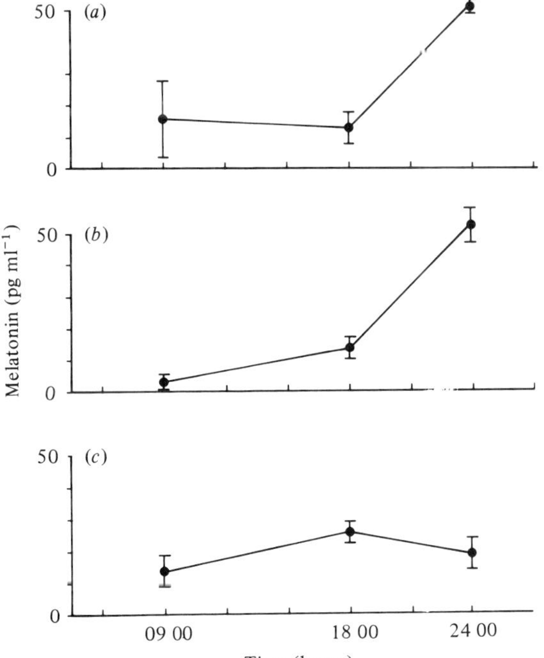

quiescence, and not subjected to any experimental photoperiod changes, were given silastic implants, which provided a continuously high level of melatonin in the circulation but had no effect on prolactin. These animals reactivated more rapidly than those in the previous study and gave birth 25–29 days after insertion of the implant (Fig. 9.1). It is clear from the results of both studies that melatonin has an important role in the control of seasonal quiescence but the difference in the response time to episodic and continual melatonin treatment suggests that the time in the 24 h when the female experiences high levels of melatonin may provide even more information about the time of the year and the appropriate reproductive response. It is of interest that the melatonin treatment that mimicked the endogenous melatonin profile after photoperiod change resulted in the same reproductive response. In the only other macropodid that displays seasonal quiescence, *M. r. rufogriseus*, Loudon, Curlewis & English (1985) induced reactivation with melatonin implants and, they also noted a lag of 6–7 days in the reactivation, and also could detect no significant change in plasma prolactin concentration as a result of the treatment.

What these results support is the idea that there are two mechanisms controlling quiescence of the corpus luteum in *M. eugenii*, and possibly also *M. rufogriseus*; the basic pituitary-dominated system with a 26–27 day response time and the over-riding photoperiod system, with a 31–33 day response, which is now shown to involve the pineal gland through its circadian secretion of melatonin. Conversely, when the melatonin profile is overwhelmed, *M. eugenii* females apparently revert to the basic pituitary-dominated system.

The next question to consider is whether the pineal gland is the main agent that controls seasonal quiescence in *M. eugenii* and, if so, how it does it. The first test of the pineal's role was reported by Renfree *et al.* (1981*a*), when they abolished seasonal quiescence by subjecting females to cervical sympathetic ganglionectomy during lactational quiescence. From August to December, when intact females are in seasonal quiescence, the ganglionectomised females produced a succession of 3 young each after removal of the preceding suckling. The interval from removal of the pouch young to birth was 27–28 days, similar to that in lactational quiescence, rather than 31–33 days as occurred in intact tammars after photoperiod change at this time of year. Conversely when McConnell & Hinds (1985) pinealectomised post-lactational females during seasonal quiescence in October, none gave birth before the summer solstice, but all did do so with the controls at the normal time in January–February. This suggested that there is either a different response to the two ways of inactivating the pineal or that there is an endogenous rhythm that comes into operation after April and continues even after pinealectomy in October. The latter idea gets some support from the fact that the animals held on summer daylength for 3 months by Sadleir & Tyndale-Biscoe (1977) gave birth at the same time as the controls held on increasing photoperiod that mimicked seasonal changes.

McConnell (1984) investigated these two possibilities by subjecting one group of females to ganglionectomy in October and other groups to pinealectomy in April, June and July. The animals ganglionectomised in October responded in the same way as those pinealectomised in October, indicating that the manner of pineal inactivation was not important. However, the results of pinealectomy in April to July were different from ganglionectomy in April–May (Renfree *et al.*, 1981*a*). Most females retained their young and, after these were removed at the equinox on 21 September, about half reactivated and produced another offspring 28–30 days later but the remainder and all the sham-operated controls gave birth in January–February. In their second year after operation, similar results were obtained with the two groups pinealectomised in June and December and one group ganglionectomised in October. Two different conclusions can be drawn from these results. The first is that pinealectomy does not abolish photosensitivity in all females, although it does abolish the diurnal profile of melatonin. There is evidence from sheep (Legan & Karsch, 1983) that total abolition of the annual cycle only occurred in blinded ewes and Dubocovich (1983) has shown that melatonin synthesised by retinal cells can affect dopamine secretion in hamsters and this may be able to mediate a photoperiod response. The alternative conclusion is that removing the pineal does abolish photosensitivity but apparently not an endogenous circannual rhythm and that, in some females, this may persist for at least 2 years after pinealectomy or ganglionectomy. With present knowledge it is not possible to differentiate between these alternatives. What is clear from the present results is that not all female *M. eugenii* respond to pinealectomy in the same way.

Stage of development of photosensitivity in the female

All present evidence supports the conclusion that male *M. eugenii* are not photosensitive or are much less sensitive than females, and that the seasonal changes in plasma testosterone and prostate weight, observed by Inns (1982) (Fig. 4.7*a*), are stimulated by changes occurring in the females with which they are associated (Fig. 6.17; Catling & Sutherland, 1980). A question one can then ask is, at what stage in sexual differentiation does photosensitivity develop in the female? Since the young females that experience their first oestrus when 8 months old in October do not complete the first pregnancy or cycle but become quiescent (see Fig. 2.26), they must already be fully sensitive to the photoperiod influences that control adult females.

This suggests that differentiation of the photosensitive centre or pathway in the brain must take place during pouch life. If exposure to light is necessary for this, differentiation could occur after 120 days of age when the eyes open or after 200 days when the young begins to put its head out of the pouch (Table 2.1). However, sexual differentiation of the brain probably occurs much earlier, since females given a testis graft 10 days after birth failed to reproduce when adult (S. J. McConnell, L. A. Hinds & C. H. Tyndale-Biscoe, unpublished results). It is also now known in *M. eugenii* that the neural pathway from the retina to the lateral geniculate nucleus is established by day 12 and to the suprachiasmatic nucleus (SCN) (Fig. 9.2) by day 65 (Wye-Dvorak, 1984). Differences have been reported in the SCN of male and female rats from birth (Le Blond *et al.*, 1982) which are thought to be related to the differentiation of the hypothalamus that occurs at an early age in this species (see Imperato-McGinley, 1983), and there is evidence that a circadian rhythm may be entrained *in utero* from the maternal rhythm (Reppert *et al.*, 1983). However, if the pattern is entrained from the female *M. eugenii* this in itself would not explain the differentiation between male and female photosensitivity.

Conclusions

The basic macropodid pattern of reproduction is controlled by lactation, although under particularly adverse conditions species, such as *M. rufus*, can become truly anoestrus (Fig. 2.25; Newsome, 1964*a*). However, in *M. eugenii* the basic pattern of lactational quiescence is only expressed when the female is experiencing shortening or uniformly short days. Under these environmental conditions ovarian activity is controlled by a sequence that begins with the frequency of the sucking stimulus to the teat, by a neural arc to the adenohypophysis, presumably via the hypothalamus, thence by means of prolactin to suppress development of the corpus luteum (Fig. 9.2). The suppressed corpus luteum, nevertheless, secretes sufficient oestradiol to so affect FSH and LH secretion by the pituitary that folliculogenesis and ovulation are inhibited. Experimental interference with this sequence allows corpus luteum reactivation with a pulse of progesterone on day 5 or 6 and birth and post-partum oestrus on day 26 or 27.

Photoperiod stimuli, associated with increasing or long daylength, prevent this response, whereas decreasing or short daylength permit the response. The pineal is involved but its role is not yet clear: some females reactivate their corpora lutea soon after denervation or removal of the pineal, but other females do not do so until the normal time after the summer solstice, so that it is necessary to invoke either an endogenous circannual rhythm which is independent of the pineal, or a residual photosensitivity, despite the lack of diurnal fluctuations in melatonin. Whatever the mechanism is, it requires about 6 days to decline to the point where the basic control sequence can begin to function and the corpus luteum be released from inhibition. Since hypophysectomy during seasonal quiescence is an effective means of inducing immediate reactivation of the corpus luteum, it must be supposed that the photoperiod-directed mechanism is ultimately operating on the pituitary. The most probable site is therefore the hypothalamus.

Differentiation of the photosensitive system must occur in females during pouch life probably before the eyes open at 120 days. The paradox in this is that the female is the homogametic sex in marsupials, as in other mammals (see p. 95), and yet this phylogenetically recent adaption is not differentiated in the male. It will be of some interest to learn whether males castrated in early pouch life become photosensitive like females do, and if females androgenised as pouch young lack photosensitivity as adults.

10

Marsupials and the evolution of mammalian reproduction

All discussions about the origin and evolution of mammalian reproduction are constrained by the time frame of the fossil record and rely to a considerable extent on comparisons between the modes of reproduction in living representatives of the Prototheria (monotremes), Metatheria (marsupials) and Eutheria (placentals). Such comparisons are limited by the degree to which the processes of reproduction are understood in each group and by the assumption that the living species resemble their remote ancestors in essential features. In comparing living mammals there is now little dissension from the view that the Monotremata share fewer characters in common with the Metatheria than the Metatheria do with the Eutheria. This applies to characters controlled by structural genes, such as serum proteins (Kirsch, 1977*a*), amino acid sequence of haemoglobin and myoglobin (Whittaker & Thompson, 1974), chromosome number and sex-determining mechanisms (VandeBerg *et al.*, 1983) which probably reveal true evolutionary affinities, as well as anatomy and reproduction (Griffiths, 1978). Where the differences of opinion occur are in how the living mammals are to be related to Mesozoic mammals and what evolutionary lines are to be drawn therefrom.

Discussion of evolutionary relationships has been enhanced by the theory of cladism but the power of cladistic analysis is only as great as the knowledge of the character states available. For fossils these are necessarily the characters controlled by regulatory genes which produce phenotypic changes in response to local environmental demands, and are limited to teeth, bones and the impressions of other organs left on them, and may be of little evolutionary significance in the cladogram, a point raised by Kemp (1983). Nevertheless, fossils provide the essential time frame for any evolutionary hypothesis.

Living species provide far more character states for analysis but the emphasis given to these may be influenced by incomplete knowledge of the character or of its distribution, and by preconceived notions of the relationships being sought. In Chapter 1 we gave an example from the nineteenth century of the resistance to an idea that contravened contemporary thinking about mammals. Analagous though less dramatic prejudices seem to colour current thinking about the relationships of living mammals today, particularly with regard to their reproduction.

The most detailed cladistic analyses based on reproductive phenomena of mammals have been made by Luckett (1977) and Marshall (1979). Luckett (1977) accepted the premise, from palaeontological evidence, of a primary dichotomy between Prototheria, of which the monotremes are the living representatives, and Theria to which the marsupials and eutherians belong. Marshall (1979) based his analysis on a number of anatomical and skeletal characters from fossil and living mammals, as well as some reproductive characters from living mammals, and concluded that they supported the primary dichotomy of Prototheria–Theria and a second dichotomy of Metatheria–Eutheria. For the reproductive character states he relied on secondary sources, particularly the reviews of Lillegraven (1969, 1975), which were written from a palaeontologist's perspective.

Since Luckett's and Marshall's papers were written, Griffiths (1978, 1984) has reviewed reproduction of monotremes and Rowlands & Weir (1984) that of non-primate eutherians, particularly the Insectivora, with which marsupials share several features of reproduction. Also, since then, there has been a burgeoning of information on the physiology of reproduction in marsupials, reviewed in the foregoing chapters, so that it seems appropriate to conclude this book by examining the reproductive relationships of living mammals in the light of this new information. We begin with a brief summary of the palaeontology of mammals, based on the recent reviews by Lillegraven, Kielan-Jaworowska & Clemens (1979), Crompton & Jenkins (1979), Crompton (1980) and Kemp (1983). This is followed by a consideration of the reproduction of living mammals, particularly those character states that were identified by Lillegraven (1975), Luckett (1977) and Marshall (1979) as being significant in cladistic analysis. We conclude with a comparison between reproduction in small and large marsupials, and reproduction in small and large eutherians.

Our thesis, briefly stated, is that the basic mode of mammalian reproduction evolved simultaneously with the origin of mammals in the Triassic and remained almost unchanged until the Tertiary because it was the most appropriate mode for small nocturnal insectivorous mammals.

The major differences seen in living mammals today, we suggest, evolved during the great adaptive radiations of the Eutheria and Metatheria in the Tertiary, in response to the metabolic requirements of increasing body size and the ecological constraints imposed by the radiation into new niches and modes of life.

The palaeontological record of mammals

The earliest fossil mammals are known from the late Triassic, 180 MY ago, of China, Europe and Southern Africa, post-dating by about 50 MY the most probable ancestors among cynodont mammal-like reptiles. Even at this early stage of mammalian evolution two orders can be recognised. The best-represented order, the Morganucodontidae, have generally been considered to be the forerunners of the late Jurassic Triconodonta, Docodonta and Multituberculata, and there is growing evidence for a close relationship between the latter order and the living Monotremata. These orders have been classified together as the Prototheria, although Griffiths (1978), Presley (1981) and Kemp (1983) disagree with this scheme. The other Triassic order, the Kuehneotheridae, is represented only by fragments of jaws and teeth but is thought on these criteria to be the forerunner of the Jurassic Symmetrodonta and Pantotheria. From within this latter order in the Cretaceous the earliest Eutheria and Metatheria are considered to have arisen, the forerunners of the two major groups of mammals alive today. These several orders are grouped together as the sub-class Theria.

According to this classification therefore, the living monotremes are only distantly related to marsupials, whereas marsupials are much more closely related to eutherian mammals. Lillegraven (1969, 1975, 1979) and Marshall (1979) concluded that the evidence from living mammals, particularly from reproduction, fully supports the primary dichotomy of Prototheria and Theria, as well as the secondary dichotomy of Metatheria and Eutheria.

However, this view of the relationships of the living and fossil mammals has been challenged by Presley (1981) and Kemp (1983). The primary dichotomy is based largely on differences in two characters, cusp patterns of the molar teeth and the construction of the side wall of the brain case. In the fossil groups included as Prototheria the side wall is formed by a large sheet continuous with the periotic posterior to the foramen ovale, whereas in therian mammals it is formed by an ascending process of the alisphenoid, which is pierced by the foramen. Both genera of living monotremes were considered to follow the prototherian pattern and on this criterion Kermack (1963) concluded that the morganucodonts, triconodonts and multituberculates formed with the monotremes a monophyletic group. However the brain case of the Kuehneotheridae is unknown and subsequent study of complete series of skulls of *Tachyglossus* by Griffiths (1978) and of *Ornithorhynchus* by Presley (1981) has shown that there is no real difference in the ossification of the side wall of the brain case between these mammals on the one hand and marsupials and eutherian mammals on the other. Therefore the morganucodontid or prototherian pattern could have been ancestral to all living mammals equally.

The second character, the pattern of molar cusps, differentiates prototherian mammals as those that have the cusps arranged linearly, or whose pattern can be derived from a linear arrangement, and therian mammals as those in which the cusps are disposed in a triangular pattern. Apart from the egg tooth, the only teeth known from living monotremes are the ephemeral milk teeth of juvenile platypus and they are said to show a linear arrangement of the two cusps. Kemp (1983) questions the logic that accepts that molar cusp pattern has been immutable through mammalian

Fig. 10.1. Phylogenetic interrelationships of the mammal groups proposed by Kemp (1983). The Eupantotheria are indicated as a paraphyletic group structurally antecedent to modern therians, monotremes, and possibly multituberculates. The main synapomorphies which support this phylogeny are as follows. (1) Elaboration of the cingulum cuspules of the molars. (2) Triangulation of the main molar cusps. (3) Ear ossicles; ascending process of the alisphenoid reduced; septomaxilla excluded from face; expanded neopallium; atlas elements fused; transverse foramina in cervical vertebrae; circular acetabulum; epiphyses. (4) Specialisation of molar teeth. (5) Fusion of anterior lamina to ventral ramus of alisphenoid and expansion of cranial process of squamosal; vertical tympanic membrane; tribosphenic molars.

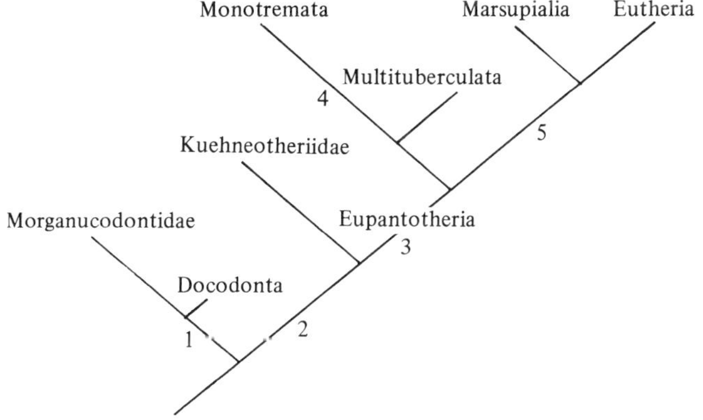

evolution but more complex structures, such as the auditory ossicles must have evolved in parallel several times (see Marshall, 1979). Kemp (1983) argues that current views overemphasise superficial dental similarities, misinterpret the structure of the mammalian braincase, and too readily accept parallel evolution. He considers the most parsimonious hypothesis at present for the relationships of fossil and living mammals is that the monotremes and modern therians form a monophyletic group, the sister group of morganucodontids and that the monotremes diverged at a later date from the therian line, possibly as (or with), the Multituberculata (Fig. 10.1). This hypothesis accords more closely with the evidence from reproduction, now to be reviewed.

Comparison of reproduction in living mammals

In this section we will compare the condition in monotremes and eutherians with that of marsupials reviewed in earlier chapters. For convenience we follow the same order (Table 10.1).

Sex chromosomes

The chromosome number of *Tachyglossus is* 64 ♀, 63 ♂ and of *Ornithorhynchus* 52 and are more similar to reptilian than to eutherian or marsupial karyotypes (Murtagh & Sharman, 1977), but the sex-determining mechanisms are similar to those of marsupials (see p. 95). Little is known of dosage compensation in monotremes but, as discussed in Chapter 3 the conclusion is that the dosage compensation has evolved independently in marsupials and eutherians (VandeBerg *et al.*, 1983), and so should be considered to be an apomorphic character for each group.

Male anatomy

All three species of monotremes are testicond but this does not necessarily imply a primitive heritage: *Ornithorhynchus* is adapted for a semi-aquatic life and *Tachyglossus* is fossorial, and many aquatic and fossorial eutherians are testicond, as is the fossorial marsupial *Notoryctes*. Although the majority of male marsupials are scrotal, the way in which the testicular and epididymal temperature differential is maintained is different from that of Eutheria, and suggests that these were independently evolved and so also may the scrotum have been. However, the pre-penial position of the scrotum in marsupials does not exclusively differentiate them from Eutheria; in lagomorphs and some myomorph rodents the scrotum is also pre-penial.

In monotremes the ureter enters the urogenital sinus beside the vas deferens, unlike in marsupials and eutherians in which the ureter enters the bladder. Monotremes and marsupials lack seminal vesicles and have disseminate prostate glands whose structure and secretory outputs are different from those of eutherians. The glans penis of monotremes, like that of sauropsids and most species of marsupial is bifid, whereas in eutherians the glans is typically undivided. The morphology of monotreme sperm is like that of birds and reptiles and distinctly different from all marsupial and eutherian sperm. Nevertheless, there are also important differences in the development of the acrosome and maturation of the sperm between those of Eutheria and of marsupials, as discussed in Chapter 4. There is also the difference of sperm conjugation in neotropical marsupials which is unknown in Australasian marsupials.

Female anatomy

The relative position of the ureters and the genital ducts is the character that most clearly separates the three groups of living mammals. In monotremes, as in sauropsids, the ureters open into the dorsal wall of the urogenital sinus opposite the urethral openings of the bladder but, in both marsupials and eutherians, the ureters migrate from a dorsal position adjacent to the Wolffian duct to a direct ventral connection with the bladder. There are only two possible routes for this; in marsupials the ureters take the medial route so that they separate the genital ducts, whereas in all eutherians they take the lateral route, so that the genital ducts of each side are contiguous.

To Buchanan & Fraser (1918) and Baxter (1935), who first described its development, and later reviewers (Eckstein & Zuckerman, 1956; Sharman, 1970; Tyndale-Biscoe, 1973) this was seen to be the most profound distinction between marsupials and eutherians, because it is exclusive. More recently, however, Lillegraven (1969, 1975), Sharman (1976) and Luckett (1977) have taken the view that the marsupial pattern is essentially the same as that of monotremes and sauropsids and only the eutherian pattern is apomorphic. They further contend that mid-line fusion of the Mullerian ducts in the Eutheria to form a single vagina and uterus was a necessary prerequisite for fetal retention in the uterus, and for its growth and eventual delivery. On this argument, with which we disagree, the marsupial condition is seen as an ineffectual attempt at mid-line fusion for retention of larger fetuses.

Against this interpretation is the fact that in many eutherian species, such as the rabbit, rat and pig, the uteri are either completely or extensively separate and each is capable of sustaining and accommodating several fetuses. Conversely in no marsupials do the uteri fuse in the mid-line even though the adjacent portions of the Mullerian ducts invariably do so to form the anterior vaginal cul de sac (see Chapter 5). Since the combined volume of the fetus and associated membranes of marsupials are comparable to the fully formed monotreme egg, which is delivered through a single oviduct, there again seems to be no strong reason for mid-line fusion to have been selected for this purpose. On the other hand, an essential adaptation for retention and nurture of the embryo was the development of a glandular endometrium capable of producing a copious secretion under the stimulus of progesterone. This is a feature shared by all three groups of mammals.

Indeed, there seems on present knowledge to be no reason for the mid-line fusion of the oviducts anterior to the ureters in all marsupials, which leads us to the conclusion that the original selection in both eutherians and marsupials was not for a reproductive but for an excretory function. For the ureters to enter the bladder directly, instead of via the urogenital sinus, the genital ducts had to be circumvented on one side or the other, a matter that was then neutral, although the subsequent consequences for reproduction were profound. This is a true dichotomy which allows for no intermediate stage, and one pattern cannot be derived from the other.

A point sometimes forgotten is that, while this dichotomy results in the bizarre vaginal anatomy of the female marsupial (Fig. 5.3), it also results in the convoluted arrangement of the vasa deferentia of the male eutherian.

Follicle and corpus luteum

Three maternal pathways are used by amniote vertebrates for transferring nutrients from the mother to the offspring: the wall of the ovarian follicle, the secretory lining of the oviduct or its derivative the uterus, and the mammary epithelium. While there has been a reduction in importance of the first route in all mammals, the latter two routes depend for their functioning on steroids secreted by the follicle or its derivative the corpus luteum, or on hormones produced by the placenta. In Eutheria the latter structure has become the predominant organ of fetal nutrition surpassing in importance the other routes.

As discussed in Chapter 3, the growth and development of the Graafian follicle is remarkably similar in all three groups of mammals, notwithstanding that the oocyte of the monotreme is about 20–40 times larger than that of the marsupial or eutherian and almost fills the antrum. Unlike sauropsids, the follicular cells of monotremes provide a secretion, which has been said to be homologous to the liquour folliculi of other mammals

(Flynn & Hill, 1939). Furthermore, not all eutherian species develop antral follicles before ovulation: in the ovaries of several species of Insectivora, e.g. *Suncus murinus*, the oocytes reach maturity and ovulate from small follicles with very small antra, which, after ovulation, form typical luteal tissue (Rowlands & Weir, 1984). We therefore disagree with Luckett (1977), who considers antral follicles to be synapomorphic for marsupials and eutherians.

Corpora lutea are formed from the ruptured follicle in all mammals and, in all species examined, they are necessary for the establishment of pregnancy. They are also necessary for the maintenance of pregnancy in some, but not all eutherian species (see Heap & Flint, 1984), although not in any marsupial species so far examined. In all marsupials examined and some eutherians the corpus luteum is necessary for parturition and mammogenesis, while in other eutherian species these functions have been assumed by the placenta. On very scant evidence the corpus luteum of monotremes is probably also necessary for intrauterine development and possibly for egg laying, since it reaches its maximum size just before this occurs. In one specimen of *Ornithorhynchus* taken just prior to egg laying progesterone and oestradiol were both elevated in the plasma (Carrick, Drinan & Cox, 1975) and, in one specimen of *Tachyglossus*, plasma progesterone was elevated during 5 days prior to egg laying and had fallen to basal levels on the day before the egg was first observed (C. H. Tyndale-Biscoe & L. A. Hinds, unpublished observations).

While the corpus luteum is not a structure unique to mammals, a bewildering variety of means to regulate its growth and decline have evolved among eutherian mammals, which attests to its central importance in the control of reproduction. However, because of this plasticity it cannot reliably be used as a character to determine evolutionary status. Rothchild (1981) has attempted to discern an underlying order in this profusion. He begins with the observation that the corpus luteum is unusual in being an ephemeral endocrine organ, which is not regulated by a negative-feedback system like other steroid secreting organs, but runs its course long or short, characteristic for each species; and that in its secretion of progesterone there is a rising phase, a plateau and a decline, which in some species is autonomous but in others may be modulated by extrinsic controls. The first phase may be deferred during embryonic diapause; the second may be truncated by luteolysis or extended by luteotrophins, but the eventual decline in all species is inevitable.

The corpora lutea of marsupials are identified as most nearly conforming to the ancestral condition because, when their lifespan is not extended by diapause, as in macropodids, they are short-lived, appear to be autonomous, and are not influenced by either luteolysins or luteotrophins. As mentioned in Chapter 6, there is now some evidence for a brief luteolysis in *M. eugenii* and possibly for a luteotrophin in *Isoodon macrourus*, but these may be minor exceptions to the general thesis. Among eutherian species, the small Insectivora appear to conform to this simple pattern also, with parturition and onset of lactation coinciding with the decline of the corpus luteum. Monoestrous Carnivora also have an autonomous corpus luteum, but its life is longer than in marsupials. In polyoestrous Eutheria, corpora lutea may be short-lived and their lifespan is always lengthened by pregnancy. Alternatively, they may be ultrashort-lived as in rodents, and the lifespan is increased by a sterile mating or a pregnancy. Thus it may be that the corpus luteum evolved in conjunction with the retention of the egg in the uterus and lactation and, in Eutheria, extrinsic controls evolved later in conjunction with the evolution of extended gestation lengths and larger fetal size. However, it should be noted in passing that there is a very poor correlation of size with gestation length. For example, hystricomorph rodents have spectacularly long gestation periods (of 60–200 days) for their size, whilst whales have gestation lengths no longer than that of the smaller herbivores (Rowlands & Weir, 1984; Short, 1984). In monotremes and marsupials, extended lactation took precedence over extended gestation and corpora lutea remained short-lived.

Intrauterine development

In the transport of the egg after ovulation there are further similarities between monotremes and marsupials, which differentiate them from eutherians. While the eggs of monotremes at ovulation are large (4.3–5 mm diameter) compared to the eggs of marsupials, they are conveyed through the oviduct as rapidly, arriving in the uterus before the completion of the first cleavage division. Fertilisation occurs in the oviduct and, as in marsupials, supernumerary sperm are trapped in the mucoid coat secreted by the oviduct. The basal layer of the shell is laid down in the oviduct but two further layers, the rodlet and outer shell are formed in the uterus (Hill, 1933; Hughes, 1977, 1984). No species of eutherian has a shell or shell membrane and only a few species have the mucoid coat. Furthermore, in most species, cleavage divisions occur in the oviduct.

Caldwell (1884, 1887) was the first to describe the meroblastic cleavage of the monotreme egg, which was amply confirmed later by Wilson & Hill (1908), Flynn & Hill (1947), Luckett (1977) and Hughes (1984). Caldwell (1887) also described meroblastic cleavage in the marsupial, *Phascolarctos cinereus* but this was refuted by Hill (1910), who showed that Caldwell had confused the polar bodies as the first blastomeres. The cleavage patterns of the much smaller eggs of marsupials and eutherians are holoblastic, although in marsupials a small amount of cytoplasm or yolk is extruded at the first two cleavage divisions. These differences between the three groups have often been citied as evidence of the evolution of the microlecithal eutherian egg from a macrolecithal sauropsid-type egg.

A distinguishing feature of all mammals is the formation of a bilaminar blastocyst (Luckett, 1977), the outer layer of which is composed of cells specialised for absorption, the trophoblast. It is incorrect to say that the trophoblast is an apomorphy of the Eutheria (see below); rather it is a synapomorphy for all mammals probably associated with reduction in yolk and development of uterine secretion. However, the bilaminar blastocyst is achieved by different means in eutherians than in monotremes and marsupials. In the latter two groups early cleavage leads to the formation of a unilaminar blastocyst, either yolk-filled or hollow, composed of uniform and apparently totipotent cells, and only after formation of the bilaminar blastocyst do the cells of the presumptive embryo differentiate.

The major distinction of eutherian early embryos is the formation of a solid morula of cells which is followed by the differentiation of an outer trophoblast and an inner cell mass, each with different developmental fates, surrounding a blastocyst cavity. These have been considered to be fundamental distinctions between marsupials and eutherians but are not so clear cut as some reviewers (e.g. Tyndale-Biscoe, 1973; Lillegraven, 1969, 1975) have suggested. The chorion is the membrane which is formed by a fusion of (extraembryonic) ectoderm and somatic mesoderm, and which later fuses with either the yolk sac or allantois to form choriovitelline or chorioallantoic placentae. The trophoblast was the name given by Hubrecht (1889) to the extraembryonic ectoderm layer of the chorion proper, and it is considered to arise from the zygote as a precociously developed ectodermal derivative (Billington, 1971). Excluded from this definition are the 'formative' blastomeres (now referred to as inner cell mass) destined to give rise to the embryo proper and the amnion. Because, as mentioned above, the marsupial has no inner cell mass in the blastocyst, McCrady (1944) suggested that the wall of the marsupial blastocyst, which gives rise to the chorion, should be termed 'protoderm'. This term has led to confusion and misinterpretation, especially in evolutionary arguments (Lillegraven, 1975; Cox, 1977), and recent workers have re-emphasised Hubrecht's (1909) homology (Taylor & Padykula, 1978) and have gone back to the term 'trophoblast' (Padykula & Taylor, 1982; Renfree, 1982).

The presence of the inner cell mass gives an easily recognisable polarity to eutherian blastocysts, but marsupials have no such marker. The question of whether the initial polarity of the preblastocyst embryo of marsupials such as *Didelphis*, *Dasyurus* and *Antechinus* is carried through to the secondary polarity at endoderm formation is still open. As mentioned in Chapter 7, McCrady (1938) did not consider the matter closed and Selwood (1982*b*) has reiterated his caution. On the other hand, the early differentiation of trophoblast and inner cell mass, said to be apomorphic for Eutheria, does not occur in all species so far examined. In several families of the Insectivora, blastocyst formation is remarkably similar to that seen in marsupials, which supports the homology. For instance in the Tenrec, *Hemicentetes semispinosus*, the first 4–8 blastomeres form a single layered blastocyst in the oviduct and, only after the embryo enters the uterus, does an inner cell mass differentiate by a proliferation of one part of the blastocyst wall (Goetz, 1938). This is similar to the early development of *Didelphis*.

The parallels with *Didelphis* are more striking in the elephant shrew *Elephantulus myurous* (van der Horst, 1942; Tripp, 1971). In this species about 50 eggs are shed at each ovulation, passage through the oviducts is rapid and cleavage occurs in the uterus. The first four blastomeres become flattened against the zona and, as in *Didelphis*, join at their margins to form a hollow blastocyst. Further cell division leads to a unilaminar blastocyst of 120 identical cells before an inner layer is formed by migratory cells with long pseudopodia, which form a reticulum beneath the outer layer. Only then does an embryonic area form by coalescence of these cells beneath the outer trophoblast. Only two of these numerous embryos implant on a special part of the endometrium. A somewhat similar pattern of early development has been described in the shrew, *Crocidura caerula* (Sansom, 1937) and in the lemur, *Galago demidoffi* (Gerard, 1932). The reason for drawing attention to these unusual patterns of development among eutherian species is to emphasise again that cladistic analyses based on generalisations may result in apparent distinctions between Eutherian and Metatheria that do not exist.

Further similarities between monotremes, marsupials and many species of eutherians are seen after formation of the unilaminar blastocyst when expansion of the embryo occurs by absorption of secretions from the glandular endometrium, stimulated, presumably, by secretions of the enlarged corpora lutea. In monotremes the embryo expands from 4–5 mm to a final diameter of 10–17 mm when the egg is laid which, in *Tachyglossus*, takes about 18 days (Griffiths, 1978). In this species and in *Ornithorhynchus* the embryo has reached the stage of 18–20 somites, which is very similar to the stage in marsupials at which the shell membrane ruptures and placental attachment takes place. In *Tachyglossus* the subsequent period of incubation lasts 10.5–11 days during which organogenesis occurs and the hatchling is very similar in size and development to the neonatal marsupial. This period is fully comparable to the last part of gestation in marsupials, except that it takes place within a shell instead of a uterus.

It is a common misconception to suppose that because monotremes deliver eggs not embryos, their reproduction is little different from that of sauropsids. This is not so. Their eggs are much smaller at ovulation than are those of equivalent-sized reptiles, and the substantial growth in the uterus depends on active absorption across the yolk sac of the living embryo. It is probably incorrect to say, as Sharman (1976) does, that 'Monotremes cannot distinguish between a fertilized and a non-fertilized egg in the uterus'. Although there is no experimental evidence at present, the difference in size would readily differentiate the two states.

Fetal membranes

The evolution of placentation in mammals has not been accompanied by the evolution of new fetal structures beyond those encountered in reptiles, so it is not surprising that parallels between marsupials and monotremes are also seen in the development and function of the fetal membranes. Compared to the reptiles, the yolk content of the monotreme egg is greatly reduced and is insufficient to maintain development. As in the pre-attachment marsupial, nutrients are supplied by the endometrial gland secretions which are presumably absorbed by the yolk sac. The endometrium shows structural similarities to the progestational condition observed in viviparous mammals during the luteal phase of the oestrous cycle. After the egg of *Tachyglossus* is laid, the allantois enlarges, makes contact with the chorion and becomes highly vascular (see Griffiths, 1978). Semon (1894) showed that it covers half the inner surface of the shell, while the vascularised yolk sac covers the other half (Fig. 1.3).

As discussed in Chapter 7 most evidence now supports Hill's (1900*b*) idea that the non-vascular (bilaminar) yolk sac is the main route of nutrient absorption and the vascular yolk sac is primarily for respiratory exchange: in *Tachyglossus* after the egg leaves the uterus a nutritional route is obviously not required but respiratory exchange is essential and the entire surface becomes vascularised, presumably for this purpose. In developing characters for his cladistic analysis, Marshall (1979) excludes monotremes from the character 'placenta', which thereby increases the apparent separation of these two groups. The functional reality is much closer; indeed one might consider, with Gregory (1947), whether the continuation of organogenesis for 10 days in the uterus or in an externally held egg is very different at all when the products are so similar. Furthermore, the fact that the egg accumulates so much nutrient material after the shell has been laid down does not conform to the definition of oviparity. It seems that on this topic terms such as 'egg', developed for another purpose, have obscured fundamental similarities and, just as Owen and Geoffroy 150 years ago could not reconcile egg bearing with lactation, so today oviparity of monotremes remains on obstacle to understanding the essential and close similarities of monotreme and marsupial reproduction.

Nevertheless, the similar altricial nature of their young suggests that physiological limits of the vascular yolk sac may have been a factor. Luckett (1977) considers that the evolution of the villous allantoic placenta by eutherians allowed for much greater exchange to take place and so opened the way for retention of the fetus during its major growth phase. The far greater penetration of maternal tissues and the intimate vascular beds are predicated on a tolerance of the fetal tissues by the maternal immune system (see Amoroso & Perry, 1975). Moors (1974) suggested that the brief period of attachment in marsupial gestation (see Fig. 7.15) was because marsupials did not evolve a trophoblast capable of masking histocompatibility antigens on its surface, and any more intimate or prolonged attachment would fail. Tyndale-Biscoe (1973) suggested that it was for this reason that the very intimate attachment of the chorioallantoic placenta of the Peramelidae is so short-lived.

Lillegraven (1975) and Cox (1977) then developed the thesis that the differentiation of a trophoblast layer capable of masking histocompatibility antigens on its surface was the major adaptation that enabled gestation to be prolonged in eutherian mammals, and that this was the major dichotomy with marsupials. As we have seen, trophoblast is not a new development of Eutheria and Kirsch (1977*b*) and Taylor & Padykula (1978) argued that the lack of extended gestation in marsupials is not intrinsically inferior to prolonged intrauterine gestation. Furthermore, the two attempts to test the hypothesis that marsupial trophoblast lacks the ability to mask histocompatibility antigens, do not support the hypothesis (see Chapter 7). Lillegraven (1979) acknowledges the first of these studies but considers that because the species used, *M. eugenii*, was derived from an island population the animals may have been closely related and hence did not provide a good test of the hypothesis. It is highly unlikely that this is so, since the island is large (4000 km^2), and the population must exceed 100000. Furthermore, in both experiments the animals rejected skin allografts in the normal time for outbred (eutherian) animals. Thus we must conclude again that this is not a significant distinction between these two groups of mammals.

Size and development at birth

The young of monotremes at hatching are remarkably similar to neonatal marsupials, not only in their minute size and early stage of development but also in their precocious and special adaptations for external life. The most pronounced of these are the well-developed forelimbs with separate digits equipped with claws, the precociously developed lungs and associated air passages, the functional mesonephric kidney, the digestive tract and anterior regions of the peripheral nervous system. There are also several special structures which are thought to have phylogenetic significance.

Prior to hatching the young of both *Tachyglossus* and *Ornithorhynchus* develop a median egg tooth, with dentine and enamel components, on the premaxillary suture, and a caruncle above the nares. These two structures are linked by a median mesenchymal strand. Hill & de Beer (1950) point out that no other amniote vertebrate has both structures; in the Squamata, which have soft-shelled eggs, the egg tooth is used to cut through the shell membrane at hatching while, in other reptiles and in birds (which have eggs with a calcareous shell), the caruncle is used to break it. The egg tooth of reptiles is homologous with the first toothlets of the jaw and, according to Hill & de Beer (1950), the monotreme egg tooth, in its formation, resembles these reptilian toothlets more than they do the alveolar teeth of mammals, including the transient teeth of juvenile *Ornithorhynchus*. If this homology is correct the monotreme egg tooth is phylogenetically ancient, pre-dating the evolution of mammalian teeth. Whatever their ultimate homology to reptiles may be shown to be, Hill & de Beer (1950) demonstrated vestiges of an homologous egg tooth in neonatal *Trichosurus vulpecula*, *Phascolarctos cinereus* and *Perameles* and a medial mesenchyme strand like that beneath the monotreme caruncle in *Didelphis marsupialis* and *Caluromys*. They considered these to have no particular function in marsupials, but to represent ontogenetic relics from their oviparous ancestors.

Neonatal marsupials have other transient structures whose function is not clear, such as the epidermal oral shield of *Didelphis virginiana*, which appears and disappears 1 day before birth (McCrady 1938), and the sternal swelling of neonatal *Dasyurus viverrinus* (Hill & Hill, 1955). Others, such as the epitrichial claws that clothe the digits of the forelimbs of neonatal marsupials and monotremes, aid in grasping the mother's hair, and the epidermal tissue that closes the ears and eyelids are presumably protective. In marsupials – but not monotremes – the lateral margins of the mouth are also sealed by the same material (see Fig. 2.23*i*) which is thought to aid in attachment to the teat and sucking of milk.

Among the Eutheria the most undeveloped or altricial young are those of species of the Insectivora, but none is so undeveloped at birth as monotreme or marsupial young. Nevertheless, Muller (1969) and Vogel (1972) have observed that similar closures of the eyelids, ears and mouth occur in the fetuses of these and other eutherian species at equivalent post-ovulatory ages to marsupials. They have postulated that these developmental stages are evidence that the ancestors of eutherians had a short gestation and delivered altricial young, resembling those of marsupials and monotremes. We will return to their ideas in the last part of the chapter, but here we conclude that the altricial young of monotremes and marsupials (including the possession of egg teeth) are sympleisiomorphic and the precocial young of Eutheria are apomorphic.

Mammary glands

During the controversy about the nature of the monotremes during the nineteenth century, the differences in the mammary apparatus were emphasised, but Bresslau (1912) recognised their essential similarity both in development and adult structure and this has been amply confirmed by Griffiths *et al.* (1972, 1973) with the light and electron microscope. Monotremes and marsupials share a common pattern of development (areolar patches, mammary hairs) different from Eutheria (mammary line); marsupials and eutherians share the character of having the galactophores gathered together in clusters opening at the ends of teats; monotremes and eutherians share the character that adult males of most species possess potentially functional mammary glands (Haacke, 1885) which, under certain hormonal conditions, can secrete milk (Griffiths, 1978; Cowie, 1984). It is not clear which of these shared characters is primitive and which is derived unless it is determined on other criteria that Eutheria are the most advanced group. The absence of all vestiges of mammary glands in adult male marsupial, is understandable because, during the first period of lactation, the young must remain permanently attached to one teat and the opportunity for the male to share lactation is excluded at this time. It is less clear why no males of any species of Monotremata or Eutheria, participate in lactation. The question has been addressed by Daly (1979) but the reasons he adduces for its absence do not seem to be sufficient to have excluded this as a reproductive strategy in all mammals, including those with exclusively monogamous mating patterns, or those such as canids in which the feeding of the young is shared by other members of the group.

Pouch and epipubic bones

The possession of epipubic bones by marsupials and monotremes and their absence from eutherians led Owen (1868) to the idea that the epipubic bones provide support for the pouch or attachment for muscles involved in pouch closure. However, as Marshall (1979) points out, epipubic bones occur in male monotremes and marsupials (most of which do not have pouches), as well as in the female *Ornithorhynchus* and the females of many small-sized marsupials, which do not develop pouches (Fig. 2.8). One cannot therefore infer from the occurrence of epipubic bones that multituberculates had pouches and carried altricial young in them. Indeed, although marsupials are identified by the pouch, it seems likely that it is a relatively modern adaptation associated with the evolution of large size and the carriage of young for a longer period (see Russell, 1982*a*).

Lactation

From the evidence presented in Chapter 8, two features of lactation in marsupials are distinctly different from eutherians: the first is the marked change in composition of the milk from early to late lactation; the second is the unusual way in which lactogenesis and the milk-ejection reflex are controlled in early lactation of macropodids. A number of species of eutherian have some changes in milk composition (Oftedal, 1984). In the giraffe, early milk is known to differ from milk secreted later in lactation (Ben Shaul, 1962) but the change occurs at 10 days, much earlier than in marsupials. Fat content rises in late lactation in the rabbit, most rodents, some carnivores, the elephant, giraffe and ruminants, but it falls in some species of seal, the pig and the horse (Oftedal, 1984). Rising fat content is accompanied by increasing protein and declining sugar levels in many species but fat and protein appear to change in opposite directions in a few (e.g. mink and pig). However none of these changes is nearly as extensive as the almost continuously altering milk composition of marsupials. For instance, the lipid content of milk of the domestic rabbit doubles, whereas that of *M. eugenii* increases by more than 400% (Green, Griffiths & Leckie, 1983). Since the profile of prolactin through lactation

Table 10.1. *A comparison of reproductive characters between the three groups of living mammals*

No.	Character	Monotremata	Metatheria	Eutheria
1	Karyotype	> 50	< 30	> 40
2	Dosage compensation	?	Paternal X inactivation	Random X inactivation
3	Bulbo-urethral glands	Present	Present	Present
4	Prostate gland	Present, disseminate	Present, disseminate	Present
5	Seminal vesicles	Absent	Absent	Present
6	Glans penis	Bifid	Bifid or single	Single
7	Scrotum	Absent	Pre-penial	Pre- or post-penial
8	Testes	Abdominal	Inguinal or scrotal	Abdominal, inguinal or scrotal
9	Testicular blood supply	Simple	Rete mirabile	Pampiniform plexus
10	Sperm head	Long, fusiform	Short	Short
11	Ureter's entry	Urogenital sinus, dorsal	Bladder, ventromedial	Bladder, ventrolateral
12	Endometrium	Secretory	Secretory	Secretory
13	Ovarian follicles	No antrum, liquour folliculi	Antral, liquour folliculi	Antral, liquour folliculi (no antrum in some species)
14	Corpora lutea	Secretory, autonomous	Secretory, autonomous	Secretory, pituitary or placenta-dependent
15	Ovum	Large, yolk-filled	Small, yolk extrusion	Very small, no yolk
16	Cleavage	Meroblastic to blastocyst	Holoblastic to blastocyst	Holoblastic to morulla (or blastocyst)
17	Bilaminar blastocyst with trophoblast	Present	Present	Present
18	Mucoid coat	Present	Present	Present in few, absent in most

19 Shell membrane	Present	Present	Absent
20 Shell	Present	Absent	Absent
21 Embryo formation	From outer layers	From outer layers	Inner cell mass in most
22 Vascularised yolk sac	Present in all	Present in all	Present in some
23 Vascularised chorioallantois	Present	Absent (except Peramelidae)	Present in all
24 Invasive villous placenta	Absent	Absent (except Peramelidae)	Present in all
25 Endocrine function of placenta	—	Slight to none	Various
26 Immunoprotection of fetus	—	Uncertain	Present
27 Delivery of young	Altricial from egg	Altricial from uterus	Altricial to precocial from uterus
28 Hair and sweat glands	Present	Present	Present
29 Mammary gland, alveolar and myoepithelial cells	Present	Present	Present
30 Mammary hairs	Present	Present	Absent
31 Mammary anlagen	Areola patches	Areola patches	Mammary lines
32 Teats	Absent	Present	Present
33 Mammary glands in males	Present	Absent	Present
34 Crural spurs and glands	Present	Absent	Absent
35 Epipubic bones	Present	Present	Absent
36 Pouch	Present/absent	Present/absent	Absent
37 Lactation	Long duration	Long duration	Short duration
38 Milk composition	Major changes through lactation	Major changes through lactation	Minor changes

is very similar in *Dasyurus viverrinus* and *T. vulpecula* to that in *M. eugenii* (see Chapter 8) it seems reasonable to conclude that the pattern described in *M. eugenii* is common to other marsupials and quite different to eutherians.

Nothing is known at present of the control of lactogenesis in monotremes except that it has been induced in male *Tachyglossus* with steroid treatment followed by ovine prolactin (Griffiths, 1978). However, like marsupials, the composition of the milk of *Tachyglossus* changes through the lengthy lactation (Griffiths, 1978) but for *Ornithorhynchus* there is no evidence that lipid and carbohydrate composition changes through lactation (Grant *et al.*, 1983; Messer *et al.*, 1983). The main feature held in common by marsupials and monotremes is the extreme length of lactation during which the young undergo much of their growth.

Conclusions

The conclusions to be drawn from this brief comparison of reproductive processes in living mammals, summarised in Table 10.1, are that many features are synapomorphic, that is to say they are derived characters held in common by all three groups of mammals. These are Graafian follicles, functional corpora lutea, bilaminar blastocysts, uterine secretion, yolk sac placentae, mammary glands and lactation. The monotremes retain some primitive (pleisiomorphic) characters such as a relatively larger egg containing yolk, egg tooth and caruncle, separate Mullerian ducts, delivery of the egg at the somite stage and lack of teats, which indicate a divergence from the stock leading to the marsupials and eutherians. Comparisons between marsupials and eutherians suggest a dichotomy in a common ancestral group, not a derivation of one from the other. This is based mainly on the evidence of the female genital tract, in which each pattern is apomorphic, and on the evidence that almost all other features are represented in some members of both groups.

The important apomorphic characters of the Eutheria are the lack of the shell membrane, early differentiation of the inner cell mass in most but not all species but in no marsupial, the precocious development of the chorioallantoic villous placenta in all species (with or without the chorio-vitelline placenta) and the assumption of a variety of extrinsic controls of the corpora lutea in several orders. All of these features are associated with the relatively longer retention of the embryo in the uterus and its delivery at a more advanced stage of development than in any marsupial or monotreme.

Conversely, the main apomorphic characters of marsupial reproduction are the fusion of the vaginae anterior to the ureters, the pseudo-vaginal canal, the pouch and the special endocrine controls of early lactation and milk secretion. Most of these are associated with greater emphasis on lactation than gestation. The question is when in the history of mammals did these major apomorphies occur.

The evolution of mammalian reproduction

Despite their wide distribution, all Triassic morganucodonts were remarkably similar in structure and level of organisation, which, according to Crompton (1980), suggests that they were able to exploit a widespread niche, hitherto left vacant by the contemporaneous and much larger herbivorous and carnivorous dinosaurs. All were very small, 10–15 cm long and 20–30 g in body weight, at least an order of magnitude smaller than any mammal-like reptile predecessor. Features of the axial skeleton and limbs are typical of small marsupials and eutherians alive today, which are arboreal or forage in forest litter. The enlarged cranial capacity anterior to the pituitary fossa suggest that they had a capacity to process more sensory information, especially from the nose and ears than mammal-like reptiles did. The separation of the incus and malleus from the articulation of the lower jaw would also have improved sound transmission to the inner ear (Kermack & Mussett, 1983). These changes, and the presumed acquisition of homeothermy, enabled the first mammals to be active at night and thus to fill the niche of nocturnal insectivore, hitherto unoccupied by heterothermic reptiles. This led to their rapid dispersal throughout the world.

The fossil record of the Jurassic and Cretaceous, some 130 MY, is very sparse but is sufficient to indicate that several orders of mammals evolved. Like the Triassic mammals before them, all were small and most were insectivorous. Only the Multituberculata showed dental specialisation for herbivory but none of them was larger than a rabbit. It is not our purpose to discuss why all Mesozoic mammals were small, but it is very relevant to consider what constraints small size may have imposed on reproduction and conversely, what constraints reproduction may have imposed on body size.

Geist (1972) and Hopson (1973) have argued that an important consequence of homeothermy and small adult size would have been the production of immature or altricial young, because of the high cost of being endothermic during the infant period, together with the development of elaborate parental care. Alternatively, for a small truly oviparous mammal to produce an independent homeothermic offspring, the egg would have to be inordinately large. While there may be no *a priori* reason against a female producing one or a few large eggs, Case (1978) points out that the relatively large amount of non-thermogenic yolk would have imposed a serious energetic burden on the maintenance of homeothermy by the female herself. On other evidence, Kielan-Jawarowska (1979) argues that multituberculates must have produced small eggs or young because the maximum diameter of the pelvic canal is about 5 mm. Griffiths (1983) however, points out that monotreme eggs can be distorted, so that a somewhat larger soft-shelled egg could have passed through such a canal. By reducing the yolk and producing altricial young from much smaller eggs, these problems would have been alleviated.

Altricial young, however, depend on parental support for food and shelter and so lactation and brood care must have evolved simultaneously with the evolution of homeothermy. Unlike the hatchlings of cynodont mammal-like reptiles that had a full set of functional teeth at hatching with which to process food, the young of Triassic mammals, like all later ones, were diphyodont. This is strong evidence that they must have been fed by the parent. Following from this premise, Pond (1977, 1983) has developed the thesis that such parental food would have been milk and that the evolution of lactation and brood care were the two most important adaptations in the evolution of mammalian reproduction, far transcending the evolution of viviparity. This led to such mammalian characteristics as rapid postnatal growth, because the young can devote a high proportion of the calories from the diet to growth and relatively little to maintenance; specialisation for particular food sources, because the definitive dentition does not form until the jaw is fully grown; and to partial independence of reproduction and food supply because essential substances can be stored in the mother's body and released during lactation. Certainly the remarkably close similarity of mammary glands (Griffiths *et al.*, 1973; Griffiths, 1978) and milk secretion (Green, 1984) in all three groups of living mammals suggests that these are shared, derived (synapomorphic) characters and supports Pond's thesis.

Another consequence of reducing the size of eggs is that the duration of folliculogenesis is much reduced and it becomes possible to produce many more eggs at the same time. Small mammals today, whether marsupial or eutherian are usually polyovular and polyoestrous and this may be the primitive or necessary condition for such mammals. In Chapter 2 we argued that since polyoestry is by far the most common condition in marsupials it is probably the primitive condition, while monoestry is the special adaptation. Apart from the Carnivora, polyoestry is also the most prevalent condition in the Eutheria.

Small mammals share several other features (Table 10.2). They are nocturnal, short-lived, and produce several litters in quick succession. The gestation period is short, the young at birth are consequently altricial, and are deposited in a nest (eutherians) or attached to the teats in a brood area or pouch (marsupials). As mentioned in Chapter 2 most species of small marsupial lack a pouch and the young are deposited in a nest after a few weeks. In both groups brain growth, homeothermy, skull ossification and differentiation of the lower jaw and auditory ossicles occur after birth. As a consequence lactation in both groups is of longer duration than gestation.

Vogel (1981) has drawn attention to the consequences of this in shrews. *Sorex* and *Neomys* conceive at a postpartum oestrus and in these species with the shortest gestation and most altricial young, implantation is delayed during early lactation. Vogel (1981) argues that this is necessary to avoid the birth of the very altricial young while the previous litter are still completing their minimum nursing period. This idea is similar to that proposed by Sharman & Berger (1969) to account for diapause in macropodid marsupials; they considered it to have evolved as an adaptation to prevent a succession of young being born during the lengthy lactation. In both cases, the effect of diapause is to space out successive litters, so that they do not overlap in the uterus (superfetation) or in the pouch respectively. Vogel (1981) further suggests that, for this reason, embryonic diapause is probably a very ancient feature of mammalian reproduction, associated with the delivery of altricial young. Other shrews, such as *Crocidura*, which deliver young at a more advanced stage after a longer gestation do not undergo diapause; in them, the lactation period is shorter than gestation. Embryonic diapause occurs in a wide range of mammals both eutherian and marsupial (Renfree & Calaby, 1981), which supports the thesis of its early evolution in mammals.

If these several aspects of reproduction reflect constraints imposed by small adult size, it is reasonable to suppose that Mesozoic mammals, which were likewise small, displayed the same characteristics of polyoestry, polyovulation, very short gestation, extremely altricial young and relatively long lactation with or without delayed implantation. Were these mammals oviparous as modern monotremes are or were they already viviparous? With Pond (1977) we consider that this change was a far less profound one than the evolution of lactation. Lactation has, after all, only evolved

Table 10.2. *A comparison of reproductive and life history characters in marsupials and eutherians of small and large body size*

	Small mammals		Large mammals	
	Eutherian	Marsupial	Marsupial	Eutherian
Activity	Nocturnal	Nocturnal	Nocturnal, crepuscular	Diurnal
Breeding	Polyoestrous	Polyoestrous	Polyoestrous	Polyoestrous/monoestrous
Ovulation rate	Polyovular	Polyovular	Monovular	Monovular/polyovular
Gestation	Short to very short	Very short	Short to very short	Long
Development at birth	Altricial	Very altricial	Very altricial	Precocial
Thermoregulation at birth	Poikilothermic	Poikilothermic	Poikilothermic	Homeothermic
Jaw development at birth	Incomplete	Incomplete	Incomplete	Complete
Brain growth	Postnatal	Postnatal	Postnatal	Pre-natal
External protection	Nest, nidicolous	Pouch or brood patch then nest	Large pouch, no nest in majority	No nest, nidifugous
Lactation	Short, > gestation	Long, > gestation	Very long, 2 phases	Short, < gestation
Delayed implantation	In some, lactational delay	In some, lactational delay and seasonal	In Macropodidae, lactational	In Carnivora, seasonal
Lifespan	1 year or less	1–2 year	More than 1 year	More than 1 year

Data for Eutheria based on Martin (1975).

in the mammals, whereas viviparity has evolved independently in every vertebrate taxon except birds and in a number of invertebrates as well (see Amoroso, Heap & Renfree, 1980). Nevertheless, if the argument is followed that reduction in egg size and vitellogenesis was a prerequisite for polyoestry and polyovulation, selection for very small yolkless eggs would be the consequence and viviparity the more probable mode.

The further question then is whether the acquisition of viviparity is a shared apomorphy of Metatheria and Eutheria or evolved independently. Resolving this question depends on the emphasis one lays on their respective reproductive characters. Lillegraven (1979), following Hill (1910), believed that the reproductive pattern of mesozoic mammals of metatherian–eutherian grade would have been marsupial-like and that all reproductive apomorphies of living Eutheria could have evolved from such a condition. From the foregoing review of these characters (Table 10.1) we are inclined to agree that the most significant apomorphies of the Eutheria (such as the development of the inner cell mass, the villous chorioallantoic placenta and the pituitary and placental control of the corpus luteum) could be so derived. However, we do not agree that the differences in the anatomy of the genital tracts can be so derived. Likewise other characters (such as the anatomy of the prostate, form of the spermatozoa and temperature control of the testis) while superficially similar, are independent apomorphies.

Thus we conclude that viviparity with delivery of highly altricial young had probably evolved before the major dichotomy of the Metatheria and Eutheria. Subsequently, at some time during the long span of the Mesozoic, the separation of the urogenital and alimentary canals occurred, for reasons unconnected with reproduction, and a consequence of this was the ventral migration of the ureters to enter the bladder either medial or lateral to the Wolffian and Mullerian ducts. Although this established the major dichotomy between Metatheria and Eutheria, the differences in reproduction between the two groups during the Cretaceous (while all species were small) would have been difficult to discern externally, as indeed they are today between the small Insectivora and small Dasyuridae and Didelphidae. It is also of interest that the present distributions of these two groups of mammals do not overlap; with a single exception, in Colombia, no species of the Insectivora occurs in South America or in Australia–New Guinea.

With the changing climates that heralded the beginning of the Tertiary and the radiations of the Angiosperms and Insecta, three major adaptive radiations of mammals occurred; eutherian herbivores and carnivores on the world continent of North America, Eurasia and Africa, eutherian herbivores and marsupials carnivores on South America, and the marsupial herbivores and carnivores on Australia–New Guinea. An important aspect of all these radiations was increase in body size of species in most of the orders of Eutheria as well as in seven families of Metatheria. The latter were the carnivorous Borhyaenidae of South America and the carnivorous Thylacinidae, Thylacoleonidae, and the herbivorous Diprotodontidae, Sthenuridae, Macropodidae and Vombatidae of Australia–New Guinea.

In recent years it has been recognised that adult body size is a major factor in determining reproductive and life history strategies (Western, 1979). We have already referred to the constraints of small size (Table 10.2). Conversely, large size correlates with increased gestation length, reduced litter size and proportionally reduced maternal investment, slower growth rates, increased age at puberty, increased longevity and reproductive effort spread over several breeding seasons (Table 10.2). Sacher & Staffeldt (1974) have argued that, since brain size also correlates with these parameters, and it is the slowest growing organ in the body, the rate of brain growth and its ultimate size are the main determinants of body size, and hence of reproductive parameters. Further, they argued that increase in brain size was the primary selective advantage in the evolution of large size.

Large mammals must provide for their young during a prolonged growth period and in eutherians this has been accommodated in most species by an extended gestation with delivery of precocious nidifugous young. However, even with extended gestation, other factors limit the size at birth; the higher primates have adopted the strategem of neoteny to allow deferment of brain growth until after birth. Neoteny has been particularly important in the evolution of humans, and has resulted in an infant which is secondarily altricial at birth. Large marsupials, as represented by the extant wombats, kangaroos and *Thylacinus* have provided for this growth phase by extended lactation and a large pouch (Russell, 1982*a*). Much of the recent work on reproduction in marsupials has been done with species of the Macropodidae that show these adaptations for large size, just as much of the recent work with eutherian species has been done on large ungulates and primates. Comparisons between these species necessarily emphasise the extremes of reproductive adaptation within the Metatheria and the Eutheria and invite comparisons of the relative merits of each mode.

Extension of the period of parental care must be accomplished by extension of intrauterine development (gestation) or extrauterine development (lactation). Amongst mammals, eutherians have elaborated the first, and marsupials the second alternative. Prolongation of gestation has involved extension of corpus luteum and placental function and both have been achieved in a variety of ways among eutherians. As mentioned earlier, Rothchild (1981) has reviewed the control of extended luteal function and emphasised the diversity even between closely related species. Equally varied patterns of placentation and placental function are also found among the Eutheria (Amoroso, 1952) and, as with the corpus luteum, it is not possible to correlate evolutionary status with placental types (Renfree, 1982). Mossman (1937) contrasted the great diversity in reproductive structures in the Mammalia with the relative uniformity of other structures and suggested that the latter are more conservative because survival depends on their constant use, whereas reproduction is an episodic function and the means for its achievement can be more varied and reflect the ecological and other constraints of the species' life. He did not mention size but its pervasive influence is clearly evident. The intimate connection of the villous placenta has involved adaptations, which still defy understanding, to mask or suppress the expression of fetal histocompatibility antigens.

Amoroso & Perry (1975) advanced the idea that capacity to secrete protein hormones by the placenta initially evolved as a means of masking fetal tissues from maternal lymphocytes and that the luteotrophic role was assumed secondarily. However, claims that human chorionic gonadotrophin suppresses lymphocyte responsiveness *in vivo* or *in vitro* have been vigorously disputed (Heap & Flint, 1984). Beer & Billingham (1979) have provided evidence that placental progesterone may have the same immunoprotective function but in some species, such as the sheep, has secondarily become the main source of progesterone for the maintenance of pregnancy. Clearly the question of immunosuppression during pregnancy is still open. Adverse consequences of extended gestation are the limits placed by the size of the pelvic canal on pre-natal brain growth, which were met in hominids by neoteny.

By contrast, the growth of the young of large marsupials during an extended period of lactation avoids these potential problems of intrauterine accommodation and brain development. The long lactation is associated with the secretion of milk of different composition to the immature and to the older young. In the macropodids this has involved complex local endocrine control to enable simultaneous secretion of two kinds of milk and adaptations of the mammary epithelial cells to change from synthesising one array of milk components to another; eutherians lack such sophisticated lactational physiology. While marsupials have not evolved luteal or placental functions that compare in complexity to those of advanced eutherians, newer evidence, which we have presented in Chapters 6 and 7, indicate that the potential for these adaptations does exist. In several macropodids, i.e. *M. giganteus*, extended gestation is associated with an extended autonomous life of the corpus luteum; in the peramelids there is a suggestion that the corpus luteum may respond to pituitary luteotrophin and in *M. eugenii* there is evidence for a luteolytic effect at the end of pregnancy. Likewise, there is evidence for steroid secretion and a local influence on the endometrium by the placenta of three species of macropodid, and the invasive chorioallantoic placenta of peramelids indicates a potential for placental adaptations that have evolved in the Eutheria.

From this we conclude that the marked differences in reproduction between the advanced Eutheria and the advanced Metatheria are the end points of two evolutionary lines with different trajectories; from similar beginnings as small, polytocous, polyoestrous, insectivorous mammals, one group exploited intrauterine nutrition for fetal growth and the other exploited the potential of the mammary epithelium to support early growth and development of the young.

Lillegraven (1975) and Cox (1977) concluded that the birth of very altricial young with precocious development of the anterior end of the body and the forelimbs for external life closed off opportunities for the full array of adaptive radiations seen among eutherian orders. They observed that no marsupials have evolved flight or a fully aquatic life and concluded that this was because the forelimbs must retain digits for travelling to the teat. Kirsch (1977*b*) countered these arguments and pointed out that the forelimb of the newborn *Notoryctes* is already highly modified for digging, and the gliding membrane and long tail of *Petauroides volans* are already evident at birth (Bancroft, 1973), which indicates that there is a potential plasticity in marsupial neonates. Lee & Cockburn (1985) further note that flying has only evolved once in the Eutheria.

Current analyses by Morton *et al.* (1982), Russell (1982*a*, *b*) and Lee & Cockburn (1985) of the ecological consequences of marsupial and eutherian reproduction have not confirmed earlier assumptions of an intrinsic inferiority of the marsupial mode and have required a reappraisal of the question why marsupials are less successful in terms of number of species and world wide distribution. Storr (1958), in a little-known article, was the first to suggest that the reason is not because they are marsupials but because they evolved on the small and isolated land masses of South America and Australia–New Guinea, whereas the dominant orders of Eutheria evolved in the more competitive arena of the world continent.

The idea was restated by Tyndale-Biscoe (1973) and more recently by Gould (1977). We hope that the perception of marsupials as mammals that took an alternative path in the evolution of those special features of mammalian reproduction – intrauterine development, lactation and maternal care – will prevail, rather than the perception of them as some irrelevant relic of the past. We hope for this, not from a misplaced chauvinism but because we believe that marsupials have much to offer in the quest for understanding of mammalian reproduction. As the contents of this book attest, marsupial models can help to elucidate the control of seasonal breeding, the control of milk synthesis and secretion, the regulation of corpus luteum function and embryonic development. The current understanding of these aspects and the development of laboratory bred species such as *Monodelphis domestica*, *Sminthopsis crassicaudata*, *Isoodon macrourus*, *Macropus eugenii* and *M. rufogriseus* now permit many of these questions to be examined with a rigour not previously possible. We confidently expect exciting advances in all these areas in the next few years.

References

Aitken, R. J. (1981). Function of the zona pellucida. *Journal of Reproduction and Fertility, Supplement*, **29**, 238–9.

Alcorn, G. T. (1975). *Development of the ovary and urino-genital ducts in the tammar wallaby* Macropus eugenii (*Desmarest, 1817*). PhD thesis, Macquarie University, Sydney.

Alcorn, G. T. & Robinson, E. S. (1983). Germ cell development in female pouch young of the tammar wallaby (*Macropus eugenii*). *Journal of Reproduction and Fertility*, **67**, 319–25.

Allen, N. T. & Bradshaw, S. D. (1980). Diurnal variation in plasma concentrations of testosterone, 5 α-dihydrotestosterone, and cortiscosteriods in the Australian brush-tailed possum, *Trichosurus vulpecula* (Kerr). *General and Comparative Endocrinology*, **40**, 455–8.

Amoroso, E. C. (1952). Placentation. In *Marshall's Physiology of Reproduction*, 3rd edn, vol. 2, ed. A. S. Parkes, pp. 127–311. London: Longmans, Green.

Amoroso, E. C. (1955). Endocrinology of pregnancy. *British Medical Bulletin*, **11**, 117–25.

Amoroso, E. C., Heap, R. B. & Renfree, M. B. (1980). Hormones and the evolution of viviparity. In *Hormones and Evolution*, vol. 2, ed. E. J. W. Barrington, pp. 925–89. London: Academic Press.

Amoroso, E. C. & Perry, J. S. (1975). The existence during gestation of an immunological buffer zone at the interface between maternal and foetal tissues. *Philosophical Transactions of the Royal Society B*, **271**, 343–61.

Andersen, D. H. (1928). Comparative anatomy of the tubo-uterine junction. Histology and physiology in the sow. *American Journal of Anatomy*, **42**, 255–305.

Andrew, D. L. & Settle, G. L. (1982). Observations on the behaviour of species of *Planigale* (Dasyuridae, Marsupialia) with particular reference to the Narrow-nosed planigale (*Planigale tenuirostris*). In *Carnivorous Marsupials*, vol. 1, ed. M. Archer, pp. 311–24. Sydney: Royal Zoological Society of New South Wales.

Andrewartha, H. G. & Barker, S. (1969). Introduction to a study of the ecology of the Kangaroo Island wallaby, *Protemnodon eugenii* (Desmarest) within Flinders Chase, Kangaroo Island, South Australia. *Transactions of the Royal Society of South Australia*, **93**, 127–32.

Archer, M. (1976). Revision of the marsupial genus *Planigale* Troughton (Dasyuridae). *Memoirs of the Queensland Museum*, **17**, 341–65.

Archer, M. (ed.) (1982). *Carnivorous Marsupials*, 2 vols. Sydney: Royal Zoological Society of New South Wales.

Armati-Gulson, P. & Lowe, J. (1985). Histology and scanning electron microscopy of the development of the reproductive tract of the common ringtail possum *Pseudocheirus peregrinus* (Pseudocheiridae: Marsupialia). *Australian Mammalogy*, **8**, 97–109.

Armstrong, D. T. & Hansel, W. (1959). Alteration of the bovine oestrous cycle with oxytocin. *Journal of Dairy Science*, **42**, 533–7.

Arnold, R. & Shorey, C. D. (1985). Structure of the oviductal epithelium of the brush-tailed possum (*Trichosurus vulpecula*). *Journal of Reproduction and Fertility*, **73**, 9–19.

Aslin, H. (1974). The behaviour of *Dasyuroides byrnei* (Marsupialia) in captivity. *Zeitschrifte für Tierpsychologie*, **35**, 187–208.

Aslin, H. J. (1975). Reproduction in *Antechinus maculatus* Gould (Dasyuridae). *Australian Wildlife Research*, **2**, 77–80.

Aslin, H. J. (1980). Biology of a laboratory colony of *Dasyuroides byrnei* (Marsupialia: Dasyuridae). *The Australian Zoologist*, **20**, 457–71.

Atherton, R. G. & Haffenden, A. T. (1982). Observations on the reproduction and growth of the long-tailed pygmy possum, *Cercartetus caudatus* (Marsupialia: Burramyidae), in captivity. *Australian Mammalogy*, **5**, 253–9.

Atramentowicz, M. (1982). Influence du milieu sur l'activité locomotrice et la reproduction de *Caluromys philander* (*L.*). *Revue d'Ecologie* (*Terre et Vie*), **36**, 373–95.

Austin, C. R. (1961). *The Mammalian Egg*. Oxford: Blackwell.

Baird, D. T. (1984). The ovary. In *Reproduction in Mammals*, Book 3. *Hormonal control of reproduction*, 2nd edn, ed. C. R. Austin & R. V. Short, pp. 91–114. Cambridge University Press.

Baldwin, J. & Temple-Smith, P. (1973). Distribution of LDH-X in mammals: presence in marsupials and absence in the monotremes platypus and echidna. *Comparative Biochemistry and Physiology*, **46B**, 805–11.

Baldwin, J., Temple-Smith, P. & Tidemann, C. (1974). Changes in testis specific lactate dehydrogenase isoenzymes during the seasonal spermatogenic cycle of the marsupial *Schoinobates volans* (Petauridae). *Biology of Reproduction*, **11**, 377–84.

Bancroft, B. J. (1973). Embryology of *Schoinobates volans* (Kerr) (Marsupialia: Petauridae). *Australian Journal of Zoology*, **21**, 33–52.

Barbour, R. A. (1963). The musculature and limb plexuses of *Trichosurus vulpecula*. *Australian Journal of Zoology*, **11**, 488–610.

Barbour, R. A. (1977). Anatomy of marsupials. In *The Biology of Marsupials*, eds B. Stonehouse & D. Gilmore, pp. 237–72. London: Macmillan.

Barnes, A. & Gemmell, R. T. (1984). Correlations between breeding activity in the marsupial bandicoots and some environmental variables. *Australian Journal of Zoology*, **32**, 219–26.

Barnes, R. D. (1977). The special anatomy of *Marmosa robinsoni*. In *The Biology of Marsupials*, ed. D. Hunsaker II, pp. 387–413. New York: Academic Press.

Barnes, R. D. & Barthold, S. W. (1969). Reproduction and breeding behaviour in an experimental colony of *Marmosa mitis* Bangs (Didelphidae). *Journal of Reproduction and Fertility, Supplement*, **6**, 477–82.

Barnes, R. D. & Wolf, H. G. (1971). The husbandry of *Marmosa mitis* as a laboratory animal. *International Zoo Yearbook*, **11**, 50–4.

Barnett, C. H. & Brazenor, C. W. (1958). The testicular rete mirabile of marsupials. *Australian Journal of Zoology*, **6**, 27–32.

Baverstock, P. R., Archer, M., Adams, M. & Richardson, B. J. (1982). Genetic relationships among 32 species of Australian Dasyurid Marsupials. In *Carnivorous Marsupials*. ed. M. Archer, pp. 641–50. Sydney: Royal Zoological Society of New South Wales.

Baxter, J. S. (1935). Development of the female genital tract in the American opossum. *Carnegie Institute Contributions to Embryology*, **25**, 15–35.

Beddard, F. E. (1891). On the pouch and brain of the male *Thylacinus*. *Proceedings of the Zoological Society of London* 1891, 138–45.

Bedford, J. M. (1978). Influence of abdominal temperature on epididymal function in the rat and rabbit. *American Journal of Anatomy*, **152**, 509–21.

Bedford, J. M., Rodger, J. C. & Breed, W. G. (1984). Why so many mammalian spermatozoa – a clue from marsupials? *Proceedings of the Royal Society B*, **221**, 221–33.

Beeck, D. M. (1955). Observations on the birth of the Grey Kangaroo (*Macropus ocydromus*). *Western Australian Naturalist*, **5**, 9.

Beer, A. E. & Billingham, R. E. (1979). Maternal immunological recognition mechanisms during pregnancy. In *Maternal Recognition of Pregnancy*, ed. J. Whelan. Ciba Foundation Symposium, New Series 64, 293–309.

Begg, R. J. (1981*a*). The small mammals of little Nourlangie Rock, N. T. II. Ecology of *Antechinus bilarni*, the sandstone antechinus (Marsupialia: Dasyuridae). *Australian Wildlife Research*, **8**, 57–72.

Begg, R. J. (1981*b*). The small mammals of little Nourlangie Rock, N. T. III. Ecology of *Dasyurus hallucatus* the northern quoll (Marsupialia: Dasyuridae). *Australian Wildlife Research*, **8**, 73–85.

Bell, B. D. (1981). Breeding and condition of possums *Trichosurus vulpecula* in the Orongorongo valley, near Wellington, New Zealand, 1966–1975. In *Marsupials in New Zealand*, ed. B. D. Bell, pp. 85–139. Victoria University of Wellington.

Belt, W. D., Cavazos, L. F., Anderson, L. L. & Melampy, R. M. (1971). Cytoplasmic granules and relaxin levels in porcine corpora lutea. *Endocrinology*, **89**, 1–10.

Bennett, J. H., Smith, M. J., Hope, R. M. & Chesson, C. M. (1982). Fat-tailed dunnart *Sminthopsis crassicaudata*: establishment and maintenance of a laboratory colony. In *The Management of Australian Mammals in Captivity*, ed. D. D. Evans, pp. 38–44. Melbourne: The Zoological Board of Victoria.

Ben Shaul, D. M. (1962). The composition of the milk of wild animals. *International Zoo Yearbook*, **4**, 333–42.

Bentley, P. J. & Shield, J. W. (1962). Metabolism and kidney function in the pouch young of the macropod marsupial *Setonix brachyurus*. *Journal of Physiology London*, **164**, 127–37.

Berger, P. J. (1966). Eleven-month 'Embryonic Diapause' in a marsupial. *Nature*, **211**, 435–6.

Berger, P. J. (1970). *Reproductive biology of the tammar wallaby*, Macropus eugenii. PhD thesis, Tulane University. *Dissertation Abstracts International*, **31B**, 3760–1.

Berger, P. J. & Sharman, G. B. (1969). Progesterone-induced development of dormant blastocysts in the tammar wallaby, *Macropus eugenii* Desmarest: Marsupialia. *Journal of Reproduction and Fertility*, **20**, 201–10.

Biggers, J. D. (1966). Reproduction in male marsupials. *Symposia of the Zoological Society of London*, **15**, 251–80.

Biggers, J. D. (1967). Notes on reproduction of the woolly opossum (*Caluromys derbianus*) in Nicaragua. *Journal of Mammalogy*, **48**, 678–80.

Biggers, J. D. & Creed, R. F. S. (1962). Conjugate spermatozoa of the North American opossum. *Nature*, **196**, 1112–13.

Biggers, J. D. & DeLamater, E. D. (1965). Marsupial spermatozoa pairing in the epididymis of American forms. *Nature*, **208**, 402–4.

Biggins, J. G. & Overstreet, D. H. (1978). Aggressive and non aggressive interactions among captive populations of the brush-tail possum, *Trichosurus vulpecula* (Marsupialia: Phalangeridae). *Journal of Mammalogy*, **59**, 149–59.

Billingham, R. E. & Head, J. R. (1983). The riddle of the allogeneic conceptus. *Transplantation proceedings*, **15**, 877–82.
Billington, W. D. (1971). Biology of the trophoblast. *Advances in Reproductive Physiology*, **5**, 28–66.
Bolliger, A. (1940). *Trichosurus* as an experimental animal. *Australian Journal of Science*, **3**, 59–61.
Bolliger, A. (1943). Functional relations between scrotum and pouch and the experimental production of a pouch like structure in the male of *Trichosurus vulpecula*. *Journal and Proceedings of the Royal Society of New South Wales*, **76**, 283–93.
Bolliger, A. (1944). The distinctive brown patch of sternal fur of *Trichosurus vulpecula* and its response to sex hormones. *Australian Journal of Science*, **6**, 181.
Bolliger, A. (1946). Some aspects of marsupial reproduction. *Journal and Proceedings of the Royal Society of New South Wales*, **80**, 2–13.
Bolliger A. & Gross, R. (1960). Nutrition of the marsupial suckling. *Australian Journal of Science*, **22**, 292–3.
Bolton, B. L., Newsome, A. E. & Merchant, J. C. (1982). Reproduction in the agile wallaby, *Macropus agilis* (Gould) in the tropical lowlands of the Northern Territory: opportunism in a seasonal environment. *Australian Journal of Ecology*, **7**, 261–77.
Bourlière, F. (1964). *The Natural History of Mammals*, 3rd edn, pp. xxxii & 387. New York: Knopf.
Bowley, E. A. (1939). Delayed fertilisation in *Dromicia*. *Journal of Mammalogy*, **20**, 499.
Bradley, A. J., McDonald, I. R. & Lee, A. K. (1980). Stress and mortality in a small marsupial (*Antechinus stuartii*, Macleay). *General and Comparative Endocrinology*, **40**, 188–200.
Bradshaw, S. D., McDonald, I. R., Hahnel, R. & Heller, H. (1975). Synthesis of progesterone by the placenta of a marsupial. *Journal of Endocrinology*, **65**, 451–2.
Braithwaite, R. W. (1974). Behavioural changes associated with the population cycle of *Antechinus stuartii* (Marsupialia). *Australian Journal of Zoology*, **22**, 45–62.
Braithwaite, R. W. & Lee, A. K. (1979) A mammalian example of semelparity. *American Naturalist*, **113**, 151–5.
Brambell, F. W. R. (1956). Ovarian changes. In *Marshall's Physiology of Reproduction*, 3rd edn, vol. 1 part 1, ed. A. S. Parkes, pp. 397–542. London: Longmans, Green.
Brambell, F. W. R. (1970). *The Transmission of Passive Immunity from Mother to Young*. Amsterdam: North Holland.
Bremer, J. L. (1904). On the lung of the opossum. *American Journal of Anatomy*, **3**, 67–73.
Bresslau, E. (1912). Die Entwickelung des Mammarapparates der Monotremen, Marsupialier und einiger Placentalier. In *Zoologische Forschungsreisen in Australien*, vol. 4, ed. R. Semon, pp. 653–874. Jena: Gustav Fischer.
Bresslau, E. (1920). *The Mammary Apparatus of the Mammalia in the light of ontogenesis and phylogenesis*. London: Methuen.
Briscoe, D. A., Murray, J. D. & Sharman, G. B. (1981). Inheritance and expression of two peptidases in the wallaroo, *Macropus robustus* (Gould). *Australian Journal of Biological Sciences*, **34**, 341–6.
Briscoe, D. A., Robinson, E. S. & Johnston, P. G. (1983). Glucose-6-phosphate dehydrogenase and lactate dehydrogenase activity in kangaroo and mouse oocytes. *Comparative Biochemistry and Physiology*, **75B**, 685–8.
Brooks, D. E., Gaughwin, M. & Mann, T. (1978). Structural and biochemical characteristics of the male accessory organs of reproduction in the hairy-nosed wombat (*Lasiorhynus latifrons*). In *Proceedings of the Royal Society of London B*, **201**, 191–207.
Broom, R. (1900). Development and morphology of the marsupial shoulder girdle. *Transactions of the Royal Society of Edinburgh*, **39**, 749–70.

Bryant, S. L. & Rose, R. W. (1986). The growth and role of the corpus luteum throughout delayed gestation in the marsupial, *Potorous tridactylus*. *Journal of Reproduction and Fertility*, **76**, 409–14.

Bryant-Greenwood, G. D., Niall, H. & Greenwood, F. (eds) (1981). *Relaxin*. Amsterdam: North Holland, Elsevier.

Buchanan, G. & Fraser, E. A. (1918). The development of the urogenital system in the Marsupialia with special reference to *T. vuplecula*. Pt 1. *Journal of Anatomy*, **53**, 35–95.

Burns, R. K. (1939*a*). The differentiation of sex in the opossum (*Didelphys virginiana*) and its modification by the male hormone testosterone propionate. *Journal of Morphology*, **65**, 79–199.

Burns, R. K. (1939*b*). Sex differentiation during the early pouch stages of the opossum (*Didelphys virginiana*) and a comparison of the anatomical changes induced by male and female sex hormones. *Journal of Morphology*, **65**, 497–547.

Burns, R. K. (1939*c*). Effect of testosterone propionate on sex differentiation in pouch young of opossum. In *Proceedings of the Society for Experimental Biology*, **41**, 60–2.

Burns, R. K. (1939*d*). Effects of female sex hormones in young opossums. In *Proceedings of the Society for Experimental Biology*, **41**, 270–2.

Burns, R. K. (1942*a*). Hormones and the growth of the parts of the urinogenital apparatus in mammalian embryos. In *Cold Spring Harbour Symposia on Quantitative Biology*, **10**, 27–34.

Burns, R. K. (1942*b*). The origin and differentiation of the epithelium of the urinogenital sinus in the opossum, with a study of the modifications induced by estrogens. *Carnegie Institute Contributions to Embryology*, **30**, 63–83.

Burns, R. K. (1945). The effects of male hormone on the differentiation of the urinogenital sinus in young opossums. *Carnegie Institute Contributions to Embryology*, **31**, 163–75.

Burns, R. K. (1956). Hormones versus constitutional factors in the growth of embryonic sex primordia in the opossum. *American Journal of Anatomy*, **98**, 35–67.

Burns, R. K. (1961). Role of hormones in the differentiation of sex. In *Sex and Internal Secretions*, vol. 1, ch. 2, ed. W. C. Young, pp. 76–158. Baltimore: Williams & Wilkins.

Burns, R. K. & Burns, L. M. (1957). Observations on the breeding of the American opossum in Florida. *Revue suisse de zoologie*, **64**, 595–605.

Byskov, A. G. (1983). Primordial germ cells and regulation of meiosis. In *Reproduction in Mammals*, 2nd edn, Book 1, ed C. R. Austin & R. V. Short, pp. 1–16. Cambridge University Press.

Cake, M. H., Owen, F. J. & Bradshaw, S. D. (1980). Difference in concentration of progesterone in the plasma between pregnant and non-pregnant quokkas (*Setonix brachyurus*). *Journal of Endocrinology*, **84**, 153–8.

Calaby, J. H. (1960). Observations on the banded ant-eater *Myrmecobius f. fasciatus* Waterhouse (Marsupialia), with particular reference to its food habits. *Proceedings of the Zoological Society of London*, **135**, 183–207.

Calaby, J. H. & Poole, W. E. (1971). Keeping kangaroos in captivity. *International Zoo Yearbook*, **11**, 5–12.

Calaby, J. H. & Taylor, J. M. (1981). Reproduction in two marsupial-mice, *Antechinus bellus* and *A. bilarni* (Dasyuridae), of Tropical Australia. *Journal of Mammalogy*, **62**, 329–41.

Caldwell, H. (1884). On the arrangement of the embryonic membranes in marsupial animals. *Quarterly Journal of Microscopical Science*, **24**, 655–8.

Caldwell, W. H. (1887). The embryology of Monotremata and Marsupialia, Pt 1. *Philosophical Transactions of the Royal Society, Series B*, **178**, 463–86.

Call, R. N., Catling, P. C. & Janssens, P. A. (1980). Development of the adrenal gland in the tammar wallaby, *Macropus eugenii* (Desmarest) (Marsupialia: Macropodidae). *Australian Journal of Zoology*, **28**, 249–59.

Callard, G. V., Petro, Z. & Tyndale-Biscoe, C. H. (1982). Aromatase activity in marsupial brain, ovaries and adrenals. *General and Comparative Endocrinology*, **46**, 541–6.

Carrick, F. N. & Cox, R. I. (1977). Testicular endocrinology of marsupials and monotremes. In *Reproduction and Evolution*, ed. J. H. Calaby & C. H. Tyndale-Biscoe, pp. 137–41. Canberra: Australian Academy of Science.

Carrick, F. N., Drinan, J. P. & Cox, R. I. (1975). Progestagens and oestrogens in peripheral plasma of the platypus, *Ornithorhynchus anatinus*. *Journal of Reproduction and Fertility*, **43**, 375–6.

Carrodus, A. & Bolliger, A. (1939). The effect of oestrogenic hormone on the prostate of the marsupial *Trichosurus vulpecula*. *Medical Journal of Australia*, **2**, 633–41.

Casanova, J. (1958). The doormouse or pigmy possum. *Walkabout*, **24**, 30–1.

Case, T. J. (1978). Endothermy and parental care in the terrestrial vertebrates. *The American Naturalist*, **112**, 861–74.

Catling, P. C. & Sutherland, R. L. (1980). Effect of gonadectomy, season, and the presence of females on plasma testosterone, luteinizing hormone, and follicle stimulating hormone levels in male tammar wallabies (*Macropus eugenii*). *Journal of Endocrinology*, **86**, 25–33.

Catling, P. C. & Vinson, G. P. (1976). Adrenocortical hormones in the neonate and pouch young of the tammar wallaby, *Macropus eugenii*. *Journal of Endocrinology*, **69**, 447–8.

Catt, D. C. (1977). The breeding biology of Bennett's wallaby (*Macropus rufogriseus fruticus*) in South Canterbury, New Zealand. *New Zealand Journal of Zoology*, **4**, 401–11.

Cerqueira, R. (1984). Reproduction de *Didelphis albiventris* dans le nord-est du Brésil (Polyprotodontia, Didelphidae). *Mammalia*, **48**, 95–104.

Chapman, H. C. (1882). On a foetal kangaroo and its membranes. In *Proceedings of the Academy of Natural Sciences Philadelphia*, 1881, 468–71.

Charles-Dominique, P. (1983). Ecology and social adaptations in didelphid marsupials: comparisons with eutherians of similar ecology. In *Advances in the Study of Mammalian Behaviour*, Special Publications American Society of Mammalogists, 7, ed. J. F. Eisenberg & D. G. Klieman, pp. 395–422. Philadelphia: American Society of Mammalogists.

Charles-Dominique, P., Atramentowicz, M., Charles-Dominique, M., Gerard, H., Hladik, A., Hladik, C. M. & Prevost, M. F. (1981). Les mammifères arboricoles nocturnes d'une forêt guyanaise: inter-relations plantes–animaux. *Revue d'Ecologie* (*Terre et Vie*), **35**, 341–435.

Chase, E. B. (1939). The reproductive system of the male opossum, *Didelphis virginiana* Kerr and its experimental modification. *Journal of Morphology*, **65**, 215–40.

Chauvet, M. T., Colne, T., Hurpet, D., Chauvet, J. & Acher, R. (1983*b*). Marsupial neurohypophysial hormones: identification of mesotocin, lysine vasopressin, and phenypressin in the quokka wallaby (*Setonix brachyurus*). *General and Comparative Endocrinology*, **52**, 309–15.

Chauvet, M. T., Colne, T., Hurpet, D., Chauvet, J. & Acher, R. (1983*c*). A multigene family for the vasopressin-like hormones? Identification of mesotocin, lysipressin and phenypressin in Australian macropods. *Biochemical and Biophysical Research Communications*, **116**, 258–63.

Chauvet, M. T., Hurpet, D., Chauvet, J. & Acher, R. (1981). A reptilian neurohypophysial hormone, mesotocin (Ile^8-oxytocin), in Australian marsupials. *FEBS Letters*, **129**, 120–2.

Chauvet, M. T., Hurpet, D., Chauvet, J. & Acher, R. (1983*a*). Identification of mesotocin, lysine vasopressin, and phenypressin in the eastern gray kangaroo (*Macropus giganteus*). *General and Comparative Endocrinology*, **49**, 63–72.

Chauvet, J., Hurpet, D., Chauvet, M. T., & Acher, R. (1984*a*). Divergent neuropeptide evolutionary drifts between American and Australian marsupials. *Bioscience Reports*, **4**, 245–52.

Chauvet, J., Hurpet, D., Colne, T., Michel, G., Chauvet, M. T. & Acher, R. (1985). Neurohypophyseal hormones as evolutionary tracers: identification of oxytocin, lysine vasopressin, and arginine vasopressin in two South American opossums (*Didelphis marsupialis* and *Philander opossum*). *General and Comparative Endocrinology*, **57**, 320–8.

Chauvet, J., Hurpet, D., Michel, G., Chauvet, M. T. & Acher, R. (1984*b*). Two multigene families for marsupial neurohypophyseal hormones? Identification of oxytocin, mesotocin, lysipressin and arginine vasopressin in the north American opossum (*Didelphis virginiana*). *Biochemical and Biophysical Research Communications*, **123**, 306–11.

Cheng, C. C. (1955). The development of the shoulder region of the opossum, *Didelphis virginiana*, with special reference to the musculature. *Journal of Morphology*, **97**, 415–71.

Chiquoine, A. D. (1954). The identification, origin and migration of the primordial germ cells in the mouse embryo. *Anatomical Record*, **118**, 135–45.

Chrétien, F. C. (1966). Etude de l'origine, de la migration et de la multiplication des cellules germinales chez l'embryon du lapin. *Journal of Embryology and Experimental Morphology*, **16**, 591–607.

Christensen, A. K. & Fawcett, D. W. (1961). The normal fine structure of opossum testicular interstitial cells. *Journal of Biophysical and Biochemical Cytology*, **9**, 653–70.

Clark, M. J. (1966). The blastocyst of the red kangaroo, *Megaleia rufa* (Desm.), during diapause. *Australian Journal of Zoology*, **14**, 19–25.

Clark, M. J. (1967). Pregnancy in the lactating pigmy possum *Cercartetus concinnus*. *Australian Journal of Zoology*, **15**, 673–83.

Clark, M. J. (1968*a*). Termination of embryonic diapause in the red kangaroo, *Megaleia rufa*, by injection of progesterone or oestrogen. *Journal of Reproduction and Fertility*, **15**, 347–55.

Clark, M. J. (1968*b*). Growth of pouch-young of the red kangaroo *Megaleia rufa* in the pouches of foster mothers of the same species. *International Zoo Yearbook*, **8**, 102–6.

Clark, M. J. & Poole, W. E. (1967). The reproductive system and embryonic diapause in the female grey kangaroo, *Macropus giganteus*. *Australian Journal of Zoology*, **15**, 441–59.

Clark, M. J. & Sharman, G. B. (1965). Failure of hysterectomy to affect the ovarian cycle of the marsupial *Trichosurus vulpecula*. *Journal of Reproduction and Fertility*, **10**, 459–61.

Cleland, K. W. & Rothschild, Lord (1959). The bandicoot spermatozoan: an electron microscope study of the tail. *Proceedings of the Royal Society B*, **150**, 24–42.

Close, R. L. (1977). Recurrence of breeding after cessation of suckling in the marsupial *Perameles nasuta*. *Australian Journal of Zoology*, **25**, 641–5.

Close, R. L. (1979). Sex chromosome mosaicism in liver, thymus, spleen and regenerating liver of *Perameles nasuta* and *Isoodon macrourus*. *Australian Journal of Biological Sciences*, **32**, 615–24.

Close, R. L. (1983). Detection of oestrus in the kowari *Dasyuroides byrnei* (Marsupialia: Dasyuridae). *Australian Mammalogy*, **6**, 41–3.

Close, R. L. (1984). Rates of sex chromosome loss during development in different tissues of the bandicoots *Perameles nasuta* and *Isoodon macrourus* (Marsupialia: Peramelidae). *Australian Journal of Biological Sciences*, **37**, 53–61.

Cockburn, A., Lee, A. K. & Martin, R. W. (1983). Macrogeographic variation in litter size in *Antechinus* (Marsupialia: Dasyuridae). *Evolution*, **37**, 86–95.
Collie, A. (1830). On some particulars connected with the natural history of the kangaroo. *Zoological Journal*, **5**, 238–41.
Collins, L. R. (1973). *Monotremes and Marsupials. A reference for Zoological Institutions.* Washington: Smithsonian Institution Press.
Cook, B., Karsch, F. J., Graber, J. W. & Nalbandov, A. V. (1977*a*). Luteolysis in the common opossum, *Didelphis marsupialis. Journal of Reproduction and Fertility*, **49**, 399–400.
Cook, B., McDonald, I. R. & Gibson, W. R. (1978). Prostatic function in the brush-tailed possum, *Trichosurus vulpecula. Journal of Reproduction and Fertility*, **53**, 369–75.
Cook, B. & Nalbandov, A. V. (1968). The effect of some pituitary hormones on progesterone synthesis *in vitro* by the luteinized ovary of the common opossum (*Didelphis marsupialis virginiana*). *Journal of Reproduction and Fertility*, **15**, 267–75.
Cook, B., Sutterlin, N. S., Graber, J. W. & Nalbandov, A. V. (1974). Gonadal steroid synthesis in the Virginian opossum, *Didelphis marsupialis. Journal of Endocrinology*, **61**, ix.
Cook, B., Sutterlin, N. S., Graber, J. W. & Nalbandov, A. V. (1977*b*). Synthesis and action of gonadal steroids in the American opossum *Didelphis marsupialis*. In *Reproduction and Evolution*. eds J. H. Calaby & C. H. Tyndale-Biscoe, pp. 253–4. Canberra: Australian Academy of Science.
Cooper, D. W., Edwards, C., James, E., Sharman, G. B., VandeBerg, J. L. & Graves, J. A. M. (1977*a*). Studies on metatherian sex chromosomes. VI. A third state of an X-linked gene: partial activity for the paternally derived Pgk-A allele in cultured fibroblasts of *Macropus giganteus* and *M. parryi. Australian Journal of Biological Sciences*, **30**, 431–43.
Cooper, D. W., Johnston, P. G., Sharman, G. B. & VandeBerg, J. L. (1977*b*). The control of gene activity on eutherian and metatherian X chromosomes: a comparison. In *Reproduction and Evolution*, eds J. H. Calaby & C. H. Tyndale-Biscoe, pp. 81–94. Canberra: Australian Academy of Science.
Cooper, D. W., VandeBerg, J. L., Sharman, G. B. & Poole, W. E. (1971). Phosphoglycerate kinase polymorphism in kangaroos provides further evidence for paternal X-inactivation. *Nature* (*New Biology*), **230**, 155–7.
Cooper, D. W., Woolley, P. A., Maynes, G. M., Sherman, F. S. & Poole, W. E. (1983). Studies on Metatherian sex chromosomes. XII. Paternal X-inactivation of α-galactosidase A in Australian marsupials. *Australian Journal of Biological Science*, **36**, 511–17.
Coulson, G. M. & Croft, D. B. (1981). Flehmen in kangaroos. *Australian Mammalogy*, **4**, 139–40.
Courot, M., Hochereau- de Reviers, M-T. & Ortavant, R. (1970). Spermatogenesis. In *The Testis*, vol. 1, ed. A. D. Johnson, W. R. Gomes & N. L. Van Demark, pp. 339–432. New York: Academic Press.
Cowie, A. T. (1984). Lactation. In *Reproduction in Mammals*. Book 3, *Hormonal Control of Reproduction* 2nd edn, C. R. Austin & R. V. Short, pp. 195–231. Cambridge University Press.
Cowie, A. T., Forsyth, I. A. & Hart, I. C. (1980). *Hormonal Control of Lactation. Monographs in Endocrinology*. Berlin: Springer-Verlag.
Cowper, W. (1704). II Carigueya, seu Marsupiale Americanum Masculum or, The Anatomy of a male opossum; in a letter to Dr Edward Tyson from Mr William Cowper. *Philosophical Transactions of the Royal Society*, **24**, 1565–90.
Cox, B. (1977). Why marsupials can't win. *Nature*, **265**, 14–15.

Crawley, M. C. (1973). A live-trapping study of Australian brush-tailed possums, *Trichosurus vulpecula* (Kerr), in the Orongorongo Valley, Wellington, New Zealand. *Australian Journal of Zoology*, **21**, 75–90.

Croft, D. B. (1982). Communication in the Dasyuridae (Marsupialia): a review. In *Carnivorous Marsupials*, vol. 1, ed. M. Archer, pp. 291–309. Sydney: Royal Zoological Society of New South Wales.

Crompton, A. W. (1980). Biology of the earliest mammals. In *Comparative Physiology: Primitive Mammals*, ed. K. Schmidt-Nielsen, L. Bolis & C. R. Taylor, pp. 1–12. Cambridge University Press.

Crompton, A. W. & Jenkins, F. A. (1979). Origin of mammals. In *Mesozoic mammals: the First Two-thirds of Mammalian History*, ed. J. A. Lillegraven, J. A. Kielen-Jawarowska & W. A. Crompton, pp. 59–73. Berkeley: University of California Press.

Cross, B. A. (1977). Comparative physiology of milk removal. *Symposium of the Zoological Society of London*, **41**, 193–210.

Crowcroft, P. & Sonderlund, R. (1977). Breeding wombats (*Lasiorhynus latifrons*) in captivity. *Der Zoologische Garten N. F. Jena*, **47**, 313–22.

Cummins, J. M. (1976). Epididymal maturation of spermatozoa in the marsupial *Trichosurus vulpecula*: changes in motility and gross morphology. *Australian Journal of Zoology*, **24**, 499–511.

Cummins, J. M. (1977). Sperm maturation in the phalanger, *Trichosurus vulpecula*. In *Reproduction and Evolution*, ed. J. H. Calaby & C. H. Tyndale-Biscoe, pp. 153–4. Canberra: Australian Academy of Science.

Cummins, J. M. (1980). Decondensation of sperm nuclei of Australian marsupials: effects of air drying and of calcium and magnesium. *Gamete Research*, **3**, 351–67.

Cummins, J. M. (1981). Sperm maturation in the possum *Trichosurus vulpecula*: a model for comparison with eutherian mammals. In *Marsupials in New Zealand*, ed. B. D. Bell, pp. 23–39. Wellington: Victoria University of Wellington.

Cummins, J. M., Temple-Smith, P. D. & Renfree, M. B. (1986). Reproduction in the male honey possum (*Tarsipes rostratus*: Marsupialia): the epididymis. *American Journal of Anatomy*, in press.

Curlewis, J. D., Axelson, M. & Stone, G. M. (1985). Identification of the major steroids in ovarian and adrenal venous plasma of the brush-tail possum (*Trichosurus vulpecula*) and changes in the peripheral plasma level of oestradiol and progesterone during the reproductive cycle. *Journal of Endocrinology*, **105**, 53–62.

Curlewis, J. & Stone, G. M. (1985*a*). Peripheral androgen levels in the male brush-tail possum (*Trichosurus vulpecula*). *Journal of Endocrinology*, **105**, 63–70.

Curlewis, J. & Stone, G. M. (1985*b*). Some effects of breeding season and castration on the prostate and epididymis of the brush-tail possum, *Trichosurus vulpecula*. *Australian Journal of Biological Sciences*, **38**, 313–26.

Curlewis, J. & Stone, G. M. (1986). Effects of oestradiol, the oestrous cycle and pregnancy on wieght, metabolism and cytosol receptors in the uterus of the brush-tail possum (*Trichosurus vulpecula*). *Journal of Endocrinology* **108**, 201–10.

Currie, W. B. & Thorburn, G. D. (1977). Parturition in goats: studies on the interactions between the foetus, placenta, prostaglandin F and progesterone before parturition, at term or at parturition induced prematurely by corticotrophin infusion of the foetus. *Journal of Endocrinology*, **73**, 263–78.

Cuttle, P. (1982). Life history strategy of the dasyurid marsupial *Phascogale tapoatafa*. In *Carnivorous Marsupials*, vol. 1, ed. M. Archer, pp. 13–22. Sydney: Royal Zoological Society of New South Wales.

Cutts, J. H., Krause, W. J. & Leeson, C. R. (1978). General observations on the growth and development of the young pouch opossum, *Didelphis virginiana*. *Biology of the Neonate*, **33**, 264–72.

Daly, M. (1979). Why don't male mammals lactate? *Journal of Theoretical Biology*, **78**, 325–45.

Davis, D. E. (1945). The annual cycle of plants, mosquitoes, birds, and mammals in two Brazilian forests. *Ecological Monographs*, **15**, 245–95.

Davis, P. J. & Jurgelski, W., Jr. (1973). Thyroid hormone-binding in opossum serum: evidence for polymorphism and relationship to haptoglobin polymorphism. *Endocrinology*, **92**, 822–32.

Dawson, T. J. & Hulbert, A. J. (1970). Standard metabolism, body temperature, and surface areas of Australian marsupials. *American Journal of Physiology*, **218**, 1233–8.

de Bavay, J. M. (1951). Notes on the female urogenital system of *Tarsipes spenserae* (Marsupialia). *Papers and Proceedings of the Royal Society of Tasmania*, **1950**, 143–50.

Del Campo, C. H. & Ginther, O. J. (1972). Vascular anatomy of the uterus and ovaries and the unilateral luteolytic effect of the uterus: guinea pigs, rats, hamsters and rabbits. *American Journal of Veterinary Research*, **33**, 2561–78.

Del Campo, C. H. & Ginther, O. J. (1973). Vascular anatomy of the uterus and ovaries and the unilateral luteolytic effect of the uterus: horses, sheep and swine. *American Journal of Veterinary Research*, **34**, 305–16.

De Melo, V. R., Orsi, A. M., Mello-Dias, S. & Oliveira, M. C. (1982). Fine structure of principal and basal cells of the cauda epididymidis of the South American opossum (*Didelphis azarae*). *Anatomischer Anzeiger Jena*, **151**, 497–502.

Denker, H. W. & Tyndale-Biscoe, C. H. (1986). Embryo implantation and proteinase activities in a marsupial (*Macropus eugenii*). *Cell and Tissue Research*, in press.

Denny, M. J. S. (1982). Review of planigale (Dasyuridae, Marsupialia) ecology. In *Carnivorous Marsupials*, vol. 1, ed. M. Archer, pp. 131–8. Sydney: Royal Zoological Society of New South Wales.

Deren, J. J., Padykula, H. A. & Wilson, T. H. (1966). Developmental structure and function in the mammalian yolk sac. II. Vitamin B12 uptake by rabbit yolk sacs. *Developmental Biology*, **13**, 349–69.

Dickman, C. R. (1982). Some ecological aspects of seasonal breeding in *Antechinus* (Dasyuridae, Marsupialia). In *Carnivorous Marsupials*. vol. 1, ed. M. Archer, pp. 139–50. Sydney: Royal Zoological Society of New South Wales.

Dimpel, H. & Calaby, J. H. (1972). Further Observations on the mountain pigmy possum (*Burramys parvus*). *Victorian Naturalist*, **89**, 101–6.

Dubocovich, M. (1983). Melatonin is a potent modulator of dopamine release in the retina. *Nature*, **306**, 782–4.

Dunnet, G. M. (1964). A field study of local populations of the brush-tailed possum *Trichosurus vulpecula* in eastern Australia. *Proceedings of the Zoological Society of London*, **142**, 665–95.

Dwyer, P. D. (1977). Notes on *Antechinus* and *Cercartetus* (Marsupialia) in the New Guinea highlands. *Proceedings of the Royal Society of Queensland*, **88**, 69–73.

Ealey, E. H. M. (1963). The ecological significance of delayed implantation in a population of the hill kangaroo (*Macropus robustus*). In *Delayed Implantation*, ed. A. C. Enders, pp. 33–47. Chicago University Press.

Ealey, E. H. M. (1967). Ecology of the euro, *Macropus robustus* (Gould) in north-western Australia. IV. Age and growth. *CSIRO Wildlife Research*, **12**, 53–65.

Eckstein, P. & Zuckerman, S. (1956). Morphology of the reproductive tract. In *Marshall's Physiology of Reproduction*, 3rd edn, vol. 1, pt. 1, ch. 2, ed. A. S. Parkes, pp. 43–155. London: Longmans, Green.

Einer-Jensen, N. & McCracken, J. A. (1981). The transfer of progesterone in the ovarian vascular pedicle of the sheep. *Endocrinology*, **109**, 685–90.

Enders, A. C., (1973). Cytology of the corpus luteum. *Biology of Reproduction*, **8**, 158–82.

Enders, A. C. & Enders, R. K. (1969). The placenta of the four-eyed opossum (*Philander opossum*). *Anatomical Record*, **165**, 431–50.

Enders, R. K. (1966). Attachment, nursing and survival of young in some didelphids. *Symposia of the Zoological Society of London*, **15**, 195–203.
Ensor, D. M. (1978). *Comparative endocrinology of prolactin*. pp. viii + 309. London: Chapman & Hall.
Evans, S. M., Tyndale-Biscoe, C. H. & Sutherland, R. L. (1980). Control of gonadotrophin secretion in the female tammar wallaby (*Macropus eugenii*). *Journal of Endocrinology*, **86**, 13–23.
Fadem, B. H. & Rayve, R. (1985). Characteristics of the oestrous cycle and influence of social factors in grey short-tailed opossums (*Monodelphis domestica*). *Journal of Reproduction and Fertility*, **73**, 337–42.
Fadem, B. H., Trupin, G. L., Maliniak, E., VandeBerg, J. L. & Hayssen, V. (1982). Care and breeding of the gray, short-tailed opossum (*Monodelphis domestica*). *Laboratory Animal Science*, **32**, 405–9.
Fanning, F. D. (1982). Reproduction, growth and development in *Ningaui* sp. (Dasyuridae, Marsupialia) from the Northern Territory. In *Carnivorous Marsupials*, vol. 1, ed. M. Archer, pp. 23–37. Sydney: Royal Zoological Society of New South Wales.
Farris, E. J. (1950). The opossum. In *The Care and Breeding of Laboratory Animals*, ed. E. J. Farris, pp. 256–67. New York: John Wiley.
Fawcett, D. W., Neaves, W. B. & Flores, M. N. (1973). Comparative observations on intertubular lymphatics and the organization of the interstitial tissue of the mammalian testis. *Biology of Reproduction*, **9**, 500–32.
Feldman, D. B. & Ross, P. W. (1975). Methods for obtaining neonates of known age from the Virginia Opossum (*Didelphis marsupialis virginiana*). *Laboratory Animal Science*, **25**, 437–9.
Findlay, L. (1982*a*). The mammary glands of the tammar wallaby (*Macropus eugenii*) during pregnancy and lactation. *Journal of Reproduction and Fertility*, **65**, 59–66.
Findlay, L. (1982*b*). *Lactation in the marsupial* Macropus eugenii. PhD thesis. Murdoch University, Perth.
Findlay, L. & Renfree, M. B. (1984). Growth, development and secretion of the mammary gland of macropodid marsupials. *Symposium of the Zoological Society of London*, **51**, 403–32.
Findlay, L., Ward, K. L & Renfree, M. B. (1983). Mammary gland lactose, plasma progesterone and lactogenesis in the marsupial *Macropus eugenii*. *Journal of Endocrinology*, **97**, 425–36.
Finkel, M. P. (1945). The relation of sex hormones to pigmentation and to testis descent in the opossum and ground squirrel. *American Journal of Anatomy*, **76**, 93–152.
Fleay, D. (1965). Australia's 'needle-in-a-haystack' marsupial. Vicissitudes in the pursuit and study of Ingram's planigale, the smallest pouch bearer. *Victorian Naturalist*, **82**, 195–204.
Fleming, D., Cinderey, R. N. & Hearn, J. P. (1983). The reproductive biology of Bennett's wallaby (*Macropus rufogriseus rufogriseus*) ranging free at Whipsnade Park. *Journal of Zoology, London*, **201**, 283–91.
Fleming, M. R. & Frey, H. (1984). Aspects of the natural history of feathertail gliders (*Acrobates pygmaeus*) in Victoria. In *Possums and Gliders*, eds A. P. Smith & I. D. Hume, pp. 403–8. Sydney: Australian Mammal Society.
Fleming, M. W. & Harder, J. D. (1981*a*). Uterine histology and reproductive cycles in pregnant and non-pregnant opossums, *Didelphis virginiana*. *Journal of Reproduction and Fertility*, **63**, 21–4.
Fleming, M. W. & Harder, J. D. (1981*b*). Effect of pregnancy on uterine constituents of the Virginia opossum. *Comparative Biochemistry and Physiology*, **69A**, 337–9.
Fleming, M. W. & Harder, J. D. (1983). Luteal and follicular populations in the ovary of

the opossum (*Didelphis virginiana*) after ovulation. *Journal of Reproduction and Fertility*, **67**, 29–34.

Fleming, M. W., Harder, J. D. & Wukie, J. J. (1981). Reproductive energetics of the Virginia opossum compared with some eutherians. *Comparative Biochemistry and Physiology*, **70B**, 645–8.

Fleming, T. H. (1973). The reproductive cycles of three species of opossums and other mammals in the Panama Canal Zone. *Journal of Mammalogy*, **54**, 439–55.

Fletcher, J. J. (1881). On the existence after parturition of a direct communication between the median vaginal cul-de-sac so-called, and the urogenital canal, in certain species of kangaroo. *Proceedings of the Linnean Society of New South Wales*, **6**, 796–811.

Fletcher, J. J. (1883). On some points in the anatomy of urogenital organs in females of certain species of kangaroos. Part 2. *Proceedings of the Linnean Society of New South Wales*, **8**, 6–11.

Fletcher, T. P. (1983). *Endocrinology of reproduction in the dasyurid marsupial* Dasyroides byrnei (*Spencer*). PhD thesis, La Trobe University, Melbourne.

Fletcher, T. P. (1985). Aspects of reproduction in the male eastern quoll, *Dasyurus viverrinus* (Shaw) (Marsupialia: Dasyuridae) with notes on polyoestry in the female. *Australian Journal of Zoology*, **33**, 101–10.

Flint, A. P. F. & Renfree, M. B. (1982). Oestradiol-17β in the circulation during seasonal reactivation of the diapausing blastocyst in a wild population of tammar wallabies, *Macropus eugenii*. *Journal of Endocrinology*, **95**, 293–300.

Flint, A. P. F. & Sheldrick, E. L. (1983). Secretion of oxytocin by the corpus luteum in sheep. In *The neurohypophysis: Structure, function and control*, ed. B. A. Cross & G. Leng, pp. 521–30. Amsterdam: Elsevier Biomedical Press.

Flynn, T. T. (1922). Notes on certain reproductive phenomena in some Tasmanian marsupials. *Annals and Magazine of Natural History*, 9th ser., **10**, 225–31.

Flynn, T. T. (1923). The yolk-sac and allantoic placenta in *Perameles*. *Quarterly Journal of Microscopical Science*, **67**, 123–82.

Flynn, T. T. (1930). The uterine cycle of pregnancy and pseudopregnancy as it is in the diprotodont marsupial *Bettongia cuniculus* with notes on other reproductive phenomena in this marsupial. *Proceedings of the Linnean Society of New South Wales*, **55**, 506–31.

Flynn, T. T. & Hill, J. P. (1939). The development of the Monotremata. IV. Growth of the ovarian ovum, fertilization and early cleavage. *Transactions of the Zoological Society of London*, **24**, 445–622.

Flynn, T. T. & Hill, J. P. (1947). The development of the Monotremata. VI. The later stages of cleavage and the formation of the primary germ layers. *Transactions of the Zoological Society of London*, **26**, 1–151.

Fox, B. J. (1982). Ecological separation and coexistence of *Sminthopsis murina* and *Antechinus sturartii* (Dasyuridae, Marsupialia): a regeneration niche? In *Carnivorous Marsupials*, vol. 1, ed. M. Archer, pp. 187–97. Sydney: Royal Zoological Society of New South Wales.

Fox, B. J. & Whitford, D. (1982). Polyoestry in a predictable coastal environment: reproduction, growth and development in *Sminthopsis murina* (Dasyuridae, Marsupialia). In *Carnivorous Maruspials*, vol. 1, ed. M. Archer, pp. 39–48. Sydney: Royal Zoological Society of New South Wales.

Fraser, E. A. (1919). The development of the urogenital system in the Marsupialia, with special reference to *Trichosurus vulpecula*. Part II. *Journal of Anatomy*, **53**, 97–129.

Fraser, E. A. & Hill, J. P. (1915). The development of the thymus, epithelial bodies and thyroid in the Marsupialia. 1. *Trichosurus vulpecula*. *Philosophical Transactions of the Royal Society B*, **207**, 1–85.

Friend, J. A. & Burrows, R. G. (1983). Bringing up young numbats. *Swans*, **13**, 3–9.

Friend, J. A. & Whitford, R. W. (1986). Captive breeding of the numbat (*Myrmecobius fasciatus*). Australian Mammal Society Proceedings of the 31st Meeting, *Australian Mammal Society Bulletin* **9**, 54.

Frith, H. J. & Sharman, G. B. (1964). Breeding in wild populations of the red kangaroo, *Megaleia rufa. CSIRO Wildlife Research*, **9**, 86–114.

Gadsby, J. E., Heap, R. B. & Burton, R. D. (1980). Oestrogen production by blastocyst and early embryonic tissue of various species. *Journal of Reproduction and Fertility*, **60**, 409–17.

Ganslosser, von U. (1979). Soziale Interaktionen des Doriabaumkänguruhs, *Dendrolagus dorianus* Ramsay, 1833 (Marsupialia, Macropodidae). *Zeitschrifte für Saugertierkunde*, **44**, 1–18.

Gardner, A. L. (1973). The systematics of the genus *Didelphis* (Marsupialia: Didelphidae) in North and middle America. *Special Publication of the Museum, Texas Technical University*, **4**, 1–81.

Gaughwin, M. D. & Wells, R. T. (1978). General features of reproduction of the hairy-nosed wombat (*Lasiorhinus latifrons*) in the Blanche town region of South Australia. *Australian Mammal Society Bulletin*, **5** (1), 46–7.

Geist, V. (1972). An ecological and behavioural explanation of mammalian characteristics. *Zeitschrifte für Saugertierkunde*, **37**, 1–15.

Gemmell, R. T. (1979). The fine structure of the luteal cells in relation to the concentration of progesterone in the plasma of the lactating bandicoot, *Isoodon macrourus* (Marsupialia: Peramelidae). *Australian Journal of Zoology*, **27**, 501–10.

Gemmell, R. T. (1981). The role of the corpus luteum of lactation in the bandicoot *Isoodon macrourus* (Marsupialia: Peramelidae). *General and Comparative Endocrinology*, **44**, 13–19.

Gemmell, R. T. (1982). Breeding bandicoots in Brisbane (*Isoodon macrourus*; Marsupialia, Peramelidae). *Australian Mammalogy*, **5**, 187–93.

Gemmell, R. T. (1984). Plasma concentrations of progesterone and 13, 14-dihydro-15-keto-prostaglandin $F_{2\alpha}$ during regression of the corpora lutea of lactation in the bandicoot (*Isoodon macrourus*). *Journal of Reproduction and Fertility*, **72**, 295–9.

Gemmell, R. T. (1985). The effect of prostaglandin $F_{2\alpha}$ analogue on the plasma concentration of progesterone in the bandicoot, *Isoodon macrourus* (Marsupialia: Peramelidae), during lactation. *General and Comparative Endocrinology*, **57**, 405–10.

Gemmell, R. T. (1986). Seasonal variations in plasma testosterone concentration in the male possum, *Trichosurus vulpecula* in captivity. Australian Mammal Society, Proceedings of the 31st meeting. *Australian Mammology*, in press.

Gemmell, R. T., Jenkin, G. & Thorburn, G. D. (1980). Plasma concentrations of progesterone and 13, 14-dihydro-15-keto-prostaglandin $F_{2\alpha}$ at parturition in the bandicoot *Isoodon macrourus. Journal of Reproduction and Fertility*, **60**, 253–6.

Gemmell, R. T. & Little, G. J. (1982). The structure of the lung of the newborn marsupial bandicoot, *Isoodon macrourus. Cell and Tissue Research*, **223**, 445–53.

Gemmell, R. T., Singh-Asa, P., Jenkin, G. & Thorburn, G. D. (1982). Ultrastructural evidence for steriod-hormone production in the adrenal of the marsupial, *Isoodon macrourus*, at birth. *Anatomical Record*, **203**, 505–12.

Gemmell, R. T., Stacy, B. D. & Thorburn, G. D. (1974). Ultrastructural study of secretory granules in the corpus luteum of the sheep during the estrous cycle. *Biology of Reproduction*, **11**, 447–62.

Geoffroy-Saint Hilaire, E. (1833). New observations on the nature of the abdominal glands of the *Ornithorhynchus. Proceedings of the Zoological Society of London*, **1**, 28–31, 91–6.

George, F. W., Hodgins, M. B. & Wilson, J. D. (1985). The synthesis and metabolism of

gonadal steroids in pouch young of the opossum, *Didelphis virginiana. Endocrinology*, **116**, 1145–50.
George, G. G. (1982). Cuscuses *Phalanger* spp.: their management in captivity. In *The Management of Australian Mammals in Captivity*, ed. D. D. Evans, pp. 67–72. Melbourne: Zoological Board of Victoria.
Gerard, P. (1932). Etudes sur l'ovogenèse et l'ontogenèse chez les lemuriens de genre *Galago. Archives de Biologie Paris*, **43**, 93–151.
Gilmore, D. P. (1965). Gynandromorphism in *Trichosurus vulpecula. Australian Journal of Science*, **28**, 165.
Gilmore, D. P. (1969). Seasonal reproductive periodicity in the male Australian brush-tailed possum (*Trichosurus vulpecula*). *Journal of Zoology, London*, **157**, 75–98.
Godfrey, G. K. (1969*a*). Reproduction in a laboratory colony of the marsupial mouse *Sminthopsis larapinta* (Marsupiala: Dasyuridae). *Australian Journal of Zoology*, **17**, 637–54.
Godfrey, G. K. (1969*b*). Influence of increased photoperiod on reproduction in the dasyurid marsupial, *Sminthopsis crassicaudata. Journal of Mammalogy*, **50**, 132–3.
Godfrey, G. K. (1975). A study of oestrus and fecundity in a laboratory colony of mouse opossums (*Marmosa robinsonii*). *Journal of Zoology, London*, **175**, 541–55.
Godfrey, G. K. & Crowcroft, P. (1971). Breeding the fat-tailed marsupial mouse *Sminthopsis crassicaudata* in captivity. *International Zoo Yearbook*, **11**, 33–8.
Godsell, J. (1982). The population ecology of the eastern quoll *Dasyurus viverrinus* (Dasyuridae, Marsupialia), in southern Tasmania. In *Carnivorous Marsupials*, vol. 1, ed. M. Archer, pp. 199–207. Sydney: Royal Zoological Society of New South Wales.
Goetz, R. H. (1938). On the early development of the Tenrecoidea (*Hemicentetes semispinosus*). *Biomorphosis*, **1**, 67–79.
Goodman, R. L., Reichert, L. E., Legan, S. J., Ryan, K. D., Foster, D. L & Karsch, F. J. (1982). Role of gonadotropins and progesterone in determining the pre-ovulatory estradiol rise in the ewe. *Biology of Reproduction*, **25**, 134–42.
Gordon, G. (1971). *A study of island populations of the short nosed bandicoot*, Isoodon macrourus *Gould*. PhD thesis, University of New South Wales, Sydney.
Gordon, G. (1974). Movements and activity of the short-nosed bandicoot *Isoodon macrourus* Gould (Marsupialia). *Mammalia*, **38**, 405–31.
Gould, S. J. (1977). Sticking up for marsupials. *Natural History*, **86** (8), 22–30.
Grant, T. R., Griffiths, M. E. & Leckie, R. M. C. (1983). Aspects of lactation in the platypus, *Ornithorhynchus anatinus* (Monotremata), in waters of eastern New South Wales. *Australian Journal of Zoology*, **31**, 881–9.
Graves, J. A. M. (1967). DNA synthesis in chromosomes of cultured leucocytes from two marsupial species. *Experimental Cell Research*, **46**, 37–57.
Green, B. (1984). Composition of milk and energetics of growth in marsupials. *Symposium of the Zoological Society of London*, **51**, 369–87.
Green, B. & Eberhard, I. (1983). Water and sodium intake, and estimated food consumption, in free-living eastern quolls, *Dasyurus viverrinus. Australian Journal of Zoology*, **31**, 871–80.
Green, B., Griffiths, M. & Leckie, R. M. C. (1983). Qualitative and quantitative changes in milk fat during lactation in the tammar wallaby (*Macropus eugenii*). *Australian Journal of Biological Sciences*, **36**, 455–61.
Green, R. H. (1967). Notes on the devil (*Sarcophilus harrisii*) and the quoll (*Dasyurus viverrinus*) in north-eastern Tasmania. *Records of the Queen Victoria Museum*, **27**, 1–13.
Green, S. W. & Renfree, M. B. (1982). Changes in the milk proteins during lactation in the tammar wallaby *Macropus eugenii. Australian Journal of Biological Sciences*, **35**, 145–52.

Gregory, W. K. (1947). The monotremes and the palimpsest theory. *Bulletin of the American Museum of Natural History*, **88**, 7–52.

Griffiths, M. (1978). *The Biology of the Monotremes*. New York: Academic Press.

Griffiths, M. (1983). Lactation in Monotremata and speculations concerning the nature of lactation in Cretaceous Multituberculata. *Acta Palaeontologica Polonica*, **28**, 93–102.

Griffiths, M. (1984). Mammals: monotremes. In *Marshall's Physiology of Reproduction*, 4th edn, vol. 1, Chap. 5, G. E. Lamming, pp. 351–85. Edinburgh: Churchill Livingstone.

Griffiths, M., Elliot, M. A., Leckie, R. M. C. & Schoefl, G. I. (1973). Observations of the comparative anatomy and ultrastructure of mammary glands, and on the fatty acids of the triglycerides in platypus and echidna milk fats. *Journal of Zoology, London*, **169**, 255–79.

Griffiths, M., McIntosh, D. L. & Leckie, R. M. C. (1972). The mammary glands of the red kangaroo, with observations on the fatty acid components of the milk triglycerides. *Journal of Zoology, London*, **166**, 265–75.

Guiler, E. R. (1960). The breeding season of *Potorous tridactylus* (Kerr). *Australian Journal of Science*, **23**, 126–7.

Guiler, E. R. (1961). Breeding season of the thylacine. *Journal of Mammology*, **42**, 396–7.

Guiler, E. R. (1970). Observations on the Tasmanian devil, *Sarcophilus harrisii* (Marsupialia, Dasyuridae). II. Reproduction, breeding, and growth of pouch young. *Australian Journal of Zoology*, **18**, 63–70.

Guiler, E. R. & Heddle, R. W. L. (1970). Testicular and body temperatures in the Tasmanian devil and three other species of marsupial. *Comparative Biochemistry and Physiology*, **33**, 881–91.

Guraya, S. S. (1968*a*). Histochemical observations on the granulosa and theca interna during follicular development and corpus luteum formation and regression in the American opossum. *Journal of Endocrinology*, **40**, 237–41.

Guraya, S. S. (1968*b*). Histochemical study of developing ovarian oocyte of the American opossum. *Acta Embryologica et Morphologica Experimentalis*, **10**, 181–91.

Haacke, W. (1885). On the marsupial ovum, the mammary pouch, and the male milk glands in *Echidna hystrix*. *Proceedings of the Royal Society of London, Series B*, **38**, 72–4.

Hall, L. S. & Hughes, R. L. (1985). The embryoogical development and cytodifferentiation of the anterior pituitary in the marsupial, *Isoodon macrourus*. *Anatomy and Embryology*, **172**, 353–63.

Hamilton, D. W. (1972). The mammalian epididymis. In *Reproductive Biology*, eds. H. Balin & S. Glasser, pp. 268–337. Amsterdam: Excerpta Medica.

Hamilton, W. J. (1958). Life history and economic relations of the opossum (*Didelphis marsupialis virginiana*) in New York State. *Memoirs of New York State College of Agriculture*, **354**, 1–48.

Hammond, J. & Marshall, F. H. A. (1925). *Reproduction in the rabbit*. Edinburgh and London: Oliver & Boyd.

Happold, M. (1972). Maternal and juvenile behaviour in the marsupial jerboa *Antechinomys spenceri* (Dasyuridae). *Australian Mammalogy*, **1**, 27–37.

Harcourt, A. H., Harvey, P. H., Larson, S. G. & Short, R. V. (1981). Testis weight, body weight and breeding system in primates. *Nature*, **293**, 55–7.

Harder, J. D. & Fleming, M. W. (1981). Estradiol and progesterone profiles indicate a lack of endocrine recognition of pregnancy in the opossum. *Science*, **212**, 1400–2.

Harder, J. D., Hinds, L. A., Horn, C. A. & Tyndale-Biscoe, C. H. (1984). Oestradiol in follicular fluid and in utero-ovarian venous and peripheral plasma during parturition and post-partum oestrus in the tammar, *Macropus eugenii*. *Journal of Reproduction and Fertility*, **72**, 551–8.

Harder, J. D., Hinds, L. A., Horn, C. A. & Tyndale-Biscoe, C. H. (1985). Effects of removal in late pregnancy of the corpus luteum, Graafian follicle or ovaries on plasma progesterone, oestradiol, LH, parturition and post-partum oestrus in the tammar, *Macropus eugenii*. *Journal of Reproduction and Fertility*, **75**, 449–59.

Harding, H. R. (1977). *Reproduction in male marsupials: a critique, with additional observations on sperm development and structure*. PhD thesis, University of New South Wales, Sydney.

Harding, H. R., Carrick, F. N. & Shorey, C. D. (1975). Ultrastructural changes in spermatozoa of the brush-tailed possum, *Trichosurus vulpecula* (Marsupialia) during epididymal transit. Part I. The flagellum. *Cell and Tissue Research*, **164**, 121–32.

Harding, H. R., Carrick, F. N. & Shorey, C. D. (1976*a*). Ultrastructural changes in spermatozoa of the brush-tailed possum *Trichosurus vulpecula* (Marsupialia), during epididymal transit. Part II. The Acrosome. *Cell and Tissue Research*, **171**, 61–73.

Harding, H. R., Carrick, F. N. & Shorey, C. D. (1976*b*). Spermiogenesis in the brush-tailed possum, *Trichosurus vulpecula* (Marsupialia): development of the acrosome. *Cell and Tissue Research*, **171**, 75–90.

Harding, H. R., Carrick, F. N. & Shorey, C. D. (1979). Special features of sperm structure and function in marsupials. In *The Spermatozoan*, ed. D. Fawcett & J. M. Bedford, pp. 289–303. Baltimore-Munich: Urban & Schwarzenberg.

Harding, H. R., Carrick, F. N. & Shorey, C. D. (1981). Marsupial phylogeny: new indications from sperm ultrastructure and development in *Tarsipes spenserae*. *Search*, **12**, 45–7.

Harding, H. R., Carrick, F. N. & Shorey, C. D. (1982*a*). Crystalloid inclusions in the Sertoli cell of the koala. *Phascolarctos cinereus* (Marsupialia). *Cell and Tissue Research*, **221**, 633–41.

Harding, H. R., Carrick, F. N. & Shorey, C. D. (1983). Acrosome development during spermiogenesis and epididymal sperm maturation in Australian marsupials. In *The Sperm Cell*, ed. J. André, pp. 411–14. The Hague: Martinus Nijhoff.

Harding, H. R., Woolley, P. A., Shorey, C. D. & Carrick, F. N. (1982*b*). Sperm ultrastructure, spermiogenesis and epididymal sperm maturation in dasyurid marsupials: phylogenetic implications. In *Carnivorous Marsupials*, vol. 2, ed. M. Archer, pp. 659–73. Sydney: Royal Zoological Society of New South Wales.

Harrison, R. G. (1949). The comparative anatomy of the blood-supply of the mammalian testis. *Proceedings of the Zoological Society of London*, **119**, 325–42.

Hart, D. B. (1909). The nature and cause of the physiological descent of the testes. *Journal of Anatomy*, **43**, 244–65.

Hartman, C. G. (1916). Studies in the development of the opossum *Didelphis virginiana* L. I. History of the early cleavage, II. Formation of the blastocyst. *Journal of Morphology*, **27**, 1–84.

Hartman, C. G. (1919). Studies in the development of the opossum, *Didelphis virginiana* L. III. Description of new material on maturation, cleavage and entoderm formation. IV. The bilaminar blastocyst. *Journal of Morphology*, **32**, 1–144.

Hartman, C. G. (1920*a*). Studies in *Didelphis*. I. The phenomena of parturition. *Anatomical Record*, **19**, 251–62.

Hartman, C. G. (1920*b*). The freemartin and its reciprocal: opossum, man, dog. *Science*, **52**, 469–71.

Hartman, C. G. (1921). Breeding habits, development, and birth of the opossum. *Annual Report of the Smithsonian Institution*, **1921**, 347–64.

Hartman, C. G. (1923*a*). The oestrous cycle in the opossum. *American Journal of Anatomy*, **32**, 353–421.

Hartman, C. G. (1923*b*). Sterility of animals under changed condition. *Anatomical Record*, **24**, 394.

Hartman, C. G. (1924). Observations on the motility of the opossum genital tract and the vaginal plug. *Anatomical Record*, **27**, 293–303.

Hartman, C. G. (1925*a*). Hysterectomy and the oestrous cycle in the opossum. *American Journal of Anatomy*, **35**, 25–9.

Hartman, C. G. (1925*b*). The interruption of pregnancy by ovariectomy in the aplacental opossum: a study in the physiology of implantation. *American Journal of Physiology*, **71**, 436–54.

Hartman, C. G. (1926). Polynuclear ova and polyovular follicles in the opossum and other mammals, with special reference to the problem of fertility. *American Journal of Anatomy*, **37**, 1–52.

Hartman, C. G. (1927). Observations on the ovary of the opossum, *Didelphis virginiana*. *Contributions to Embryology of the Carnegie Institute*, **19**, 285–300.

Hartman, C. G. (1928). The breeding season of the opossum (*Didelphis virginiana*) and the rate of intrauterine and postnatal development. *Journal of Morphology and Physiology*, **46**, 143–215.

Hartman, C. G. (1952). *Possums*. Austin: University of Texas Press.

Hartman, C. G. & League, B. (1925). Description of a sex-intergrade opossum, with an analysis of the constituents of its gonads. *Anatomical Record*, **29**, 283–98.

Hayman, D. L., Ashworth, L. K. & Carrano, A. U. (1982). The relative DNA contents of the eutherian and marsupial X chromosomes. *Cytogenics and Cell Genetics*, **34**, 265–70.

Hayman, D. L. & Martin, P. G. (1965*a*). An autoradiographic study of DNA synthesis in the sex chromosomes of two marsupials with an XX/XY_1Y_2 sex chromosome mechanism. *Cytogenetics*, **4**, 209–18.

Hayman, D. L. & Martin, P. G. (1965*b*). Sex chromosome mosaicism in the marsupial genera *Isoodon* and *Perameles*. *Genetics*, **52**, 1201–6.

Hayman, D. L. & Martin, P. G. (1974). Mammalia I: Monotremata and Marsupialia. In *Animal Cytogenetics*, vol. 4, Chordata 4, ed. B. John, pp. 1–110. Berlin: Gebruder Borntraeger.

Hayman, D. L. & Rofe, R. H. (1977). Marsupial sex chromosomes. In *Reproduction and Evolution*, ed. J. H. Calaby & C. H. Tyndale-Biscoe, pp. 69–79. Canberra: Australian Academy of Science.

Hayman, D. & Sharp, P. (1981). Verification of the structure of the complex sex chromosome system in *Lagorchestes conspicillatus* Gould (Marsupialia: Mammalia). *Chromosoma*, **83**, 263–74.

Heap, R. B. & Flint, A. P. F. (1984). Pregnancy. In *Reproduction in Mammals*, 2nd edn, Book 3, ed. C. R. Austin & R. V. Short, pp. 153–94. Cambridge University Press.

Heap, R. B., Renfree, M. B. & Burton, R. D. (1980). Steroid metabolism in the yolk sac placenta and endometrium of the tammar wallaby, *Macropus eugenii*. *Journal of Endocrinology*, **87**, 339–49.

Hearn, J. P. (1972*a*). *The Pituitary Gland and Reproduction in the Marsupial*, Macropus eugenii (*Desmarest*). PhD thesis, Australian National University, Canberra.

Hearn, J. P. (1972*b*). Effect of advanced photoperiod on termination of embryonic diapause in the marsupial, *Macropus eugenii* (Macropodidae). *Australian Mammalogy*, **1**, 40–2.

Hearn, J. P. (1973). Pituitary inhibition of pregnancy. *Nature*, **241**, 207–8.

Hearn, J. P. (1974). The pituitary gland and implantation in the tammar wallaby, *Macropus eugenii*. *Journal of Reproduction and Fertility*, **39**, 235–41.

Hearn, J. P. (1975*a*). The role of the pituitary in the reproduction of the male tammar wallaby, *Macropus eugenii*. *Journal of Reproduction and Fertility*, **42**, 399–402.

Hearn, J. P. (1975*b*). Hypophysectomy of the tammar wallaby, *Macropus eugenii*: surgical approach and general effects. *Journal of Endocrinology*, **64**, 403–16.

Hearn, J. P., Short, R. V. & Baird, D. T. (1977). Evolution of the luteotrophic control of the mammalian corpus luteum. In *Reproduction and Evolution*, ed. J. H. Calaby & C. H. Tyndale-Biscoe, pp. 255–64. Canberra: Australian Academy of Science.

Heddle, R. W. L. & Guiler, E. R. (1970). The form and function of the testicular rete mirabile of marsupials. *Comparative Biochemistry and Physiology*, **35**, 415–25.

Heeres, J. E. (1899). The part borne by the Dutch in the discovery of Australia 1606–1765. Leiden and London: Royal Dutch Geographical Society.

Heinsohn, G. E. (1966). Ecology and reproduction of the Tasmanian bandicoots (*Perameles gunnii* and *Isoodon obesulus*). *University of California Publications in Zoology*, **80**, 1–107.

Heinsohn, G. E. (1968). Habitat requirements and reproductive potential of the macropod marsupial *Potorous tridactylus* in Tasmania. *Mammalia*, **32**, 30–43.

Heinsohn, G. E. (1970). World's smallest marsupial. The flat headed marsupial mouse. *Animals*, **13**, 220–2.

Henry, S. R. (1984). The social organisation of the greater glider in Victoria. In *Possums and Gliders*, ed. A. P. Smith & I. D. Hume, pp. 221–8. Sydney: Australian Mammal Society.

Henry, S. R. & Craig, S. A. (1984). Diet, ranging behaviour and social organization of the yellow-bellied glider in Victoria. In *Possums and Gliders*, ed. A. P. Smith & I. D. Hume, pp. 331–41. Sydney: Australian Mammal Society.

Hershkovitz, P. (1972). The recent mammals of the neotropical region: a zoogeographic and ecological review. In *Evolution, Mammals, and Southern Continents*, ed. A. Keast, F. C. Erk & B. Glass, pp. 311–432. Albany: State University of New York Press.

Hill, C. J. (1933). The development of the Monotremata. I. The histology of the oviduct during gestation. *Transactions of the Zoological Society of London*, **21**, 413–43.

Hill, J. P. (1895). Preliminary note on the occurrence of a placental connection in *Perameles obesula*, and on the foetal membranes of certain macropods. *Proceedings of the Linnean Society of New South Wales*, **10**, 578–81.

Hill, J. P. (1898). Contributions to the embryology of the Marsupialia. I. The placentation of *Perameles*. *Quarterly Journal of Microscopical Science*, **40**, 385–446.

Hill, J. P. (1899). Contributions to the morphology and development of the female urogenital organs in the Marsupialia. I. On the female urogenital organs of *Perameles*, with an account of the phenomenon of parturition. *Proceedings of the Linnean Society of New South Wales*, **24**, 42–82.

Hill, J. P. (1900*a*). Contributions to the embryology of the Marsupialia. 2. On a further stage of placentation of *Perameles*. 3. On the foetal membranes of *Macropus parma*. *Quarterly Journal of Microscopical Science*, **43**, 1–22.

Hill, J. P. (1900*b*). On the foetal membranes, placentation and parturition, of the native cat (*Dasyurus viverrinus*). *Anatomischer Anzeiger*, **18**, 364–73.

Hill, J. P. (1900*c*). Contributions to the morphology and development of the female urogenital organs in the Marsupialia. *Proceedings of the Linnean Society of New South Wales*, **25**, 519–32.

Hill, J. P. (1910). Contributions to the embryology of the Marsupialia. 4. The early development of the Marsupialia with special reference to the native cat (*Dasyurus viverrinus*). *Quarterly Journal of Microscopical Science*, **56**, 1–134.

Hill, J. P. (1918). Some observations on the early development of *Didelphys aurita*. *Quarterly Journal of Microscopical Science*, **63**, 91–139.

Hill, J. P. & de Beer, G. R. (1950). Development of the Monotremata. Part VII. The development and structure of the egg-tooth and the caruncle in the monotremes and on the occurrence of vestiges of the egg-tooth and caruncle in marsupials. *Transactions of the Zoological Society of London*, **26**, 503–44.

Hill, J. P. & Fraser, E. (1925). Some observations on the female urogenital organs of Didelphyidae. *Proceedings of the Zoological Society of London*, **1925**, 189–219.

Hill, J. P. & Hill, W. C. O. (1955). The growth stages of the pouch young of the native cat (*Dasyurus viverrinus*) together with observations on the anatomy of the new born young. *Transactions of the Zoological Society of London*, **28**, 349–453.

Hill, J. P. & O'Donoghue, C. H. (1913). The reproductive cycle in the marsupial *Dasyurus viverrinus*. *Quarterly Journal of Microscopical Science*, **59**, 133–174.

Hinds, L. A. (1983). *Progesterone and prolactin in marsupial reproduction*. PhD thesis, Australian National University, Canberra.

Hinds, L. A. & den Ottolander, R. C. (1983). Effect of changing photoperiod on peripheral plasma prolactin and progesterone concentrations in the tammar wallaby (*Macropus eugenii*). *Journal of Reproduction and Fertility*, **69**, 631–9.

Hinds, L. A., Evans, S. M. & Tyndale-Biscoe, C. H. (1983). *In-vitro* secretion of progesterone by the corpus luteum of the tammar wallaby, *Macropus eugenii*. *Journal of Reproduction and Fertility*, **67**, 57–63.

Hinds, L. A. & Janssens, P. A. (1986). Changes in prolactin in peripheral plasma during lactation in the brushtail possum *Trichosurus vulpecula*. *Australian Journal of Biological Sciences*, **39**, in press.

Hinds, L. A. & Merchant, J. C. (1986). Plasma prolactin concentrations throughout lactation in the eastern quoll, *Dasyurus viverrinus* (Marsupialia: Dasyuridae). *Australian Journal of Biological Sciences*, **39**, in press.

Hinds, L. A. & Tyndale-Biscoe, C. H. (1982*a*). Plasma progesterone levels in the pregnant and non-pregnant tammar, *Macropus eugenii*. *Journal of Endocrinology*, **93**, 99–107.

Hinds, L. A. & Tyndale-Biscoe, C. H. (1982*b*). Prolactin in the marsupial, *Macropus eugenii* during the estrous cycle, pregnancy and lactation. *Biology of Reproduction*, **26**, 391–8.

Hinds, L. A. & Tyndale-Biscoe, C. H. (1985). Seasonal and circadian patterns of circulating prolactin during lactation and seasonal quiescence in the tammar, *Macropus eugenii*. *Journal of Reproduction and Fertility*, **74**, 173–83.

Hobson, B. M. (1983). An appraisal of the mouse uterine weight assay for the bioassay of chorionic gonadotrophin in the macaque term placenta. *Journal of Reproduction and Fertility*, **68**, 457–63.

Hollis, D. E. & Lyne, A. G. (1977). Endoderm formation in blastocysts of the marsupials *Isoodon macrourus* and *Perameles nasuta*. *Australian Journal of Zoology*, **25**, 207–23.

Hollis, D. E. & Lyne, A. G. (1980). Ultrastructure of luteal cells in fully-formed and regressing corpora lutea during pregnancy and lactation in the marsupials *Isoodon macrourus* and *Perameles nasuta*. *Australian Journal of Zoology*, **28**, 195–211.

Holmes, P. V. & Dickson, A. D. (1975). Temporal and spatial aspects of oestrogen-induced RNA, protein and DNA synthesis in delayed implantation mouse blastocysts. *Journal of Anatomy*, **119**, 453–9.

Holmes, R. S., Cooper, D. W. & VandeBerg, J. L. (1973). Marsupial and monotreme lactate dehydrogenase isozymes: phylogeny, ontogeny, and homology with eutherian mammals. *Journal of Experimental Zoology*, **184**, 127–48.

Holstein, A. F. (1965). Elektronmikroskopische untersuchungen am spermatozoon des opossum (*Didelphys virginiana* Kerr). *Zeitschrifte für Zellforschungen*, **65**, 904–14.

Home, E. (1795). Some observations on the mode of generation of the kangaroo: a particular description of the organs themselves. *Philosophical Transactions of the Royal Society*, **85**, 221–30.

Hopson, J. A. (1973). Endothermy, small size, and the origin of mammalian reproduction. *American Naturalist*, **107**, 446–52.

Horn, C. A., Fletcher, T. P. & Carpenter, S. (1985). Effects of oestradiol-17β on peripheral plasma concentrations of LH and FSH in ovariectomised tammars (*Macropus eugenii*). *Journal of Reproduction and Fertility*, **73**, 585–92.

Horst, C. J. van der (1942). Early stages in the embryonic development of *Elephantulus*. *South African Journal of Medical Science*, **7**, 55–65.

How, R. A. (1976). Reproduction, growth and survival of young in the mountain possum, *Trichosurus caninus* (Marsupialia). *Australian Journal of Zoology*, **24**, 189–99.

How, R. A. (1981). Population parameters of two congeneric possums, *Trichosurus* spp., in north-eastern New South Wales. *Australian Journal of Zoology*, **29**, 205–15.

How, R. A., Barnett, J. L., Bradley, A. J., Humphreys, W. F. & Martin, R. (1984). The population biology of *Pseudocheirus peregrinus* in a *Leptospermum laevigatum* thicket. In *Possums and Gliders*, ed. A. P. Smith & I. D. Hume, pp. 261–88. Sydney: Australian Mammal Society.

Hruban, Z., Martan, J., Slesers, A., Steiner, D. F., Lubran, M. & Richeigl, M. (1965). Fine structure of the prostatic epithelium of the opossum (*Didelphis virginiana*). *Journal of Experimental Zoology*, **160**, 81–106.

Hsu, T. L. & Benirschke, K. (1977). *Atlas of Mammalian Chromosomes*. New York: Springer-Verlag.

Hubrecht, A. A. W. (1889). Studies in mammalian embryology. I. The placentation of *Erinaceus europaeus*, with remarks on the phylogeny of the placenta. *Quarterly Journal of Microscopical Science*, **30**, 283–404.

Hubrecht, A. A. W. (1909). Early ontogenetic phenomena in mammals and their bearing on our interpretation of the phylogeny of vertebrates. *Quarterly Journal of Microscopical Science*, **53**, 1–181.

Hughes, R. L. (1962*a*). Reproduction in the macropod marsupial *Potorous tridactylus* (Kerr). *Australian Journal of Zoology*, **10**, 193–224.

Hughes, R. L. (1962*b*). Role of the corpus luteum in marsupial reproduction. *Nature*, **194**, 890–1.

Hughes, R. L. (1965). Comparative morphology of spermatozoa from five marsupial families. *Australian Journal of Zoology*, **13**, 533–43.

Hughes, R. L. (1974). Morphological studies on implantation in marsupials. *Journal of Reproduction and Fertility*, **39**, 173–86.

Hughes, R. L. (1977). Egg membranes and ovarian function during pregnancy in monotremes and marsupials. In *Reproduction and Evolution*, ed. J. H. Calaby & C. H. Tyndale-Biscoe, pp. 281–91. Canberra: Australian Academy of Science.

Hughes, R. L. (1982). Reproduction in the Tasmanian devil, *Sarcophilus harrisii* (Dasyuridae, Marsupialia). In *Carnivorous Marsupials*, vol. 1, ed. M. Archer, pp. 49–63. Sydney: Royal Zoological Society of New South Wales.

Hughes, R. L. (1984). Structural adaptations of the eggs and the fetal membranes of monotremes and marsupials for respiration and metabolic exchange. In *Respiration and Metabolism of Embryonic Vertebrates*, ed. R. S. Seymour, pp. 389–421. Doordrecht: Junk.

Hughes, R. L. & Hall, L. S. (1984). Embryonic development in the common brushtail possum *Trichosurus vulpecula*. In *Possums and Gliders*, ed. A. P. Smith & I. D. Hume, pp. 197–212. Sydney: Australian Mammal Society.

Hughes, R. L. & Rodger, J. C. (1971). Studies on the vaginal mucus of the marsupial *Trichosurus vulpecula*. *Australian Journal of Zoology*, **19**, 19–33.

Hughes, R. L. & Shorey, C. D. (1973). Observations on the permeability properties of the egg membranes of the marsupial, *Trichosurus vulpecula*. *Journal of Reproduction and Fertility*, **32**, 25–32.

Hughes, R. L., Thomson, J. A. & Owen, W. H. (1965). Reproduction in natural populations of the Australian ringtail possum, *Pseudocheirus peregrinus* (Marsupialia: Phalangeridae), in Victoria. *Australian Journal of Zoology*, **13**, 383–406.

Hulbert, A. J. (1972). Growth and development of pouch young in the rabbit-eared bandicoot, *Macrotis lagotis* (Peramelidae). *Australian Mammalogy*, **1**, 38–9.

Hume, I. D. (1982). *Digestive physiology and nutrition of marsupials*. Cambridge University Press.

Hunsaker, D., II (1977). Ecology of New World marsupials. In *The Biology of Marsupials*, ed. D. Hunsaker II, pp. 95–156. New York: Academic Press.

Hunsaker, D., II & Shupe, D. (1977). Behaviour of New World marsupials. In *The Biology of Marsupials*, ed. D. Hunsaker II, pp. 279–347. New York: Academic Press.

Hurpet, D., Chauvet, M. T., Chauvet, J. & Acher, R. (1982). Marsupial hypothalamo-neurohypophyseal hormones. The brushtailed possum (*Trichosurus vulpecula*) active peptides. *International Journal of Peptide and Protein Research*, **19**, 366–71.

Hutson, G. D. (1976). Grooming behaviour and birth in the dasyurid marsupial *Dasyuroides byrnei*. *Australian Journal of Zoology*, **24**, 277–82.

Imperato-McGinley, J. (1983). Sexual differentiation: normal and abnormal. In *Current Topics in Experimental Endocrinology*, vol. 5, ed. L. Martini & V. H. T. James, pp. 232–309. New York and London: Academic Press.

Inns, R. W. (1976). Some seasonal changes in *Antechinus flavipes* (Marsupialia: Dasyuridae). *Australian Journal of Zoology*, **24**, 523–31.

Inns, R. W. (1982). Seasonal changes in the accessory reproductive system and plasma testosterone levels of the male tammar wallaby, *Macropus eugenii*, in the wild. *Journal of Reproduction and Fertility*, **66**, 675–80.

Jacobs, H. T. T. M., S. J. (1971). *A treatise on the Moluccas* (*c. 1544*). Rome: Jesuit Historical Institute.

Janssens, P. A., Jenkinson, L. A., Paton, B. C. & Whitelaw, E. (1977). The regulation of gluconeogenesis in pouch young of the tammar wallaby, *Macropus eugenii* (Desmarest). *Australian Journal of Biological Sciences*, **30**, 183–95.

Jarman, P. (1983). Mating system and sexual dimorphism in large, terrestrial, mammalian herbivores. *Biological Reviews*, **58**, 485–520.

Johnson, K. A. (1977). *Ecology and management of the red-necked pademelon*, Thylogale thetis, *on the Dorrigo plateau of northern New South Wales*. PhD thesis, University of New England, Armidale.

Johnson, P. M. (1978). Studies of Macropodidae in Queensland. 9. Reproduction of the rufous rat-kangaroo (*Aepyprymnus rufescens* (Gray)) in captivity with age estimation of pouch young. *Queensland Journal of Agricultural and Animal Sciences*, **35**, 69–72.

Johnson, P. M. (1979). Reproduction in the plain rock-wallaby *Petrogale penicillata inornata* Gould, in captivity, with age estimation of the pouch young. *Australian Wildlife Research*, **6**, 1–4.

Johnston, P. G. & Robinson, E. S. (1985). Glucose-6-phosphate dehydrogenase expression in heterozygous kangaroo embryos and extra-embryonic membranes. *Genetical Research, Cambridge*, **45**, 205–8.

Johnston, P. G., Robinson, E. S. & Sharman, G. B. (1976). X chromosome activity in oocytes of kangaroo pouch young. *Nature*, **264**, 359–60.

Johnston, P. G. & Sharman, G. B. (1975). Studies on metatherian sex chromosomes. I. Inheritance and inactivation of sex-linked allelic genes determining glucose-6-phosphate dehydrogenase variation in kangaroos. *Australian Journal of Biological Sciences*, **28**, 567–74.

Jones, R. C., Hinds, L. A. & Tyndale-Biscoe, C. H. (1984). Ultrastructure of the epididymis of the tammar, *Macropus eugenii*, and its relationship to sperm maturation. *Cell and Tissue Research*, **237**, 525–35.

Jost, A. (1953). Problems of fetal endocrinology: the gonadal and hypophyseal hormones. *Recent Progress in Hormone Research*, **8**, 379–418.

Jurgelski, W. & Porter, M. E. (1974). The opossum (*Didelphis virginiana* Kerr) as a biomedical model. III. Breeding the opossum in captivity: methods. *Laboratory Animal Science*, **24**, 412–25.

Karsch, F. J., Bittman, E. L., Foster, D. L., Goodman, R. L., Legan, S. J. & Robinson, J. E. (1984). Neuroendocrine basis of seasonal reproduction. *Recent Progress in Hormone Research*, **40**, 185–225.

Kean, R. I. (1975). Growth of opossums (*Trichosurus vulpecula*) in the Orongorongo valley, Wellington, New Zealand, 1953–61. *New Zealand Journal of Zoology*, **2**, 435–44.

Kean, R. I., Marryatt, R. G. & Carroll, A. L. K. (1964). The female urogenital system of *Trichosurus vulpecula* (Marsupialia). *Australian Journal of Zoology*, **12**, 18–41.

Kemp, T. S. (1983). The relationships of mammals. *Zoological Journal of the Linnean Society*, **77**, 353–84.

Kennaway, D. J., Frith, R. G., Phillipou, G., Mathews, C. D. & Seamark, R. F. (1977). A specific radioimmunoassay for melatonin in biological tissue and fluids and its validation by gas chromatography – mass spectrometry. *Endocrinology*, **101**, 119–27.

Kerle, J. A. (1984). Growth and development of *Burramys parvus* in captivity. In *Possums and Gliders*, ed. A. P. Smith & I. D. Hume, pp. 409–12. Sydney: Australian Mammal Society.

Kermack, K. A. (1963). The cranial structure of the triconodonts. *Philosophical Transactions of the Royal Society B*, **246**, 83–103.

Kermack, K. A. & Mussett, F. (1983). The ear in mammal-like reptiles and early mammals. *Acta Palaeontologica Polonica*, **28**, 147–58.

Kerr, J. B. & Hedger, M. P. (1983). Spontaneous spermatogenic failure in the marsupial mouse *Antechinus stuartii* Macleay (Dasyuridae: Marsupialia). *Australian Journal of Zoology*, **31**, 445–66.

Kerr, T. (1935). Blastocysts of *Potorous tridactylus* and *Bettongia cuniculus*. *Quarterly Journal of Microscopical Science*, **77**, 305–15.

Kerr, T. (1936). On the primitive streak and associated structures in the marsupial *Bettongia cuniculus*. *Quarterly Journal of Microscopical Science*, **78**, 687–715.

Khin Aye Than & McDonald, I. R. (1976). Corticosteriod binding by plasma proteins and the effects of oestrogens in the marsupial brush-tailed possum, *Trichosurus vulpecula* (Kerr). *Journal of Endocrinology*, **68**, 31–41.

Kielan-Jaworowska, Z. (1979). Pelvic structure and nature of reproduction in Multituberculata. *Nature*, **277**, 402–3.

Kirkpatrick, T. H. (1965). Studies of Macropodidae in Queensland. 3. Reproduction in the grey kangaroo (*Macropus major*) in southern Queensland. *Queensland Journal of Agricultural and Animal Sciences*, **22**, 319–28.

Kirkpatrick, T. H. & Johnson, P. M. (1969). Studies on Macropodidae in Queensland. 7. Age estimation and reproduction in the agile wallaby (*Wallabia agilis* (Gould)). *Queensland Journal of Agricultural and Animal Sciences*, **26**, 691–8.

Kirsch, J. A. W. (1977*a*). The comparative serology of Marsupialia, and a classification of marsupials. *Australian Journal of Zoology, Supplementary Series* **52**, 1–152.

Kirsch, J. A. W. (1977*b*). Biological aspects of the marsupial-placental dichotomy: a reply to Lillegraven. *Evolution*, **31**, 898–900.

Kirsch, J. A. W. & Calaby, J. H. (1977). The species of living marsupials: an annotated list. In *The Biology of Marsupials*, ed. B. Stonehouse & D. Gilmore, pp. 9–26. London: Macmillan.

Kirsch, J. A. W. & Poole, W. E. (1972). Taxonomy and distribution of the grey kangaroos, *Macropus giganteus* Shaw and *Macropus fuliginosus* (Desmarest), and their subspecies (Marsupialia: Macropodidae). *Australian Journal of Zoology*, **20**, 315–39.

Kirsch, J. A. W. & Waller, P. F. (1979). Notes on the trapping and behavior of the Caenolestidae (Marsupialia). *Journal of Mammalogy*, **60**, 390–5.

Kitchener, D. J. (1981). Breeding, diet and habitat preference of *Phascogale calura* (Gould, 1844) (Marsupialia: Dasyuridae) in the southern wheat belt, Western Australia. *Records of the Western Australian Museum*, **9**, 173–86.

Krause, W. J. & Cutts, J. H. (1979). Pairing of spermatozoa in the epididymis of the opossum (*Didelphis virginiana*): a scanning electron microscopic study. *Archivum histologicum japonicum*, **42**, 181–90.

Krause, W. J. & Cutts, J. H. (1983). Ultrastructural observations on the shell membrane of the North American opossum (*Didelphis virginiana*). *Anatomical Record*, **207**, 335–8.

Krause, W. J. & Cutts, J. H. (1984). Scanning electron microscopic observations on the opossum yolk sac chorion immediately prior to uterine attachment. *Journal of Anatomy*, **138**, 189–91.

Krause, W. J. & Cutts, J. H. (1985). Placentation in the opossum, *Didelphis virginiana. Acta Anatomica*, **123**, 156–71.

Krause, W. J., Cutts, J. H. & Leeson, C. R. (1976). Type II pulmonary epithelial cells of the newborn opossum lung. *American Journal of Anatomy*, **146**, 181–8.

Krause, W. J., Cutts, J. H. & Lesson, C. R. (1979*a*). Morphological observations on the mesonephros in the postnatal opossum, *Didelphis virginiana. Journal of Anatomy*, **129**, 377–97.

Krause, W. J., Cutts, J. H. and Leeson, C. R. (1979*b*). Morphological observations on the metanephros in the postnatal opossum, *Didelphis virginiana. Journal of Anatomy*, **129**, 459–77.

Krause, W. J. & Leeson, C. R. (1973). The postnatal development of the respiratory system of the opossum. I. Light and scanning electron microscopy. *American Journal of Anatomy*, **137**, 337–56.

Kuhn, N. J. (1977). Lactogenesis: the search for trigger mechanisms in different species. *Symposium of the Zoological Society of London*, **41**, 165–92.

Ladman, A. J. (1967). Fine structure of the ductuli efferentes of the opossum. *Anatomical Record*, **157**, 559–75.

Lay, D. W. (1942). Ecology of the opossum in eastern Texas. *Journal of Mammalogy*, **23**, 147–59.

Leatherland, J. F. & Renfree, M. B. (1982). Ultrastructure of the non-granulated cells and morphology of the extracellular spaces in the pars distalis of adult and pouch-young tammar wallabies (*Macropus eugenii*). *Cell and Tissue Research*, **227**, 439–50.

Leatherland, J. F. & Renfree, M. B. (1983). Structure of the pars distalis in pouch-young tammar wallabies (*Macropus eugenii*). *Cell and Tissue Research*, **230**, 587–603.

Le Blond, C. B., Morris, S., Karakuilakis, G., Powell, R. & Thomas, P. J. (1982). Development of sexual dimorphism in the suprachiasmatic nucleus of the rat. *Journal of Endocrinology*, **95**, 137–45.

Lee, A. K., Bradley, A. J. & Braithwaite, R. W. (1977). Corticosteroid levels and male mortality in *Antechinus stuartii*. In *The Biology of Marsupials*, ed. B. Stonehouse & D. Gilmore, pp. 209–20. London: Macmillan.

Lee, A. K. & Cockburn, A. (1985). *The Evolutionary Ecology of Marsupials*. Cambridge University Press.

Lee, A. K., Woolley, P. & Braithwaite, R. W. (1982). Life history strategies of dasyurid marsupials. In *Carnivorous Marsupials*, vol. 1, ed. M. Archer, pp. 1–11. Sydney: Royal Zoological Society of New South Wales.

Lee, C. S. & O'Shea, J. D. (1977). Observations on the vasculature of the reproductive tract in some Australian marsupials. *Journal of Morphology*, **154**, 95–114.

Legan, S. J. & Karsch, F. J. (1983). Importance of retinal photoreceptors to the photoperiodic control of seasonal breeding in the ewe. *Biology of Reproduction*, **29**, 316–25.

Legan, S. J., Karsch, F. J. & Foster, D. L. (1977). The endocrine control of seasonal reproductive function in the ewe: a marked change in response to the negative feedback action of estradiol on luteinizing hormone secretion. *Endocrinology*, **101**, 818–24.

Lemon, M. (1972). Peripheral plasma progesterone during pregnancy and the oestrous cycle in the tammar wallaby *Macropus eugenii*. *Journal of Endocrinology*, **55**, 63–71.

Lemon, M. & Barker, S. (1967). Changes in milk composition of the red kangaroo, *Megaleia rufa* (Desmarest), during lactation. *Australian Journal of Experimental Biology and Medical Science*, **45**, 213–19.

Lemon, M. & Poole, W. E. (1969). Specific proteins in the whey from milk of the grey kangaroo. *Australian Journal of Experimental Biology and Medical Science*, **47**, 283–5.

Lewis, P. R., Fletcher, T. P. & Renfree, M. B. (1986). Prostaglandin in the peripheral plasma of tammar wallabies during parturition. *Journal of Endocrinology*, **110**, in press.

Lierse, W. (1965). The wall construction and blood vessel system of the uterus of the opossum. *Acta Anatomica*, **60**, 152–63.

Liggins, G. C., Fairclough, R. J., Grieves, S. A., Forster, C. S. & Knox, B. S. (1977). Parturition in the sheep. In *The Fetus and Birth*. Ciba Foundation Symposium 47, ed. J. Knight & M. O'Connor, pp. 5–25. Amsterdam: Elsevier/Excerpta Medica.

Lillegraven, J. A. (1969). Latest cretaceous mammals of upper part of Edmonton formation of Alberta, Canada and review of marsupial–placental dichotomy in mammalian evolution. *Paleontological Contributions from the University of Kansas*, **50**, 1–122.

Lillegraven, J. A. (1975). Biological considerations of the marsupial–placental dichotomy. *Evolution*, **29**, 707–22.

Lillegraven, J. A. (1979). Reproduction in Mesozoic mammals. In *Mesozoic Mammals: The First Two-thirds of Mammalian History*, eds J. A. Lillegraven, Z. Kielan-Jaworowska & W. A. Clemens, Chapter 13, pp. 259–76. Berkeley: University of California Press.

Lillegraven, J. A., Kielan-Jaworowska, Z. & Clemens, W. A. (Eds) (1979). *Mesozoic Mammals: The First Two-thirds of Mammalian History*. Berkeley: University of California Press.

Lillie, F. R. (1917). The freemartin: a study of the action of sex hormones in the fetal life of cattle. *Journal of Experimental Zoology*, **23**, 371–452.

Lincoln, D. W. & Paisley, A. C. (1982). Neuroendocrine control of milk ejection. *Journal of Reproduction and Fertility*, **65**, 571–86.

Lincoln, D. W. & Renfree, M. B. (1981*a*). Milk ejection in a marsupial, *Macropus agilis*. *Nature*, **289**, 504–6.

Lincoln, D. W. & Renfree, M. B. (1981*b*). Mammary gland growth and milk ejection in the agile wallaby *Macropus agilis*, displaying concurrent asynchronous lactation. *Journal of Reproduction and Fertility*, **63**, 193–203.

Lincoln, G. A. (1976). Seasonal variation in the episodic secretion of luteinizing hormone and testosterone in the ram. *Journal of Endocrinology*, **69**, 213–26.

Lincoln, G. A. (1978). Plasma testosterone profiles in male macropodid marsupials. *Journal of Endocrinology*, **77**, 347–51.

Lincoln, G. A., Peet, M. J. & Cunningham, R. A. (1977). Seasonal and circadian changes in the episodic release of follicle-stimulating hormone, luteinizing hormone and testosterone in rams exposed to artificial photoperiods. *Journal of Endocrinology*, **72**, 337–49.

Lincoln, G. A. & Short, R. V. (1980). Seasonal breeding: nature's contraceptive. *Recent Progress in Hormone Research*, **36**, 1–52.

Lindner, H. R. & Sharman, G. B. (1966). The pregnancy hormone in the red kangaroo (*Megaleia rufa* Desm.) *Second International Congress on Hormonal Steroids*, p. 371. Amsterdam: Excerpta Medica.

Lintern-Moore, S., Moore, G.P. M., Tyndale-Biscoe, C. H. & Poole, W. E. (1976). The growth of oocyte and follicle in the ovaries of monotremes and marsupials. *The Anatomical Record*, **185**, 325–32.

Lintern-Moore, S. & Moore, G. P. M. (1977). Comparative aspects of oocyte growth in mammals. In *Reproduction and Evolution*, ed. J. H. Calaby & C. H. Tyndale-Biscoe, pp. 215–19. Canberra: Australian Academy of Science.

Lister, J. J. & Fletcher, J. J. (1881). On the condition of the median portion of the vaginal apparatus in the Macropodidae. *Proceedings of the Zoological Society of London*, **1881.** 976–96.

Loeb, L. (1923). The effect of extirpation of the uterus on the life and function of the corpus luteum in the guinea pig. *Proceedings of the Society for Experimental Biology, New York*, **20**, 441–3.

Long, J. A. & Evans, H. M. (1922). The oestrous cycle of the rat and its associated phenomena. *Memoirs of the University of California*, **6**, 1–148.

Loudon, A. S. I., Curlewis, J. D. & English, J. (1985). The effect of melatonin on the seasonal embryonic diapause of the Bennett's wallaby (*Macropus rufogriseus rufogriseus*). *Journal of Zoology*, **206**, 35–9.

Low, B. S. (1978). Environmental uncertainty and the parental strategies of marsupials and placentals. *American Naturalist*, **112**, 197–213.

Luckett, W. P. (1977). Ontogeny of amniote fetal membranes and their application to phylogeny. In *Major Patterns in Vertebrate Evolution*, ed. M. K. Hecht, P. C. Goody & B. M. Hecht, pp. 439–516. New York: Plenum Press.

Lustick, S. & Lustick, D. D. (1972). Energetics in the opossum, *Didelphis marsupialis virginiana. Comparative Biochemistry and Physiology*, **43A**, 643–7.

Lyne, A. G. (1964). Observations on the breeding and growth of the marsupial *Perameles nasuta* Geoffroy, with notes on other bandicoots. *Australian Journal of Zoology*, **12**, 322–39.

Lyne, A. G. (1974). Gestation period and birth in the marsupial *Isoodon macrourus. Australian Journal of Zoology*, **22**, 303–9.

Lyne, A. G. (1976). Observations on oestrus and the oestrous cycle in the marsupials *Isoodon macrourus* and *Perameles nasuta. Australian Journal of Zoology*, **24**, 513–21.

Lyne, A. G. (1982). The bandicoots *Isoodon macrourus* and *Perameles nasuta*: their maintenance and breeding in captivity. In *The Management of Australian Mammals in Captivity*, ed. D. D. Evans, pp. 47–52. Melbourne: The Zoological Board of Victoria.

Lyne, A. G. & Hollis, D. E. (1976). Early embryology of the marsupials *Isoodon macrourus* and *Perameles nasuta. Australian Journal of Zoology*, **24**, 361–82.

Lyne, A. G. & Hollis, D. E. (1977*a*). An unusual blastocyst stage of the marsupial *Isoodon macrourus. Australian Journal of Zoology*, **25**, 225–31.

Lyne, A. G. & Hollis, D. E. (1977*b*). The early development of marsupials, with special reference to bandicoots. In *Reproduction and Evolution*, ed. J. H. Calaby & C. H. Tyndale-Biscoe, pp. 293–302. Canberra: Australian Academy of Science.

Lyne, A. G. & Hollis, D. E. (1979). Observations on the corpus luteum during pregnancy and lactation in the marsupials *Isoodon macrourus* and *Perameles nasuta. Australian Journal of Zoology*, **27**, 881–99.

Lyne, A. G. & Hollis, D. E. (1982). Observations on the lateral vaginae and birth canals in the marsupials *Isoodon macrourus* and *Perameles nasuta* (Mammalia). *Journal of Zoology, London*, **198**, 263–77.

Lyne, A. G. & Hollis, D. E. (1983). Observations on Graafian follicles and their oocytes during lactation and after the removal of pouch young in the marsupials *Isoodon macrourus* and *Perameles nasuta. American Journal of Anatomy*, **166**, 41–61.

Lyne, A. G., Pilton, P. E. & Sharman, G. B. (1959). Oestrous,cycle, gestation period and parturition in the marsupial *Trichosurus vulpecula. Nature*, **183**, 622–3.

Lyne, A. G. & Verhagen, A. M. W. (1957). Growth of the marsupial *Trichosurus vulpecula* and a comparison with some higher mammals. *Growth*, **21**, 167–95.
Lyon, M. F. (1972). X chromosome inactivation and developmental patterns in mammals. *Biological Reviews*, **47**, 1–35.
McConnell, S. J. (1984). *The pineal gland and reproduction in the tammar wallaby,* Macropus eugenii. PhD thesis, Australian National University, Canberra.
McConnell, S. J. (1986). Seasonal changes in the circadian plasma profile of the tammar, *Macropus eugenii. Journal of Pineal Research*, **3**, in press.
McConnell, S. J. & Hinds, L. A. (1985). Effect of pinealectomy on plasma melatonin, prolactin and progesterone concentrations during seasonal reproductive quiescence in the tammar, *Macropus eugenii. Journal of Reproduction and Fertility*, **7**, 433–40.
McConnell, S. J. & Tyndale-Biscoe, C. H. (1985). Response to photoperiod in peripheral plasma melatonin and the effects of exogenous melatonin on seasonal quiescence in the tammar, *Macropus eugenii. Journal of Reproduction and Fertility*, **73**, 529–38.
McCracken, H. E. (1986). Observations on the oestrous cycle and gestation period of the greater bilby, *Macrotis lagotis* (Reid) (Marsupialia: Thylacomyidae). *Australian Mammalogy*, **9**, 5–16.
McCracken, J. A., Baird, D. T. & Goding, J. R. (1971). Factors affecting the secretion of steroids from the transplanted ovary of the sheep. *Recent Progress in Hormone Research*, **27**, 537–82.
McCrady, E. (1938). The embryology of the opossum. *American Anatomical Memoirs*, **16**, 1–233.
McCrady, E. (1944). The evolution and significance of the germ layers. *Journal of the Tennessee Academy of Science*, **19**, 240–51
McDonald, I. R., Lee, A. K., Bradley, A. J. & Than, K. A. (1981). Endocrine changes in dasyurid marsupials with differing mortality patterns. *General and Comparative Endocrinology*, **44**, 292–301.
McDonald, I. R. & Waring, H. (1982). Hormones of marsupials and monotremes. In *Hormones and Evolution*, ed. E. J. W. Barrington, vol. 2, pp. 873–924. London: Academic Press.
McFarlane, J. R., Carrick, F. N. & Brown, A. S. (1986). Seasonal changes in the size of the prostate in the brushtail possum *Trichosurus vulpecula*. Australian Mammal Society Proceedings of the 31st Meeting. *Australian Mammalogy*, in press.
McIlroy, J. C. (1973). *Aspects of the ecology of the common wombat*, Vombatus ursinus (*Shaw*, 1800). PhD thesis, Australian National University, Canberra.
McKay, G. M., McQuade, L. R., Murray, J. D. & Von Sturmer, S. R. (1984). Sex-chromosome mosaicism in the lemur-like possum *Hemibelideus lemuroides* (Marsupialia: Petauridae). *Australian Journal of Biological Sciences*, **37**, 131–5.
McNeilly, A. S. & Land, R. B. (1979). Effect of suppression of plasma prolactin on ovulation, plasma gonadotrophins and corpus luteum function in LH-RH-treated anoestrous ewes. *Journal of Reproduction and Fertility*, **56**, 601–9.
Mack, G. (1961). Mammals from southwestern Queensland. *Memoirs of the Queensland Museum*, **13**, 213–28.
Mann, F. G. (1958). Reproduccion de *Dromiciops australis* (Marsupialia, Didelphydae). *Investigaciones Zoologicas Chilenas*, **4**, 209–13.
Mann, T. (1964). *The Biochemistry of Semen and of the Male Reproductive Tract.* London: Methuen.
Marlow, B. J. (1961). Reproductive behaviour of the marsupial mouse, *Antechinus flavipes* (Waterhouse) (Marsupialia) and the development of the pouch young. *Australian Journal of Zoology*, **9**, 203–18.
Marshall, L. G. (1979). Evolution of metatherian and eutherian (Mammalian) characters: a review based on cladistic methodology. *Zoological Journal of the Linnean Society*, **66**, 369–410.

Martan, J. & Allen, J. M. (1965). The cytological and chemical organization of the prostatic epithelium of *Didelphis virginiana* Kerr. *The Journal of Experimental Zoology*, **159**, 209–30.

Martan, J., Hruban, Z. & Slesers, A. (1967). Cytological studies on the ductuli efferentes of the opossum. *Journal of Morphology*, **121**, 81–101.

Martin, P. G. & Hayman, D. L. (1966). A complex sex-chromosome system in the hare-wallaby *Lagorchestes conspicillatus* Gould. *Chromosoma*, **19**, 159–75.

Martin, R. W. & Lee, A. (1984). The koala, *Phascolarctos cinereus*, the largest marsupial folivore. In *Possums and Gliders*, ed. A. P. Smith & I. D. Hume, pp. 463–7. Sydney: Australian Mammal Society.

Martinez-Esteve, P. (1937). Le cycle sexuel vaginal chez le marsupial *Didelphys azarae*. *Comptes Rendus de Societé de Biologie, Paris*, **124**, 502–4.

Martinez-Esteve, P. (1942). Observations on the histology of the opossum ovary. *Carnegie Institute Contributions to Embryology* **30**, 17–26.

Maynes, G. M. (1973*a*). Reproduction in the parma wallaby, *Macropus parma* Waterhouse. *Australian Journal of Zoology*, **21**, 331–51.

Maynes, G. M. (1973*b*). Aspects of reproduction in the whiptail wallaby *Macropus parryi*. *Australian Zoologist*, **18**, 43–6.

Maynes, G. M. (1977). Breeding and age structure of the population of *Macropus parma* on Kawau Island, New Zealand. *Australian Journal of Ecology*, **2**, 207–14.

Meites, J., Nicoll, C. S. & Talwalker, P. K. (1959). The effects of reserpine and serotonin on milk secretion and mammary growth in the rat. *Proceedings of the Society for Experimental Biology*, **101**, 563–5.

Merchant, J. C. (1976). Breeding biology of the agile wallaby, *Macropus agilis* (Gould) (Marsupialia: Macropodidae), in captivity. *Australian Wildlife Research*, **3**, 93–103.

Merchant, J. C. (1979). The effect of pregnancy on the interval between one oestrus and the next in the tammar wallaby, *Macropus eugenii*. *Journal of Reproduction and Fertility*, **56**, 459–63.

Merchant, J. C. & Calaby, J. H. (1981). Reproductive biology of the red-necked wallaby (*Macropus rufogriseus banksianus*) and Bennett's wallaby (*M.r. rufogriseus*) in captivity. *Journal of Zoology, London*, **194**, 203–17.

Merchant, J. C., Newgrain, K. & Green, B. (1984). Growth of the eastern quoll, *Dasyurus viverrinus* (Shaw) (Marsupialia) in captivity. *Australian Wildlife Research*, **11**, 21–9.

Merchant, J. C. & Sharman, G. B. (1966). Observations on the attachment of marsupial pouch young to the teats and on the rearing of pouch young by foster-mothers of the same or different species. *Australian Journal of Zoology*, **14**, 593–609.

Merry, D. E., Pathak, S. & VandeBerg, J. L. (1983). Differential NOR activities in somatic and germ cells of *Monodelphis domestica* (Marsupialia: Mammalia). *Cytogenetics and Cell Genetics*, **35**, 244–51.

Messer, M., Gadiel, P. A., Ralston, G. B. & Griffiths, M. (1983). Carbohydrates of the milk of the platypus. *Australian Journal of Biological Sciences*, **36**, 129–37.

Michener, G. R. (1969). Notes on the breeding and young of the crest-tailed marsupial mouse, *Dasycercus cristicauda*. *Journal of Mammalogy*, **50**, 633–5.

Miegs, C. D. (1847). Memoirs on the reproduction of the opossum *Didelphys virginiana*. *Proceedings of the American Philosophical Society*, **4**, 327–30.

Mintz, B. (1957). Germ cell origin and history in the mouse: genetic and histochemical evidence. *Anatomical Record*, **127**, 335–6.

Mohla, S. & Prasad, M. R. N. (1971). Early action of oestrogen on the incorporation of [^{3}H]uridine in the blastocyst and uterus of rat during delayed implantation. *Journal of Endocrinology*, **49**, 87–92.

Moore, C. R. (1939). Modification of sexual development in the opossum by sex hormones. *Proceedings of the Society for Experimental Biology*, **40**, 544–6.

Moore, C. R. (1941*a*). On the role of sex hormones in sex differentiation in the opossum (Didelphis virginiana). *Physiological Zoology*, **14**, 1–47.

Moore, C. R. (1941*b*). Embryonic differentiation of the opossum prostrate following castration and responses of the juvenile gland to hormones. *Anatomical Record*, **80**, 315–27.

Moore, C. R. (1943). Sexual differentiation in the opossum after early gonadectomy. *Journal of Experimental Zoology*, **94**, 415–62.

Moore, C. R. (1945). Prostate gland induction in the female opossum by hormones and the capacity of the gland for development. *American Journal of Anatomy*, **76**, 1–31.

Moore, C. R. & Morgan, C. F. (1942). Responses of the testis to androgenic treatments. *Endocrinology*, **30**, 990–9.

Moore, C. R. & Morgan, C. F. (1943). First response of developing opossum gonads to equine gonadotropic treatment. *Endocrinology*, **32**, 17–26.

Moore, G. P. M. (1978). Embryonic diapause in the marsupial *Macropus eugenii.* Stimulation of nuclear RNA polymerase activity in the blastocyst during resumption of development. *Journal of Cellular Physiology*, **94**, 31–6.

Moore, G. P. M., Lintern-Moore, S., Peters, H. & Faber, M. (1974). RNA synthesis in the mouse oocyte. *Journal of Cell Biology*, **60**, 416–22.

Moore, G. P. M. & Lintern-Moore, S. (1978). Transcription of the mouse oocyte genome. *Biology of Reproduction*, **18**, 865–70.

Moors, P. J. (1974). The foeto-maternal relationship and its significance in marsupial reproduction: a unifying hypothesis. *Australian Mammalogy*, **1**, 263–6.

Moors, P. J. (1975). The urogenital system and notes on the reproductive biology of the female rufous rat-kangaroo, *Aepyprymnus rufescens* (Gray) (Macropodidae). *Australian Journal of Zoology*, **23**, 355–61.

Morgan, C. F. (1943). The normal development of the ovary of the opossum from birth to maturity and its reactions to sex hormones. *Journal of Morphology*, **72**, 27–85.

Morgan, J. (1829). Description of the mammary organs of the kangaroo. *Transactions of the Linnean Society of London*, **16**, 61–84

Morton, S. R. (1978). An ecological study of *Sminthopsis crassicaudata* (Marsupialia: Dasyuridae). III. Reproduction and life history. *Australian Wildlife Research*, **5**, 183–211.

Morton, S. R., Recher, H. F., Thomspon, S. D. & Braithwaite, R. W. (1982). Comments on the relative advantages of marsupial and eutherian reproduction. *American Naturalist*, **120**, 128–34.

Mossman, H. W. (1937). Comparative morphogenesis of the fetal membranes and accessory uterine structures. *Carnegie Institute Contributions to Embryology*. **26**, 129–246.

Mossman, H. W. & Duke, K. L. (1973). *Comparative Morphology of the Mammalian ovary*. Madison: University of Wisconsin Press.

Muller, F. (1969). Zur fruhen Evolution der Saugerontogenesetypen: Versuch einer Rekonstruktion Aufgrund der Ontogenese-verhaltnisse bei den *Marsupialia. Acta Anatomica*, **74**, 297–404.

Murphy, C. R. & Smith, J. R. (1970). Age determination of pouch young and juvenile Kangaroo Island wallabies. *Transactions of the Royal Society of South Australia*, **94**, 15–20.

Murray, J. D., McKay, G. M. & Sharman, G. B. (1979). Studies on metatherian sex chromosomes. IX. Sex chromosomes of the greater glider (Marsupialia: Petauridae). *Australian Journal of Biological Sciences*, **32**, 375–86.

Murtagh, C. E. (1977). A unique cytogenetic system in monotremes. *Chromosoma (Berlin)*, **65**, 37–57.

Murtagh, C. E. & Sharman, G. B. (1977). Chromosomal sex determination in monotremes. In *Reproduction and Evolution*, ed. J. H. Calaby & C. H. Tyndale-Biscoe, pp. 61–6. Canberra: Australian Academy of Science.

Nalbandov, A. V. (1969). Comparative morphology and anatomy of the oviduct. In *The Mammalian Oviduct*, ed. E. S. E. Hafez & R. J. Blandau, pp. 48–56. Chicago and London: University of Chicago Press.

Nelsen, O. E. & Maxwell, N. (1942). The structure and function of the urogenital region in the female opossum compared with the same region in other marsupials. *Journal of Morphology*, **71**, 463–92.

Nelsen, O. E. & Swain, E. (1942). The prepubertal origin of germ cells in the ovary of the opossum (*Didelphys virginiana*). *Journal of Morphology*, **71**, 335–56.

Nelson, J. E. (Ed.) (1978). The phylogeny and evolution of the Macropodidae. *Australian Mammalogy*, **2**, 1–80.

Nelson, J. E. & Smith, G. (1971). Notes on growth rates in native cats of the family Dasyuridae. *International Zoo Yearbook*, **11**, 38–41.

New, D. A. T. & Mizell, M. (1972). Opossum fetuses grown in culture. *Science*, **175**, 533–6.

New, D. A. T., Mizell, M. & Cockroft, D. L. (1977). Growth of opossum embryos *in vitro* during organogenesis. *Journal of Experimental Morphology*, **41**, 111–23.

Newsome, A. E. (1964*a*). Anoestrus in the red kangaroo *Megaleia rufa* (Desmarest). *Australian Journal of Zoology*, **12**, 9–17.

Newsome, A. E. (1964*b*). Oestrus in the lactating red kangaroo, *Megaleia rufa* (Desmarest). *Australian Journal of Zoology*, **12**, 315–21.

Newsome, A. E. (1965). Reproduction in natural populations of the red kangaroo, *Megaleia rufa* (Desmarest), in Central Australia. *Australian Journal of Zoology*, **13**, 735–59.

Newsome, A. E. (1966). The influence of food on breeding in the red kangaroo in Central Australia. *CSIRO Wildlife Research*, **11**, 187–96.

Newsome, A. E. (1973). Cellular degeneration in the testis of red kangaroos during hot weather and drought in Central Australia. *Journal of Reproduction and Fertility, Supplement* **19**, 191–201.

Newsome, A. E. (1975). An ecological comparison of the two arid-zone kangaroos of Australia, and their anomalous prosperity since the introduction of ruminant stock to their environment. *The Quarterly Review of Biology*, **50**, 389–424.

Nicholas, K. R. & Tyndale-Biscoe, C. H. (1985). Prolactin-dependent accumulation of α-lactalbumin in mammary gland explants from the pregnant tammar (*Macropus eugenii*). *Journal of Endocrinology*, **106**, 337–42.

Nierembergius, J. E. (1635). *Historia naturae, maxime peregrinae*. Antwerp: Plantiniana B. Moreti.

Nilsson, O. & Reinius, S. (1969). Light and electron microscopic structure of the oviduct. In *The Mammalian Oviduct*, ed. E. S. E. Hafez & R. J. Blandau, pp. 57–84. Chicago and London: University of Chicago Press.

Noqueira, J. C., Godinho, H. P. & Cardoso, F. M. (1977). Microscopic anatomy of the scrotum, testis with its excurrent duct system and spermatic cord of *Didelphis azarae*. *Acta Anatomica*, **99**, 209–19.

O'Donoghue, C. H. (1911). The growth-changes in the mammary apparatus of *Dasyurus* and the relation of the corpora lutea thereto. *Quarterly Journal of Microscopical Science*, **57**, 187–234.

O'Donoghue, C. H. (1912). The corpus luteum in the non-pregnant *Dasyurus* and polyovular follicles in *Dasyurus*. *Anatomischer Anzeiger*, **41**, 353–68.

O'Donoghue, C. H. (1914). Uber die corpora lutea bei einigen Beuteltieren. *Archiv für Mikroskopische Anatomie*, **84**, 1–47.

O'Donoghue, C. H. (1916). On the corpora lutea and interstitial tissue of the ovary in the Marsupialia. *Quarterly Journal of Microscopical Science*, **61**, 433–73.

Oftedal, O. T. (1984). Milk composition, milk yield and energy output at peak lactation: a comparative review. *Symposia of the Zoological Society of London*, **51**, 33–85.

Ohno, S., Kaplan, W. D. & Kinosita, R. (1960). The basis of nuclear sex difference in somatic cells of the opossum *Didelphis virginiana*. *Experimental Cell Research* **19**, 417–20.

Olson, G. (1975). Observations on the ultrastructure of a fiber network in the flagellum of sperm of the brush-tailed phalanger, *Trichosurus vulpecula*. *Journal of Ultrastructural Research*, **50**, 193–8.

Olson, G. E. (1980). Changes in intramembranous particle distribution in the plasma membrane of *Didelphis virginiana* spermatozoa during maturation in the epididymis. *Anatomical Record*, **197**, 471–88.

Olson, G. E. & Hamilton, D. W. (1976). Morphological changes in the midpiece of woolly opossum spermatozoa during epididymal transit. *Anatomical Record*, **186**, 387–404.

Olson, G. E., Lifsics, M., Fawcett, D. & Hamilton, D. W. (1977). Structural specializations in the flagellar plasma membrane of opossum spermatozoa. *Journal of Ultrastructural Research*, **59**, 207–21.

Orsi, A. M., Ferreira, A. L. & De Melo, V. R. (1979). On the morphology of the rete testis of the opossum (*Didelphis azarae*). *Ciencia e Cultura*, **31**, 1029–31.

Orsi, A. M., Ferreira, A. L., De Melo, V. R. & Oliveira, M. C. (1981). Regional histology of the epididymis in the South American opossum. Light microscope study. *Anatomischer Anzeiger, Jena*, **150**, 521–8.

Osgood, W. H. (1921). A monographic study of the American marsupial *Caenolestes*. *Field Museum of Natural History Zoological Series*, **14**, 1–162.

Owen, F. J. (1984). *The role of progesterone in the reproduction of* Setonix brachyurus (*quokka*). MSc thesis. University of Western Australia, Perth.

Owen, F. J., Cake, M. H. & Bradshaw, S. D. (1982). Characterization and properties of a progesterone receptor in the uterus of the quokka (*Setonix brachyurus*). *Journal of Endocrinology*, **93**, 17–24.

Owen, R. (1832). On the Mammary glands of the *Ornithorhynchus paradoxus*. *Philosophical Transactions of the Royal Society*, **1832**, 517–38.

Owen, R. (1834). On the generation of the marsupial animals, with a description of the impregnated uterus of the kangaroo. *Philosophical Transactions of the Royal Society*, **1834**, 333–64.

Owen, R. (1852). Notes on the anatomy of the tree kangaroo (*Dendrolagus inustus*, Gould). *Proceedings of the Zoological Society of London*, **20**, 103–7.

Owen, R. (1868). *On the Comparative Anatomy and Physiology of Vertebrates*, Vol. III *Mammals*, pp. 760 – 75. London: Longmans, Green.

Owen, R. (1880). On the ova of the *Echidna Hystrix*. *Philosophical Transactions of the Royal Society of London*, **1880**, 1051–4.

Owen, R. (1887). Description of a newly-excluded young of the *Ornithorhynchus paradoxus*. *Proceedings of the Royal Society*, **42**, 391.

Padykula, H. A., Deren, J. J. & Wilson, T. H. (1966). Development of structure and function in the mammalian yolk sac. I. Developmental morphology and Vitamin B12 uptake of the rat yolk sac. *Developmental Biology*, **13**, 311–48.

Padykula, H. A. & Taylor, J. M. (1976*a*). Cellular mechanisms involved in cyclic stromal renewal of the uterus. I. The opossum, *Didelphis virginiana*. *Anatomical Record*, **184**, 5–26.

Padykula, H. A. & Taylor, J. M. (1976*b*). Ultrastructural evidence for loss of the trophoblastic layer in the chorioallantoic placenta of Australian bandicoots (Marsupialia: Peramelidae). *Anatomical Record*, **186**, 357–85.

Padykula, H. A. & Taylor, J. M. (1977). Uniqueness of the bandicoot chorioallantoic placenta (Marsupialia: Peramelidae). Cytological and evolutionary interpretations. In *Reproduction and Evolution*, ed. J. H. Calaby & C. H. Tyndale-Biscoe, pp. 303–23. Canberra: Australian Academy of Science.

Padykula, H. A. & Taylor, J. M. (1982). Marsupial placentation and its evolutionary significance. *Journal of Reproduction and Fertility, Supplement* **31**, 95–104.

Painter, T. S. (1922). Studies in mammalian spermatogenesis. I. The spermatogenesis of the opossum. *Journal of Experimental Zoology*, **35**, 13–45.

Panyaniti, W., Carpenter, S. M. & Tyndale-Biscoe, C. H. (1985). Effects of hypophysectomy on folliculogenesis in the tammer *Macropus eugenii* (Marsupialia: Macropodidae). *Australian Journal of Zoology*, **33**, 303–11.

Parker, K. M. (1917). The development of the hypophysis cerebri, preoral gut, and related structures in the Marsupialia. *Journal of Anatomy, London*, **51**, 181–249.

Parker, P. (1977). An ecological comparison of marsupial and placental patterns of reproduction. In *The Biology of Marsupials*, ed. B. Stonehouse & D. Gilmore, pp. 273–86. London: Macmillan.

Pearson, J. (1945). The female urogenital system of the Marsupialia with special reference to the vaginal complex. *Papers and Proceedings of the Royal Society of Tasmania*, **1944**, 71–98.

Pearson, J. (1946). The affinities of the rat kangaroos (Marsupialia) as revealed by a comparative study of the female urogenital system. *Papers and Proceedings of the Royal Society of Tasmania*, **1945**, 13–25.

Pearson, J. & de Bavay, J. (1951). The female urogenital system of *Antechinus* (Marsupialia). *Papers and Proceedings of the Royal Society of Tasmania*, **143**, 137–44.

Pedersen, T. & Peters, H. (1968). Proposal for a classification of oocytes and follicles in the mouse ovary. *Journal of Reproduction and Fertility*, **17**, 555–7.

Perondini, A. L. P. & Perondini, D. R. (1966). Sex chromatin in somatic cells of the Philander opossum quica (Temminck 1827, Marsupialia). *Cytogenetics*, **5**, 28–33.

Peters, D. G. & Rose, R. W. (1979). The oestrus cycle and basal body temperature in the common wombat (*Vombatus ursinus*). *Journal of Reproduction and Fertility*, **57**, 453–60.

Phillips, C. J. & Jones, J. K. (1968). Additional comments on reproduction in the woolly opossum *Caluromys derbianus* in Nicaragua. *Journal of Mammalogy*, **49**, 320–1.

Phillips, C. J. & Jones, J. K. (1969). Notes on reproduction and development in the four-eyed opossum, *Philander opossum*, in Nicaragua. *Journal of Mammalogy*, **50**, 345–8.

Phillips, D. M. (1970). Development of spermatozoa in the woolly opossum with special reference to the shaping of the sperm head. *Journal of Ultrastructural Research*, **33**, 369–80.

Pike, I. L. (1981). Comparative studies of embryo metabolism in early pregnancy. *Journal of Reproduction and Fertility, Supplement* **29**, 203–13.

Pilton, P. E. (1961). Reproduction in the great grey kangaroo. *Nature*, **189**, 984–5.

Pilton, P. E. & Sharman, G. B. (1962). Reproduction in the marsupial *Trichosurus vulpecula. Journal of Endocrinology*, **25**, 119–36.

Piso, W. (1648). *Historia naturalis Brasiliae*. Amsterdam: L. Elzevirium.

Pocock, R. J. (1926). The external characters of *Thylacinus*, *Sarcophilus*, and some related mammals. *Proceedings of the Zoological Society of London*, **1926**, 1037–84.

Poelman, C. (1851). Description des organes de la génération chez le *Macropus (Halmaturus) bennettii* femelle. *Bulletin de l'Academie Royale des Sciences de Belgique*, **18**, 595–9.

Pond, C. M. (1977). The significance of lactation in the evolution of mammals. *Evolution*, **31**, 177–99.

Pond, C. M. (1983). Parental feeding as a determinant of ecological relationships in mesozoic terrestrial ecosystems. *Acta Palaeontologica Polonica*, **28**, 215–24.
Poole, W. E. (1973). A study of breeding in grey kangaroos, *Macropus giganteus* Shaw and *M. fuliginosus* (Desmarest), in central New South Wales. *Australian Journal of Zoology*, **21**, 183–212.
Poole, W. E. (1975). Reproduction in the two species of grey kangaroos, *Macropus giganteus* Shaw, and *M. fuliginosus* (Desmarest). II. Gestation, parturition and pouch life. *Australian Journal of Zoology*, **23**, 333–53.
Poole, W. E. (1976). Breeding biology and current status of the grey kangaroos, *Macropus fuliginosus fuliginosus*, of Kangaroo Island, South Australia. *Australian Journal of Zoology*, **24**, 169–87.
Poole, W. E. & Catling, P. C. (1974). Reproduction in the two species of grey kangaroo, *Macropus giganteus* Shaw and *M. fuliginosus* (Desmarest). I. Sexual maturity and oestrus. *Australian Journal of Zoology*, **22**, 277–302.
Poole, W. E. & Pilton, P. E. (1964). Reproduction in the grey kangaroo, *Macropus canguru*, in captivity. *CSIRO Wildlife Research*, **9**, 218–34.
Presley, R. (1981). Alisphenoid equivalents in placentals, marsupials, monotremes and fossils. *Nature*, **294**, 668–70.
Rafferty-Machlis, G. R. & Hartman, C. G. (1953). Early death of the ovum in the opossum with observations on moribund mouse eggs. *Journal of Morphology*, **92**, 455–84.
Rattner, J. B. (1972). Nuclear shaping in marsupial spermatids. *Journal of Ultrastructural Research*, **40**, 498–512.
Read, D. (1982). Observations on the movements of two arid zone planigales (Dasyuridae: Marsupialia). In *Carnivorous Marsupials*, vol. 1, ed. M. Archer, pp. 227–31. Sydney: Royal Zoological Society of New South Wales.
Reiter, R. (Ed.) (1981). *The Pineal Gland*. Florida: CRC Press.
Renfree, M. B. (1970). Protein, amino acids and glucose in the yolk-sac fluids and maternal blood sera of the tammar wallaby, *Macropus eugenii* (Desmarest). *Journal of Reproduction and Fertility*, **22**, 483–92.
Renfree, M. B. (1972*a*). *Embryo–maternal relationships in the tammar wallaby*, Macropus eugenii. PhD thesis, Australian National University, Canberra.
Renfree, M. B. (1972*b*). Influence of the embryo on the marsupial uterus. *Nature*, **240**, 475–7.
Renfree, M. B. (1973*a*). Proteins in the uterine secretions of the marsupial *Macropus eugenii*. *Developmental Biology*, **32**, 41–9.
Renfree, M. B. (1973*b*). The composition of fetal fluids of the marsupial *Macropus eugenii*. *Developmental Biology*, **33**, 62–79.
Renfree, M. B. (1974*a*). Ovariectomy during gestation in the American opossum, *Didelphis marsupialis virginiana*. *Journal of Reproduction and Fertility*, **39**, 127–30.
Renfree, M. B. (1974*b*). Yolk sac fluid and yolk sac membrane enzymes in the marsupial, *Macropus eugenii*. *Comparative Biochemistry and Physiology*, **49B**, 273–9.
Renfree, M. B. (1975). Uterine proteins in the marsupial, *Didelphis marsupialis virginiana*, during gestation. *Journal of Reproduction and Fertility*, **42**, 163–6.
Renfree, M. B. (1977). Feto-placental influences in marsupial gestation. In *Reproduction and Evolution*, ed. J. H. Calaby & C. H. Tyndale-Biscoe, pp. 303–11. Canberra: Australian Academy of Science.
Renfree, M. B. (1978). Embryonic diapause in mammals – a developmental strategy. In *Dormancy and Developmental Arrest: Experimental Analysis in Plants and Animals*, ed. M. E. Clutter, pp. 1–46. New York: Academic Press.

Renfree, M. B. (1979). Initiation of development of diapausing embryo by mammary denervation during lactation in a marsupial. *Nature*, **278**, 549–51.
Renfree, M. B. (1980*a*). Embryonic diapause in the honey possum *Tarsipes spencerae. Search*, **11**, 81.
Renfree, M. B. (1980*b*). Placental function and embryonic development in marsupials. In *Comparative Physiology: Primitive Mammals*, ed. K. Schmidt-Nielsen, L. Bolis & C.R. Taylor, pp. 269–84. Cambridge University Press.
Renfree, M. B. (1980*c*). Endocrine activity in marsupial placentation. In *Endocrinology 1980*, ed. I. A. Cumming, J. W. Funder & F. A. O. Mendelsohn, pp. 83–6. Canberra: Australian Academy of Science.
Renfree, M. B. (1981*a*). Embryonic diapause in marsupials. *Journal of Reproduction and Fertility, Supplement* **29**, 67–78.
Renfree, M. B. (1981*b*). Marsupials: alternative mammals. *Nature*, **293**, 100–1.
Renfree, M. B. (1982). Implantation and placentation. In *Reproduction in Mammals*, 2nd edn. Book 2, ed. C. R. Austin & R. V. Short, *Embryonic and Fetal Development*, pp. 26–69. Cambridge University Press.
Renfree, M. B. (1983). Marsupial reproduction: the choice between placentation and lactation. In *Oxford Reviews of Reproductive Biology 5*, ed. C. A. Finn. Oxford University Press.
Renfree, M. B. & Calaby, J. H. (1981). Background to delayed implantation and embryonic diapause. *Journal of Reproduction and Fertility, Supplement* **29**, 1–9.
Renfree, M. B., Flint, A. P. F., Green, S. W. & Heap, R. B. (1984*b*). Ovarian steroid metabolism and luteal oestrogens in the corpus luteum of the tammar wallaby. *Journal of Endocrinology*, **101**, 231–40.
Renfree, M. B. & Fox, D. J. (1975). Pre- and postnatal development of lactate and malate dehydrogenases in the marsupial *Didelphis marsupialis virginiana. Comparative Biochemistry and Physiology*, **52B**, 347–50.
Renfree, M. B., Green, S. W. & Young, I. R. (1979). Growth of the corpus luteum and its progesterone content during pregnancy in the tammar wallaby, *Macropus eugenii. Journal of Reproduction and Fertility*, **57**, 131–6.
Renfree, M. B. & Heap, R. B. (1977). Steroid metabolism by the placenta, corpus luteum and endometrium during pregnancy in the marsupial *Macropus eugenii. Theriogenology*, **8**, 164.
Renfree, M. B., Holt, A. B., Green, S. W., Carr, J. & Cheek, D. B. (1982*a*). Ontogeny of the brain in a marsupial, *Macropus eugenii*, throughout pouch life. I. Brain growth and structure. *Brain, Behaviour and Evolution*, **20**, 57–71.
Renfree, M. B., Lincoln, D. W., Almeida, O. F. X. & Short, R. V. (1981*a*). Abolition of seasonal embryonic diapause in a wallaby by pineal denervation. *Nature*, **293**, 138–9.
Renfree, M. B., Meier, P., Teng, C. & Battaglia, F. C. (1981*b*). Relationship between amino acid intake and accretion in a marsupial, *Macropus eugenii. Biology of the Neonate*, **40**, 29–37.
Renfree, M. B., Russell, E. M. & Wooller, R. D. (1984*a*). Reproduction and life history of the honey possum, *Tarsipes rostratus*. In *Possums and Gliders*, ed. A. P. Smith & I. D. Hume, pp. 427–37. Sydney: Australian Mammal Society.
Renfree, M. B. & Short, R. V. (1984). Seasonal reproduction in marsupials. In *Proceedings of the 7th International Congress of Endocrinology*, ed. F. Labrie & L. Proulx, pp. 789–92. Excerpta Medica International Congress Series. Amsterdam: Elsevier Science Publishers.
Renfree, M. B. & Tyndale-Biscoe, C. H. (1973*a*). Intra-uterine development after diapause in the marsupial *Macropus eugenii. Developmental Biology*, **32**, 28–40.
Renfree, M. B. & Tyndale-Biscoe, C. H. (1973*b*). Transferrin variation between mother

and fetus in the marsupial *Macropus eugenii*. *Journal of Reproduction and Fertility*, **32**, 113–15.

Renfree, M. B. & Tyndale-Biscoe, C. H. (1978). Manipulation of marsupial embryos and pouch young. In *Methods in Mammalian Reproduction*, Chapter 15, ed. J. C. Daniel, pp. 307–31. New York: Academic Press.

Renfree, M. B. Wallace, G. I. & Young, I. R. (1982*b*). Effects of progesterone, oestradiol-17β and androstenedione on follicular growth after removal of the corpus luteum during lactational and seasonal quiescence in the tammar wallaby. *Journal of Endocrinology*, **92**, 397–403.

Renfree, M. B. & Young, I. R. (1979). Steriods in pregnancy and parturition in the marsupial, *Macropus eugenii*. *Journal of Steroid Biochemistry*, **11**, 515–22.

Reppert, S. M. & Schwartz, W. J. (1983). Maternal coordination of the fetal biological clock *in vitro*. *Science*, **220**, 969–71.

Reynolds, H. C. (1952). Studies on reproduction in the opossum (*Didelphis virginiana virginiana*). *University of California Publications in Zoology*, **52**, 223–84.

Richardson, B. J., Czuppon, A. B. & Sharman, G. B. (1971). Inheritance of glucose-6-phosphate dehydrogenase variation in kangaroos. *Nature*, *New Biology*, **230**, 154–5.

Risman, G. C. (1947). The effects of oestradiol and progesterone on the reproductive tract of the opossum and their possible relation to parturition. *Journal of Morphology*, **81**, 343–97.

Robinson, E. S., Johnston, P. G. & Sharman, G. B. (1977). X chromosome activity in germ cells of female kangaroos. In *Reproduction and Evolution*, ed. J. H. Calaby & C. H. Tyndale-Biscoe, pp. 89–94. Canberra: Australian Academy of Science.

Rodger, J. C. (1978). Male reproduction – its usefulness in discussions of Macropodidae evolution. *Australian Mammalogy*, **2**, 73–80.

Rodger, J. C. (1982). The testis and its excurrent ducts in American caenolestid and didelphid marsupials. *The American Journal of Anatomy*, **163**, 269–82.

Rodger, J. C. & Bedford, J. M. (1982*a*). Induction of oestrus, recovery of gametes, and the timing of fertilization events in the opossum, *Didelphis virginiana*. *Journal of Reproduction and Fertility*, **64**, 159–69.

Rodger, J. C. & Bedford, J. M. (1982*b*). Separation of sperm pairs and sperm-egg interaction in the opossum, *Didelphis virginiana*. *Journal of Reproduction and Fertility*, **64**, 171–9.

Rodger, J. C., Fletcher, T. P. & Tyndale-Biscoe, C. H. (1985). Active anti-paternal immunization does not affect the success of marsupial pregnancy. *Journal of Reproductive Immunology*, **8** 249–56.

Rodger, J. C. & Hughes, R. L. (1973). Studies of the accessory glands of male marsupials. *Australian Journal of Zoology*, **21**, 303–20.

Rodger, J. C. & White, I. G. (1975). Electro-ejaculation of Australian marsupials and analyses of the sugars in the seminal plasma from three macropod species. *Journal of Reproduction and Fertility*, **43**, 233–9.

Rodger, J. C. & White I. G. (1976*a*). Source of seminal *N*-acetylglucosamine in Australian marsupials and further studies of free sugars of the marsupial prostate gland. *Journal of Reproduction and Fertility*, **46**, 467–9.

Rodger, J. C. & White, I. G. (1976*b*). Comparison of sperm numbers released in the urine and in ejaculates of the brush-tailed possum. *Journal of Reproduction and Fertility*, **46**, 502–3.

Rodger, J. C. & White, I. G. (1980). Glycogen not *N*-acetylglucosamine the prostatic carbohydrate of three Australian and American marsupials, and patterns of these sugars in Marsupialia. *Comparative Biochemistry and Physiology*, **67B**, 109–13.

Rodger, J. C. & Young, R. J. (1981). Glycosidase and cumulus dispersal activities of acrosomal extracts from opossum (marsupial) and rabbit (eutherian) spermatozoa. *Gamete Research*, **4**, 507–14.

Rose, R. W. (1978). Reproduction and evolution in female Macropodidae. *Australian Mammalogy*, **2**, 65–72.

Rose, R. W. (1985). *Reproductive biology of the Tasmanian bettong*, Bettongia gaimardi. PhD thesis, University of Tasmania, Hobart.

Rose, R. W. & McCartney, D. J. (1982*a*). Reproduction of the red-bellied pademelon, *Thylogale billardierii* (Marsupialia). *Australian Wildlife Research*, **9**, 27–32.

Rose, R. W. & McCartney, D. J. (1982*b*). Growth of the red-bellied pademelon, *Thylogale billardierii*, and age estimation of pouch young. *Australian Wildlife Research*, **9**, 33–8.

Rotenberg, D. (1928). Notes on the male generative apparatus of *Tarsipes spencerae*. *Journal of the Royal Society of Western Australia*, **15**, 9–17.

Rothchild, I. (1981). The regulation of the mammalian corpus luteum. *Recent Progress in Hormone Research*, **37**, 183–298.

Rowlands, I. W. & Weir, B. J. (1984). Mammals: non-primate eutherians. In *Marshall's Physiology of Reproduction*, 4th edn. vol. 1, Chap. 7, ed. G. E. Lamming, pp. 455–658. Edinburgh: Churchill Livingstone.

Rubin, D. (1943). Embryonic differentiation of Cowper's and Bartholins glands of the opossum following castration and ovariectomy. *Journal of Experimental Zoology*, **94**, 463–73.

Rubin, D. (1944). The relation of hormones to the development of Cowper's and Bartholin's glands in the opossum (*Didelphys virginiana*). *Journal of Morphology*, **74**, 213–78.

Russell, E. M. (1974). The biology of kangaroos (Marsupialia-Macropodidae). *Mammal Reviews*, **4**, 1–59.

Russell, E. M. (1982*a*). Patterns of parental care and parental investment in marsupials. *Biological Reviews*, **57**, 423–86.

Russell, E. M. (1982*b*). Parental investment and desertion of young in marupials. *American Naturalist*, **119**, 744–8.

Russell, E. M. (1984*a*). Social behaviour and social organisation of marsupials. *Mammal Reviews*, **14**, 101–54.

Russell, R. (1984*b*) Social behaviour of the yellow-bellied glider *Petaurus australis* in north Queensland. In *Possums and Gliders*, ed. A. P. Smith & I. D. Hume, pp. 343–53. Sydney: Australian Mammal Society.

Sacher, G. A. & Staffeldt, E. F. (1974). Relation of gestation time to brain weight for placental mammals: Implications for the theory of vertebrate growth. *The American Naturalist*, **108**, 593–615.

Sadleir, R. M. F. S. (1965). Reproduction in two species of kangaroo *Macropus robustus* and *Megaleia rufa* in the arid Pilbara region of Western Australia. *Proceedings of the Zoological Society of London*, **145**, 239–61.

Sadleir, R. M. F. S. & Tyndale-Biscoe, C. H. (1977). Photoperiod and the termination of embryonic diapause in the marsupial *Macropus eugenii*. *Biology of Reproduction*, **16**, 605–8.

Sandes, F. P. (1903). The corpus luteum of *Dasyurus viverrinus* with observations on the growth and atrophy of the Graafian follicle. *Proceedings of the Linnean Society of New South Wales*, **28**, 364–405.

Sansom, G. S. (1937). The placentation of the Indian musk-shrew (*Crocidura caerula*). *Transactions of the Zoological Society of London*, **23**, 267–314.

Sapsford, C. S. & Rae, C. A. (1969). Ultrastructural studies on Sertoli cells and

spermatids in the bandicoot and ram during the movement of mature spermatids into the lumen of the seminferous tubule. *Australian Journal of Zoology*, **17**, 415–45.
Sapsford, C. S., Rae, C. A. & Cleland, K. W. (1967). Ultrastructural studies on spermatids and Sertoli cells during early spermiogenesis in the bandicoot *Perameles nasuta* Geoffroy (Marsupialia). *Australian Journal of Zoology*, **15**, 881–909.
Sapsford, C. S., Rae, C. A. & Cleland, K. W. (1969*a*) Ultrastructural studies on maturing spermatids and on Sertoli cells in the bandicoot *Perameles nasuta* Geoffroy (Marsupialia). *Australian Journal of Zoology*, **17**, 195–292.
Sapsford, C. S., Rae, C. A. & Cleland K. W. (1969*b*). The fate of residual bodies and degenerating germ cells and the lipid cycle in Sertoli cells in the bandicoot *Perameles nasuta* Geoffroy (Marsupialia). *Australian Journal of Zoology*, **17**, 729–53.
Sapsford, C. S., Rae, C. A. & Cleland, K. W. (1970). Ultrastructural studies on the development and form of the principal piece sheath of the bandicoot spermatozoon. *Australian Journal of Zoology*, **18**, 21–48.
Schaller, O., Habel, R. E. & Frewein, J. (1972). *Nomina Anatomica Veterinaria*, 2nd edn. Vienna: International Committee on Veterinary Anatomical Nomenclature.
Schmidt, G. H. (1971). *Biology of Lactation*. San Francisco: W. H. Freeman.
Schneider, L. K. (1970). RNA synthesis in the sex chromosomes of the opossum. *Didelphis virginiana*. 1. Female. *Experientia*, **26**, 914–16.
Schultze–Westrum, T. G. (1969). Social communication by chemical signals in flying phalangers (*Petaurus breviceps papuanus*). In *Olfaction and Taste*, ed. C. Pfaffmann, pp. 268–77. New York: Rockefeller University Press.
Seiler, B. W. (1828). Einige Bemerkungen über die arste Geburt des Kangeru-Embryo und seine Ernährung in dem Beutel. *Isis von Oken*, **12**, 475–7.
Selenka, E. (1887). *Studien uber Entwickelungsgeschichte der Thiere. 4. Das opossum, Didelphis virginiana*. Wiesbaden: C. W. Kreidel.
Selenka, E. (1892). *Studien uber Entwickelungsgeschichte der Thiere. 5. Beutelfuchs und Kanguruhratte*. Wiesbaden: C. W. Kreidel.
Selwood, L. (1980). A timetable of embryonic development of the dasyurid marsupial *Antechinus stuartii* (Macleay). *Australian Journal of Zoology*, **28**, 649–68.
Selwood, L. (1981). Delayed embryonic development in the dasyurid marsupial, *Antechinus stuartii*. *Journal of Reproduction and Fertility, Supplement* **29**, 79–82.
Selwood, L. (1982*a*). Brown antechinus *Antechinus stuartii*: management of breeding colonies to obtain embryonic material and pouch young. In *The Management of Australian Mammals in Captivity*, ed. D. D. Evans, pp. 31–7. Melbourne: Zoological Board of Victoria.
Selwood, L. (1982*b*). A review of maturation and fertilization in marsupials with special reference to the dasyurid: *Antechinus stuartii*. In *Carnivorous Marsupials*, ed. M. Archer, pp. 65–76. Sydney: Royal Zoological Society of New South Wales.
Selwood, L. (1983). Factors influencing pre-natal fertility in the brown marsupial mouse, *Antechinus stuartii*. *Journal of Reproduction and Fertility*, **68**, 317–24.
Selwood, L. & Young, G. J. (1983). Cleavage *in vivo* and in culture in the dasyurid marsupial *Antechinus stuartii* (Macleay). *Journal of Morphology*, **176**, 43–60.
Semon, R. (1894). Die Embryonalhullen der Monotremen und Marsupialier. In *Zoologische Forschungsreisen in Australien und dem Malayischen Archipel*, vol. 2, ed. R. Semon, pp. 19–58. Jena: Gustav Fischer.
Sernia, C., Bradley, A. J. & McDonald, I. R. (1979). High affinity binding of adrenocortical and gonadal steroids by plasma proteins of Australian marsupials. *General and Comparative Endocrinology*, **38**, 496–503.
Sernia, C., Hinds, L. & Tyndale-Biscoe, C. H. (1980). Progesterone metabolism during embryonic diapause in the tammar wallaby, *Macropus eugenii*. *Journal of Reproduction and Fertility*, **60**, 139–47.

Sernia, C. & Tyndale-Biscoe, C. H. (1979). Prolactin receptors in the mammary gland, corpus luteum and other tissues of the tammar wallaby, *Macropus eugenii*. *Journal of Endocrinology*, **83**, 79–89.

Setchell, B. P. (1970*a*). Testicular blood supply, lymphatic drainage, and secretion of fluid. In *The Testis*, ed. A. D. Johnson, W. R. Gomes, & N. L. Van Demark, pp. 101–239. New York: Academic Press.

Setchell, B. P. (1970*b*). Fluid secretion by the testes of an Australian marsupial, *Macropus eugenii*. *Comparative Biochemistry and Physiology*, **36**, 411–14.

Setchell, B. P. (1974). Secretions of the testis and epididymis. *Journal of Reproduction and Fertility*, **37**, 165–77.

Setchell, B. P. (1977). Reproduction in male marsupials. In *The Biology of Marsupials*, ed. B. Stonehouse & D. Gilmore, pp. 411–57. London: Macmillan.

Setchell, B. P. & Carrick, F. N. (1973). Spermatogenesis in some Australian marsupials. *Australian Journal of Zoology*, **21**, 491–9.

Setchell, B. P. & Thorburn, G. D. (1969). Effect of local heating on blood flow through the testes of some Australian marsupials. *Comparative Biochemistry and Physiology*, **31**, 675–7.

Setchell, B. P. & Thorburn, G. D. (1971). The effect of artificial cryptorchidism on the testis and on testicular blood flow in an Australian marsupial *Macropus eugenii*. *Comparative Biochemistry and Physiology*, **38A**, 705–8.

Setchell, B. P. & Waites, G. M. H. (1969). Pulse attenuation and countercurrent heat exchange in the internal spermatic artery of some Australian marsupials. *Journal of Reproduction and Fertility*, **20**, 165–9.

Setchell, B. P., Waites, G. M. H. & Lindner, H. R. (1965). Effect of undernutrition on testicular blood flow and metabolism and the output of testosterone in the ram. *Journal of Reproduction and Fertility*, **9**, 149–62.

Settle, G. A. (1978). The quiddity of tiger quolls. *Australian Natural History*, **19**, 164–9.

Sharman, G. B. (1954). Reproduction in marsupials. *Nature*, **173**, 302–3.

Sharman, G. B. (1955*a*). Studies on marsupial reproduction. II. The oestrous cycle of *Setonix brachyrus*. *Australian Journal of Zoology*, **3**, 44–55.

Sharman, G. B. (1955*b*). Studies on marsupial reproduction. III. Normal and delayed pregnancy in *Setonix brachyurus*. *Australian Journal of Zoology*, **3**, 56–70.

Sharman, G. B. (1955*c*). Studies on marsupial reproduction. IV. Delayed birth in *Protemnodon eugenii*. *Australian Journal of Zoology*, **3**, 156–61.

Sharman, G. B. (1959). Marsupial reproduction. In *Biogeography and Ecology in Australia*. *Monographiae Biologicae*, **8**, 332–68.

Sharman, G. B. (1961*a*). The mitotic chromosomes of marsupials and their bearing on taxonomy and phylogeny. *Australian Journal of Zoology*, **9**, 38–60.

Sharman, G. B. (1961*b*). The embryonic membranes and placentation in five genera of diprotodont marsupials. *Proceedings of the Zoological Society of London*, **137**, 197–220.

Sharman, G. B. (1962). The initiation and maintenance of lactation in the marsupial, *Trichosurus vulpecula*. *Journal of Endocrinology*, **25**, 375–85.

Sharman, G. B. (1963). Delayed implantation in marsupials. In *Delayed Implantation*, ed. A. C. Enders, pp. 3–14. Chicago University Press.

Sharman, G. B. (1965*a*). The effects of the suckling stimulus and oxytocin injection on the corpus luteum of delayed implantation in the red kangaroo. *Excerpta Medica*, **83**, 669–74.

Sharman, G. B. (1965*b*). The effects of suckling on normal and delayed cycles of reproduction in the red kangaroo. *Sonderdruck aus Zeitschrift für Saugetierkunde*, **30**, 10–20.

Sharman, G. B. (1965*c*). Marsupials and the evolution of viviparity. *Viewpoints in Biology*, **4**, 1–28.

Sharman, G. B. (1970). Reproductive physiology of marsupials. *Science, New York*, **167**, 1221–8.
Sharman, G. B. (1971). Late DNA replication in the paternally derived X chromosome of female kangaroos. *Nature*, **230**, 231–2.
Sharman, G. B. (1973*a*). The chromosomes of non-eutherian mammals. In *Cytotaxonomy and Vertebrate Evolution*, ed. A. B. Chiavelli & E. Capanna, pp. 485–527. New York: Academic Press.
Sharman, G. B. (1973*b*). Adaptations of marsupial pouch young for extra-uterine existence. In *The Mammalian Fetus in Vitro*, ed. C. R. Austin, pp. 67–90. London: Chapman & Hall.
Sharman, G. B. (1976). Evolution of viviparity in mammals. In *Reproduction in Mammals*. Book 6. ed C. R. Austin & R. V. Short, *The Evolution of Reproduction*, pp. 32–70. Cambridge University Press,
Sharman, G. B. (1982). Karyotypic similarities between *Dromiciops australis* (Microbiotheriidae, Marsupialia) and some Australian marsupials. In *Carnivorous Marsupials*, vol. 2, ed. M. Archer, pp. 711–14. Sydney: Royal Zoological Society of New South Wales.
Sharman, G. B. & Barber, H. N. (1952). Multiple sex-chromosomes in the marsupial *Potorous*. *Heredity*, **6**, 345–55.
Sharman, G. B. & Berger, P. J. (1969). Embryonic diapause in marsupials. *Advances in Reproductive Physiology*, **4**, 211–40.
Sharman, G. B. & Calaby, J. H. (1964). Reproductive behaviour in the red kangaroo, *Megaleia rufa*, in captivity. *CSIRO Wildlife Research*, **9**, 58–85.
Sharman, G. B., Calaby, J. H. & Poole, W. E. (1966). Patterns of reproduction in female diprotodont marsupials. *Symposium of the Zoological Society of London*, **15**, 205–32.
Sharman, G. B. & Clark, M. J. (1967). Inhibition of ovulation by the corpus luteum in the red kangaroo, *Megaleia rufa*. *Journal of Reproduction and Fertility*, **14**, 129–37.
Sharman, G. B. & Pilton, P. E. (1964). The life history and reproduction of the red kangaroo (*Megaleia rufa*). *Proceedings of the Zoological Society of London*, **142**, 29–48.
Sharman, G. B., Robinson, E. S., Walton, S. M. & Berger, P. J. (1970). Sex chromosomes and reproductive anatomy of some intersexual marsupials. *Journal of Reproduction and Fertility*, **21**, 57–68.
Shaw, G. (1983*a*). Effect of $PGF_{2\alpha}$ on uterine activity, and concentrations of 13, 14-dihydro-15-keto-$PGF_{2\alpha}$ in peripheral plasma during parturition in the tammar wallaby (*Macropus eugenii*). *Journal of Reproduction and Fertility*, **69**, 429–36.
Shaw, G. (1983*b*). *Pregnancy after diapause in the tammar wallaby*, Macropus eugenii. PhD thesis, Murdoch University, Western Australia.
Shaw, G. & Renfree, M. B. (1984). Concentrations of oestradiol-17β in plasma and corpora lutea throughout pregnancy in the tammar, *Macropus eugenii*. *Journal of Reproduction and Fertility*, **72**, 29–37.
Shaw, G. & Renfree, M. B. (1986). Uterine and embryonic metabolism after diapause in the tammar wallaby, *Macropus eugenii*. *Journal of Reproduction and Fertility*, **76**, 339–47.
Shaw, G. & Rose, R. W. (1979). Delayed gestation in the potoroo *Potorous tridactylus* (Kerr). *Australian Journal of Zoology*, **27**, 901–12.
Shield, J. (1964). A breeding season difference in two populations of the Australian macropod marsupial (*Setonix brachyurus*). *Journal of Mammalogy*, **45**, 616–25.
Shield, J. (1968). Reproduction of the quokka, *Setonix brachyurus*, in captivity. *Journal of Zoology, London*, **155**, 427–44.
Shield, J. W. & Woolley, P. (1960). Gestation time for delayed birth in the quokka. *Nature*, **188**, 163–4.

Shield, J. W. & Woolley, P. (1963). Population aspects of delayed birth in the quokka (*Setonix brachyurus*). *Proceedings of the Zoological Society of London*, **141**, 783–90.
Shorey, C. D. & Hughes, R. L. (1972). Uterine glandular regeneration during the follicular phase in the marsupial *Trichosurus vulpecula*. *Australian Journal of Zoology*, **20**, 235–47.
Shorey, C. D. & Hughes, R. L. (1973*a*). Cyclical changes in the uterine endometrium and peripheral plasma concentrations of progesterone in the marsupial *Trichosurus vulpecula*. *Australian Journal of Zoology*, **21**, 1–19.
Shorey, C. D. & Hughes, R. L. (1973*b*). Development, function and regression of the corpus luteum in the marsupial *Trichosurus vulpecula*. *Australian Journal of Zoology*, **21**, 477–89.
Shorey, C. D. & Hughes, R. L. (1975). Uterine response to ovariectomy during the proliferative and luteal phases in the marsupial, *Trichosurus vulpecula*. *Journal of Reproduction and Fertility*, **42**, 221–8.
Short, R. V. (1984). Species differences in reproductive mechanisms. In *Reproduction in Mammals*, 2nd edn. Book 4, ed. C. R. Austin & R. V. Short, *Reproductive Fitness*, pp. 24–61. Cambridge University Press.
Short, R. V., Flint, A. P. F. & Renfree, M. B. (1985). Influence of passive immunization against GnRH on pregnancy and parturition in the tammar wallaby, *Macropus eugenii*. *Journal of Reproduction and Fertility*, **75**, 567–75.
Smith, A. P. (1982). Diet and feeding strategies of the marsupial sugar glider in temperate Australia. *Journal of Animal Ecology*, **51**, 149–66.
Smith, A. P. (1984*a*). Demographic consequences of reproduction, dispersal and social interaction in a population of Leadbeater's possum. In *Possums and Gliders*, ed. A. P. Smith & I. D. Hume, pp. 359–73. Sydney: Australian Mammal Society.
Smith, A. P. & Hume, I. D. (Eds) (1984). *Possums and Gilders*, Sydney: Surrey Beatty and Australian Mammal Society.
Smith, G. C. (1984*b*). The biology of the yellow-footed antechinus, *Antechinus flavipes* (Marsupialia: Dasyuridae), in a swamp forest on Kinaba Island, Cooloola, Queensland. *Australian Wildlife Research*, **11**, 465–80.
Smith, Malcolm (1979*a*). Notes on reproduction and growth in the koala, *Phascolarctos cinereus* (Goldfuss). *Australian Wildlife Research*, **6**, 5–12.
Smith, M. J. (1971). Breeding the sugar-glider *Petaurus breviceps* in captivity; and growth of pouch-young. *International Zoo Yearbook*, **11**, 26–8.
Smith, M. J. (1973). *Petaurus breviceps*. *Mammalian Species*, **30**, 1–5.
Smith, M. J. (1979*b*). Observations on growth of *Petaurus breviceps* and *P. norfolcensis* (Petauridae: Marsupialia) in captivity. *Australian Wildlife Research*, **6**, 141–50.
Smith, M. J. (1981). Morphological observations on the diapausing blastocyst of some macropodid marsupials. *Journal of Reproduction and Fertility*, **61**, 483–6.
Smith, M. J., Bennett, J. H. & Chesson, C. M. (1978). Photoperiod and some other factors affecting reproduction in female *Sminthopsis crassicaudata* (Gould) (Marsupialia: Dasyuridae) in captivity. *Australian Journal of Zoology*, **26**, 449–63.
Smith, M. J., Brown, B. K. & Frith, H. J. (1969). Breeding of the brush-tailed possum, *Trichosurus vulpecula* (Kerr), in N.S.W. *CSIRO Wildlife Research*, **14**, 181–93.
Smith, M. J. & Godfrey, G. K. (1970). Ovulation induced by gonadotrophins in the marsupial, *Sminthopsis crassicaudata* (Gould). *Journal of Reproduction and Fertility*, **22**, 41–7.
Smith, M. J., Hayman, D. L. & Hope, R. M. (1979). Observations on the chromosomes and reproductive systems of four macropodine interspecific hybrids (Marsupialia: Macropodidae). *Australian Journal of Zoology*, **27**, 959–72.
Smith, M. J. & How, C. A. (1973). Reproduction in the mountain possum *Trichosurus caninus* (Ogilby), in captivity. *Australian Journal of Zoology*, **21**, 321–9.

Smith, M. J. & Sharman, G. B. (1969). Development of dormant blastocysts induced by oestrogen in the ovariectomised marsupial, *Macropus eugenii*. *Australian Journal of Biological Sciences*, **22**, 171–80.

Smith, R. F. C. (1969). Studies on the marsupial glider, *Schoinobates volans* (Kerr). I. Reproduction. *Australian Journal of Zoology*, **17**, 625–36.

Sorenson, M. W. (1970). Observations on the behaviour of *Dasycercus cristicauda* and *Dasyuroides byrnei* in captivity. *Journal of Mammalogy*, **51**, 123–30.

Stewart, F. (1984). Mammogenesis and changing prolactin receptor concentrations in the mammary glands of the tammar wallaby (*Macropus eugenii*). *Journal of Reproduction and Fertility*, **71**, 141–8.

Stewart, F., Sutherland, R. L. & Tyndale-Biscoe, C. H. (1981). Macropodid marsupial testicular gonadotrophin receptors and their use in assays for marsupial gonadotrophins. *Journal of Endocrinology*, **89**, 213–23.

Stewart, F. & Tyndale-Biscoe, C. H. (1982). Prolactin and luteinizing hormone receptors in marsupial corpora lutea: relationship to control of luteal function. *Journal of Endocrinology*, **92**, 63–72.

Stewart, F. & Tyndale-Biscoe, C. H. (1983). Pregnancy and lactation in marsupials. *Current Topics in Experimental Endocrinology*, **4**, 1–33.

Stirling, E. C. (1889). On some points in the anatomy of the female organs of generation of the kangaroo, especially in relation to the acts of impregnation and parturition. *Proceedings of the Zoological Society of London*, **1889**, 433–40.

Stirling, E. C. (1891). Description of a new genus and species of Marsupialia, '*Notoryctes typhlops*'. *Transactions of the Royal Society of South Australia*, **14**, 154–87.

Stirton, R. A., Tedford, R. H. & Woodburne, M. O. (1968). Australian Tertiary deposits containing terrestrial mammals. *University of California Publications in Geology*, **77**, 1–30.

Stockard, C. R. & Papanicolau, G. N. (1917). The existence of a typical oestrous cycle in the guinea pig with a study of its histological and physiological changes. *American Journal of Anatomy*, **22**, 225–84.

Stodart, E. (1966*a*). Management and behaviour of breeding groups of the marsupial *Perameles nasuta* Geoffroy in captivity. *Australian Journal of Zoology*, **14**, 611–23.

Stodart, E. (1966*b*). Observations on the behaviour of the marsupial *Bettongia lesueuri* (Quoy and Gaimard) in an enclosure. *CSIRO Wildlife Research*, **11**, 91–9.

Stodart, E. (1977). Breeding and behaviour of Australian bandicoots. In *The Biology of Marsupials*, ed. B. Stonehouse & D. Gilmore, pp. 179–91. London: Macmillan.

Stoddart, D. M. & Braithwaite, R. W. (1979). A strategy for utilization of regenerating heathland habitat by the brown bandicoot (*Isoodon obesulus*; Marsupialia, Peramelidae). *Journal of Animal Ecology*, **48**, 165–79.

Stokes, J. L. (1846). *Discoveries in Australia, With an Account of the Coasts and Rivers Explored and Surveyed During the Voyage of H.M.S. Beagle in the Years* 1837, 38, 39, 40, 41, 42, 43. London: T. & W. Boone.

Storr, G. M. (1958). Are marsupials 'second-class' mammals? *Western Australian Naturalist*, **6**, 179–83.

Strahan, R. (Ed) (1983). *The Australian Museum Complete book of Australian Mammals*. Sydney: Angus & Robertson.

Streilein, K. E. (1982). The ecology of small mammals in the semi-arid Brazilian Caatinga. III. Reproductive biology and population ecology. *Annals of the Carnegie Museum*, **51**, 251–9.

Suckling, G. C. (1984). Population ecology of the sugar glider *Petaurus breviceps* in a system of fragmented habitat. *Australian Wildlife Research*, **11**, 49–76.

Surani, M. A. H. (1977). Cellular and molecular approaches to blastocyst uterine interactions at implantation. In *Development in Mammals*, vol. 1, ed. M. H. Johnson, pp. 245–306. Amsterdam: North Holland.

Sutherland, R. L., Evans, S. M. & Tyndale-Biscoe, C. H. (1980). Macropodid marsupial luteinising hormone: validation of assay procedures and changes in plasma levels during the oestrous cycle in the female tammar wallaby (*Macropus eugenii*). *Journal of Endocrinology*, **86**, 1–12.

Sweet, G. (1907). Contribution to our knowledge of the anatomy of *Notoryctes typhlops* Stirling. V. The reproductive organs of *Notoryctes*. *Quarterly Journal of Microscopical Science*, **51**, 333–44.

Talbot, P. & Dicarlantonio, G. (1984). Ultrastructure of opossum oocyte investing coats and their sensitivity to trysin and hyaluronidase. *Developmental Biology*, **103**, 159–67.

Talice, R. V. & Lagomarsino, J. C. (1961). Comportamiento sexual y naciementos en cautividad de la 'Comadreja overa': *Didelphis azarae*. *Congresso Sudamericana de Zoologica I*, **5**, 81–98.

Taplin, L. E. (1980). Some observations on the reproductive biology of *Sminthopsis virginiae* (Tarragon), (Marsupialia: Dasyuridae). *Australian Zoologist*, **20**, 407–18.

Taylor, J. M., Calaby, J. H. & Redhead, T. D. (1982). Breeding in wild populations of the marsupial-mouse *Planigale maculata sinualis* (Dasyuridae, Marsupialia). In *Carnivorous Marsupials*, vol. 1, ed. M. Archer, pp. 83–7. Sydney: Royal Zoological Society of New South Wales.

Taylor, J. M. & Horner, B. E. (1970). Gonadal activity in the marsupial mouse, *Antechinus bellus*, with notes on other species of the genus (Marsupialia: Dasyuridae). *Journal of Mammalogy*, **51**, 659–68.

Taylor, J. M. & Padykula, H. A. (1978). Marsupial trophoblast and mammalian evolution. *Nature*, **271**, 588.

Temple-Smith, P. D. (1984*a*). Phagocytosis of sperm cytoplasmic droplets by a specialized region in the epididymis of the brushtailed possum, *Trichosurus vulpecula*. *Biology of Reproduction*, **30**, 707–20.

Temple-Smith, P. D. (1984*b*). Reproductive structures and strategies in male possums and gliders. In *Possums and Gliders*, ed. A. P. Smith & I. D Hume, pp. 89–106. Sydney: Australian Mammal Society.

Temple-Smith, P. D. & Bedford, J. M. (1976). The features of sperm maturation in the epididymis of a marsupial, the brushtailed possum *Trichosurus vulpecula*. *American Journal of Anatomy*, **147**, 471–500.

Temple-Smith, P. D. & Bedford, J. M. (1980). Sperm maturation and the formation of sperm pairs in the epididymis of the opossum, *Didelphis virginiana*. *Journal of Experimental Zoology*, **214**, 161–71.

Temple-Smith, P. D. & Grant, T. R. (1986). Sperm structure and marsupial phylogeny. In *Possums and Opossums: Studies in Evolution*, ed. M. Archer, in press. Sydney: Australian Mammal Society.

Thomson, J. A. & Owen, W. H. (1964). A field study of the Australian ringtail possum *Pseudocheirus peregrinus* (Marsupialia: Phalangeridae). *Ecological Monographs*, **34**, 27–52.

Thorburn, G. D., Challis, J. R. G. & Robinson, J. S. (1977). Endocrine control of parturition. In *Biology of the Uterus*, ed. R. M. Wynn. pp. 653–732. London: Plenum Press.

Thorburn, G. D., Cox, R. I. & Shorey, C. D. (1971). Ovarian steroid secretion rates in the marsupial *Trichosurus vulpecula*. *Journal of Reproduction and Fertility*, **24**, 139.

Thornber, E. J., Renfree, M. B. & Wallace, G. I. (1981). Biochemical studies of intrauterine components of the tammar wallaby *Macropus eugenii* during pregnancy. *Journal of Embryology and Experimental Morphology*, **62**, 325–38.

Topper, Y. J. & Freeman, C. S. (1980). Multiple hormone interactions in the development biology of the mammary gland. *Physiological Reviews*, **60**, 1049–1106.

Towers, P. A., Shaw, G. & Renfree, M. B. (1986). Urogenital vasculature and local steroid concentrations in the uterine branch of the ovarian vein of the female tammar wallaby (*Macropus eugenii*). *Journal of Reproduction and Fertility*, **78**, in press.

Tribe, M. (1918). The development of the pancreas, the pancreatic and hepatic ducts, in *Trichosurus vulpecula. Philosophical Transactions of the Royal Society B*, **208**, 307–50.

Tribe, M. (1923). The development of the hepatic venous system and the post caval vein in the Marsupialia. *Philosophical Transactions of the Royal Society B*, **212**, 147–207.

Tripp, H. R. H. (1971). Reproduction in elephant shrews (Macroscelidae) with special reference to ovulation and implantation. *Journal of Reproduction and Fertility*, **26**, 149–59.

Trupin, G. L. & Fadem, B. H. (1982). Sexual behaviour of the Grey short-tailed opossum *Monodelphis domestica. Journal of Mammalogy*, **63**, 409–14.

Turnbull, K. E., Mattner, P. E. & Hughes, R. L. (1981). Testicular descent in the marsupial *Trichosurus vulpecula* (Kerr). *Australian Journal of Zoology*, **29**, 189–98.

Tyndale-Biscoe, C. H. (1955). Observations on the reproduction and ecology of the brush tailed possum *Trichosurus vulpecula* (Marsupialia). *Australian Journal of Zoology*, **3**, 162–84.

Tyndale-Biscoe, C. H. (1963a). The role of the corpus luteum in the delayed implantation in marsupials. In *Delayed Implantation*, ed. A. C. Enders, pp. 15–32. Chicago University Press.

Tyndale-Biscoe, C. H. (1963*b*). Effects of ovariectomy in the marsupial, *Setonix brachyurus. Journal of Reproduction and Fertility*, **6**, 25–40.

Tyndale-Biscoe, C. H. (1965). The female urogenital system and reproduction of the marsupial *Lagostrophus fasciatus. Australian Journal of Zoology*, **13**, 255–67.

Tyndale-Biscoe, C. H. (1966). The marsupial birth canal. *Symposium of the Zoological Society of London*, **15**, 233–50.

Tyndale-Biscoe, C. H. (1968). Reproduction and post-natal development in the marsupial *Bettongia lesueur* (Quoy & Gaimard). *Australian Journal of Zoology*, **16**, 577–602.

Tyndale-Biscoe, C. H. (1969). Relaxin activity during the oestrous cycle of the marsupial, *Trichosurus vulpecula* (Kerr). *Journal of Reproduction and Fertility*, **19**, 191–3.

Tyndale-Biscoe, C. H. (1970). Resumption of development by quiescent blastocysts transferred to primed, ovariectomised recipients in the marsupial, *Macropus eugenii. Journal of Reproduction and Fertility*, **23**, 25–32.

Tyndale-Biscoe, C. H. (1973). *Life of Marsupials*. London: Arnold.

Tyndale-Biscoe, C. H. (1979). Hormonal control of embryonic diapause and reactivation in the tammar wallaby. In *Maternal Recognition of Pregnancy*. Ciba Foundation Symposium 64 (new series), pp. 173–90. Amsterdam: Excerpta Medica.

Tyndale-Biscoe, C. H. (1980*a*). Observations on the biology of marsupials in Colombia and Venezuela. *National Geographic Society Research Reports*, **12**, 711–20.

Tyndale-Biscoe, C. H. (1980*b*). Photoperiod and the control of seasonal reproduction in marsupials. In *Proceedings of the VI International Congress of Endocrinology*, ed. I. A. Cumming, J. W. Funder & F. A. O. Mendelsohn, pp. 277–82. Canberra: Australian Academy of Science.

Tyndale-Biscoe, C. H. (1981). Evidence for relaxin in marsupials. In *Relaxin*, ed. G. Bryant-Greenwood, H. Niall & F. Greenwood, pp. 225–32. Amsterdam; North Holland/Elsevier Biomedical Press.

Tyndale-Biscoe, C. H. (1984). Mammals-Marsupials. In *Marshall's Physiology of Reproduction*, 4th Edn, vol. 1, Chapter 5, ed. G. E. Lamming. pp. 386–454. Edinburgh: Churchill Livingstone.

Tyndale-Biscoe, C. H. & Hawkins, J. (1977). The corpora lutea of marsupials, aspects of function and control. In *Reproduction and Evolution*, ed. J. H. Calaby & C. H. Tyndale-Biscoe, pp. 245–52. Canberra: Australian Academy of Science.

Tyndale-Biscoe, C. H. & Hearn, J. P. (1981). Pituitary and ovarian factors associated with seasonal quiescence of the tammar wallaby, *Macropus eugenii. Journal of Reproduction and Fertility*, **63**, 225–30.

Tyndale-Biscoe, C. H., Hearn, J. P. & Renfree, M. B. (1974). Control of reproduction in macropodid marsupials. *Journal of Endocrinology*, **63**, 589–614.

Tyndale-Biscoe, C. H. & Hinds, L. (1981). Hormonal control of the corpus luteum and embryonic diapause in macropodid marsupials. *Journal of Reproduction and Fertility, Supplement* **29**, 111–17.

Tyndale-Biscoe, C. H. & Hinds, L. A. (1984). Seasonal patterns of circulating progesterone and prolactin and response to bromocriptine in the female tammar, *Macropus eugenii. General and Comparative Endocrinology*, **53**, 58–68.

Tyndale-Biscoe, C. H., Hinds, L. A., Horn, C. A. & Jenkin, G. (1983). Hormonal changes at oestrus, parturition and post-partum oestrus in the tammar wallaby (*Macropus eugenii*). *Journal of Endocrinology*, **96**, 155–61.

Tyndale-Biscoe, C. H. & Mackenzie, R. B. (1976). Reproduction in *Didelphis marsupialis* and *D. albiventris* in Colombia. *Journal of Mammalogy*, **57**, 249–65.

Tyndale-Biscoe, C. H. & Rodger, J. C. (1978). Differential transport of spermatozoa into the two sides of the genital tract of a monovular marsupial, the tammar wallaby (*Macropus eugenii*). *Journal of Reproduction and Fertility*, **52**, 37–43.

Tyndale-Biscoe, C. H. & Smith, R. F. C. (1969). Studies on the marsupial glider *Schoinobates volans* (Kerr). II. Population structure and regulatory mechanisms. *Journal of Animal Ecology*, **38**, 637–50.

Tyndale-Biscoe, C. H., Stewart, F. & Hinds, L. A. (1984). Some factors in the initiation and control of lactation in the tammar wallaby. *Symposium of the Zoological Society of London*, **51**, 389–401.

Tyson, E. (1698). Carigueya, seu Marsupiale Americanum or the anatomy of an opossum dissected at Gresham college. *Philosophical Transactions of the Royal Society*, **20**, 105–64.

Ullmann, S. L. (1978). Observations on the primordial oocyte of the bandicoot *Isoodon macrourus* (Peramelidae, Marsupialia). *Journal of Anatomy*, **128**, 619–31.

Ullmann, S. L. (1981*a*). Observations on the primordial germ cells of bandicoots (Peramelidae, Marsupialia). *Journal of Anatomy*, **132**, 581–95.

Ullmann, S. L. (1981*b*). Sexual differentiation of the gonads in bandicoots (Peramelidae, Marsupialia). *Proceedings of Vth workshop on the Development and Function of the Reproductive Organs*, International Congress Series No. 559, ed. A. G. Byskov & H. Peters, pp. 41–50. Amsterdam: Excerpta Medica.

Ullmann, S. L. (1984). Early differentiation of the testis in the native cat, *Dasyurus viverrinus* (Marsupialia). *Journal of Anatomy*, **138**, 675–88.

Valentyne, F. (1726). *Oud en nieuw Oost-Indien*. Doordrecht: J. Van Braam & G. O. den Linden.

VandeBerg, J. L. (1983*a*). Developmental aspects of X chromosome inactivation in eutherian and metatherian mammals. *Journal of Experimental Zoology*, **228**, 271–86.

VandeBerg, J. L. (1983*b*). The grey short-tailed opossum: A new laboratory animal. *Institute of Laboratory Animal Resources News*, **26**, 9–12.

VandeBerg, J. L., Cooper, D. W., Sharman, G. B. & Poole, W. E. (1977). Studies on metatherian sex chromosomes. IV. X linkage of PGK-A with paternal X inactivation confirmed in erythrocytes of grey kangaroos by pedigree analysis. *Australian Journal of Biological Sciences*, **30**, 115–25.

VandeBerg, J. L., Johnston, P. G., Cooper, D. W. & Robinson, E. W. (1983). X-chromosome inactivation and evolution in marsupials and other mammals. In *Isozymes: Current Topics in Biological and Medical Research*, vol. 9, *Gene Expression and Development*, ed. M. C. Rattazi, J. G. Scandalios & G. S. Whitt. pp. 201–18. New York: Alan R. Liss Inc.

VandeBerg, J. L., Thiel, J. E., Hope, R. M. & Cooper, D. W. (1979). Expression of PGK-A in the Australian brush-tailed possum, *Trichosurus vulpecula* (Kerr), consistent with paternal X inactivation. *Biochemical Genetics*, **17**, 325–32.

Van Dyck, S. (1979). Behaviour in captive individuals of the dasyurid marsupial *Planigale maculata* (Gould, 1851). *Memoirs of the Queensland Museum*, **19**, 413–29.

Van Dyck, S. (1982). The relationships of *Antechinus stuartii* and *A. flavipes* (Dasyuridae, Marsupialia) with special reference to Queensland. In *Carnivorous Marsupials*, ed. M. Archer, pp. 723–66. Sydney: Royal Zoological Society of New South Wales.

Vogel, P. (1972). Vergleichende Untersuchung zum Ontogenesemodus einheimischer Soriciden (*Crocidura russula, Sorex araneus* und *neomys fodiens*). *Revue Suisse de Zoologie*, **79**, 1201–332.

Vogel, P. (1981). Occurrence and interpretation of delayed implantation in insectivores. *Journal of Reproduction and Fertility, Supplement* **29**, 51–60.

Von der Borch, S. M. (1963). Unilateral hormone effect in the marsupial *Trichosurus vulpecula. Journal of Reproduction and Fertility*, **5**, 447–9.

Von Korff, K. (1902). Zur Histogenese der Spermien von *Phalangista vulpina. Archiv für Mikroscopische Anatomie*, **60**, 232–60.

Wainer, J. W. (1976). Studies of an island population of *Antechinus minimus* (Marsupialia, Dasyuridae). *Australian Journal of Zoology* **19**, 1–7.

Wakefield, N. A. & Warneke, R. M. (1963). Some revision in *Antechinus* (Marsupialia). *Victorian Naturalist*, **80**, 194–219.

Walker, K. Z. & Tyndale-Biscoe, C. H. (1978). Immunological aspects of gestation in the tammar wallaby, *Macropus eugenii. Australian Journal of Biological Sciences*, **31**, 173–82.

Walker, M. T. (1983). *Gestational hormones: embryonic maintenance in the marsupial* Macropus rufogriseus. PhD thesis, University of Queensland, Brisbane.

Walker, M. T. & Gemmell, R. T. (1983*a*). Plasma concentrations of progesterone, oestradiol-17β and 13, 14-dihydro-15-oxo-prostaglandin $F_{2\alpha}$ in the pregnant wallaby (*Macropus rufogriseus rufogriseus*). *Journal of Endocrinology*, **97**, 369–77.

Walker, M. T. & Gemmell, R. T. (1983*b*). Organogenesis of the pituitary, adrenal, and lung at birth in the wallaby, *Macropus rufogriseus. American Journal of Anatomy*, **168**, 331–44.

Walker, M. T., Gemmell, R. T. & Hughes, R. L. (1983). Ultrastructure of the reactivated corpus luteum after embryonic diapause in the marsupial, *Macropus rufogriseus banksianus. Cell and Tissue Research*, **228**, 61–74.

Walker, M. T. & Rose, R. (1981). Prenatal development after diapause in the marsupial *Macropus rufogriseus. Australian Journal of Zoology*, **29**, 167–87.

Wallace, G. I. (1981). Uterine factors and delayed implantation in macropodid marsupials. *Journal of Reproduction and Fertility, Supplement* **29**, 173–81.

Walton, S. M. (1971). Sex-chromosome mosaicism in pouch young marsupials *Perameles* and *Isoodon. Cytogenetics*, **10**, 115–20.

Ward, K. L. & Renfree, M. B. (1984). Effects of progesterone on parturition in the tammar, *Macropus eugenii. Journal of Reproduction and Fertility*, **72**, 21–8.

Waring, H., Moir, R. J. & Tyndale-Biscoe, C. H. (1966). Comparative physiology of marsupials. *Advances in Comparative Physiology and Biochemistry*, **2**, 237–376.

Waring, H., Sharman, G. B., Lovat, D. & Kahan, M. (1955). Studies on marsupial reproduction. 1. General features and techniques. *Australian Journal of Zoology*, **3**, 34–43.

Watson, D. M. S. (1917). The evolution of the tetrapod shoulder girdle and forelimb. *Journal of Anatomy*, **52**, 1–63.

Webster, G. M. & Haresign, W. (1983). Seasonal changes in LH and prolactin concentrations in ewes of two breeds. *Journal of Reproduction and Fertility*, **67**, 465–71.

Western, D. (1979). Size life history and ecology in mammals. *African Journal of Ecology*, **17**, 185–204.

Whitford, D., Fanning, F. D. & White, A. W. (1982). Some information on reproduction, growth and development in *Planigale gilesi* (Dasyuridae, Marsupialia). In *Carnivorous Marsupials*, ed. M. Archer, pp. 77–81. Sydney: Royal Zoological Society of New South Wales.

Whittaker, R. G. & Thompson, E. O. P. (1974). Studies on monotreme proteins. V. Amino acid sequence of the α-chain of haemoglobin from the platypus, *Ornithorhynchus anatinus*. *Australian Journal of Biological Sciences*, **27**, 111–15.

Wilkes, G. E. (1984). *The development of structure and function of the kidney in the tammar wallaby*, Macropus eugenii. PhD thesis, Australian National University, Canberra.

Williams, R. & Williams, A. (1982). The life cycle of *Antechinus swainsonii* (Dasyuridae: Marsupialia). In *Carnivorous Marsupials*, ed. M. Archer, pp. 89–95. Sydney: Royal Zoological Society of New South Wales.

Wilson, B. A. (1986). Reproduction in the female dasyurid *Antechinus minimus maritimus* (Marsupialia: Dasyuridae). *Australian Journal of Zoology*, **34**, 189–97.

Wilson, B. A. & Bourne, A. R. (1984). Reproduction in the male dasyurid *Antechinus minimus maritimus* (Marsupialia: Dasyuridae). *Australian Journal of Zoology*, **32**, 311–18.

Wilson, J. D., Griffin, J. E., George, F. W. & Leshin, M. (1983). The endocrine control of male phenotypic development. *Australian Journal of Biological Sciences*, **36**, 101–28.

Wilson, J. T. & Hill, J. P. (1908). Observations on the development of *Ornithorhynchus*. *Philosophical Transactions of the Royal Society B*, **199**, 31–168.

Wimsatt, W. A. (1975). Some comparative aspects of implantation. *Biology of Reproduction*, **12**, 1–40.

Winter, J. W. (1978). *The behaviour and social organisation of the brush-tail possum (Trichosurus vulpecula: Kerr)*. PhD thesis, University of Queensland, Brisbane.

Witschi, E. (1956). *Development of Vertebrates*. Philadelphia: Saunders.

Wood, D. H. (1970). An ecological study of *Antechinus stuartii* (Marsupialia) in a south-east Queensland rain forest. *Australian Journal of Zoology*, **18**, 185–207.

Wooller, R. D., Renfree, M. B., Russell E. M., Dunning, A., Green, S. W. & Duncan, P. (1981). Seasonal changes in a population of the nectar-feeding marsupial *Tarsipes spencerae* (Marsupialia: Tarsipedidae). *Journal of Zoology, London*, **195**, 267–79.

Woolley, P. (1966*a*). Reproduction in *Antechinus* spp. and other dasyurid marsupials. *Symposium of the Zoological Society of London*, **15**, 281–94.

Woolley, P. (1966*b*). *Reproductive biology of* Antechinus stuartii *Macleay (Marsupialia: Dasyuridae)*. PhD thesis, Australian National University, Canberra.

Woolley, P. (1971*a*). Maintenance and breeding of laboratory colonies, *Dasyuroides byrnei* and *Dasycercus cristicauda*. *International Zoo Yearbook*, **11**, 351–4.

Woolley, P. (1971*b*). Observations on the reproductive biology of the dibbler, *Antechinus apicalis* (Marsupialia: Dasyuridae). *Journal of The Royal Society of Western Australia*, **54**, 99–102.

Woolley, P. (1973). Breeding patterns, and the breeding and laboratory maintenance of dasyurid marsupials. *Experimental Animals*, **22**, 161–72.

Woolley, P. (1974). The pouch of *Planigale subtilissima* and other dasyurid marsupials. *Journal of the Royal Society of Western Australia*, **57**, 11–15.

Woolley, P. (1975*a*). The seminiferous tubules in dasyurid marsupials. *Journal of Reproduction and Fertility*, **45**, 255–61.

Woolley, P. A. (1982). Phallic morphology of the Australian species of *Antechinus* (Dasyuridae, Marsupialia): a new taxonomic tool? In *Carnivorous Marsupials*, ed. M. Archer, pp. 767–81. Sydney: Royal Zological Society of New South Wales.

Woolley, P. A. (1984). Reproduction in *Antechinomys laniger* ('*spenceri*' form) (Marsupialia: Dasyuridae): field and laboratory investigations. *Australian Wildlife Research*, **11**, 481–9.

Woolley, P. A. (1986). The seminiferous tubules, rete testis and efferent ducts in didelphid, caenolestid and microbiotheriid marsupials. In *Possums and Opossums: Studies in Evolution*, ed. M. Archer, in press. Sydney: Australian Mammal Society.

Woolley, P. A. & Ahern, L. D. (1983). Observations on the ecology and reproductive biology of *Sminthopsis leucopus* (Marsupialia: Dasyuridae). *Proceedings of the Royal Society of Victoria*, **95**, 169–80.

Woolley, P. A. & Watson, M. R. (1984). Observations on a captive outdoor breeding colony of a small dasyurid marsupial, *Sminthopsis crassicaudata*. *Australian Wildlife Research*, **11**, 249–54.

Woolley, P. A. & Webb, S. J. (1977). The penis of dasyurid marsupials. In *The Biology of Marsupials*, ed. B. Stonehouse & D. P. Gilmore, pp. 306–26. London: Macmillan.

Wye-Dvorak, J. (1984). Postnatal development of primary visual projections in the tammar wallaby (*Macropus eugenii*). *Journal of Comparative Neurology*, **288**, 491–508.

Young, A. H. (1879). On the male generative organs of the koala (*Phascolarctos cinereus*). *Journal of Anatomy and Physiology*, **13**, 305–17.

Young, C. E. & McDonald, I. R. (1982). Oestrogen receptors in the genital tract of the Australian marsupial *Trichosurus vulpecula*. *General and Comparative Endocrinology*, **46**, 417–27.

Young, I. R. & Renfree, M. B. (1979). The effects of corpus luteum removal during gestation on parturition in the tammar wallaby (*Macropus eugenii*). *Journal of Reproduction and Fertility*, **56**, 249–54.

Ziegler, A. C. (1977). Evolution of New Guinea's marsupial fauna in response to a forested environment. In *The Biology of Marsupials*, ed. B. Stonehouse & D. P. Gilmore, pp. 117–38. London: Macmillan.

Index

Page numbers in **bold type** refer to illustrations

Acrobates pygmaeus, 18, 19, 62, 127, 128, **176**
acrosome, **152**, **155**, 156, 157, **261**, 263
adrenal glands, development of, 308
Aepyprymnus rufescens, **10**, 18, 19, 23, 87, 95, 243, 311
agile wallaby, *see Macropus agilis*
allantois, **78**, 80, 81, 305, **306**, **312**, **315**, **321**
amnion, **78**, **79**, 80, 81, 303, **306**, **312**
anoestrus, 67, **68**
Antechinomys laniger, 16, 17, 42
Antechinomys spenceri, **39**
Antechinus apicalis, **146**
Antechinus bellus, 44, 159
Antechinus flavipes, 16, 17, 43, 103, 143, 159, 170
Antechinus melanurus, 44
Antechinus minimus, 43, 137, **141**, 143, 145, 159, 201
Antechinus naso, 44
Antechinus rosamondae, **146**
Antechinus stuartii,
  embryo development, **259**, 264, 265, 267, 271, **272**, 274, 276–8, 294, **295**, 296, 297, 299, 303, **304**
  life history of, 42–3
  plasma progesterone in, 220
  reproduction in female, 16, 17, 22, **34**, 178, **179**, 198, 199, 201, 204, 210, 211, 218
  reproduction in male, 42–3, 123, **125**, 137, **140**, 145, **146**, 159, **162**
Antechinus swainsonii, 16, 17, 43, 143, 159
aromatase activity in immature ovary, **118**

basal body temperature, during reproductive cycle, **65**, 241

Bennett's wallaby, *see Macropus rufogriseus*
bettong, *see Bettongia* spp.
Bettongia gaimardi, 18, 19, 23, **60**, 81, 87, 187, 241, **298**, 299, 311, 316, 328
Bettongia lesueur, 18, 19, 23, **74**, 80, 81, 82, 87, 125, **180**
Bettongia penicillata, 18, 19, 23, 87
bilaminar blastocyst, 299, 405
blastocyst,
  development of, **61**, **259**, **272**, 278, 279
  metabolism of, 287–90, **290**
body size,
  and reproduction, **32**, 89, 92, 416, 417, 418, 420
  of Mesozoic mammals, 415
body weight,
  changes related to oestrous cycle, 34, 218, **219**
  in relation to breeding success, **93**
breeding season, 17, 19, 21, 26–30, **28**, **29**, 40–7, 52–3, 57, 83–8, **84**, **86**
bromocriptine, 382, 383
brush-tailed possum, *see Trichosurus vulpecula*
bulbo-urethral or Cowper's glands, 124, **125**
Burramys parvus, 18, 19, 60, 124, 128

Caenolestes obscurus, 31, 124, 128, 135, 147, **155**, 156, 169, 170, **176**, 181
Caluromys derbianus, 29, 168
Caluromys philander, 16, 17, **29**, 30, 151, **155**
Caluromys sp., 145
caruncle, 409
castration effect on gonadotrophins, 139

Cercartetus caudatus, 61
Cercartetus concinnus, 18, 19, 60–1, **61**, 127, 199
Cercartetus lepidus, 128
Cercartetus nanus, 18, 19, 61, 128
cervical sympathetic ganglionectomy, effect on quiescence, 392, 393
Chironectes minimus, 121, 156
chorio-allantoic placenta, 313, 319, **321** **322**, 332
chorio-vitelline placenta, 311–3, **312**, **315**, **322**
chorion, **312**, 405
chromosome number, 95–6, **96**
cladistic analysis, 396, 397
common ringtail possum, *see Pseudocheirus peregrinus*
copulation, 36, 49, **50**, 71–3, **74**, 260, 264
corpus albicans, 211
corpus luteum,
  effects of bromocriptine on, 380, 382
  endocrine control of, 233, 235, **235**, 249–54, **375**, 379, 382, 404
  endocrine function, 9, 221–3, **222**, 238, **238**, 239, 335–6
  in Eutheria, 403, 421
  in monotremes, 403
  prolactin receptors on, **252**, 253, 379, 383
  reactivation after summer solstice, 384, **386**
  role of, 207, 208, 228–30, 242–5, **245**, 248, **248**, 249, 282–4, 357
  structure of, 208–11, **209**, 223, 225, 256
corticosteroid binding globulin (CBG), 212
cortisol, 308
Crocidura cacrula, 406
cuscus, *see Phalanger orientalis*

Dasycercus cristicauda, 16, 17, 45
Dasykaluta rosamondae, 127
Dasypus novemcinctus, **6**
Dasyuridae, breeding seasons of, 40–7
Dasyuroides byrnei,
  reproduction in female, 16, 17, 22, 34, 44, 200, 204, 208, 218, **219**, 332
  reproduction in male, **134**, 139, 143, 170
Dasyurus hallucatus, 16, 17, 45, 127, 159
Dasyurus maculatus, 16, 17
Dasyurus viverrinus,
  development, 16, 17, 45, 51, 106, 108–10, 112, **113**, 215–7, **268**, 269, 271, 274, 296, 297, 299, 303, **312**, 313, 314, **363**, 409
  lactation, 46, 47, 93, 345, 351, **362**, 363
  reproduction in female, 22, 33, 45, **179**, 186, 198, 200, 201, 204, 210, 211, 214–7, **217**, 217–8
  reproduction in male, 144, 154, 170
delayed oestrous cycle, in macropodids, 233, 234, 237
delayed pregnancy, in macropodids, 233, 234, 235, **236**
Didelphis albiventris, 2, 16, 17, 123, 127, 128, 131, 210, 212
Didelphis marsupialis, 16, 17, 27, **28**, 123, 145, 199, 200, **270**, 271, 274, 275
Didelphis virginiana, **12**
  chromosomes in, 104
  development, 7, 25–7, 98, 106, 108, **109**, 112, 114, 115, **115**, 116, 122, 258, **261**, **262**, 260–4, 265, 267, **270**, 271, 273, 274, 275, 279, **295**, 296, 297, 299, **304**, 305, 307, 308, 311, 314, **318**, 324, 327, 409
  lactation, 26, 93, 345, 351
  reproduction in female, 3, **12**, 15, 16, 17, 20–1, 22, 25, 173, 174, 175, **178**, 181, **182**, 183, 186, **187**, 196, 198, 200, 201, 212, 214, **215**, 254, 255, 330, 332, 336
  reproduction in male, 3, **4**, 125, **125**, 127, 128, 135, 136, 140, 145, 151, **155**, 156, 165, **166**, 167–9, **167**, **169**
  steroids in, 117, 118, **118**, **119**, 120, **120**, 121, 212–4, **213**
dihydrotestosterone, effect on urogenital system, 121
5 alpha-dihydrotestosterone, 136–7, 143–4
Distoechurus pennatus, 62
Dromiciops australis, 30, 155, 156, 168

eastern grey kangaroo, *see Macropus giganteus*
efferent duct of testis, **126**, **130**, 131
egg cleavage, **259**, 276–8
  in monotremes, 404
egg tooth, 409
egg,
  dimensions of, 271
  of monotremes, 8
Elephantulus myurus, embryo development in, 406
embryo attachment or implantation, **304**, 314, **318**,
embryo culture, 277, 278, 279
embryo development, of monotremes, 9
embryo vesicle, **281**, **300**
embryogenesis, 215–7, **281**, **295**, 299, 301, 303, 304
embryonic diapause,
  in marsupials, 11, 61, 62, 63, 67, 69, 91–2, 279–80
  occurrence in mammals, 417
endoderm formation, **270**, 294, **298**, 299
endometrium,
  secretions of 324, **326**
  weight changes in, 214, **215**, 328–30, **329**

epididymis, **126**, **130**, 131, 133, **139**, **141**, **164**, **165**, **167**, 170
epipubic bones, 411
euro, *see Macropus robustus*
Eutheria, 398, **399**
altricial young of, 410
reproductive characters of, 412–3
extra-embryonic coelom, **302**, **312**

feathertail glider, *see Acrobates pygmaeus*
female urogenital system, **6**, **173**, **176**, **180**, 190–5, **192**, **193**
fertilisation, **259**, **261**, **262**, 263–4
in monotremes, 404
fetal membranes,
of marsupials, **10**, **306**, **312**, **315**, 311–4
of monotremes, **10**, 407
of rabbits, **10**
feto-maternal contact, 314–23, **318**, **321**, **322**
fetus, **281**, **306**
role in parturition, 332, **338**, 339, 340
flehman, 249
follicle stimulating hormones, plasma concentration, **136**, 206, 246–8, 374, **375**
follicular phase of oestrous cycle, 22–3

galactophores, 344, 346, **368**, **369**, 435
alpha-galactosidase A, on X chromosomes, 99
Galago demidoffi, embryo development in, 406
germ cells, **101**, 105, 106, **106**, **107**, 108
gestation length, 22–3, 36, 75, 234
effect of anti-paternal sensitisation on, 333
effect of fetal genotype on, 332
gestation, duration of delayed, 282, 285, 286
glucose, concentration in fetal fluids, 324, **325**, 327
glucose-6-phosphate dehydrogenase (G6PD), on X chromosomes, 99–102
gonad, development of in marsupials, 108–14
gonadotrophin receptors, 137, 251, **252**
gonadotrophin releasing hormone, 137, 206, 246, 251, 338, 341, 376
Graafian follicle, 194–201, **197**, **205**, 206–8, **245**, 246, 251, **252**, 402
in Eutheria, 198, 403
in monotremes, 196–8, 402
greater glider, *see Petauroides volans*
Gymnobelideus leadbeateri, 18, 19, 22, 59–60

hairy-nosed wombat, *see Lasiorhinus latifrons*
Hemibelideus lemuroides, 58, 104,
Hemicentetes semispinosus, 406
heterokaryons of placenta, 319, **321**, **322**, 323
homokaryons of placenta, 317, 319, **320**, **322**
honey possum, *see Tarsipes rostratus*
hypophysectomy, 139, 200, 206, **250**, 252–3, 377
hypothalamo-pituitary axis, 246–8, 376, **378**
hypoxanthine-guanine phosphoribosyl-transferase (HGPT), on X chromosomes, 99
Hypsiprymnodon moschatus, 18, 19, 66, 81
hysterectomy, effects on corpus luteum, 254, 255

inner cell mass, in eutherian embryo, 405
Insectivora, fetal and neonatal development of, 410
intersex, reproductive system of, 96, **97**, 98, 122
interstitial tissue of ovary, 201–2, **209**
Isoodon macrourus,
corpus luteum of, 210, 211, 224, 225, **226**, 244, 254
development, 18, 19, 103, 105, **106**, 108, 109, **111**, 271, 275, **295**, 299, 308, 309
placenta of, 313, 317, 319, **320**, **321**, **322**
prostaglandin at parturition in, 256, 339
reproduction in female, 22, 52, 53
spermatozoon of, **154**, 170
Isoodon obesulus, 18, 19, 52, 53, 103, 105, 127, 170, 210, 224, 313

koala, *see Phascolarctos cinereus*
Kuehneotheridae, 398, **399**

alpha-lactalbumin, 360
lactational quiescence, 67, **68**, 69, 242, 243, 244, 250
control of, 377, **378**
lactation, 356, 358, 361
cost of, 27, **29**, 30, 52, 64, 92–4, **347**, 348
evolution of in mammals, 416
sequential in macropodids, 81–4, **82**, **84**, **354**, 355, 356, **368**
water and sodium influxes during, 46, 47
lactic acid dehydrogenase, 159
Lagorchestes conspicillatus, 96
Lagostrophus fasciatus, 177
Lasiorhinus latifrons, 18, 19, 65–6, 124, 125, 128, 147, 151, 170
lateral vaginae, **179**, **180**, 181, **183**, 189
Leydig cell, 112, 135, 136
litter size, 16, 18, 20, 26–7, 37, 39, 52–3, 81–2, 89
liver, development of, 308, 327

lung, development of, **302**, 307
lutectomy, effects of, 207, 229, 230, 242–4, **245**, **281**, 283, 284, 336–7
luteinising hormone, plasma concentration, **136**, **205**, 206–8, 246–8, **338**, 374, **375**
  stimulation of testosterone synthesis, 137
luteinising hormone receptors, 251, **252**, 254

m. cremaster, 122, 129, 364
m. ilio-marsupialis, 122, 348, 364
Macropus agilis, 20–1, 23, 75, 84, 127, 191, 193, 194, 351, 367, **368**, **369**, 370
Macropus eugenii, 2, 20, 21, **71**, 72, **74**, 82, 96, **97**, 98, 99, 122, **354**, **375**
  corpus luteum of, 231, **238**, 238–9, 242–7, **245**, 283, 284, 335, 336, 386
  embryo development of, **101**, 102, 105, **107**, 108, 110, 112, 114, 260, 265, **266**, **268**, 270, 271, **272**, 273, 276, 278, 280, **281**, 283, 286, 287–90, **290**, **295**, 296, 297, **300**, 301, **302**, **304**, 305, **306**, 307–9
  lactation in, 346, 347, 348, 350, 351, **351**, **352**, 357, **357**, 358, 359, **359**, 361, 380
  ovary of, 196, **197**, 198, 199, 200, 201–2, 204–8, **209**, 210
  parturition in, **78**, **79**, 80, 336, 337–41, **338**, **340**
  pineal and melatonin in, **390**, **391**, 392, 393
  placenta of, 313, 317, **318**, 324–7, **325**, **326**, 330, **331**, 333
  reproduction in female, 173, **173**, **183**, 184, 190, 191, 192, **193**, 194, 195, 290–3, **292**, 328, **329**, 334, **335**
  reproduction in male, 69, **125**, **126**, 127, 128, 129, 131, 132, 135, **136**, **138**, 139, 140, 143, **148**, **150**, 151, **158**, 159, 167, 169, **248**, 249
  reproductive cycle of, 23, 75, 231, **232**, 233, 234, **236**, **239**, 239–240, **250**, 251, 255, 256, 282, 283, 381, 382, 384, **386**
  seasonal breeding in, 85–7, **86**, **138**, 374, 384, 385, 387, 388, 394
Macropus fuliginosus, 20, 21, 23, 75, 80, 81, 87, 99, 127, 128, 170, 244, 279, 332, 338
Macropus giganteus,
  development, 20, 21, **76**, 81, **268**, 311, 332
  maternal investment in, 92
  reproduction in female, **6**, 23, **68**, 73, 75, 80, 81, 87, 88, 177, 188, 191, 193, 199, 210, 211, 241, 243, 244, 328, 338
  reproduction in male, 127, 128, 143
Macropus parma, 20, 21, 23, 72, 73, 75, 88, 125, 271, 313, 316
Macropus parryi, 20, 21, 23, 88, 100
Macropus robustus, 20, 21, **76**, 84, 125, 137, 142, 311
  X chromosomes of, 98, 99–103, **100**, 102
Macropus rufogriseus
  development, 296, 297, 299, 301, 303, 305, 308–9, 311, 316, 317
  reproduction in female, 20, 21, 23, 72, 75, 80, 81, 230, 231, 233, 234, 235, **236**, 237, 240–1, 339
  reproduction in male, 69, 127, 143, 156, 159, 161, 170, 210
Macropus rufogriseus banksianus, seasonal breeding in, 84
Macropus rufogriseus rufogriseus, seasonal breeding in, 85–6, 392
Macropus rufus, **76**, 99, **100**, 114
  corpus luteum of, 235, 243, 244, 254
  lactation, 82, **349**, **354**
  reproduction in female, 20, 21, 23, **70**, 75, 76, **77**, **78**, **79**, 80, 81, 83–4, **84**, **179**, 191, 193, 207, **272**, 285, 286, 338
  reproduction in male, **72**, 73, **74**, 128, 129, 137, 142, 143, 170
Macrotis lagotis, 18, 19, 22, 53, 95, 125, 145
male behaviour, 70–3, **71**, **72**
male die-off, 42–3
mammals, palaeontological record of, 398
mammary area, 37, **38**, 39, **40**, **48**,
mammary gland,
  alpha lactalbumin synthesis in, 359, **359**, 360
  development and growth, 26, 45, 345, 347, 348, **349**, 350, **350**, **351**, 355, 356
  of monotremes, 8, 410
  prolactin receptors on, **252**, **357**, 358, 361
  response to electrical stimulation of hypothalamus, **369**
mammary hair, 344, 345, 410
  of monotremes, 345, 419
Marmosa mitis, 151, 156
Marmosa murina, **6**, 127, 128
Marmosa robinsoni, 16, 17, 20, 22, **24**, 25, **28**, 30, 199, 200, 345
Marmosa sp., 145
maternal recognition of pregnancy, 327–32
medullary plate, **298**, 299, **300**, 301
melatonin,
  effects of treatment on reactivation, **390**, 392
  plasma concentration, 389, **390**
mesonephros, **109**, **110**, **302**, 305
Metachirus sp., 156
Metatheria, 398, 412–3
midpiece of spermatozoon, **152**, **153**, 154, 156

milk composition,
in eutherians, 353, 411
in marsupials, 352, **352**, 353
in monotremes, 414
milk ejection reflex, 365, 367, **368**, **369**, 370
molar teeth, in Prototheria, 398, 399
Monodelphis domestica, 16, 17, 20, 22, **24**, 25, 104, 169, 170
Monodelphis sp., 156
monoestry, 88, 89
Monotremata, 8, **10**, 96, 196–8, 343, 398–401, 403–4, 406–14
moon phase, effect on time of reactivation, 387
Morganucodontidae, 398, **399**, 415
mucoid coat, **262**, 264, **266**, 267, **268**, 269, **270**, 271
Mullerian duct, 114–6, **119**, 401, 402
Multituberculata, 398, **399**
musky rat kangaroo, *see Hypsiprymnodon moschatus*
Myrmecobius fasciatus, 16, 17, 47–8, **48**

neonatus, 16, 18, 20, **78**, **79**, 80, 309–10
neuropeptides,
in eutherians, 366
in marsupials, 365, 366, 367
in monotremes, 366
Ningaui ridei, 16, 17, **33**, **35**, 42, 92
Notoryctes typhlops, 16, 17, 124
numbat, *see Myrmecobius fasciatus*, 47

oestradiol,
effects of, **115**, 117, 118, **119**, 120, 121, **245**, 246, 255, 285, 286, 376
plasma concentration, 64, 205–8, **205**, 213, **213**, 214, 223, **239**, 239–40, 384, **386**
receptors in urogenital tract, 188, 189, 190
oestrogen: progesterone ratio, at parturition, 337–8
oestrous behaviour, 21, 35, 36, 49, 70
oestrous cycle, 20–3, 34, 69–70, **70**, 73, 75
influence of pregnancy on, 75, 234, 332
oestrus, hormonal control of, 204–8, **205**
oocyte,
of Eutheria, 196, 198
of marsupials, 196–9, **197**
of monotremes, 196, 197
Ornithorhynchus anatinus, 8, 96, 196, 400
ovarian quiescence, 67
ovariectomy,
effects on gonadotrophins, 376
effects on pregnancy, 214
ovary,
blood supply of, 191, **192**, **193**, 194, 195
differentiation of, 112, 114
oviduct, **169**, 172–4, **174**, 189
Ovis aries, 142
ovulation, hormone control of, 204–8, **205**
ovulation rate, 22–3, 37, 200, 264, 265
oxytocin, 251, **253**, 255, 334, **335**, 365–7

pampiniform plexus of Eutheria, 129
pancreas, development of, 308
Parantechinus apicalis, 16, 17, 44
Parantechinus bilarni, 16, 17, 44
parental investment, **32**, 92
parma wallaby, *see Macropus parma*
parturition, 5, 7, 25, 37, **50**, 51, 56, 76, **77**, **78**, **79**, 80, 81
hormone changes at, 337–41, **338**, **340**
penis,
accessory erectile body of dasyurids, **146**
of monotremes, 401
Peradorcas concinna, 20, 21
Perameles gunnii, 18, 19, 49, 52, 53, 125, 313
Perameles nasuta,
development, 18, 19, 105, 108, 271, 275, 276, 299, 303, **304**
placenta of, **312**, 313, 314, 316, 317, 319, **322**, 323, 324
reproduction in female, 22, 49, **50**, 51, 52, 53, **179**, 210, 211, 224, 225
reproduction in male, **125**, 137, 142, 151, 157
sex chromosome elimination in, 103
Petauroides volans, 18, 19, 58–9, 104, 159, 199, 276, 299, 303, 305, 311
Petaurus australis, 18, 19, 59, 123
Petaurus breviceps, 18, 19, 22, 59, 123
Petaurus norfolcensis, 311
Petrogale penicillata, 20, 21, 23, 84
Phalanger gymnotis, 58
Phalanger maculatus, 58
Phalanger orientalis, 2, 58
Phascogale calura, 44
Phascogale tapoatafa, 16, 17, **40**, 44
Phascolarctos cinereus, **10**, 18, 19, 64–5, 92, 145, 153, 165, 167, 204, 210, 211, **312**, 313, 314
Philander opossum, 16, 17, 30, 104, 127, 128, 145, 156, 225, 274, 305, 311, 314, 315, **315**, 316, 317
phosphoglycerate kinase-A (PGK-A), on X chromosomes, 99–100
photoperiod effect on breeding season, 41, 42, 47, 53, 57, 85–7, **86**, 381, 384–93, **386**, **390**
pineal gland, **387**, 389, **391**
pinealectomy, 392, 393
pituitary gland, 198, 200, 206, 208, 246, 247, 254, 309, 340, 341, 358, 361, 374, **378**, 379

placenta, 51, 310–11, **312**, 313, **315**, 330–2, **331**
Planigale gilesi, 16, 17, 42
Planigale ingrami, 16, 17, 42
Planigale maculatus, 16, 17, 42
Planigale tenuirostris, 16, 17, 42
platypus, *see Ornithorhynchus anatinus*
polyoestry, 88–9
post-partum oestrus, 75, 204–5, **205**, 232, 234, 237, **338**
potoroo, *see Potorous tridactylus*
Potorous tridactylus, 18, 19, 23, 51, 72, 73, 75, 80, 81, 87, 95, 99, 145, 178, 188, 194, 199, 235, 279, 283, 284, **304**, 316, 330
pouch, 37–9, **38**, **40**, **50**, **82**, 121–3, **354**, 411
pregnenolone, 117
primitive groove, **300**, 301, **302**
proamnion, **300**, **302**, 303, **312**
progesterone,
  effects of, **245**, 246, 283, 285, 286, **329**, 336
  in corpus luteum, 212
  plasma concentration during cycle, 64, **205**, 206–8, 213, **213**, **217**, 218, **219**, 220, **221**, **232**, **236**, 237, **375**
  plasma concentration during pregnancy, 213, **213**, 218, **219**, 220, **221**, **226**, 227, 228, **232**, **236**, 237–9, **250**, 251, 337, **338**
  pulse, **250**, 251, 280, **281**, 282, 286, 380, 381, 384, 388
  receptors for, 188, 189, 190, 329–30
  synthesis, **118**, 330, **331**
prolactin, 251, **253**, **338**, 340, 341, **357**, 360, 361, **362**, **375**
prolactin receptor concentration, **252**, 252–4, **357**, 358, 361, 379, 383
prostaglandin, 227, 255, 256, 334, **335**, **338**, 339, **340**
prostate gland, **138**, **139**, **141**, 144, 147, **148**, 149–51, **150**
  in monotremes, 401
prostatic sugars, **150**, 151
protein concentration in fetal fluids, 324, 325, **326**, 327
Prototheria, 397, 398
Pseudantechinus macdonnellensis, 127
Pseudocheirus archeri, 58
Pseudocheirus dahli, 58
Pseudocheirus herbertensis, 58
Pseudocheirus peregrinus, 18, 19, **55**, 58, 142, 147, **148**, **174**, **179**, 196, 198–200, 303, 311, 316
pseudovaginal canal, 7, 9, 177, 178, **179**, **180**, 181, 334–7
pygmy possum, *see Cercartetus*
quokka, *see Setonix brachyurus*

rabbit-eared bandicoot, *see Macrotis lagotis*
red kangaroo, *see Macropus rufus*
5 alpha-reductase activity, **120**
relaxin, 223, 335–6
reproductive patterns, classification of in marsupials, 90–1
reproductive strategies, comparison between Metatheria and Eutheria, 420–3
rete mirabile, **126**, 129
rete testis, **113**, 129, **134**, 135
rete testis fluid, 135
RNA synthesis by developing blastocyst, 287–90

Sarcophilus harrisii, 16, 17, 45, 47, 98, 122, 128, 200, 271, 275
Schoinobates, see Petauroides
scrotum, 121–3, 129, **130**
seasonal quiescence, 67, 69, 243, 244, **245**, **378**, 382, 384
semelparity, 43, 44
seminiferous tubules, **113**, **134**, **140**, **141**
Sertoli cell, **111**, 156
Setonix brachyurus,
  corpus luteum of 231, 233, 235, 237, 242, 243, 245, 279–80, 283, 284
  development, 273, 303, 305
  placenta of, 32, **312**, 313, 316
  reproduction in female, 20, 21, 23, 80, 85, 187, 188, 190, 191, 193, 194, 199, 200, 210, **236**, 240, 330
sex chromosome mosaic, 98
sex chromosomes,
  in monotremes, 400
  in marsupials, 95–6, 98
sex hormone binding globulin (SHBG), 212
sexual dimorphism, 73, **76**, 123, **123**
sheep, seasonal breeding in, 373
shell, of monotreme egg, 404
shell membrane, **259**, **266**, **268**, 269, **270**, 271, **272**, 273, 274, **315**
short-nosed bandicoot, *see Isoodon macrourus*
sinus terminalis, **312**, **315**, **320**
Sminthopsis crassicaudata, 16, 17, 22, 41, 42, 143, 170, 313
Sminthopsis leucopis, 42
Sminthopsis macroura, 16, 17, 22, 42, 170, 198–200, 210
Sminthopsis murina, 16, 17, 42
Sminthopsis virginae, 170
sperm pairing, 163

spermatogenesis, 156, 157, **158**, 159, 161, **162**, 163
spermatogonia, 156
spermatozoa, 36, 135, **152**, **153**, **154**, **160**, 163–7, **164**, **165**, **166**, **167**, 168–70, 260–1, 265, 271
  of monotremes, 401
spermatozoon pairs, in American marsupials, **155**, 156, **166**, **167**, 168–70
spermiogenesis, 157
summer solstice, relation to season of birth, 85, 384–7
Suncus murinus, oocyte maturation in, 403
surfactant, 307
swamp wallaby, *see Wallabia bicolor*

Tachyglossus aculeatus, **10**, 96, 400, 407
tammar wallaby, *see Macropus eugenii*
Tarsipes rostratus, 18, 19, **61**, 62, 63, **63**, 64, 123, 127, 128, **130**, 131, 145, 151, **153**, 154, **154**, 165, **176**, 177, 178, **179**, 181, 201, 211, 294, 314, 334
Tasmanian devil, *see Sarcophilus harrisii*
teat, 43, **51**, 52, 82, **82**, 344, 346, **354**, 355, 356
testicular descent, 125
testis, **111**, 112, **113**, 124, 125, 127, 128–9, 132, **133**
testosterone, **115**, **118**, 136, **136**, 137, **138**, 139, **140**, 142–5, **248**, 249
  effect of breeding females on, 144–5
  effects of, 117, 118, **119**, 120, 121
Theria, 397
Thylacinus cynocephalus, 47, 121
Thylogale billardierii, 20, 21, 23, 87, 191, 243
Thylogale brunii, 3
Thylogale thetis, 20, 21, 143
tongue, development of, 308
Trichosurus arnhemensis, 55, 58
Trichosurus caninus, 18, 19, 22, 55, 57, 211
Trichosurus vulpecula,
  corpus luteum of, 220–3, 242, 243, 247, 253–4, 335–6
  embryo development of, 57, **110**, 114, 265, 267, 269, 271, 273, 276, **295**, 296, 297, **304**, 305, 308
  lactation in, 348, **362**, 363
  luteinising hormone receptors on corpus luteum, 254
  reproduction in female, 18, 19, 22, 55–7, 93–4, **93**, 103, 173, 174, **176**, **179**, 181, 183, 184, **185**, 186, **187**, 188, 189, 190, 191, 192, **192**, 194, 198, 199, 200, 208, 211, 311, 314, 316, 324, 327, 328, 330, 341
  reproduction in male, 57, 98, 106, 122, 123, 125, **125**, 127, 128, 129, 131, 132, **133**, 135, 136, 137, **139**, 142, 143, 144, 147, **148**, 151, **152**, 159, **160**, 161, 163–7, **164**, **165**, 170
  seasonal distribution of births, **139**
  shell membrane of, 273
trophoblast, 288, 289, **315**, **318**, 319, **322**, 333, 405, 406, 408
  ability to mask histocompatibility antigens, 408
tunica vaginalis, **126**, 127, 128

urea concentration in fetal fluids, 324, **325**
ureter, 114–5, 400, 401
urogenital system, **107**, 114–6, 117
uterotubal junction, 173–4, 269
uterus, **68**, 175, **178**, 184, **185**, 186, **187**, 189, 191, **192**, **193**, 194, 195
  myometrial activity during pregnancy, 334, **335**

vas deferens, 135
Virginian opossum, *see Didelphis virginiana*
vitelline membrane, 265
viviparity, evolution of in mammals, 419
vocalisation, in dasyurids during breeding season, 35
Vombatus ursinus, 18, 19, 65, **65**, 92, 125, 191, 193, 194, 313, 314

Wallabia bicolor, 20, 21, 23, **76**, 95, 99, 170
water opossum, *see Chironectes minimus*
weaning, age at in dasyurids, **32**
western grey kangaroo, *see Macropus fuliginosus*
Wolffian duct, 114–6, **119**
wombat, *see Vombatus ursinus*

X chromosome, 98–104

yellow-bellied glider, *see Petaurus australis*
yolk extrusion, **259**, **270**, **272**, 274, 275
yolk sac, **306**, **312**, **315**, **318**, **320**, **322**
yolk sac fluid, **78**, 80, 324, **325**, **326**, 327

zona pellucida, **261**, **262**, 263, 265–7, **266**, **268**, 271